ELEMENTARY MATRICES

AND SOME APPLICATIONS TO DYNAMICS AND DIFFERENTIAL EQUATIONS

by

R. A. FRAZER, D.Sc., F.R.Ae.S., F.I.Ae.S., F.R.S.
Formerly Deputy-Chief Scientific Officer in the Aerodynamics Division, the National Physical Laboratory

W. J. DUNCAN, C.B.E., D.Sc., F.R.S.
Mechan Professor of Aeronautics and Fluid Mechanics in the University of Glasgow, Fellow of University College London

AND

A. R. COLLAR, M.A., D.Sc., F.R.Ae.S.
Sir George White Professor of Aeronautical Engineering in the University of Bristol

CAMBRIDGE
AT THE UNIVERSITY PRESS
1963

CAMBRIDGE UNIVERSITY PRESS
Cambridge, New York, Melbourne, Madrid, Cape Town, Singapore, São Paulo

Cambridge University Press
The Edinburgh Building, Cambridge CB2 8RU, UK

Published in the United States of America by Cambridge University Press, New York

www.cambridge.org
Information on this title: www.cambridge.org/9780521091558

First published 1938
Reprinted 1963
Re-issued in this digitally printed version 2007

A catalogue record for this publication is available from the British Library

ISBN 978-0-521-09155-8 paperback

CONTENTS

CHAPTER I

FUNDAMENTAL DEFINITIONS AND ELEMENTARY PROPERTIES

CHAPTER II

POWERS OF MATRICES, SERIES, AND INFINITESIMAL CALCULUS

CHAPTER IV

MISCELLANEOUS NUMERICAL METHODS

CHAPTER V

LINEAR ORDINARY DIFFERENTIAL EQUATIONS WITH CONSTANT COEFFICIENTS

Part I. *General Properties*

Part II. *Construction of the Complementary Function and of a Particular Integral*

CHAPTER VI

LINEAR ORDINARY DIFFERENTIAL EQUATIONS WITH CONSTANT COEFFICIENTS (*continued*)

Part I. *Boundary Problems*

PART II. *Systems of First Order*

CHAPTER VII

NUMERICAL SOLUTIONS OF LINEAR ORDINARY DIFFERENTIAL EQUATIONS WITH VARIABLE COEFFICIENTS

CHAPTER VIII

KINEMATICS AND DYNAMICS OF SYSTEMS

PART I. *Frames of Reference and Kinematics*

PART II. *Statics and Dynamics of Systems*

CHAPTER IX

SYSTEMS WITH LINEAR DYNAMICAL EQUATIONS

CHAPTER X

ITERATIVE NUMERICAL SOLUTIONS OF LINEAR DYNAMICAL PROBLEMS

CHAPTER XI

DYNAMICAL SYSTEMS WITH SOLID FRICTION

CHAPTER XII

ILLUSTRATIVE APPLICATIONS OF FRICTION THEORY TO FLUTTER PROBLEMS

CHAPTER XIII

PITCHING OSCILLATIONS OF A FRICTIONALLY CONSTRAINED AEROFOIL

ADDENDA ET CORRIGENDA

Additional Definitions.

The *trace* of a square matrix is the sum of the elements in the principal diagonal. It is equal to the sum of the latent roots.

Characteristic vector or *proper vector* is equivalent to "modal column" or "modal row", though less explicit. The term "eigenvector" is sometimes used but is to be strongly deprecated.

p. 33. *Special types of square matrix.* Add

"Matrix is *unitary* when $u^{-1}=\bar{u}'$.

A real unitary matrix is orthogonal but an orthogonal matrix need not be real.

All the latent roots of a unitary matrix have unit modulus.

Matrix is *persymmetric* when the value of u_{ij} depends only on $(i+j)$."

p. 110. Insert dagger † against footnote.

p. 120, 2nd table, 2nd row, for "$r_6 - 2r_9$" read "$r_6 - 3r_9$".

p. 121, line 1. Delete "method of".

p. 121, § 4 12, lines 2, 3. For "the first or the third" read "any".

p. 144, para. beginning at line 8 should read

"If there are p distinct dominant roots $\lambda_1 \lambda_2 \ldots \lambda_p$ and if $\kappa_1 \kappa_2 \ldots \kappa_p$ are the corresponding modal rows, the procedure is as follows. Partition the $(\bar{p}, n)$ matrix $\{\kappa_1, \kappa_2, \ldots, \kappa_p\}$ in the form $[\alpha, \beta]$, where α is a (p, p) submatrix, assumed to be non-singular (rearrangement of the rows of u and columns of $[\alpha, \beta]$ may be necessary to satisfy this condition). In this case the required matrix w is constructed in the partitioned form

$$w = \begin{bmatrix} I, & \alpha^{-1}\beta \\ 0, & 0 \end{bmatrix}$$

and then

$$v = u(I - w) = u\begin{bmatrix} 0, & -\alpha^{-1}\beta \\ 0, & I \end{bmatrix}.$$

Evidently v has p zero columns and hence p zero latent roots. If rearrangement has been required, u must be in the corresponding rearranged form.

The choice of a non-singular submatrix α is a generalization of the choice of a non-zero element κ_{r1} in the elimination of a single dominant root.

This process is in effect that which is applied in the numerical example on p. 330."

p. 150, § 4·21, line 9. For "a machine could no doubt be" read "machines have been" and in line 10 delete "most of".

p. 152, line 5. For "changed" read "reversed".

p. 176, equation (4), denominator of third fraction, for

"$\Delta^{(1)}(\lambda_r)(\lambda - \lambda_r)$" read "$\lambda_r \Delta^{(1)}(\lambda_r)(\lambda - \lambda_r)$".

p. 195, § 6·5, line 7. For "intial" read "initial".

p. 252, equation at bottom, interchange first and third matrices on the right-hand side.

p. 277. The symbol α stands for a set of parameters *and* for the components of the total acceleration of P. One of these should be represented by β, say.

p. 291, § 9·9. The following is a simple alternative proof of the reality of the roots of the determinantal equation $\Delta_m(z) = 0$ when A and E are real and symmetrical.

Let z, k respectively denote any root and its associated modal column, and let $\bar{z}$, $\bar{k}$ be the corresponding conjugates (see § 1·17). Then

$$zAk = Ek. \qquad \ldots\ldots(1)$$

Premultiplication by $\bar{k}'$ yields

$$z\bar{k}'Ak = \bar{k}'Ek, \qquad \ldots\ldots(2)$$

and by transposition

$$zk'A\bar{k} = k'Ek.$$

The conjugate relation is

$$\bar{z}\bar{k}'Ak = \bar{k}'Ek. \qquad \text{......(3)}$$

Comparison of (2) and (3) gives $z = \bar{z}$, which shows that z is real. Thus by (1) k is real, and by (2) z is positive when the potential energy function is positive and definite.

p. 296, equation (7). An alternative is $Q = k_0'Eq$.

p. 309, equation in (*b*). We may replace $k_r'A$ by $k_r'E$ which may be simpler.

p. 310, § 10·2 (*e*), second sentence should read "The principle shows that first order errors in the mode yield only second order errors in the frequency as calculated by the equation of energy".

Also line 10 should read "used, and when U happens to be symmetrical, a convenient...".

p. 315, line 9 from bottom, for "Rayleigh's principle will next be applied" read "Since U is symmetrical, the extension of Rayleigh's principle given in § 10·2 (*e*) can be applied...".

p. 363, § 12·3, line 4, for "given" read "are given"

p. 396, line above the diagram. For "0·84 degree" read "84 degrees per lb.ft."

PREFACE

The number of published treatises on matrices is not large, and so far as we are aware this is the first which develops the subject with special reference to its applications to differential equations and classical mechanics. The book is written primarily for students of applied mathematics who have no previous knowledge of matrices, and we hope that it will help to bring about a wider appreciation of the conciseness and power of matrices and of their convenience in computation. The general scope of the book is elementary, but occasional discussions of advanced questions are not avoided. The sections containing these discussions, which may with advantage be omitted at the first reading, are distinguished by an asterisk.

The first four chapters give an account of those properties of matrices which are required later for the applications. Chapters I to III introduce the general theory of matrices, while Chapter IV is devoted to various numerical processes, such as the reciprocation of matrices, the solution of algebraic equations, and the calculation of latent roots of matrices by iterative methods.

The remainder of the book is concerned with applications. Chapters V and VI deal in some detail with systems of linear ordinary differential equations with constant coefficients, and Chapter VII contains examples of numerical solutions of systems of linear differential equations with variable coefficients. The last six chapters take up the subject of mechanics. They include an account of the kinematics and dynamics of systems, a separate discussion of motions governed by linear differential equations, illustrations of iterative methods of numerical solution, and a treatment of simple dynamical systems involving solid friction. The part played by friction in the motions of dynamical systems is as yet very incompletely understood, and we have considered it useful to include a very brief description of some experimental tests of the theory.

A considerable number of worked numerical examples has been included. It is our experience that the practical mathematician, whose requirements we have mainly considered, is often able to grasp the significance of a general algebraic theorem more thoroughly when it is illustrated in terms of actual numbers. For examples of applications of dynamical theory we have usually chosen problems relating to the oscillations of aeroplanes or aeroplane structures. Such problems conveniently illustrate the properties of dissipative dynamical systems, and they have a considerable practical importance.

A word of explanation is necessary in regard to the scheme of numbering adopted for paragraphs, equations, tables, and diagrams. The fourth paragraph of Chapter I, for example, is denoted by § 1·4. The two equations introduced in § 1·4 are numbered (1) and (2), but when it is necessary in later paragraphs to refer back to these equations they are described, respectively, as equations (1·4·1) and (1·4·2). Tables and diagrams are numbered in each paragraph in serial order: thus, the two consecutive tables which appear in § 7·13 are called Tables 7·13·1 and 7·13·2, while the single diagram introduced is Fig. 7·13·1.

The list of references makes no pretence to be complete, and in the case of theorems which are now so well established as to be almost classical, historical notices are not attempted. We believe that much of the subject-matter—particularly that relating to the applications—presents new features and has not appeared before in text-books. However, in a field so extensive and so widely explored as the theory of matrices, it would be rash to claim complete novelty for any particular theorem or method.

The parts of the book dealing with applications are based very largely on various mathematical investigations carried out by us during the last seven years for the Aeronautical Research Committee. We wish to express our great indebtedness to that Committee and to the Executive Committee of the National Physical Laboratory for permission to refer to, and expand, a number of unpublished reports, and for granting many other facilities in the preparation of the book. We wish also to record our appreciation of the care which the Staff of the Cambridge University Press has devoted to the printing.

Our thanks are also due to Miss Sylvia W. Skan of the Aerodynamics Department of the National Physical Laboratory for considerable assistance in computation and in the reading of proofs.

R. A. F.
W. J. D.
A. R. C.

March 1938

CHAPTER I

FUNDAMENTAL DEFINITIONS AND ELEMENTARY PROPERTIES

1·1. Preliminary Remarks. Matrices are sets of numbers or other elements which are arranged in rows and columns as in a double entry table and which obey certain rules of addition and multiplication. These rules will be explained in §§ 1·3, 1·4.

Rectangular arrays of numbers are of course very familiar in geometry and physics. For example, an ordinary three-dimensional vector is represented by three numbers called its components arranged in one row, while the state of stress at a point in a medium can be represented by nine numbers arranged in three rows and three columns. However, two points must be emphasised in relation to matrices. Firstly, the idea of a matrix implies the treatment of its elements taken as a whole and in their proper arrangement. Secondly, matrices are something more than the mere arrays of their elements, in view of the rules for their addition and multiplication.

1·2. Notation and Principal Types of Matrix. (*a*) *Rectangular Matrices.* The usual method of representing a matrix is to enclose the array of its elements within brackets, and in general square brackets are used for this purpose.* For instance, the matrix formed from the array

$$\begin{matrix} 1 & 12 & 0 \\ 5 & 6 & 1 \end{matrix}$$

is represented by

$$\begin{bmatrix} 1 & 12 & 0 \\ 5 & 6 & 1 \end{bmatrix}.$$

The meaning of other special brackets will be explained later. If a matrix contains lengthy numbers or complicated algebraic expressions, the elements in the rows can be shown separated by commas to avoid confusion.

The typical element of a matrix such as

$$\begin{bmatrix} A_{11} & A_{12} & \cdots & A_{1n} \\ A_{21} & A_{22} & \cdots & A_{2n} \\ \cdots & \cdots & \cdots & \cdots \\ A_{m1} & A_{m2} & \cdots & A_{mn} \end{bmatrix}$$

* Some writers employ thick round brackets or double lines.

can be denoted by A_{ij}, where the suffices i and j are understood to range from 1 to m and from 1 to n, respectively. A convenient abbreviated notation for the complete matrix is then $[A_{ij}]$, but in cases where no confusion can arise it is preferable to omit the matrix brackets and the suffices altogether and to write the matrix simply as A.

The letters i, j are generally used in the sense just explained as suffices for a typical element of a matrix. Specific elements will generally have other suffices, such as m, n, r, s.

(*b*) *Order*. A matrix having m rows and n columns is said to be of order m by n. For greater brevity, such a matrix will usually be referred to as an $(\overline{m}, n)$ matrix; the bar shows which of the two numbers m, n relates to the rows.*

(*c*) *Column Matrices and Row Matrices*. A matrix having m elements arranged in a single column—namely, an $(\overline{m}, 1)$ matrix—will be called a *column matrix*. A column of numbers occupies much vertical space, and it is often preferable to adopt the convention that a single row of elements enclosed within braces represents a column matrix. For instance,

$$\{x_1, x_2, x_3\} \equiv \begin{bmatrix} x_1 \\ x_2 \\ x_3 \end{bmatrix}.$$

A literal matrix such as the above can be written in the abbreviated form $\{x_i\}$.

In the same way a matrix with only a single row of elements will be spoken of as a *row matrix*.† When it is necessary to write a row matrix at length, the usual square brackets will be employed; but the special brackets $\lfloor\ \rfloor$ will be used to denote a literal row matrix in the abbreviated form. For example,

$$\lfloor y_j \rfloor \equiv [y_1, y_2, y_3].$$

In accordance with the foregoing conventions, the matrix formed from the rth column of an $(\overline{m}, n)$ matrix $[A_{ij}]$ is

$$\{A_{1r}, A_{2r}, \ldots, A_{mr}\},$$

and this can be represented as $\{A_{ir}\}$, provided that i is always taken to be the typical suffix and r the specific suffix. In the same way the matrix formed from the sth row of $[A_{ij}]$ is

$$[A_{s1}, A_{s2}, \ldots, A_{sn}],$$

and this can be expressed as $\lfloor A_{sj} \rfloor$.

* An alternative notation, which is in current use, is $[A]_m^n$.

† A row matrix is often called a *line matrix*, a *vector of the first kind*, or a *prime*; while a column matrix is referred to as a *vector of the second kind*, or a *point*.

The most concise notation for column and row matrices is, as with matrices of a general order, by means of single letters. The particular type of matrix represented by a single letter will always be clear from the context.

(*d*) *Transposition of Matrices.* The *transposed* A' of a matrix A is defined to be the matrix which has rows identical with the columns of A. Thus if $A = [A_{ij}]$, then $A' = [A_{ji}]$. In particular the transposed of a column matrix is a row matrix, and *vice versa.*

In this book an accent applied to a matrix will always denote the transposition of that matrix.

(*e*) *Square, Diagonal, and Unit Matrices.* When the numbers of the rows and columns are equal, say n, the matrix is said to be *square* and of order n: the elements of type A_{ii} then lie in the *principal diagonal.* If all the elements other than those in the principal diagonal are zero, the matrix is called a *diagonal matrix.* The *unit matrix* of order n is defined to be the diagonal matrix of order n which has units for all its principal diagonal elements. It is denoted by I_n, or more simply by I when the order is apparent.*

(*f*) *Symmetrical and Skew Matrices.* When $A_{ij} = A_{ji}$ the matrix A is said to be *symmetrical,* and it is then identical with its transposed. If $A_{ij} = -A_{ji}$, whereas the elements of type A_{ii} are not all zero, the matrix is *skew*; but if both $A_{ij} = -A_{ji}$ and $A_{ii} = 0$ the matrix is *skew symmetric* or *alternate.* Both symmetrical and skew matrices are necessarily square.

(*g*) *Null Matrices.* A matrix of which the elements are all zero is called a *null matrix,* and is represented by 0.

EXAMPLES

(i) $(\bar{3}, 2)$ *Matrix.*
$$\begin{bmatrix} 1 & 2 \\ -1 & 0 \\ 3 & -1 \end{bmatrix}.$$

(ii) *Row Matrix.* $[0, 1, -3, 0]$.

(iii) *Column Matrix.* $\{2, -1, -3, 1\}$.

(iv) *Symmetrical Square Matrix.*
$$\begin{bmatrix} 1 & 2 & 0 \\ 2 & 0 & -1 \\ 0 & -1 & -2 \end{bmatrix}.$$

* On the Continent the symbol commonly used for the unit matrix is E.

(v) *Diagonal Matrix.* $\begin{bmatrix} 2 & 0 & 0 \\ 0 & 1 & 0 \\ 0 & 0 & -3 \end{bmatrix}.$

(vi) *Unit Matrix I_2.* $\begin{bmatrix} 1 & 0 \\ 0 & 1 \end{bmatrix}.$

(vii) *Transposed Matrices.*

If $A = \begin{bmatrix} 1 & 2 \\ -1 & 0 \\ 3 & -1 \end{bmatrix}$, then $A' = \begin{bmatrix} 1 & -1 & 3 \\ 2 & 0 & -1 \end{bmatrix}.$

(viii) *Null Matrices.* $\{0, 0, 0\}$; $\begin{bmatrix} 0 & 0 & 0 \\ 0 & 0 & 0 \\ 0 & 0 & 0 \end{bmatrix}.$

1·3. Summation of Matrices and Scalar Multipliers. Operations with matrices involve operations with the elements of which they are composed. Unless the contrary is stated, these elements will always be understood to be numbers, real or complex, which obey the laws of ordinary algebra (i.e. scalars). It is, however, sometimes useful to consider matrices the elements of which are not ordinary numbers. For example, the elements may themselves be matrices (see § 1·7).

(*a*) *Equality of Matrices.* Equal matrices are necessarily of the same order and have their corresponding elements equal. Thus $A = B$ if $A_{ij} = B_{ij}$.

The equality of two matrices of order m by n implies by definition the satisfaction of mn ordinary equations between their elements. Conversely, a set of ordinary equations can always be represented by a single equation between matrices.

(*b*) *Addition and Subtraction of Matrices.* These operations can only be performed on matrices of the same order.

The *sum* of two such matrices A and B is defined to be the matrix C the typical element C_{ij} of which is $A_{ij} + B_{ij}$. Then

$$A + B = C. \qquad \text{......(1)}$$

The *difference* of A and B is similarly the matrix D the typical element D_{ij} of which is $A_{ij} - B_{ij}$. Then

$$A - B = D. \qquad \text{......(2)}$$

Since any element in a matrix sum is equal to the algebraic sum of the corresponding elements in the summed matrices, we see that the addition of matrices is subject to the same laws as the addition of scalars. Thus the associative and commutative laws of addition hold good.

(c) *Scalar Multipliers.* If $A = B$, it is natural to write equation (1) as $2A = C$, with $C_{ij} = 2A_{ij}$. More generally, the convention is adopted that multiplication of a matrix by a *scalar* coefficient, say l, written either before or after the matrix, is equivalent to multiplication of every element by l. Thus, if

$$lA = Al = C,$$

then

$$C_{ij} = lA_{ij}.$$

The foregoing definitions and conventions are sufficient for the interpretation and reduction of any expression which is linear and homogeneous in a set of matrices of the same order.

Examples

(i) *Matrix Equation Expressed as Scalar Equations.* The single matrix equation

$$\begin{bmatrix} a_{11} & a_{12} \\ a_{21} & a_{22} \end{bmatrix} = \begin{bmatrix} b_{11} & b_{12} \\ b_{21} & b_{22} \end{bmatrix}$$

yields the four scalar equations

$$a_{11} = b_{11};\ a_{12} = b_{12};\ a_{21} = b_{21};\ a_{22} = b_{22}.$$

(ii) *Scalar Equations Expressed as Matrix Equation.* The four scalar equations

$$a_1 = b_1;\ a_2 = b_2;\ a_3 = b_3;\ a_4 = b_4$$

are contained in each of the matrix equations

$$\begin{bmatrix} a_1 & a_2 \\ a_3 & a_4 \end{bmatrix} = \begin{bmatrix} b_1 & b_2 \\ b_3 & b_4 \end{bmatrix},$$

$$[a_1, a_2, a_3, a_4] = [b_1, b_2, b_3, b_4],$$

$$\{a_1, a_2, a_3, a_4\} = \{b_1, b_2, b_3, b_4\}.$$

(iii) *Sum of Matrices.*

$$\begin{bmatrix} 1 & 2 & 3 \\ 2 & 3 & 4 \end{bmatrix} + \begin{bmatrix} 4 & 5 & 6 \\ 5 & 6 & 7 \end{bmatrix} = \begin{bmatrix} 5 & 7 & 9 \\ 7 & 9 & 11 \end{bmatrix}.$$

(iv) *Difference of Matrices.*

$$\begin{bmatrix} 1 & 0 & -1 \\ 2 & -7 & 3 \end{bmatrix} - \begin{bmatrix} -1 & 1 & 2 \\ 1 & -2 & 0 \end{bmatrix} = \begin{bmatrix} 2 & -1 & -3 \\ 1 & -5 & 3 \end{bmatrix}.$$

(v) *Scalar Multipliers.*

$$2\begin{bmatrix} 1 & 2 \\ 3 & 4 \end{bmatrix} + 3\begin{bmatrix} -2 & 0 \\ 0 & -1 \end{bmatrix} - 5\begin{bmatrix} -2 & 7 \\ 2 & 1 \end{bmatrix} = \begin{bmatrix} 6 & -31 \\ -4 & 0 \end{bmatrix}.$$

1·4. Multiplication of Matrices. With matrix multiplication two essential facts must be borne in mind. Firstly, matrices are in general not commutative in multiplication. Secondly, two matrices can only be multiplied in a given order provided that a certain condition is satisfied. If the number of columns in B is equal to the number of rows in A, the two matrices are described as *conformable*, and they can then be multiplied in the order $B \times A$. Specifically, if B is a $(\bar{q}, n)$ matrix and A is an $(\bar{n}, p)$ matrix, then the product BA is a $(\bar{q}, p)$ matrix. A scheme which expresses this rule very simply is

$$(\bar{q}, n) \times (\bar{n}, p) = (\bar{q}, p). \qquad \text{......(1)}$$

The product BA is referred to either as A *premultiplied* by B, or as B *postmultiplied* by A.

We may now define the process of multiplication. To obtain the ith element in the jth column of the product $P \equiv BA$, select the ith row of B and the jth column of A, and sum the products of their corresponding elements, beginning at the left-hand end and the top, respectively: thus

$$P_{ij} = \sum_{r=1}^{n} B_{ir} A_{rj}. \qquad \text{......(2)}$$

As a particular case assume in (1) that $q = 1$ and $p = 1$, so that the first matrix has merely a single row, say $\lfloor B_j \rfloor$, of n elements, while the second has a single column, say $\{A_i\}$, of n elements. The product $\lfloor B_j \rfloor \{A_i\}$ in this case is a $(\bar{1}, 1)$ matrix, or a scalar*, which is given by (2) as the sum of the products of the corresponding elements in $\lfloor B_j \rfloor$ and $\{A_i\}$. The general process of multiplication of two matrices may accordingly be interpreted as follows: To obtain the typical element P_{ij} of the product BA, postmultiply the ith row of B by the jth column of A.

From (1) it is seen that two matrices B and A can be multiplied in both the orders BA and AB only provided they are of the types $(\bar{p}, n)$ and $(\bar{n}, p)$: the products are then of the types $(\bar{p}, p)$ and $(\bar{n}, n)$, respectively. In particular this condition is satisfied for square matrices of equal order: however, even in this case the two products are usually not the same. Two matrices having the special property that $BA = AB$ are said to *commute* or to be *permutable*. The unit matrix I, for instance, commutes with any square matrix of the same order.

* The caution should be added that, although a $(\bar{1}, 1)$ matrix can always be treated as a scalar, the converse is only true when conformability allows.

Examples

(i) *Product of Rectangular Matrices.*

$$\begin{bmatrix} 4 & 2 & -1 & 2 \\ 3 & -7 & 1 & -8 \\ 2 & 4 & -3 & 1 \end{bmatrix} \begin{bmatrix} 2 & 3 \\ -3 & 0 \\ 1 & 5 \\ 3 & 1 \end{bmatrix}$$

$$= \begin{bmatrix} (4\times 2)-(2\times 3)-(1\times 1)+(2\times 3), & (4\times 3)+(2\times 0)-(1\times 5)+(2\times 1) \\ (3\times 2)+(7\times 3)+(1\times 1)-(8\times 3), & (3\times 3)-(7\times 0)+(1\times 5)-(8\times 1) \\ (2\times 2)-(4\times 3)-(3\times 1)+(1\times 3), & (2\times 3)+(4\times 0)-(3\times 5)+(1\times 1) \end{bmatrix}$$

$$= \begin{bmatrix} 7 & 9 \\ 4 & 6 \\ -8 & -8 \end{bmatrix}.$$

The rule (1) here gives $(\bar{3}, 4) \times (\bar{4}, 2) = (\bar{3}, 2)$. The matrices are not conformable when taken in the reverse order.

(ii) *Products of Square Matrices.*

$$\begin{bmatrix} 3 & 4 \\ -2 & -1 \end{bmatrix} \begin{bmatrix} 1 & 2 \\ 2 & 5 \end{bmatrix} = \begin{bmatrix} (3\times 1)+(4\times 2), & (3\times 2)+(4\times 5) \\ -(2\times 1)-(1\times 2), & -(2\times 2)-(1\times 5) \end{bmatrix} = \begin{bmatrix} 11, & 26 \\ -4, & -9 \end{bmatrix}.$$

When the matrices are multiplied in the reverse order the product is

$$\begin{bmatrix} 1 & 2 \\ 2 & 5 \end{bmatrix} \begin{bmatrix} 3 & 4 \\ -2 & -1 \end{bmatrix} = \begin{bmatrix} (1\times 3)-(2\times 2), & (1\times 4)-(2\times 1) \\ (2\times 3)-(5\times 2), & (2\times 4)-(5\times 1) \end{bmatrix} = \begin{bmatrix} -1 & 2 \\ -4 & 3 \end{bmatrix}.$$

Another illustration is provided by the pair of products

$$\begin{bmatrix} 3 & 4 & 2 \\ -2 & -1 & -1 \\ -1 & -3 & -1 \end{bmatrix} \begin{bmatrix} 1 & 1 & 1 \\ 2 & 2 & 2 \\ 5 & 5 & 5 \end{bmatrix} = \begin{bmatrix} 21 & 21 & 21 \\ -9 & -9 & -9 \\ -12 & -12 & -12 \end{bmatrix},$$

$$\begin{bmatrix} 1 & 1 & 1 \\ 2 & 2 & 2 \\ 5 & 5 & 5 \end{bmatrix} \begin{bmatrix} 3 & 4 & 2 \\ -2 & -1 & -1 \\ -1 & -3 & -1 \end{bmatrix} = 0.$$

(iii) *Products of Permutable Matrices.*

$$\begin{bmatrix} 0 & -3 & 1 \\ 2 & -1 & 1 \\ 2 & -1 & 1 \end{bmatrix} \begin{bmatrix} 0 & -1 & 1 \\ 0 & 1 & -1 \\ 0 & 3 & -3 \end{bmatrix} = \begin{bmatrix} 0 & -1 & 1 \\ 0 & 1 & -1 \\ 0 & 3 & -3 \end{bmatrix} \begin{bmatrix} 0 & -3 & 1 \\ 2 & -1 & 1 \\ 2 & -1 & 1 \end{bmatrix} = 0,$$

$$\begin{bmatrix} a & b \\ c & d \end{bmatrix} \begin{bmatrix} 1 & 0 \\ 0 & 1 \end{bmatrix} = \begin{bmatrix} 1 & 0 \\ 0 & 1 \end{bmatrix} \begin{bmatrix} a & b \\ c & d \end{bmatrix} = \begin{bmatrix} a & b \\ c & d \end{bmatrix}.$$

(iv) *Column Matrix Premultiplied by Row Matrix.*

$$[5\ \ 2\ \ -3]\{2\ \ -1\ \ 4\} \equiv [5\ \ 2\ \ -3]\begin{bmatrix} 2 \\ -1 \\ 4 \end{bmatrix} = -4.$$

The rule (1) here gives $(\bar{1}, 3) \times (\bar{3}, 1) = (\bar{1}, 1)$; hence the product is a scalar.

(v) *Row Matrix Premultiplied by Column Matrix.*

$$\{2\ \ -1\ \ 4\}[5\ \ 2\ \ -3] \equiv \begin{bmatrix} 2 \\ -1 \\ 4 \end{bmatrix}[5\ \ 2\ \ -3] = \begin{bmatrix} 10 & 4 & -6 \\ -5 & -2 & 3 \\ 20 & 8 & -12 \end{bmatrix}.$$

In this case the rule (1) gives $(\bar{3}, 1) \times (\bar{1}, 3) = (\bar{3}, 3)$. Note that the product here is of a very special type. The elements in any row (or column) are proportional to the corresponding elements in any other row (or column), so that the product has in fact only a single linearly independent row (or column). Examples (ii) and (iii) contain other illustrations of square matrices with this property. More generally, an $(\bar{m}, n)$ matrix with only a single linearly independent row (or column) is always expressible as a product of the form $\{A_i\}\lfloor B_j\rfloor$, where $\{A_i\}$ is a column of m elements and $\lfloor B_j \rfloor$ is a row of n elements. For instance,

$$\begin{bmatrix} 10, & -5, & 20 \\ 4, & -2, & 8 \end{bmatrix} = \begin{bmatrix} 5 \\ 2 \end{bmatrix}[2, -1, 4] \equiv \{5, 2\}[2, -1, 4].$$

(vi) *Square Matrix Postmultiplied by Column Matrix.*

$$ax \equiv \begin{bmatrix} a_{11} & a_{12} & a_{13} \\ a_{21} & a_{22} & a_{23} \\ a_{31} & a_{32} & a_{33} \end{bmatrix}\begin{bmatrix} x_1 \\ x_2 \\ x_3 \end{bmatrix} = \begin{bmatrix} a_{11}x_1 + a_{12}x_2 + a_{13}x_3 \\ a_{21}x_1 + a_{22}x_2 + a_{23}x_3 \\ a_{31}x_1 + a_{32}x_2 + a_{33}x_3 \end{bmatrix}.$$

The system of linear algebraic equations

$$a_{11}x_1 + a_{12}x_2 + a_{13}x_3 = b_1,$$
$$a_{21}x_1 + a_{22}x_2 + a_{23}x_3 = b_2,$$
$$a_{31}x_1 + a_{32}x_2 + a_{33}x_3 = b_3$$

is accordingly concisely expressible as the matrix equation

$$ax = b.$$

(vii) *Abbreviated Rules for Products of Special Matrices.*

Matrix × column = column,
Row × matrix = row,
Row × column = scalar,
Column × row = matrix with proportional rows and proportional columns.

1·5. Continued Products of Matrices. A continued product of matrices, such as CBA, is to be interpreted as follows. First premultiply A by B, and then premultiply the product BA by C. This process will of course not be possible unless B is conformable with A and C with BA.

In the foregoing definition of a continued product a specific order of multiplication is laid down. However, it will now be shown that the associative law holds good for matrix multiplication, so that the factors in a product may be grouped in any convenient manner, provided that the order of multiplication is not altered.

The associative law requires that if $Y = CB$ and $X = BA$, then

$$CBA = YA = CX. \qquad \text{......(1)}$$

To prove formally that this is a consequence of the definitions of § 1·4, we note firstly that equation (1·4·2) gives for the typical element of X

$$X_{ij} = \sum_{r=1}^{n} B_{ir} A_{rj},$$

where n is the number of columns in B and of rows in A. Hence the (k,j)th element of CX is

$$\sum_{i=1}^{m} C_{ki} X_{ij} = \sum_{i=1}^{m} \sum_{r=1}^{n} C_{ki} B_{ir} A_{rj}, \qquad \text{......(2)}$$

where m is the number of columns in C. Similarly, since

$$Y_{kr} = \sum_{i=1}^{m} C_{ki} B_{ir},$$

the (k,j)th element of YA is

$$\sum_{r=1}^{n} Y_{kr} A_{rj} = \sum_{i=1}^{m} \sum_{r=1}^{n} C_{ki} B_{ir} A_{rj}, \qquad \text{......(3)}$$

in agreement with (2). This proves the truth of (1).

In view of the associative law, it will now be clear that a continued product, or product chain, such as

$$A_m A_{m-1} \dots A_2 A_1$$

will only have a meaning provided that adjacent matrices A_s, A_{s-1} in the chain are conformable. Thus, if r_s and c_s denote, respectively, the number of rows and columns in A_s, the conditions to be satisfied are

$$c_s = r_{s-1}$$

for $s = m, m-1, \dots, 2$. The scheme of multiplication in this case may be represented by

$$(\bar{r}_m, r_{m-1}) \times (\bar{r}_{m-1}, r_{m-2}) \times \dots \times (\bar{r}_2, r_1) \times (\bar{r}_1, c_1) = (\bar{r}_m, c_1).$$

The product, accordingly, is a matrix having r_m rows and c_1 columns.

For brevity, the product of two equal square matrices A is written A^2, and a similar notation is adopted for the other positive integral powers.

The distributive law also holds good for matrix multiplication. For example,

$$E(A+B)F = EAF+EBF. \qquad \text{......(4)}$$

The correctness of this follows at once from the formula (2) and the definition of addition.

Examples

(i) *Associative Law.* Compare

$$\begin{bmatrix} 3 & 4 \\ -2 & -1 \end{bmatrix}\begin{bmatrix} 1 & 2 \\ 2 & 5 \end{bmatrix}\times\begin{bmatrix} 2 & -1 \\ 0 & 3 \end{bmatrix} = \begin{bmatrix} 11 & 26 \\ -4 & -9 \end{bmatrix}\begin{bmatrix} 2 & -1 \\ 0 & 3 \end{bmatrix} = \begin{bmatrix} 22 & 67 \\ -8 & -23 \end{bmatrix} \qquad \text{......(5)}$$

with

$$\begin{bmatrix} 3 & 4 \\ -2 & -1 \end{bmatrix}\times\begin{bmatrix} 1 & 2 \\ 2 & 5 \end{bmatrix}\begin{bmatrix} 2 & -1 \\ 0 & 3 \end{bmatrix} = \begin{bmatrix} 3 & 4 \\ -2 & -1 \end{bmatrix}\begin{bmatrix} 2 & 5 \\ 4 & 13 \end{bmatrix} = \begin{bmatrix} 22 & 67 \\ -8 & -23 \end{bmatrix}. \qquad \text{......(6)}$$

(ii) *Rule for Product.* With

$$P = \{2 \ \ -1 \ \ 4\}[5 \ \ 2 \ \ -3]\{1 \ \ 0 \ \ 2\}[-1 \ \ 2], \qquad \text{......(7)}$$

the rule for the product gives

$$(\bar{3},1)\times(\bar{1},3)\times(\bar{3},1)\times(\bar{1},2) = (\bar{3},2).$$

The product thus exists, and is a matrix with three rows and two columns. To evaluate P, note that the part-product $[5 \ \ 2 \ \ -3]\{1 \ \ 0 \ \ 2\}$ yields the scalar -1, which may be brought to the front: hence

$$P = -1\times\{2 \ \ -1 \ \ 4\}[-1 \ \ 2] = \begin{bmatrix} 2 & -4 \\ -1 & 2 \\ 4 & -8 \end{bmatrix}.$$

(iii) *Positive Powers of a Square Matrix.*

$$\begin{bmatrix} 3 & 4 \\ -2 & 1 \end{bmatrix}^2 \equiv \begin{bmatrix} 3 & 4 \\ -2 & 1 \end{bmatrix}\begin{bmatrix} 3 & 4 \\ -2 & 1 \end{bmatrix} = \begin{bmatrix} 1 & 16 \\ -8 & -7 \end{bmatrix},$$

$$\begin{bmatrix} 3 & 4 \\ -2 & 1 \end{bmatrix}^3 = \begin{bmatrix} 3 & 4 \\ -2 & 1 \end{bmatrix}\begin{bmatrix} 1 & 16 \\ -8 & -7 \end{bmatrix} = \begin{bmatrix} 1 & 16 \\ -8 & -7 \end{bmatrix}\begin{bmatrix} 3 & 4 \\ -2 & 1 \end{bmatrix} = \begin{bmatrix} -29 & 20 \\ -10 & -39 \end{bmatrix}.$$

Note that positive integral powers of a square matrix are permutable.

(iv) *Computation from "Right to Left".* In this method the complete product is evaluated by successive premultiplications. For instance, if the product required is (7),

$$\{1\ \ 0\ \ 2\}[-1\ \ 2] = \begin{bmatrix} -1 & 2 \\ 0 & 0 \\ -2 & 4 \end{bmatrix},$$

$$[5\ \ 2\ \ -3]\begin{bmatrix} -1 & 2 \\ 0 & 0 \\ -2 & 4 \end{bmatrix} = [1\ \ -2],$$

and finally $$P = \{2\ \ -1\ \ 4\}[1\ \ -2] = \begin{bmatrix} 2 & -4 \\ -1 & 2 \\ 4 & -8 \end{bmatrix}.$$

Equation (6) is a further simple illustration of computation from right to left.

(v) *Computation from "Left to Right"*. Here successive post-multiplications are used. Thus, if the complete product is (7),

$$\{2\ \ -1\ \ 4\}[5\ \ 2\ \ -3] = \begin{bmatrix} 10 & 4 & -6 \\ -5 & -2 & 3 \\ 20 & 8 & -12 \end{bmatrix},$$

$$\begin{bmatrix} 10 & 4 & -6 \\ -5 & -2 & 3 \\ 20 & 8 & -12 \end{bmatrix}\{1\ \ 0\ \ 2\} = \{-2\ \ 1\ \ -4\},$$

and lastly, $$P = \{-2\ \ 1\ \ -4\}[-1\ \ 2] = \begin{bmatrix} 2 & -4 \\ -1 & 2 \\ 4 & -8 \end{bmatrix}.$$

Equation (5) provides another example.

(vi) *Use of Subproducts*. In this treatment the complete product chain is first split into a convenient set of subproducts, which are separately computed. The calculation is completed by multiplication of the chain of subproducts, or by separation into further subproducts. For instance,

$$P = \begin{bmatrix} 3 & 4 \\ -2 & -1 \end{bmatrix}\begin{bmatrix} 1 & 2 \\ 2 & 5 \end{bmatrix} \times \begin{bmatrix} 1 & 2 \\ -1 & -1 \end{bmatrix}\begin{bmatrix} 0 & 1 \\ 1 & -5 \end{bmatrix} \times \begin{bmatrix} 5 & 1 \\ 4 & 1 \end{bmatrix}\begin{bmatrix} -1 & 1 \\ 5 & 0 \end{bmatrix}$$

$$= \begin{bmatrix} 11 & 26 \\ -4 & -9 \end{bmatrix} \times \begin{bmatrix} 2 & -9 \\ -1 & 4 \end{bmatrix}\begin{bmatrix} 0 & 5 \\ 1 & 4 \end{bmatrix}$$

$$= \begin{bmatrix} 11 & 26 \\ -4 & -9 \end{bmatrix}\begin{bmatrix} -9 & -26 \\ 4 & 11 \end{bmatrix} = 5I_2.$$

Some of the methods of numerical solution of differential equations to be described in Chapter VII involve the computation of lengthy product chains of square matrices. With such product chains it is often preferable to adopt the method of subproducts rather than to compute the complete chain directly from left to right, or right to left; the use of subproducts facilitates the correction of errors. For instance, suppose the second matrix from the left, namely $\begin{bmatrix} 1 & 2 \\ 2 & 5 \end{bmatrix}$, to have been incorrectly entered as $\begin{bmatrix} 1 & 2 \\ 2 & 3 \end{bmatrix}$. The rectification of this error involves a recalculation of the first subproduct but no recalculation of the other subproducts. On the other hand, if the chain is computed from left to right, the correction is much more troublesome.

A further important case is that in which a chain of square matrices is postmultiplied by a column matrix. For instance, consider the product

$$\begin{bmatrix} 1 & 2 \\ -1 & -1 \end{bmatrix}\begin{bmatrix} 0 & 1 \\ 1 & -5 \end{bmatrix}\begin{bmatrix} 5 & 1 \\ 4 & 1 \end{bmatrix}\begin{bmatrix} -1 & 1 \\ 5 & 0 \end{bmatrix}\begin{bmatrix} 3 \\ 4 \end{bmatrix}.$$

Here computation from right to left obviously requires the least labour, since at each stage merely the postmultiplication of a square matrix by a column matrix is involved.

1·6. Properties of Diagonal and Unit Matrices. Suppose A to be a square matrix of order n, and B to be a diagonal* matrix of the same order. If

$$P = BA,$$

then

$$P_{ij} = \sum_{r=1}^{n} B_{ir} A_{rj} = B_{ii} A_{ij}, \qquad \text{......(1)}$$

since in the present case $B_{ir} = 0$ unless $r = i$. Hence premultiplication by a diagonal matrix has the effect of multiplying every element in any row of a matrix by a constant. Similarly, postmultiplication by a diagonal matrix results in the multiplication of every element in any column of a matrix by a constant.

Diagonal matrices are clearly permutable with one another, and the product is always another diagonal matrix. In the particular case where $A_{ii} = B_{ii}^{-1}$,

$$BA = AB = I.$$

These relations are characteristic of *reciprocal* or *inverse* matrices (see § 1·11).

* For definition of diagonal matrix see § 1·2 (*e*).

From the definition of the unit matrix it follows that if A is any square matrix

$$IA = AI = A, \qquad \text{......(2)}$$

and that

$$I^m = I. \qquad \text{......(3)}$$

It is on account of the properties expressed by (2) and (3) that the name unit matrix is justified.

A diagonal matrix whose diagonal elements are all equal is called a *scalar matrix*. It is equal to the unit matrix multiplied by a scalar.

Examples

(i) *Premultiplication by Diagonal Matrix.*

$$\begin{bmatrix} 1 & 0 & 0 \\ 0 & -4 & 0 \\ 0 & 0 & 3 \end{bmatrix} \begin{bmatrix} 2 & 0 & 7 \\ -1 & 4 & 5 \\ 3 & 1 & 2 \end{bmatrix} = \begin{bmatrix} 2 & 0 & 7 \\ 4 & -16 & -20 \\ 9 & 3 & 6 \end{bmatrix}.$$

(ii) *Postmultiplication by Diagonal Matrix.*

$$\begin{bmatrix} 2 & 0 & 7 \\ -1 & 4 & 5 \\ 3 & 1 & 2 \end{bmatrix} \begin{bmatrix} 1 & 0 & 0 \\ 0 & -4 & 0 \\ 0 & 0 & 3 \end{bmatrix} = \begin{bmatrix} 2 & 0 & 21 \\ -1 & -16 & 15 \\ 3 & -4 & 6 \end{bmatrix}.$$

(iii) *Commutative Properties.*

$$\begin{bmatrix} 1 & 0 & 0 \\ 0 & -4 & 0 \\ 0 & 0 & 3 \end{bmatrix} \begin{bmatrix} 2 & 0 & 0 \\ 0 & 4 & 0 \\ 0 & 0 & 2 \end{bmatrix} = \begin{bmatrix} 2 & 0 & 0 \\ 0 & 4 & 0 \\ 0 & 0 & 2 \end{bmatrix} \begin{bmatrix} 1 & 0 & 0 \\ 0 & -4 & 0 \\ 0 & 0 & 3 \end{bmatrix} = \begin{bmatrix} 2 & 0 & 0 \\ 0 & -16 & 0 \\ 0 & 0 & 6 \end{bmatrix}.$$

(iv) *Reciprocal Diagonal Matrices.*

$$\begin{bmatrix} 0{\cdot}1 & 0 & 0 \\ 0 & -0{\cdot}2 & 0 \\ 0 & 0 & -4 \end{bmatrix} \begin{bmatrix} 10 & 0 & 0 \\ 0 & -5 & 0 \\ 0 & 0 & -0{\cdot}25 \end{bmatrix} = I_3.$$

(v) *Scalar Matrices.*

$$\begin{bmatrix} -2 & 0 & 0 \\ 0 & -2 & 0 \\ 0 & 0 & -2 \end{bmatrix} \begin{bmatrix} 3 & 0 & 0 \\ 0 & 3 & 0 \\ 0 & 0 & 3 \end{bmatrix} = -6I_3.$$

1·7. Partitioning of Matrices into Submatrices. The array of a given matrix may be divided into smaller arrays by horizontal and vertical lines, and the matrix is then said to be *partitioned* into *submatrices*. The partitioning lines are usually shown dotted, as in

$$\left[\begin{array}{c:cc} 3 & 0 & 2 \\ -2 & -1 & -1 \\ \hdashline -1 & -3 & 5 \end{array}\right].$$

Each of the submatrices in a partitioned matrix may be represented by a single symbol, and the original matrix becomes a matrix of matrices; in this case the partitioning lines will usually be omitted, or perhaps indicated by commas. With certain fairly obvious restrictions partitioned matrices can be added and multiplied as if the submatrices were ordinary matrix elements, and the result of any such operation is the same as if the operation were performed on the original unpartitioned matrices.

Suppose two matrices of the same order to be partitioned in a corresponding manner. Then the submatrices occupying corresponding positions will be of the same order, and may be added. Since each element in the sum is the sum of the individual elements, it is clear that the sum of the partitioned matrices is equal to the sum of the original matrices.

Multiplication of partitioned matrices requires more detailed consideration. Let $BA = P$, where B is of type $(\overline{m}, p)$ and A of type $(\overline{p}, n)$. Suppose now that B is partitioned by, say, two vertical lines into three submatrices B_1, B_2, B_3; thus $B \equiv [B_1, B_2, B_3]$. Then since B now has three (matrix) elements arranged in a line, conformability requires that A shall be partitioned into three (matrix) elements A_1, A_2, A_3 arranged in a column; also the products B_1A_1, B_2A_2, B_3A_3 must be conformable. Hence the p columns of B and the p rows of A must be correspondingly partitioned by vertical and horizontal lines, respectively. Clearly the values of P given by BA and by $B_1A_1 + B_2A_2 + B_3A_3$ will be identical, for the partitioning has only the effect of splitting the typical sum $\sum_{r=1}^{p} B_{ir}A_{rj}$ into three portions which are subsequently added. Since the numbers of rows in B and of columns in A are quite unrelated, any horizontal partitioning of B and vertical partitioning of A may be quite arbitrary: the conformability of the submatrices will not be affected.

In general, for the multiplication of two partitioned matrices, it is necessary and sufficient that for every partitioning line between columns of the matrix on the left there shall be a partitioning line between the corresponding rows of the matrix on the right, and no further horizontal partitioning on the right.

EXAMPLES

(i) *Addition.*

$$\begin{bmatrix} 2 & 0 & 4 \\ 0 & 2 & 7 \\ -3 & 8 & -1 \end{bmatrix} + \begin{bmatrix} 3 & 0 & 1 \\ 0 & 3 & -2 \\ 8 & -3 & 6 \end{bmatrix} = \left[\begin{array}{cc|c} \multicolumn{2}{c|}{\multirow{2}{*}{$2I_2$}} & 4 \\ & & 7 \\ \hdashline -3 & 8 & -1 \end{array}\right] + \left[\begin{array}{cc|c} \multicolumn{2}{c|}{\multirow{2}{*}{$3I_2$}} & 1 \\ & & -2 \\ \hdashline 8 & -3 & 6 \end{array}\right] = 5\left[\begin{array}{cc|c} \multicolumn{2}{c|}{\multirow{2}{*}{I_2}} & 1 \\ & & 1 \\ \hdashline 1 & 1 & 1 \end{array}\right].$$

(ii) *Multiplication.*

$$\begin{bmatrix} 3 & 0 & 2 \\ -2 & -1 & -1 \\ -1 & -3 & 5 \end{bmatrix}\begin{bmatrix} 1 & -1 & 4 \\ 2 & 3 & 0 \\ 5 & 0 & 2 \end{bmatrix} = \left[\begin{array}{c|c|c} 3 & 0 & 2 \\ -2 & -1 & -1 \\ -1 & -3 & 5 \end{array}\right]\left[\begin{array}{ccc} 1 & -1 & 4 \\ \hdashline 2 & 3 & 0 \\ \hdashline 5 & 0 & 2 \end{array}\right]$$

$$= \{3, -2, -1\}[1, -1, 4] + \{0, -1, -3\}[2, 3, 0] + \{2, -1, 5\}[5, 0, 2]$$

$$= \begin{bmatrix} 3 & -3 & 12 \\ -2 & 2 & -8 \\ -1 & 1 & -4 \end{bmatrix} + \begin{bmatrix} 0 & 0 & 0 \\ -2 & -3 & 0 \\ -6 & -9 & 0 \end{bmatrix} + \begin{bmatrix} 10 & 0 & 4 \\ -5 & 0 & -2 \\ 25 & 0 & 10 \end{bmatrix}$$

$$= \begin{bmatrix} 13 & -3 & 16 \\ -9 & -1 & -10 \\ 18 & -8 & 6 \end{bmatrix}.$$

Another scheme of partitioning for the same product is

$$\left[\begin{array}{cc|c} 3 & 0 & 2 \\ -2 & -1 & -1 \\ \hdashline -1 & -3 & 5 \end{array}\right]\left[\begin{array}{ccc} 1 & -1 & 4 \\ 2 & 3 & 0 \\ \hdashline 5 & 0 & 2 \end{array}\right]$$

$$= \left[\begin{array}{c} \begin{bmatrix} 3 & 0 \\ -2 & -1 \end{bmatrix}\begin{bmatrix} 1 & -1 & 4 \\ 2 & 3 & 0 \end{bmatrix} + \begin{bmatrix} 2 \\ -1 \end{bmatrix}[5 \;\; 0 \;\; 2] \\ \hdashline [-1 \;\; -3]\begin{bmatrix} 1 & -1 & 4 \\ 2 & 3 & 0 \end{bmatrix} + 5[5 \;\; 0 \;\; 2] \end{array}\right]$$

$$= \left[\begin{array}{c} \begin{bmatrix} 3 & -3 & 12 \\ -4 & -1 & -8 \end{bmatrix} + \begin{bmatrix} 10 & 0 & 4 \\ -5 & 0 & -2 \end{bmatrix} \\ \hdashline [-7 \;\; -8 \;\; -4] + [25 \;\; 0 \;\; 10] \end{array}\right] = \begin{bmatrix} 13 & -3 & 16 \\ -9 & -1 & -10 \\ 18 & -8 & 6 \end{bmatrix}.$$

The convenience of partitioning when the matrices include many zero elements is illustrated by

$$\left[\begin{array}{cc|cc} 5 & 2 & 0 & 0 \\ 2 & 1 & 0 & 0 \\ \hdashline 0 & 0 & 8 & 3 \\ 0 & 0 & 5 & 2 \end{array}\right]\left[\begin{array}{cc|cc} 1 & -2 & 0 & 0 \\ -2 & 5 & 0 & 0 \\ \hdashline 0 & 0 & 2 & -3 \\ 0 & 0 & -5 & 8 \end{array}\right] = \left[\begin{array}{cc|cc} \multicolumn{2}{c|}{\begin{bmatrix} 5, & 2 \\ 2, & 1 \end{bmatrix}\begin{bmatrix} 1, & -2 \\ -2, & 5 \end{bmatrix}} & 0 & 0 \\ & & 0 & 0 \\ \hdashline 0 & 0 & \multicolumn{2}{c}{\begin{bmatrix} 8, & 3 \\ 5, & 2 \end{bmatrix}\begin{bmatrix} 2, & -3 \\ -5, & 8 \end{bmatrix}} \\ 0 & 0 & & \end{array}\right] = I_4.$$

1·8. Determinants of Square Matrices. The determinant of a square matrix A is the determinant whose array of elements is identical with that of A, and it is represented by $|A|$. It is shown in treatises on determinants that the product of two determinants $|B|$ and $|A|$ of the same order m can be represented by another determinant of order m. The multiplication rule for determinants can be expressed in several different ways, but one of them is as follows:

$$|B| \times |A| = |C|, \qquad \text{......(1)}$$

if

$$C_{ij} = \sum_{r=1}^{m} B_{ir} A_{rj}.$$

The array of the elements in $|C|$ is thus identical with that in the matrix product BA. Hence the determinant of the product of two matrices equals the product of their determinants. It should be noted particularly that if c is a scalar and A is a square matrix of order m, then

$$|cA| = c^m |A|. \qquad \text{......(2)}$$

Square matrices are also associated with minor determinants and cofactors. It may be recalled that the first minor of the determinant $|A|$, corresponding to the element A_{ij}, is defined to be the determinant obtained by omission of the ith row and the jth column of $|A|$; while the cofactor of A_{ij} is this minor multiplied by $(-1)^{i+j}$. If $|A|$ is of order m, any first minor is of order $m-1$. Similarly, the determinant obtained by omission of any s rows and s columns from $|A|$ is called an sth minor, or a minor of order $m-s$.

Methods for the computation of determinants and cofactors by the use of matrices are described in Chapter IV. A few special properties of determinants to which reference will be made later are given in the examples which follow.

EXAMPLES

(i) *Differentiation of Determinants.* Suppose $\Delta(\lambda)$ to denote a determinant of order m, the elements of which are polynomials of a parameter λ: for instance,

$$\Delta(\lambda) = \begin{vmatrix} f_{11}(\lambda) & f_{12}(\lambda) & \cdots & f_{1m}(\lambda) \\ f_{21}(\lambda) & f_{22}(\lambda) & \cdots & f_{2m}(\lambda) \\ \cdots & \cdots & \cdots & \cdots \\ f_{m1}(\lambda) & f_{m2}(\lambda) & \cdots & f_{mm}(\lambda) \end{vmatrix}.$$

Then by the usual rule for the differentiation of determinants $\frac{d\Delta}{d\lambda}$ is the sum of the determinants obtained when any one column is differentiated and the remaining columns are taken as they stand. Since each such determinant is expansible in terms of first minors of $\Delta(\lambda)$, it follows that $\dfrac{d\Delta}{d\lambda}$ is a linear homogeneous function of the first minors Δ_1 of $\Delta(\lambda)$. Hence also $\dfrac{d^2\Delta}{d\lambda^2}$ is a similar function of the second minors Δ_2 of $\Delta(\lambda)$, and more generally $\dfrac{d^p\Delta}{d\lambda^p}$ is linear in the pth minors Δ_p.

(ii) *Determinants whose Minors of a Certain Order all Vanish.* Suppose that, when $\lambda = \lambda_s$, at least one of the qth minors Δ_q of $\Delta(\lambda)$ does not vanish, whereas all the $(q-1)$th minors Δ_{q-1} do vanish, so that $\lambda - \lambda_s$ is a factor of every minor Δ_{q-1}. Now, by the theorem proved in example (i), $\dfrac{d\Delta_{q-2}}{d\lambda}$ is linear in the first minors of Δ_{q-2}—that is, in the minors Δ_{q-1} of $\Delta(\lambda)$. Hence $\lambda - \lambda_s$ is certainly a factor of $\dfrac{d\Delta_{q-2}}{d\lambda}$, so that $(\lambda - \lambda_s)^2$ at least is a factor of every minor Δ_{q-2}. In the same way $\dfrac{d\Delta_{q-3}}{d\lambda}$ is linear in the minors Δ_{q-2}, and thus contains the factor $(\lambda - \lambda_s)^2$. Accordingly $(\lambda - \lambda_s)^3$ at least is a factor of every minor Δ_{q-3}. Proceeding in this way we see finally that $(\lambda - \lambda_s)^{q-1}$ at least is a factor of every first minor Δ_1 of $\Delta(\lambda)$, while $(\lambda - \lambda_s)^q$ at least is a factor of $\Delta(\lambda)$ itself.

For instance, suppose

$$\Delta(\lambda) = \begin{vmatrix} 1 & 0 & 0 & 0 \\ 0 & \lambda & 0 & 0 \\ 0 & 0 & \lambda & 0 \\ 0 & 0 & 0 & \lambda \end{vmatrix}.$$

Here, when $\lambda = 0$, all the second minors of $\Delta(\lambda)$ are zero, but one third minor is not zero. Clearly λ^2 is a factor of every first minor, while λ^3 is a factor of $\Delta(\lambda)$.

Again, if
$$\Delta(\lambda) = \begin{vmatrix} 1 & 0 & 0 & 0 \\ 0 & \lambda & 0 & 0 \\ 0 & 0 & \lambda^2 & 0 \\ 0 & 0 & 0 & \lambda^3 \end{vmatrix},$$

we see that as before, when $\lambda = 0$, all second minors vanish whereas one third minor does not. In this case λ^3 is a factor of every first minor and λ^6 is a factor of $\Delta(\lambda)$.

1·9. Singular Matrices, Degeneracy, and Rank. In general the rows of a square matrix are linearly independent, but examples have already been given of matrices which do not satisfy this condition. If the determinant of a square matrix vanishes—so that the rows are not linearly independent—the matrix is said to be *singular*.

The rows of a singular matrix may be linearly connected by only a single relation, in which case the matrix is said to be *simply degenerate*, or to have *degeneracy** 1. If the rows are linearly connected by more than a single relation, the matrix is *multiply degenerate*, and in fact the degeneracy is q if there are q such relations. In accordance with this definition the rows of a square matrix of order m and degeneracy q will all be expressible as linear combinations of $m-q$ linearly distinct rows of elements. The quantity $m-q$ is usually spoken of as the *rank* of the matrix. The preceding considerations are, of course, true equally for the rows and the columns. The formal definitions of degeneracy and rank are as follows: A square matrix of order m is of degeneracy q when at least one of its qth minor determinants does not vanish, whereas all its $(q-1)$th minor determinants do vanish. The rank then is $m-q$.

From the preceding remarks it will be clear that the question of the degeneracy of a square matrix a of order m is intimately bound up with the number of linearly independent sets of quantities x which can satisfy the m homogeneous scalar equations contained in the matrix equation $ax = 0$ (see example (vi) of § 1·4). Here x denotes a column of m unknowns. When a is non-singular $x = 0$ is the only solution. When a is simply degenerate a non-zero column, say $x = \boldsymbol{x}$, can be found to satisfy the equation, and the most general solution then is an arbitrary multiple of $\boldsymbol{x}$. Again, when a is doubly degenerate two distinct non-zero columns can be found, say $\boldsymbol{x}(1)$ and $\boldsymbol{x}(2)$, and the most general solution is an arbitrary linear combination of these. More generally, when a has degeneracy q there are q linearly independent non-zero solutions.

Examples

(i) *Non-Singular Matrix.* The matrix

$$a \equiv \begin{bmatrix} 1 & 2 & 3 \\ -1 & 0 & 2 \\ 0 & 1 & 0 \end{bmatrix}$$

* Also called *nullity* by some writers.

is non-singular, since $|a| \neq 0$. Also $x = 0$ is the only solution of the matrix equation $ax = 0$, i.e. of the three scalar equations

$$\begin{aligned} x_1 + 2x_2 + 3x_3 &= 0, \\ -x_1 \qquad + 2x_3 &= 0, \\ x_2 \qquad &= 0. \end{aligned}$$

(ii) *Simply Degenerate Matrix.* The matrix

$$a \equiv \begin{bmatrix} 1 & 2 & 9 \\ 2 & 0 & 2 \\ 3 & -2 & -5 \end{bmatrix}$$

is of degeneracy 1 (or rank $3-1=2$), since at least one first minor of $|a|$ does not vanish whereas $|a| = 0$. Hence a single linear relation connects the rows (or the columns). In fact

$$[1, 2, 9] - 2[2, 0, 2] + [3, -2, -5] = 0,$$

and $$\{1, 2, 3\} + 4\{2, 0, -2\} - \{9, 2, -5\} = 0.$$

From the second of these relations we see that, in the present case, the equations $ax = 0$, or

$$\begin{aligned} x_1 + 2x_2 + 9x_3 &= 0, \\ 2x_1 \qquad + 2x_3 &= 0, \\ 3x_1 - 2x_2 - 5x_3 &= 0, \end{aligned}$$

have the particular solution $x = \{1, 4, -1\} \equiv \boldsymbol{x}$, and the general solution $x = c\boldsymbol{x}$, where c is an arbitrary constant.

(iii) *Multiply Degenerate Matrices.* The matrix

$$a \equiv \begin{bmatrix} 0 & 0 & 0 & 0 \\ 0 & 0 & 0 & 0 \\ 0 & 0 & 1 & 0 \\ 0 & 0 & 0 & 1 \end{bmatrix}$$

is of degeneracy 2 and rank 2. The equations $ax = 0$ can here be satisfied by $\boldsymbol{x}(1) = \{1, 0, 0, 0\}$ and $\boldsymbol{x}(2) = \{0, 1, 0, 0\}$, and the most general solution is $x = c_1\boldsymbol{x}(1) + c_2\boldsymbol{x}(2)$.

As another illustration we may take

$$a \equiv \begin{bmatrix} 1 & 2 & 3 & -1 \\ 2 & 4 & 6 & -2 \\ -1 & -2 & -3 & 1 \\ 0 & 0 & 0 & 0 \end{bmatrix}.$$

This matrix is of degeneracy 3 and rank 1. The equations $ax = 0$ are satisfied by the three columns

$$\boldsymbol{x}(1) = \{2, -1, 0, 0\},$$
$$\boldsymbol{x}(2) = \{3, 0, -1, 0\},$$
$$\boldsymbol{x}(3) = \{1, 0, 0, 1\},$$

and the most general solution is

$$x = c_1\boldsymbol{x}(1) + c_2\boldsymbol{x}(2) + c_3\boldsymbol{x}(3).$$

(iv) *Singular Matrices Expressed as Products.* A square matrix a which has proportional columns and rows is of rank 1, and conversely. Such a matrix has effectively one linearly independent column and one linearly independent row, and is expressible as a product of the type $\{b_i\}\lfloor\beta_j\rfloor$. For instance, the second matrix in example (iii) can be written as the product

$$\{1, 2, -1, 0\}[1, 2, 3, -1].$$

In the same way a square matrix a of rank 2 is expressible as a product of the type $\boldsymbol{b}\boldsymbol{\beta}$, in which $\boldsymbol{b}$ consists of two independent columns and $\boldsymbol{\beta}$ consists of two independent rows. The columns of a are all linear combinations of the two columns of $\boldsymbol{b}$, and the rows of a are all linear combinations of the rows of $\boldsymbol{\beta}$. For instance, the first matrix in example (iii) can be written as

$$\begin{bmatrix} 0 & 0 \\ 0 & 0 \\ 1 & 0 \\ 0 & 1 \end{bmatrix} \begin{bmatrix} 0 & 0 & 1 & 0 \\ 0 & 0 & 0 & 1 \end{bmatrix},$$

while similarly the matrix a of rank 2 in example (ii) can be represented in the following ways:

$$\begin{bmatrix} 1 & 2 & 9 \\ 2 & 0 & 2 \\ 3 & -2 & -5 \end{bmatrix} = \begin{bmatrix} 1 & 2 \\ 2 & 0 \\ 3 & -2 \end{bmatrix} \begin{bmatrix} 1 & 0 & 1 \\ 0 & 1 & 4 \end{bmatrix} = \tfrac{1}{5}\begin{bmatrix} 1 & 11 \\ 2 & 2 \\ 3 & -7 \end{bmatrix} \begin{bmatrix} 5 & -1 & 1 \\ 0 & 1 & 4 \end{bmatrix}.$$

More generally a square matrix a of order n and rank r can be represented by

$$a = \boldsymbol{b}\boldsymbol{\beta},$$

in which $\boldsymbol{b}$ is an $(\overline{n}, r)$ matrix with linearly distinct columns and $\boldsymbol{\beta}$ is an $(\overline{r}, n)$ matrix with linearly distinct rows.

1·10. Adjoint Matrices. If $[a_{ij}]$ is a square matrix and A_{ij} is the cofactor of a_{ij} in $|a|$, the matrix $[A_{ji}]$ is said to be the *adjoint* of $[a_{ij}]$. It should be noted carefully that the adjoint of a matrix a is the *transposed* of the matrix of the cofactors of a.

Since by the properties of determinants

$$\sum_{r=1}^{n} a_{ir}A_{jr} = 0 \quad \text{if } i \neq j,$$

and

$$\sum_{r=1}^{n} a_{ir}A_{ir} = |a|,$$

it follows that

$$[a_{ij}][A_{ji}] = |a| I. \qquad \text{......(1)}$$

Similarly,

$$[A_{ji}][a_{ij}] = |a| I. \qquad \text{......(2)}$$

From equations (1·8·2) and (1) we see also that

$$|A_{ji}| = |a|^{n-1}. \qquad \text{......(3)}$$

The properties of the adjoint matrix expressed by equations (1) and (2) are of great importance. In the special case where a is singular, so that $|a| = 0$,

$$aA = Aa = 0. \qquad \text{......(4)}$$

Anticipating results given in § 1·12 regarding square matrices having a null product, we can deduce from (4) that the adjoint of a simply degenerate matrix a necessarily has unit rank. If a has multiple degeneracy, the adjoint is by definition null.

Examples

(i) *Adjoint of Non-Singular Matrix.*

$$[a_{ij}] = \begin{bmatrix} 2 & 0 & 7 \\ -1 & 4 & 5 \\ 3 & 1 & 2 \end{bmatrix}; \qquad [A_{ji}] = \begin{bmatrix} 3 & 7 & -28 \\ 17 & -17 & -17 \\ -13 & -2 & 8 \end{bmatrix}.$$

The product of these two matrices in either order is

$$\begin{bmatrix} -85 & 0 & 0 \\ 0 & -85 & 0 \\ 0 & 0 & -85 \end{bmatrix} = -85I_3 = \begin{vmatrix} 2 & 0 & 7 \\ -1 & 4 & 5 \\ 3 & 1 & 2 \end{vmatrix} I_3.$$

(ii) *Adjoint of Simply Degenerate Matrix.*

$$[a_{ij}] = \begin{bmatrix} 2 & 0 & 2 \\ -1 & 4 & 3 \\ 3 & 1 & 4 \end{bmatrix}; \qquad [A_{ji}] = \begin{bmatrix} 13 & 2 & -8 \\ 13 & 2 & -8 \\ -13 & -2 & 8 \end{bmatrix}.$$

In this case both products aA and Aa vanish, and A has unit rank.

(iii) *Adjoint of Multiply Degenerate Matrix.*

$$[a_{ij}] = \begin{bmatrix} 0 & 0 & 0 & 0 \\ 0 & 0 & 0 & 0 \\ 0 & 0 & 1 & 0 \\ 0 & 0 & 0 & 2 \end{bmatrix}; \qquad [A_{ji}] = 0.$$

Note that here a is of rank 2 and degeneracy 2, and that A is null.

1·11. Reciprocal Matrices and Division. If a is a non-singular square matrix, the elements of the adjoint A may be divided by $|a|$. The matrix so obtained is called the *reciprocal* or *inverse* of a, and is written a^{-1}. By (1·10·1) and (1·10·2) it follows that

$$aa^{-1} = a^{-1}a = I. \qquad \text{......(1)}$$

It is easy to see that if a is given, the reciprocal of a is the only square matrix x which satisfies the equation $ax = I$. Similarly, a^{-1} is the only matrix which satisfies $xa = I$.

If a square matrix is not singular it possesses a reciprocal, and multiplication by this reciprocal is in many ways analogous to division in ordinary algebra. In conformity with the terms adopted for matrix multiplication, we may refer to $b^{-1}a$ (when it exists) as a predivided by b, and to ab^{-1} as a postdivided by b.

Methods for the computation of reciprocal matrices will be described in Chapter IV.

EXAMPLES

(i) *Reciprocal Matrices.*

$$a = \begin{bmatrix} 3 & -2 & 0 & -1 \\ 0 & 2 & 2 & 1 \\ 1 & -2 & -3 & -2 \\ 0 & 1 & 2 & 1 \end{bmatrix}; \qquad a^{-1} = \begin{bmatrix} 1 & 1 & -2 & -4 \\ 0 & 1 & 0 & -1 \\ -1 & -1 & 3 & 6 \\ 2 & 1 & -6 & -10 \end{bmatrix}.$$

The product of these two matrices can be verified to be I_4.

(ii) *Predivision and Postdivision.*

$$a = \begin{bmatrix} 1 & 2 \\ -1 & 3 \end{bmatrix}; \qquad b = \begin{bmatrix} 2 & 5 \\ 1 & 3 \end{bmatrix}; \qquad b^{-1} = \begin{bmatrix} 3 & -5 \\ -1 & 2 \end{bmatrix}.$$

Predivision of a by b gives

$$\begin{bmatrix} 3 & -5 \\ -1 & 2 \end{bmatrix}\begin{bmatrix} 1 & 2 \\ -1 & 3 \end{bmatrix} = \begin{bmatrix} 8 & -9 \\ -3 & 4 \end{bmatrix}.$$

Postdivision of a by b gives

$$\begin{bmatrix} 1 & 2 \\ -1 & 3 \end{bmatrix}\begin{bmatrix} 3 & -5 \\ -1 & 2 \end{bmatrix} = \begin{bmatrix} 1 & -1 \\ -6 & 11 \end{bmatrix}.$$

1·12. Square Matrices with Null Product. In scalar algebra the equation $ab = 0$ can only be satisfied when at least one of the factors a and b vanishes. However, if a and b are matrices, their product can be null even though neither factor is null. If one factor is null, then obviously the other is arbitrary; but if neither factor is null, then both factors must be singular.

Let $ab = 0$, where the matrices are square and of order n. Since all the elements in the pth column of the product vanish, it follows that

$$a_{11}b_{1p}+a_{12}b_{2p}+\ldots+a_{1n}b_{np} = 0,$$
$$a_{21}b_{1p}+a_{22}b_{2p}+\ldots+a_{2n}b_{np} = 0,$$
$$\cdots\cdots\cdots\cdots\cdots\cdots\cdots\cdots\cdots\cdots$$
$$a_{n1}b_{1p}+a_{n2}b_{2p}+\ldots+a_{nn}b_{np} = 0.$$

Firstly, suppose that $|a|$ is not zero. Then the foregoing equations can only be satisfied if all the elements of the pth column of b vanish. Hence b is null, i.e. it is of rank 0. The rank of a is n, so that in this case the sum of the ranks of a and b is n. Next suppose that $|a|$ vanishes, while the cofactors of its elements are not all zero, i.e. let a be of rank $n-1$. The equations can now be satisfied by non-zero values for some, at least, of the elements b_{ip}, and the ratios of these elements are uniquely determined. But the elements b_{iq} in the qth column of b must also satisfy the equations and must therefore be proportional to the corresponding elements of the pth column. Hence b is of rank 1. Evidently b can also be null, i.e. of rank 0; hence the sum of the ranks of a and of b is in this case either n or $n-1$.

More generally it can be shown that if the product of two square matrices is null the sum of their ranks cannot exceed their order. This is a particular case of Sylvester's law of degeneracy,* which states that the degeneracy of the product of two matrices is at least as great as the degeneracy of either factor, and at most as great as the sum of the degeneracies of the factors (see example (ii)). This law is illustrated in a simple way when both matrices concerned are diagonal, with certain diagonal elements zero. Suppose the degeneracies of a, b are p, q, respectively; this implies that p diagonal elements of a, and q of b, are zero. Evidently if $p \geqslant q$ and all the q diagonal ciphers of b are in positions corresponding to ciphers in the diagonal of a, then the product will contain only p diagonal ciphers, i.e. the degeneracy of the product is p. On the other hand, if none of the q ciphers in b is in a position

* Often called Sylvester's law of nullity. A short proof is given on p. 78 of Ref. 1.

corresponding to a cipher in a (which will be possible provided $p+q \ngtr n$), then the product will contain $p+q$ ciphers; i.e. the degeneracy will be $p+q$.

Examples

(i) *Square Matrices with Null Product.* Let

$$a = \begin{bmatrix} 1 & 0 & 0 & 0 \\ 0 & 1 & 0 & 0 \\ 0 & 0 & 0 & 0 \\ 0 & 0 & 0 & 0 \end{bmatrix}, \quad b = \begin{bmatrix} 0 & 0 & 0 & 0 \\ 0 & 0 & 0 & 0 \\ 0 & 0 & 1 & 0 \\ 0 & 0 & 0 & 1 \end{bmatrix}.$$

Here $ab = 0$, both matrices are of rank 2, and $2+2 \ngtr 4$. Note that if the rank of either matrix is increased by the addition of a further unit in the principal diagonal, the product cannot vanish.

Again, if $a = \begin{bmatrix} 1 & 1 & 1 \\ 2 & 2 & 2 \\ 5 & 5 & 5 \end{bmatrix}, \quad b = \begin{bmatrix} 3 & 4 & 2 \\ -2 & -1 & -1 \\ -1 & -3 & -1 \end{bmatrix},$

then $ab = 0$. Here a is of rank 1, b is of rank 2, and $1+2 \ngtr 3$.

(ii) *Sylvester's Law of Degeneracy.* The validity of this law is illustrated by the following considerations. Let a, b have degeneracies p, q, respectively. Then it is possible to represent a by a product $A\alpha$, where A, α are $(\overline{n}, n-p)$, $(\overline{n-p}, n)$ matrices, respectively (see § 1·9, example (iv)). Similarly, b can be expressed as $B\beta$, where B, β are $(\overline{n}, n-q)$, $(\overline{n-q}, n)$ matrices, respectively. Moreover, the columns in A and in B, and the rows in α and in β, severally, will be linearly independent. Hence we may write

$$ab = A\alpha B\beta = A\gamma,$$

where

$$\gamma = \alpha B\beta.$$

For definiteness suppose $p > q$. Then since A has only $n-p$ linearly independent columns, and since the rows of γ may not be linearly independent, the product has at most $n-p$ linearly independent rows and columns, and its degeneracy is thus at least p. Similarly, if $q > p$, the degeneracy is at least q.

Next consider the product in the form $aB\beta$. Since a has degeneracy p, there are p relations connecting the columns of a; i.e. there are p columns x such that $ax = 0$ (see § 1·9). In the extreme case, therefore, it is possible for p of the $n-q$ columns of B to give null columns when premultiplied by a, and the remaining $n-p-q$ columns will in fact be linearly independent. Hence aB has $n-p-q$ linearly independent columns, and the degeneracy of the product ab in this extreme case is therefore $p+q$.

1·13. Reversal of Order in Products when Matrices are Transposed or Reciprocated. (*a*) *Transposition.* Let A be a $(\overline{p}, n)$ matrix and B an $(\overline{m}, p)$ matrix. Then the product $P = BA$ is an $(\overline{m}, n)$ matrix of which the typical element is

$$P_{ij} = \sum_{r=1}^{p} B_{ir} A_{rj}.$$

When transposed, A and B become, respectively, $(\overline{n}, p)$ and $(\overline{p}, m)$ matrices; they are now conformable when multiplied in the order $A'B'$. This product is an $(\overline{n}, m)$ matrix, which is readily seen to be the transposed of P, since the typical element is

$$\sum_{r=1}^{p} A_{ri} B_{jr} = P_{ji}.$$

Hence when a matrix product is transposed, the order of the matrices forming the product must be reversed. Similarly, if

$$CBA = CP = R, \quad \text{then} \quad R' = P'C' = A'B'C'.$$

It is evident that the reversal rule holds for any number of factors.

(*b*) *Reciprocation.* Suppose that in the equation $P = BA$ the matrices are square and non-singular. Premultiply both sides of the equation by $A^{-1}B^{-1}$ and postmultiply by P^{-1}; then $A^{-1}B^{-1} = P^{-1}$. Similarly, if $R = CBA$, then $R^{-1} = A^{-1}B^{-1}C^{-1}$. The reversal rule again applies for any number of factors.

EXAMPLES

(i) *Transposition of Product of Square Matrix and Column.* If $ax = b$, then $x'a' = b'$. For example, if

$$\begin{bmatrix} 7 & 4 & 2 \\ 5 & 3 & 1 \\ 3 & 2 & 1 \end{bmatrix} \begin{bmatrix} x_1 \\ x_2 \\ x_3 \end{bmatrix} = \begin{bmatrix} 21 \\ 14 \\ 10 \end{bmatrix},$$

then
$$[x_1 \quad x_2 \quad x_3] \begin{bmatrix} 7 & 5 & 3 \\ 4 & 3 & 2 \\ 2 & 1 & 1 \end{bmatrix} = [21 \quad 14 \quad 10].$$

Both of these matrix equations represent the same set of three scalar equations.

(ii) *If A, B are Symmetrical, then the Continued Product $ABA \dots BA$ is Symmetrical.* For if $ABA \dots BA = Q$, then

$$Q' = A'B' \dots A'B'A' = AB \dots ABA = Q.$$

For example

$$\begin{bmatrix} 2 & 1 \\ 1 & 1 \end{bmatrix}\begin{bmatrix} 5 & 2 \\ 2 & 1 \end{bmatrix}\begin{bmatrix} 2 & 1 \\ 1 & 1 \end{bmatrix} = \begin{bmatrix} 12 & 5 \\ 7 & 3 \end{bmatrix}\begin{bmatrix} 2 & 1 \\ 1 & 1 \end{bmatrix} = \begin{bmatrix} 29 & 17 \\ 17 & 10 \end{bmatrix}.$$

(iii) *The Reciprocal of a Symmetrical Matrix is also Symmetrical.* If α is the reciprocal of a symmetrical matrix A, then $A\alpha = I$. By the reversal rule $\alpha'A' = \alpha'A = I' = I$, and postmultiplication by α yields $\alpha = \alpha'$: hence α is symmetrical. For example,

$$\begin{bmatrix} 4 & 3 & 2 \\ 3 & 2 & 1 \\ 2 & 1 & 1 \end{bmatrix}^{-1} = \begin{bmatrix} -1 & 1 & 1 \\ 1 & 0 & -2 \\ 1 & -2 & 1 \end{bmatrix}.$$

(iv) *The Reciprocal of a Skew Symmetric Matrix of Even Order is also Skew Symmetric.* If A is a skew symmetric matrix of order n, then $A = -A'$. Hence $|A| = (-1)^n |A'| = (-1)^n |A|$. When n is odd, then $|A| = 0$, so that a skew symmetric matrix of odd order has no reciprocal. When A is of even order, let $A\alpha = I$. By the reversal rule $\alpha'A' = -\alpha'A = I' = I$. Postmultiply by α: then $\alpha' = -\alpha$, so that α is also skew symmetric. For example,

$$\begin{bmatrix} 0 & 1 \\ -1 & 0 \end{bmatrix}^{-1} = \begin{bmatrix} 0 & -1 \\ 1 & 0 \end{bmatrix}.$$

(v) *The Product of any Matrix and its Transposed is Symmetrical.* If $P = u'u$, then $P' = u'u = P$. For instance,

$$\begin{bmatrix} 4 & 2 & 1 \\ 1 & 7 & -3 \end{bmatrix}\begin{bmatrix} 4 & 1 \\ 2 & 7 \\ 1 & -3 \end{bmatrix} = \begin{bmatrix} 21 & 15 \\ 15 & 59 \end{bmatrix}.$$

An obvious extension is that if a is a symmetrical matrix, then the product $u'au$ is symmetrical.

1·14. Linear Substitutions. Suppose that

$$\begin{aligned} y_1 &= u_{11}x_1 + u_{12}x_2 + \dots + u_{1n}x_n, \\ y_2 &= u_{21}x_1 + u_{22}x_2 + \dots + u_{2n}x_n, \\ &\dots\dots\dots\dots\dots\dots\dots\dots \\ y_n &= u_{n1}x_1 + u_{n2}x_2 + \dots + u_{nn}x_n. \end{aligned}$$

Then the set of variables y is said to be derived from the set x by a *linear transformation*, and the equations are also said to define a *linear substitution*. The whole set of equations can be represented by the single matrix equation

$$y = ux, \qquad \ldots\ldots(1)$$

where x and y are column matrices. The square matrix u is called the matrix of the transformation.

Now suppose that a third set of variables z is derived from the set y by the transformation whose matrix is v. Then $z = vy = vux$. Thus the transformation of x into y and the subsequent transformation of y into z is equivalent to a certain direct transformation of x into z. This is called the product of the transformations, and its matrix is the product of the matrices of the successive transformations taken in the proper order.

When u is non-singular, equation (1) may be multiplied by u^{-1} to give

$$x = u^{-1}y. \qquad \ldots\ldots(2)$$

The matrix of the inverse transformation is therefore the reciprocal of the matrix of the transformation. If the quantities x_i in (1) are regarded as unknowns, then their values are given explicitly by (2). Hence the solution of a set of linear equations may be found by the calculation of the reciprocal of the matrix of the coefficients, provided this matrix is not singular.

Examples

(i) *Linear Substitutions*. If

$$\begin{aligned} y_1 &= 2x_1 - 2x_2 + 3x_3, \\ y_2 &= x_1 + x_2 + x_3, \\ y_3 &= x_1 + 3x_2 - x_3, \end{aligned}$$

and

$$\begin{aligned} z_1 &= y_1 \qquad + 2y_3, \\ z_2 &= \quad - y_2 + y_3, \\ z_3 &= 2y_1 + 3y_2 + 4y_3, \end{aligned}$$

then

$$z = \begin{bmatrix} 1 & 0 & 2 \\ 0 & -1 & 1 \\ 2 & 3 & 4 \end{bmatrix} \begin{bmatrix} 2 & -2 & 3 \\ 1 & 1 & 1 \\ 1 & 3 & -1 \end{bmatrix} x = \begin{bmatrix} 4 & 4 & 1 \\ 0 & 2 & -2 \\ 11 & 11 & 5 \end{bmatrix} x.$$

(ii) *Solution of Linear Algebraic Equations*. If

$$\begin{aligned} 3x_1 - 2x_2 \qquad - x_4 &= 7, \\ 2x_2 + 2x_3 + x_4 &= 5, \\ x_1 - 2x_2 - 3x_3 - 2x_4 &= -1, \\ x_2 + 2x_3 + x_4 &= 6, \end{aligned}$$

then
$$\begin{bmatrix} 3 & -2 & 0 & -1 \\ 0 & 2 & 2 & 1 \\ 1 & -2 & -3 & -2 \\ 0 & 1 & 2 & 1 \end{bmatrix} \begin{bmatrix} x_1 \\ x_2 \\ x_3 \\ x_4 \end{bmatrix} = \begin{bmatrix} 7 \\ 5 \\ -1 \\ 6 \end{bmatrix}.$$

Hence (see example (i) of § 1·11)
$$\begin{bmatrix} x_1 \\ x_2 \\ x_3 \\ x_4 \end{bmatrix} = \begin{bmatrix} 1 & 1 & -2 & -4 \\ 0 & 1 & 0 & -1 \\ -1 & -1 & 3 & 6 \\ 2 & 1 & -6 & -10 \end{bmatrix} \begin{bmatrix} 7 \\ 5 \\ -1 \\ 6 \end{bmatrix} = \begin{bmatrix} -10 \\ -1 \\ 21 \\ -35 \end{bmatrix},$$

or
$$x_1 = -10, \quad x_2 = -1, \quad x_3 = 21, \quad x_4 = -35.$$

1·15. Bilinear and Quadratic Forms. If x, y are two sets of n variables, a function $A(y, x)$ which is linear and homogeneous in the variables of each set separately is called a *bilinear form*. Evidently such a function has n^2 coefficients and can be written
$$A(y, x) = \sum_{r=1}^{n} y_r(a_{r1}x_1 + a_{r2}x_2 + \dots + a_{rn}x_n)$$
$$= [y_1 \quad y_2 \quad \dots \quad y_n] \begin{bmatrix} a_{11} & a_{12} & \dots & a_{1n} \\ a_{21} & a_{22} & \dots & a_{2n} \\ \dots & \dots & \dots & \dots \\ a_{n1} & a_{n2} & \dots & a_{nn} \end{bmatrix} \begin{bmatrix} x_1 \\ x_2 \\ \dots \\ x_n \end{bmatrix}.$$

A bilinear form can therefore be expressed concisely as
$$A(y, x) = yax = x'a'y',$$
in which y is a row matrix and x is a column matrix.

When the sets of variables are identical, so that $y = x'$, the bilinear form becomes a *quadratic form*, i.e. a homogeneous function of the second degree. The coefficient of $x_i x_j$ $(i \neq j)$ in the quadratic form is $a_{ij} + a_{ji}$, and this will be unaltered if a_{ij} and a_{ji} are both changed to $\frac{1}{2}(a_{ij} + a_{ji})$. Hence a convenient expression for a quadratic form is
$$A(x, x) = x'ax,$$
where a is a symmetrical matrix.

If we make the substitution $x = hz$, where h is a non-singular square matrix of constants, then
$$A(x, x) = z'h'ahz = z'bz,$$
where b is the symmetrical matrix $h'ah$. When two square matrices a and b (not necessarily symmetrical) are related by an equation

of the type $b = p'ap$, where p is non-singular, they are said to be connected by a *congruent transformation*.

The representation of bilinear and quadratic forms* by matrices has two great advantages. It is concise, and it emphasises the fact that any special properties of the form correspond to special properties of the matrix of its coefficients. Examples of its use in relation to dynamics will be given in Chapter VIII.

EXAMPLES

(i) *Bilinear Forms having Unequal Numbers of Variables.* If there are n variables y and m variables x the appropriate bilinear form is $A(y,x) = yax = x'a'y'$, where a is an $(\bar{n}, m)$ matrix. For instance,

$$A(y,x) = [y_1, y_2, y_3]\begin{bmatrix}1, & 2\\ 3, & 0\\ 0, & -1\end{bmatrix}\begin{bmatrix}x_1\\ x_2\end{bmatrix} = [x_1, x_2]\begin{bmatrix}1, & 3, & 0\\ 2, & 0, & -1\end{bmatrix}\begin{bmatrix}y_1\\ y_2\\ y_3\end{bmatrix}$$

$$= y_1(x_1+2x_2)+3y_2x_1-y_3x_2.$$

(ii) *Partitioning of Bilinear Forms.* The usual rules for the partitioning of matrices can be applied with bilinear forms. As an illustration suppose

$$A(y,x) = [y_1, y_2, y_3]\begin{bmatrix}a_{11} & a_{12} & a_{13}\\ a_{21} & a_{22} & a_{23}\\ a_{31} & a_{32} & a_{33}\end{bmatrix}\begin{bmatrix}x_1\\ x_2\\ x_3\end{bmatrix}.$$

If $[y_1, y_2] = \boldsymbol{y}$ and $\{x_1, x_2\} = \boldsymbol{x}$, we may write

$$A(y,x) = [\boldsymbol{y}, y_3]\left[\begin{array}{cc|c}a_{11} & a_{12} & a_{13}\\ a_{21} & a_{22} & a_{23}\\ \hline a_{31} & a_{32} & a_{33}\end{array}\right]\begin{bmatrix}\boldsymbol{x}\\ x_3\end{bmatrix},$$

or say

$$A(y,x) = [\boldsymbol{y}, y_3]\begin{bmatrix}\alpha, & \beta\\ \gamma, & \delta\end{bmatrix}\begin{bmatrix}\boldsymbol{x}\\ x_3\end{bmatrix}$$

$$= [\boldsymbol{y}, y_3]\begin{bmatrix}\alpha\boldsymbol{x}+\beta x_3\\ \gamma\boldsymbol{x}+\delta x_3\end{bmatrix}$$

$$= \boldsymbol{y}\alpha\boldsymbol{x}+\boldsymbol{y}\beta x_3+y_3\gamma\boldsymbol{x}+y_3\delta x_3.$$

As a further example consider a quadratic form

$$A(x,x) = x'ax.$$

* A simple account of the properties of bilinear and quadratic forms is given in Chaps. VIII to XI of Ref. 1.

Then if we suppose x to be divided into m variables y and k variables η, so that $x=\{y,\eta\}$, we may write

$$A(x,x)=[y',\eta']\begin{bmatrix}\alpha, & \beta\\ \gamma, & \delta\end{bmatrix}\begin{bmatrix}y\\ \eta\end{bmatrix}.$$

Since a is symmetrical, the submatrices α, δ are square and symmetrical, and are of the orders m and k, respectively. Further, the rectangular submatrices β, γ are, respectively, of the types $(\overline{m},k)$ and $(\overline{k},m)$, while $\beta'=\gamma$. The expanded expression is

$$\begin{aligned}A(x,x)&=[y',\eta']\begin{bmatrix}\alpha y+\beta\eta\\ \gamma y+\delta\eta\end{bmatrix}\\ &=y'\alpha y+y'\beta\eta+\eta'\gamma y+\eta'\delta\eta\\ &=y'\alpha y+\eta'(\beta'+\gamma)\,y+\eta'\delta\eta\\ &=y'\alpha y+2\eta'\gamma y+\eta'\delta\eta.\end{aligned}$$

1·16. Discriminants and One-Signed Quadratic Forms. If $A(x,x)\equiv x'ax$ is a quadratic form of m variables x, then the determinant $\Delta_m\equiv|a|$ is usually termed the *discriminant* of the form. The conditions that a given quadratic form $A(x,x)$ shall be one-signed* (say positive) for all real values of its variables are of great importance in dynamics, and they are connected with the signs of Δ_m and of the discriminants of the forms derived from $A(x,x)$ by omission of any number of the variables x. A brief discussion of this question follows.

For simplicity assume, firstly, that none of the discriminants in question is zero. This implies, in particular, that none of the principal diagonal elements a_{ii} of a vanishes. In this case all the terms involving the variable x_1 can be collected together as a perfect square, and the quadratic form can be rewritten as

$$A(x,x)=a_{11}\left(x_1+\frac{a_{12}}{a_{11}}x_2+\ldots+\frac{a_{1m}}{a_{11}}x_m\right)^2+B,$$

where B is a quadratic form involving all the variables with the exception of x_1. Treating B in a similar manner, and continuing the process, we finally express $A(x,x)$ as the sum of m perfect squares. Thus, say,

$$A(x,x)=E_1\xi_1^2+E_2\xi_2^2+\ldots+E_m\xi_m^2\equiv\xi'E\xi, \qquad \ldots\ldots(1)$$

where E is a diagonal matrix the elements of which are all clearly rational functions of the elements of a. Moreover, the variables ξ_i and

* A quadratic form which is positive or zero for all real values of the variables is spoken of as a *positive definite* form.

x_i will be connected by a linear transformation of the special "triangular" type

$$\xi = ux = \begin{bmatrix} 1 & u_{12} & u_{13} & \dots & u_{1m} \\ 0 & 1 & u_{23} & \dots & u_{2m} \\ 0 & 0 & 1 & \dots & u_{3m} \\ \dots & \dots & \dots & \dots & \dots \\ 0 & 0 & 0 & \dots & 1 \end{bmatrix} x,$$

and a will be derived from E by the congruent transformation $a = u'Eu$ (see § 1·15). Since $|u| = |u'| = 1$, it follows that

$$E_1 E_2 \dots E_m = \Delta_m.$$

Suppose next that $x_m = 0$, so that also $\xi_m = 0$. Then, as previously, we can prove that

$$E_1 E_2 \dots E_{m-1} = \Delta_{m-1},$$

where Δ_{m-1} is the discriminant of the quadratic form obtained from $A(x, x)$ when $x_m = 0$. More generally, if $x_m = x_{m-1} = \dots = x_{j+1} = 0$, then

$$E_1 E_2 \dots E_j = \Delta_j.$$

Hence we find that

$$E_1 = \Delta_1; \quad E_2 = \frac{\Delta_2}{\Delta_1}; \quad E_3 = \frac{\Delta_3}{\Delta_2}; \quad \dots; \quad E_m = \frac{\Delta_m}{\Delta_{m-1}}. \quad \dots(2)$$

From (1) it is clear that the necessary and sufficient conditions that $A(x, x)$ shall be a positive function are that all the coefficients E shall be positive. Hence by (2) all the discriminants $\Delta_m, \Delta_{m-1}, \dots, \Delta_1$ must be positive. We could, of course, have reduced $A(x, x)$ to a sum of squares such as (1) by collecting together the variables in quite a different order. Each such method of reduction would lead to a different set of discriminants, but the final criteria are in fact all equivalent.

The argument given above requires modification if any of the discriminants is zero. For example, if $a_{11} = 0$ the first step in the reduction of $A(x, x)$ would fail. In this case, leaving x_1 and x_r general and choosing the remaining variables to be zero, we have $A(x, x) = a_{rr}x_r^2 + 2a_{1r}x_1x_r$. This expression will not be one-signed for all values of x_1 and x_r unless $a_{1r} = 0$. Since this conclusion applies for all elements a_{1r}, we see that if $a_{11} = 0$ the form cannot be one-signed unless x_1 is completely absent from $A(x, x)$. More generally, we may immediately rule out all quadratic forms having any zero principal diagonal elements.

Suppose next that, while $a_{11} \neq 0$, yet B contains some zero principal diagonal elements. By the same argument as before B (and therefore

A) will not necessarily be positive unless the variables in B corresponding to its zero principal diagonal elements are all absent. It is easy to show that in order that, say, x_s may be completely absent from B the sth row (or column) of elements in a must be proportional to the first row (or column). The procedure then is to retain the first row and first column of a and all other rows and columns not proportional to the first. The discriminants actually to be used are then constructed as before from the matrix a as thus reduced. In these circumstances the total number of squares involved in (1) will be less than m. Illustrations of some of the possible cases which can arise are included in the examples which follow.

Examples

(i) *Discriminants all Positive.*

$$A(x,x) = [x_1, x_2, x_3, x_4]\begin{bmatrix} 1 & -1 & 2 & 1 \\ -1 & 3 & 0 & -3 \\ 2 & 0 & 9 & -6 \\ 1 & -3 & -6 & 19 \end{bmatrix}\begin{bmatrix} x_1 \\ x_2 \\ x_3 \\ x_4 \end{bmatrix}.$$

Here we may take $\Delta_1 = 1$ and

$$\Delta_2 = \begin{vmatrix} 1 & -1 \\ -1 & 3 \end{vmatrix} = 2; \qquad \Delta_3 = \begin{vmatrix} 1 & -1 & 2 \\ -1 & 3 & 0 \\ 2 & 0 & 9 \end{vmatrix} = 6;$$

$$\Delta_4 = \begin{vmatrix} 1 & -1 & 2 & 1 \\ -1 & 3 & 0 & -3 \\ 2 & 0 & 9 & -6 \\ 1 & -3 & -6 & 19 \end{vmatrix} = 24.$$

These particular discriminants are all positive, and $A(x,x)$ is accordingly a positive function. It can be verified that $A(x,x)$ is expressible as

$$A(x,x) = (x_1 - x_2 + 2x_3 + x_4)^2 + 2(x_2 + x_3 - x_4)^2 + 3(x_3 - 2x_4)^2 + 4x_4^2.$$

(ii) *Case of Proportional Rows and Columns.*

$$A(x,x) = [x_1, x_2, x_3, x_4]\begin{bmatrix} 1 & -1 & 2 & 1 \\ -1 & 1 & -2 & -1 \\ 2 & -2 & 7 & -4 \\ 1 & -1 & -4 & 17 \end{bmatrix}\begin{bmatrix} x_1 \\ x_2 \\ x_3 \\ x_4 \end{bmatrix}.$$

Since the second row of a is proportional to the first, the variable x_2 will be absent from B. Omitting the second row and column, and distinguishing the discriminants for the modified matrix by clarendon type, we have

$$\Delta_1 = 1; \qquad \boldsymbol{\Delta}_2 = \begin{vmatrix} 1 & 2 \\ 2 & 7 \end{vmatrix} = 3; \qquad \boldsymbol{\Delta}_3 = \begin{vmatrix} 1 & 2 & 1 \\ 2 & 7 & -4 \\ 1 & -4 & 17 \end{vmatrix} = 12.$$

The form in question is accordingly positive and expressible as the sum of three squares. It can be verified that

$$A(x, x) = (x_1 - x_2 + 2x_3 + x_4)^2 + 3(x_3 - 2x_4)^2 + 4x_4^2.$$

(iii) *If $a = u'u$, where u is Real, the Discriminants of $x'ax$ are all Positive.* Let $y = ux$, where x, y are real columns. Then

$$x'ax = x'u'ux = y'y.$$

The product $y'y$ is a sum of squares and is therefore always positive. Hence $x'ax$ is a positive form and the discriminants are all positive.

More generally, if d is a diagonal matrix of positive quantities, the discriminants of $x'u'dux$ are all positive.

1·17. Special Types of Square Matrix. For convenience of reference the formal definitions of one or two particularly important types of square matrix will now be given. Some of these types have already been referred to incidentally, and exemplified, in preceding pages. In the list which follows the matrix concerned is denoted simply as u, and as usual an accent means transposition. If the elements of u are complex numbers, the matrix with the corresponding conjugate complex elements—that is to say the *conjugate** of u—is denoted as $\bar{u}$.

$u = u'$	... ...	*Symmetrical*
$u = -u'$	... ...	*Skew Symmetric*, or *Alternate*
$u^{-1} = u'$	... ...	*Orthogonal*
$\bar{u} = u'$	... ...	*Hermitian*

A few additional properties† are given in the examples which follow.

Examples

(i) *Symmetrical Matrices and Determinants.* It is to be noted that the product of two symmetrical matrices is, in general, not symmetrical. For instance

$$\begin{bmatrix} 1 & 2 \\ 2 & 3 \end{bmatrix}\begin{bmatrix} 4 & 5 \\ 5 & 6 \end{bmatrix} = \begin{bmatrix} 14 & 17 \\ 23 & 28 \end{bmatrix}.$$

* The term "conjugate" is used by some writers as meaning "transposed".

† For a more complete account see, for example, Chap. IV of Ref. 2.

If $[a_{ij}]$ is a symmetrical matrix of order n, the adjoint $[A_{ji}]$ is also symmetrical. Since $\sum_{r=1}^{n} a_{ir}A_{jr} = 0$ and $\sum_{r=1}^{n} a_{ir}A_{ir} = |a|$, we have by direct multiplication

$$\begin{bmatrix} 1 & 0 & 0 & \dots & 0 \\ 0 & 1 & 0 & \dots & 0 \\ a_{31} & a_{32} & a_{33} & \dots & a_{3n} \\ a_{41} & a_{42} & a_{43} & \dots & a_{4n} \\ \dots & \dots & \dots & \dots & \dots \\ a_{n1} & a_{n2} & a_{n3} & \dots & a_{nn} \end{bmatrix} [A_{ji}] = \begin{bmatrix} A_{11} & A_{21} & A_{31} & \dots & A_{n1} \\ A_{12} & A_{22} & A_{32} & \dots & A_{n2} \\ 0 & 0 & |a| & \dots & 0 \\ \dots & \dots & \dots & \dots & \dots \\ \dots & \dots & \dots & \dots & \dots \\ 0 & 0 & 0 & \dots & |a| \end{bmatrix}.$$

Moreover, $|A_{ji}| = |a|^{n-1}$. Hence

$$|a| \begin{vmatrix} a_{33} & a_{34} & \dots & a_{3n} \\ a_{43} & a_{44} & \dots & a_{4n} \\ \dots & \dots & \dots & \dots \\ a_{n3} & a_{n4} & \dots & a_{nn} \end{vmatrix} = A_{11}A_{22} - A_{12}^2. \qquad \dots\dots(1)$$

It follows from this equation that if $A_{11} = 0$ the two determinants on the left have opposite signs. Again, if both $|a|$ and A_{11} vanish, then also $A_{12} = 0$: similarly, all other first minors of type A_{1r} vanish. Accordingly, if a symmetrical determinant and its leading first minor both vanish, then the first minors of all elements in the first row (or column) also vanish. An application of this property to dynamics is given in § 9·9.

(ii) *Orthogonal Matrices.* The matrix of the direction cosines of three mutually perpendicular axes OX_1, OX_2, OX_3 referred to three fixed axes is orthogonal. For, let (l_{11}, l_{12}, l_{13}), (l_{21}, l_{22}, l_{23}), (l_{31}, l_{32}, l_{33}) be the direction cosines of the three axes, respectively. Then

$$l = \begin{bmatrix} l_{11} & l_{12} & l_{13} \\ l_{21} & l_{22} & l_{23} \\ l_{31} & l_{32} & l_{33} \end{bmatrix} \quad \text{and} \quad l' = \begin{bmatrix} l_{11} & l_{21} & l_{31} \\ l_{12} & l_{22} & l_{32} \\ l_{13} & l_{23} & l_{33} \end{bmatrix}.$$

Now by the properties of direction cosines

$$l_{11}^2 + l_{12}^2 + l_{13}^2 = 1, \quad l_{11}l_{21} + l_{12}l_{22} + l_{13}l_{23} = 0, \quad \text{etc.},$$

so that $ll' = I$. Hence $l^{-1} = l'$, and l is therefore orthogonal. The application of orthogonal matrices to the subject of kinematics will be considered in Chapter VIII.

(iii) *Matrices Corresponding to Complex Scalars and Quaternions.* If I denotes the second order unit matrix, and if $\boldsymbol{I}$ is the orthogonal matrix

$$\boldsymbol{I} = \begin{bmatrix} 0 & 1 \\ -1 & 0 \end{bmatrix}, \qquad \text{......(2)}$$

then evidently $I^2 = I$, $\boldsymbol{I}^2 = -I$, $I\boldsymbol{I} = \boldsymbol{I}I = \boldsymbol{I}$. Moreover, if

$$aI + b\boldsymbol{I} = a'I + b'\boldsymbol{I},$$

where a, a', b, b' are scalar multipliers, then $a = a'$ and $b = b'$. The algebra of matrices of the type $aI + b\boldsymbol{I}$, where a and b are scalars, is thus identical with that of the scalar complexes $a + ib$. The complex quantity $a + ib$ is in fact here replaced by the real matrix $\begin{bmatrix} a & b \\ -b & a \end{bmatrix}$. It will be seen later that the "latent roots" (see § 3·6) of this matrix are $a \pm ib$.

Next suppose I to denote the fourth order unit matrix, and in addition introduce the orthogonal matrices

$$\boldsymbol{I} = \begin{bmatrix} 0 & 1 & 0 & 0 \\ -1 & 0 & 0 & 0 \\ 0 & 0 & 0 & -1 \\ 0 & 0 & 1 & 0 \end{bmatrix}; \quad \boldsymbol{J} = \begin{bmatrix} 0 & 0 & 1 & 0 \\ 0 & 0 & 0 & 1 \\ -1 & 0 & 0 & 0 \\ 0 & -1 & 0 & 0 \end{bmatrix}; \quad \boldsymbol{K} = \begin{bmatrix} 0 & 0 & 0 & 1 \\ 0 & 0 & -1 & 0 \\ 0 & 1 & 0 & 0 \\ -1 & 0 & 0 & 0 \end{bmatrix}. \qquad \text{......(3)}$$

Then

$$\left. \begin{aligned} I^2 = I, \quad & \boldsymbol{I}^2 = \boldsymbol{J}^2 = \boldsymbol{K}^2 = \boldsymbol{IJK} = -I, \\ I\boldsymbol{I} = \boldsymbol{I}I = \boldsymbol{I}, \quad & I\boldsymbol{J} = \boldsymbol{J}I = \boldsymbol{J}, \quad I\boldsymbol{K} = \boldsymbol{K}I = \boldsymbol{K}. \end{aligned} \right\} \qquad \text{......(4)}$$

These relations provide a basis for the algebra of matrices of the type $wI + x\boldsymbol{I} + y\boldsymbol{J} + z\boldsymbol{K}$, in which w, x, y, z are scalars.

The calculus of quaternions* is concerned with expressions of the type $1w + ix + jy + kz$, in which 1 is the scalar unit and i, j, k are three mutually perpendicular unit vectors. The laws of combination postulated for 1, i, j, k are precisely those given by (4), so that the two algebras considered formally correspond. A quaternion is thus completely represented by the real matrix

$$q = \begin{bmatrix} w & x & y & z \\ -x & w & -z & y \\ -y & z & w & -x \\ -z & -y & x & w \end{bmatrix}.$$

The "latent roots" of q are $w \pm i\sqrt{x^2 + y^2 + z^2}$, repeated. If

$$w^2 + x^2 + y^2 + z^2 \equiv T^2, \quad \text{then} \quad q'q = qq' = T^2 I.$$

* See, for example, Ref. 3.

In the algebra of quaternions q and q' would be described as "conjugates", while T would be called the "tensor" of q and q'.

It may be noted that the four matrices of fourth order used in the preceding representation of a quaternion can, when partitioned, be expressed in terms of the second order unit matrix and the matrix $\boldsymbol{I}$ defined by (2). Hence a quaternion can also be represented as a linear combination of the four second order matrices

$$\begin{bmatrix} 1 & 0 \\ 0 & 1 \end{bmatrix}, \quad \begin{bmatrix} i & 0 \\ 0 & -i \end{bmatrix}, \quad \begin{bmatrix} 0 & 1 \\ -1 & 0 \end{bmatrix}, \quad \begin{bmatrix} 0 & i \\ i & 0 \end{bmatrix},$$

where $i = \sqrt{-1}$.

CHAPTER II

POWERS OF MATRICES, SERIES, AND INFINITESIMAL CALCULUS

2·1. Introductory. In the present chapter we shall consider some of the properties of matrices which are expressible as functions of a given matrix or which have elements functionally dependent on real or complex parameters. In the latter connection the ideas of differentiation and integration of matrices will be developed. We shall also deal with certain types of infinite series of matrices (e.g. that defining the exponential function) which are of importance in the infinitesimal calculus.

2·2. Powers of Matrices. If u is a square matrix of order n, then the continued product $uuu \dots u$ to m factors is written u^m. In addition u^0 is interpreted to mean the unit matrix I_n. By the associative law of multiplication,

$$u^l \times u^m = u^m \times u^l = u^{m+l},$$

and

$$(u^m)^l = (u^l)^m = u^{ml}.$$

Hence in the multiplication of powers of matrices the usual index laws of scalar algebra hold for positive integral and zero indices.

The reciprocal matrix u^{-1} exists provided u is not singular, and the higher negative integral powers of u are then defined as the powers of the reciprocal. Thus

$$u^{-m} = (u^{-1})^m.$$

It readily follows that, provided u is not singular, the index laws are applicable for all integral indices.

EXAMPLES

(i) *Integral Powers.* If

$$u = \begin{bmatrix} 3 & -4 \\ 1 & -1 \end{bmatrix}, \qquad \dots\dots(1)$$

then

$$u^2 = \begin{bmatrix} 3 & -4 \\ 1 & -1 \end{bmatrix}\begin{bmatrix} 3 & -4 \\ 1 & -1 \end{bmatrix} = \begin{bmatrix} 5 & -8 \\ 2 & -3 \end{bmatrix}.$$

Hence also

$$u^3 = u^2u = uu^2 = \begin{bmatrix} 5 & -8 \\ 2 & -3 \end{bmatrix}\begin{bmatrix} 3 & -4 \\ 1 & -1 \end{bmatrix} = \begin{bmatrix} 3 & -4 \\ 1 & -1 \end{bmatrix}\begin{bmatrix} 5 & -8 \\ 2 & -3 \end{bmatrix} = \begin{bmatrix} 7 & -12 \\ 3 & -5 \end{bmatrix},$$

and

$$u^6 = (u^3)^2 = \begin{bmatrix} 7 & -12 \\ 3 & -5 \end{bmatrix}\begin{bmatrix} 7 & -12 \\ 3 & -5 \end{bmatrix} = \begin{bmatrix} 13 & -24 \\ 6 & -11 \end{bmatrix}.$$

Again, the reciprocal of u is $\begin{bmatrix} -1 & 4 \\ -1 & 3 \end{bmatrix}$, so that

$$u^{-3} \equiv (u^{-1})^3 = \begin{bmatrix} -1 & 4 \\ -1 & 3 \end{bmatrix}\begin{bmatrix} -1 & 4 \\ -1 & 3 \end{bmatrix}\begin{bmatrix} -1 & 4 \\ -1 & 3 \end{bmatrix}$$

$$= \begin{bmatrix} -3 & 8 \\ -2 & 5 \end{bmatrix}\begin{bmatrix} -1 & 4 \\ -1 & 3 \end{bmatrix} = \begin{bmatrix} -5 & 12 \\ -3 & 7 \end{bmatrix}.$$

It is easy to verify by induction, or otherwise, that when s is any integer

$$u^s = \begin{bmatrix} (1+2s), & -4s \\ s, & (1-2s) \end{bmatrix}. \qquad \ldots\ldots(2)$$

(ii) *Fractional Powers.* If m is an integer and if v is any square matrix such that $v^m = u$, then we may write $v = u^{1/m}$ and refer to v as an mth root of u. The number of mth roots which a given square matrix possesses depends on the nature of the matrix, and in special cases there may be an infinite number corresponding to any assigned value of m. Two simple illustrations will make this clear. Firstly, suppose u to be the unit matrix I_2, and assume a square root to be

$$v = \begin{bmatrix} \alpha & \beta \\ \gamma & \delta \end{bmatrix}.$$

Then we require

$$v^2 = \begin{bmatrix} \alpha^2+\beta\gamma & \beta(\alpha+\delta) \\ \gamma(\alpha+\delta) & \beta\gamma+\delta^2 \end{bmatrix} = \begin{bmatrix} 1 & 0 \\ 0 & 1 \end{bmatrix},$$

or $\quad \beta\gamma = 1-\alpha^2 = 1-\delta^2 \quad$ with $\quad \gamma(\alpha+\delta) = 0, \ \beta(\alpha+\delta) = 0.$

The possible square roots are readily seen to be $\pm I_2$ and

$$\begin{bmatrix} \alpha & \beta \\ \gamma & -\alpha \end{bmatrix},$$

with α, β, γ related by the condition $\beta\gamma = 1-\alpha^2$ but otherwise arbitrary. Hence in this case there is a doubly infinite number of square roots.

Next let

$$u = \begin{bmatrix} 3 & -4 \\ 1 & -1 \end{bmatrix}.$$

Then it can be verified by a similar method that the only possible distinct square roots of u are the two matrices

$$\pm\begin{bmatrix} 2 & -2 \\ \frac{1}{2} & 0 \end{bmatrix}.$$

It may be noted that one of these square roots is given by the formula (2) with $s = \frac{1}{2}$.

A fractional power of a matrix is commutative with any other power, and the usual index laws are applicable to non-singular matrices for all real indices. The subject of fractional powers will be discussed further in example (iv) of § 3·9.

2·3. Polynomials of Matrices. The typical polynomial of a square matrix u of order n is

$$P(u) = p_0 u^m + p_1 u^{m-1} + \dots + p_{m-1} u + p_m I_n, \qquad \dots\dots(1)$$

where m is a positive integer (the degree of the polynomial) and the coefficients p are scalar constants. It is to be noted that the final coefficient p_m is multiplied by the unit matrix. The corresponding polynomial of a scalar variable x may be denoted by

$$P(x) = p_0 x^m + p_1 x^{m-1} + \dots + p_{m-1} x + p_m. \qquad \dots\dots(2)$$

Evidently an identity such as

$$P_1(x)\, P_2(x) \equiv P_3(x), \qquad \dots\dots(3)$$

involving scalar polynomials, necessarily implies a corresponding matrix identity

$$P_1(u)\, P_2(u) \equiv P_3(u), \qquad \dots\dots(4)$$

for all the coefficients of the powers of x in the expanded form of (3) can be identified term by term with the coefficients of the corresponding powers of u in (4), on account of the properties given in § 2·2. As a corollary, the multiplication of expressions containing only a single matrix is commutative. It therefore appears that the algebras of scalar polynomials of a single variable and of polynomials of a single square matrix are completely analogous.

Examples

(i) *Factorised Form of a Matrix Polynomial.* If $\alpha_1, \alpha_2, \dots, \alpha_m$ are the roots of $P(x) = 0$, then $P(x) \equiv p_0(x-\alpha_1)(x-\alpha_2)\dots(x-\alpha_m)$. The corresponding matrix polynomial may accordingly be written in the factorised form $P(u) \equiv p_0(u-\alpha_1 I)(u-\alpha_2 I)\dots(u-\alpha_m I)$.

For instance, let $u = \begin{bmatrix} 3, & -4 \\ 1, & -1 \end{bmatrix}$ and suppose

$$\begin{aligned} P(u) &= 3u^2 - 9u + 6I \\ &= 3\begin{bmatrix} 5, & -8 \\ 2, & -3 \end{bmatrix} - 9\begin{bmatrix} 3, & -4 \\ 1, & -1 \end{bmatrix} + 6\begin{bmatrix} 1, & 0 \\ 0, & 1 \end{bmatrix} = \begin{bmatrix} -6, & 12 \\ -3, & 6 \end{bmatrix}. \end{aligned}$$

In this case $P(x) = 3x^2 - 9x + 6 = 3(x-1)(x-2)$. Hence also

$$P(u) = 3(u-I)(u-2I)$$
$$= 3\begin{bmatrix} 2, & -4 \\ 1, & -2 \end{bmatrix}\begin{bmatrix} 1, & -4 \\ 1, & -3 \end{bmatrix} = 3\begin{bmatrix} -2, & 4 \\ -1, & 2 \end{bmatrix} = \begin{bmatrix} -6, & 12 \\ -3, & 6 \end{bmatrix}.$$

(ii) *Theorem on Factorisation of* $P(\lambda)I - P(u)$. If $P(x)$ is the polynomial (2) and if λ is any constant, then

$$P(\lambda) - P(x) \equiv \sum_{r=0}^{m} p_r(\lambda^{m-r} - x^{m-r}).$$

The expression on the right is exactly divisible by $(\lambda - x)$, so that

$$P(\lambda) - P(x) = (\lambda - x)\, S(\lambda, x),$$

where S is of degree $m-1$ in both x and λ. It follows that

$$P(\lambda)I - P(u) \equiv P(\lambda I) - P(u) = (\lambda I - u)\, S(\lambda I, u).$$

Hence $\lambda I - u$ is a factor of $P(\lambda)I - P(u)$.

(iii) *Lagrange's Interpolation Formula*. If $a_1, a_2, \ldots, a_n$ are distinct but otherwise arbitrary constants, then, provided that the degree of $P(x)$ does not exceed $n-1$,

$$P(x) \equiv \sum_{r=1}^{n} P(a_r) \frac{\prod_{s \neq r} (a_s - x)}{\prod_{s \neq r} (a_s - a_r)}.$$

This scalar identity is usually known as the Lagrange interpolation formula. It gives rise to the corresponding matrix identity

$$P(u) \equiv \sum_{r=1}^{n} P(a_r) \frac{\prod_{s \neq r} (a_s I - u)}{\prod_{s \neq r} (a_s - a_r)}, \qquad \ldots\ldots(5)$$

which holds good for all distinct values of the constants a_r provided that the degree of $P(u)$ does not exceed $n-1$.

2·4. Infinite Series of Matrices. Let there be a sequence of matrices $u_0, u_1, u_2, \ldots$, *ad inf.*, which are all of the same order, and let $S_p = \sum_{r=0}^{p} u_r$. Then the infinite series $\sum_{r=0}^{\infty} u_r$ is defined to be convergent if every element in the matrix S_p converges to a not infinite limit as p tends to infinity.

An important case is that in which the series is of the type

$$S = \sum_{r=0}^{\infty} \alpha_r u^r, \qquad \ldots\ldots(1)$$

where u is a given square matrix of order n, and the coefficients α are all scalar. Methods for the discussion of the convergence of such power series and for their summation will be given in § 3·9. Sufficient, but not necessary, conditions for the convergence can readily be deduced from elementary considerations as follows. Suppose U to denote the greatest modulus of any of the elements of u. Then clearly the modulus of an element of u^2 cannot exceed nU^2. Similarly, n^2U^3 is an upper bound for the modulus of any element of u^3, and generally $n^{r-1}U^r$ is an upper bound for the modulus of every element of u^r. Hence the series whose terms are the moduli of any homologous elements in the successive terms of the power series (1) is not greater than

$$\begin{aligned}\sigma &= A_0+A_1U+nA_2U^2+n^2A_3U^3+\dots\\ &= \left(\frac{n-1}{n}\right)A_0+\frac{1}{n}(A_0+A_1\theta+A_2\theta^2+\dots), \qquad \text{......(2)}\end{aligned}$$

where $\theta = nU$, and A_r is the modulus of α_r. Accordingly, the matrix series (1) will certainly converge if the corresponding scalar series (2) converges.

EXAMPLE

Suppose
$$u = \begin{bmatrix} 0{\cdot}15, & -0{\cdot}01 \\ -0{\cdot}25, & 0{\cdot}15 \end{bmatrix},$$
and
$$S = I+u+u^2+u^3+\dots.$$
Here $n = 2$ and $U = 0{\cdot}25$, so that

$$\begin{aligned}2\sigma &= 1+1+(0{\cdot}5)+(0{\cdot}5)^2+\dots\\ &= 1+(1-0{\cdot}5)^{-1} = 3.\end{aligned}$$

Hence no element of S can exceed 1·5. The exact sum is evidently given by $S(I-u) = I$, or

$$S = (I-u)^{-1} = \begin{bmatrix} 0{\cdot}85 & 0{\cdot}01 \\ 0{\cdot}25 & 0{\cdot}85 \end{bmatrix}^{-1} = \frac{1}{0{\cdot}72}\begin{bmatrix} 0{\cdot}85 & -0{\cdot}01 \\ -0{\cdot}25 & 0{\cdot}85 \end{bmatrix}.$$

The greatest element in this is 85/72 (< 1·5).

2·5. The Exponential Function. The exponential function of a square matrix is defined by the same power series as the exponential function of a scalar. Thus

$$\exp u \equiv e^u \equiv I+u+\frac{u^2}{\underline{|2}}+\frac{u^3}{\underline{|3}}+\dots, \qquad \text{......(1)}$$

and

$$\exp(-u) \equiv e^{-u} \equiv I-u+\frac{u^2}{\underline{|2}}-\frac{u^3}{\underline{|3}}+\dots.$$

On application of (2·4·2) it is seen that a dominant series for each of the series which constitute the elements of $\exp u$ is

$$\sigma = \frac{n-1}{n} + \frac{1}{n}\exp(nU),$$

where U is the greatest modulus in u. The exponential series thus converges for all square matrices u.

The scalar exponential function has the property

$$e^{x+y} = e^x e^y,$$

which is proved by direct multiplication of the two series on the right and identification of the product with the series on the left. Clearly, the result of the multiplication of the matrix exponentials $\exp u$ and $\exp v$ will correspond to that given by the multiplication of the scalar functions provided that u and v are permutable in multiplication. Hence

$$e^u e^v = e^v e^u = e^{u+v} \qquad \text{......(2)}$$

whenever u and v commute. In particular, this condition will be satisfied if u and v are polynomials of a given matrix. It should be noted, however, that (2) will not be valid if u and v are general matrices of the same order.

From (2) it follows that

$$e^u e^{-u} = e^{-u} e^u = I.$$

Hence $\exp u$ and $\exp(-u)$ are reciprocal.

Example

Remainder after s Terms. An expression for the remainder after s terms of the exponential series can be derived from the known scalar identity

$$e^x = 1 + x + \frac{x^2}{\lfloor 2} + \dots + \frac{x^{s-1}}{\lfloor s-1} + \frac{x^s}{\lfloor s-1}\int_0^1 (1-z)^{s-1} e^{xz}\,dz.$$

The corresponding identity for a matrix of order n is

$$e^u = I + u + \frac{u^2}{\lfloor 2} + \dots + \frac{u^{s-1}}{\lfloor s-1} + R_s,$$

where*

$$R_s = \frac{u^s}{\lfloor s-1}\int_0^1 (1-z)^{s-1} e^{uz}\,dz.$$

If U is the greatest modulus of any of the elements of u, and if [1] denotes the square matrix of order n having all its elements unity, then

* See § 2·10 for definition of the integral of a matrix.

clearly $\exp(nU)[1] > \exp(uz)$ for all values of z within the range of integration $z = 0$ to 1. Further, $n^{s-1}U^s$ is an upper bound for any element of u^s. Hence an upper bound for the value of R_s is

$$\frac{n^{s-1}U^s}{\lfloor s-1}[1]e^{nU}[1]\int_0^1 (1-z)^{s-1}\,dz = \frac{(nU)^s}{\lfloor s} e^{nU}[1].$$

2·6. Differentiation of Matrices. The elements of a matrix are often functions of a variable, say t. When it is necessary to exhibit the functional dependence of u on t, the matrix is written $u(t)$ and the typical element is $u_{ij}(t)$. However, the simpler notation u and u_{ij} is usually sufficient, since the functional nature of the matrix will be clear from the context.

If t receives the increment δt, there will be corresponding increments of the elements of u, and the matrix of the increments may be written δu. The differential coefficient (or derivative) of u with respect to t is then defined by the equation

$$\frac{du}{dt} = \lim_{\delta t \to 0} \frac{\delta u}{\delta t}.$$

Hence the elements of du/dt are the differential coefficients of the corresponding elements of u. An immediate deduction is that

$$\frac{d}{dt}(u_1 + u_2) = \frac{du_1}{dt} + \frac{du_2}{dt}.$$

If $w = u_1 u_2$, then

$$\begin{aligned} w + \delta w &= (u_1 + \delta u_1)(u_2 + \delta u_2) \\ &= u_1 u_2 + (\delta u_1) u_2 + u_1(\delta u_2) + (\delta u_1)(\delta u_2), \end{aligned}$$

and on neglect of the second order term, this gives

$$\frac{d}{dt}(u_1 u_2) = u_1\left(\frac{du_2}{dt}\right) + \left(\frac{du_1}{dt}\right) u_2.$$

More generally the derivative of a continued product is formed as for scalar expressions, *except that the original order of the factors must be preserved throughout*. For instance,

$$\frac{d}{dt}u^3 = \frac{du}{dt}u^2 + u\frac{du}{dt}u + u^2\frac{du}{dt},$$

and there is a similar expression for the derivative of any positive integral power of a matrix. Again $u^{-1}u = I$, and it readily follows that

$$\frac{du^{-1}}{dt} = -u^{-1}\frac{du}{dt}u^{-1}.$$

As for scalar functions, the symbol D will sometimes be used as an abbreviation for a differential operator such as $\frac{d}{dt}$. Thus $\frac{du}{dt} \equiv Du$, and more generally $\frac{d^s u}{dt^s} \equiv D^s u$.

The sth derivative will also often be denoted by $\overset{(s)}{u}$, and in particular $Du \equiv \overset{(1)}{u}$. The accent notation for derivatives will be avoided, as it might be confused with the operation of transposition.

In the multiplication of matrix products which involve the differential operator D, care should always be taken to make sure which of the matrix products concerned are subject to the differentiation. For instance, both sides of the matrix equation $Du = v$ can be premultiplied by a conformable matrix w to give $wDu = wv$. On the other hand, if both sides are postmultiplied by w, the result should be written $(Du)w = vw$ and not $Duw = vw$.

Examples

(i) *Differentiation of a Product.* Let

$$u_1 = \begin{bmatrix} \cos a, & \sin a \\ \cos t, & \sin t \end{bmatrix} \quad \text{and} \quad u_2 = \begin{bmatrix} \cos t, & \cos a \\ \sin t, & \sin a \end{bmatrix}.$$

Then
$$\overset{(1)}{u}_1 = \begin{bmatrix} 0, & 0 \\ -\sin t, & \cos t \end{bmatrix} \quad \text{and} \quad \overset{(1)}{u}_2 = \begin{bmatrix} -\sin t, & 0 \\ \cos t, & 0 \end{bmatrix}.$$

Hence, if $w = u_1 u_2$, we have

$$\begin{aligned} \overset{(1)}{w} &= \overset{(1)}{u}_1 u_2 + u_1 \overset{(1)}{u}_2 \\ &= \begin{bmatrix} \cos a, & \sin a \\ \cos t, & \sin t \end{bmatrix} \begin{bmatrix} -\sin t, & 0 \\ \cos t, & 0 \end{bmatrix} + \begin{bmatrix} 0, & 0 \\ -\sin t, & \cos t \end{bmatrix} \begin{bmatrix} \cos t, & \cos a \\ \sin t, & \sin a \end{bmatrix} \\ &= \begin{bmatrix} \sin(a-t), & 0 \\ 0, & 0 \end{bmatrix} + \begin{bmatrix} 0, & 0 \\ 0, & \sin(a-t) \end{bmatrix} = I \sin(a-t). \end{aligned}$$

This may be verified by differentiation of

$$w = \begin{bmatrix} \cos a, & \sin a \\ \cos t, & \sin t \end{bmatrix} \begin{bmatrix} \cos t, & \cos a \\ \sin t, & \sin a \end{bmatrix} = \begin{bmatrix} \cos(a-t), & 1 \\ 1, & \cos(a-t) \end{bmatrix}.$$

(ii) *Taylor's Theorem for Matrices.* If u and v are permutable square matrices, the algebra of polynomials involving no other matrices than these (and the unit matrix) is formally identical with scalar algebra.

In this case if $P(u)$ is a polynomial of degree m,

$$P(u+v) = P(u) + v\overset{(1)}{P}(u) + \frac{v^2}{\underline{|2}}\overset{(2)}{P}(u) + \ldots + \frac{v^m}{\underline{|m}}\overset{(m)}{P}(u),$$

where $\overset{(i)}{P}(u)$ is the usual ith derived function of $P(u)$, namely, the matrix polynomial obtained by substitution of u for x in the scalar polynomial $\frac{d^i}{dx^i}P(x)$. In particular, if h is a scalar

$$P(u+hI) = P(u) + h\overset{(1)}{P}(u) + \frac{h^2}{\underline{|2}}\overset{(2)}{P}(u) + \ldots + \frac{h^m}{\underline{|m}}\overset{(m)}{P}(u).$$

(iii) *Differentiation of a Function of a Matrix with respect to the Matrix.* The differentiation of a matrix function $f(u)$ with respect to u, which has already been illustrated in example (ii), can be defined more generally as

$$\lim_{h\to 0}\left[\frac{f(u+hv)-f(u)}{h}\right]v^{-1},$$

where v is permutable with u and h is a scalar. If Taylor's theorem can be applied, then

$$\begin{aligned}\lim_{h\to 0}\left[\frac{f(u+hv)-f(u)}{h}\right]v^{-1} &= \lim_{h\to 0}\left[v\overset{(1)}{f}(u) + \frac{hv^2}{\underline{|2}}\overset{(2)}{f}(u) + \ldots\right]v^{-1} \\ &= \overset{(1)}{f}(u).\end{aligned}$$

2·7. Differentiation of the Exponential Function. If t is a scalar variable (real or complex) and u is a matrix of constants, the function exp (ut), defined similarly to (2·5·1), is

$$\exp(ut) \equiv e^{ut} = I + ut + \frac{u^2t^2}{\underline{|2}} + \ldots. \qquad \ldots\ldots(1)$$

An application of the formula (2·4·2) shows that the foregoing series is absolutely and uniformly convergent for all values of t. Now the series obtained by differentiation of (1) term by term with respect to t is

$$u + u^2t + \frac{u^3t^2}{\underline{|2}} + \ldots = ue^{ut}.$$

Since this series is absolutely and uniformly convergent, it represents the differential coefficient of exp (ut). Hence

$$De^{ut} \equiv \frac{d}{dt}(e^{ut}) = ue^{ut} = e^{ut}u. \qquad \ldots\ldots(2)$$

More generally $D^m e^{ut} = u^m e^{ut} = e^{ut}u^m$, so that if $P(D)$ is any polynomial of the differential operator D,

$$P(D)\,e^{ut} = P(u)\,e^{ut} = e^{ut}P(u).$$

In the calculus of scalar functions

$$P(D)\,(e^{at}X) = e^{at}P(D+a)\,X,$$

where $P(D)$ is any polynomial of D, and X is any function of t. It will be useful to consider whether a similar rule is applicable for matrices. Suppose $v(t)$ to be a matrix, not necessarily permutable with u. Then

$$D(e^{ut}v) = e^{ut}Dv + e^{ut}uv = e^{ut}(ID+u)\,v.$$

Again, $$D^2(e^{ut}v) = De^{ut}(ID+u)\,v = e^{ut}(ID+u)\,(ID+u)\,v.$$

The result may be written

$$D^2(e^{ut}v) = e^{ut}(ID+u)^2\,v.$$

Similarly, $$D^m(e^{ut}v) = e^{ut}(ID+u)^m\,v,$$

and more generally $$P(D)\,(e^{ut}v) = e^{ut}P(ID+u)\,v.$$

This result is valid when u is a matrix of constants, and $v(t)$ is a matrix not necessarily permutable with u. When the order of the matrices is reversed to give $P(D)\,(ve^{ut})$ there is no corresponding simple formula.

2·8. Matrices of Differential Operators. The use of matrices of differential operators is often a valuable aid to conciseness. As examples of branches of mathematics which naturally invite treatment by operational matrices we may refer to the theory of linear differential equations (see Chapters V, VI, VII) and to analytical dynamics (see Chapters VIII, IX).

To show how such matrices arise, let us suppose that we have to deal with a pair of first-order linear ordinary differential equations such as

$$v_{11}\frac{dy_1}{dt} + u_{11}y_1 + v_{12}\frac{dy_2}{dt} + u_{12}y_2 = \eta_1,$$

$$v_{21}\frac{dy_1}{dt} + u_{21}y_1 + v_{22}\frac{dy_2}{dt} + u_{22}y_2 = \eta_2,$$

in which the coefficients v_{ij}, u_{ij} are given constants and η_1, η_2 are functions of t. These equations can be expressed without ambiguity as

$$\begin{bmatrix} v_{11}D+u_{11}, & v_{12}D+u_{12} \\ v_{21}D+u_{21}, & v_{22}D+u_{22} \end{bmatrix} \begin{bmatrix} y_1 \\ y_2 \end{bmatrix} = \begin{bmatrix} \eta_1 \\ \eta_2 \end{bmatrix}.$$

The operational matrix on the left may be denoted simply by $f(D)$, and the equation can be abbreviated to

$$f(D)\,y = \eta.$$

Similarly, if the differential equations were of order higher than the first, the appropriate matrix $f(D)$ would have for elements polynomials of the operator D.

Matrices of partial differential operators often arise in connection with problems involving several independent variables. If the independent variables are denoted by $x_1, x_2, \ldots, x_n$, the simplest operational matrices of this type are the column

$$\left\{\frac{\partial}{\partial x_i}\right\} \equiv \left\{\frac{\partial}{\partial x_1}, \frac{\partial}{\partial x_2}, \ldots, \frac{\partial}{\partial x_n}\right\},$$

and the row

$$\left\lfloor\frac{\partial}{\partial x_j}\right\rfloor \equiv \left\lfloor\frac{\partial}{\partial x_1}, \frac{\partial}{\partial x_2}, \ldots, \frac{\partial}{\partial x_n}\right\rfloor.$$

Examples

(i) *Matrix Differentiation of a Matrix Product.* As an illustration of the matrix differentiation of a matrix product, suppose $\phi(D)$ to be an operational square matrix of order 2, and let

$$k \equiv \begin{bmatrix} k_{11} & k_{12} \\ k_{21} & k_{22} \end{bmatrix} \quad \text{and} \quad M(t) \equiv \begin{bmatrix} e^{\lambda_1 t} & 0 \\ 0 & e^{\lambda_2 t} \end{bmatrix},$$

where λ_1, λ_2, and the elements k_{ij}, are constants. Then

$$\phi(D)\,(kM) = \begin{bmatrix} \phi_{11}(D) & \phi_{12}(D) \\ \phi_{21}(D) & \phi_{22}(D) \end{bmatrix} \begin{bmatrix} e^{\lambda_1 t}k_{11} & e^{\lambda_2 t}k_{12} \\ e^{\lambda_1 t}k_{21} & e^{\lambda_2 t}k_{22} \end{bmatrix}.$$

By the usual properties of differential operators this product reduces to

$$\begin{bmatrix} \phi_{11}(\lambda_1)\,e^{\lambda_1 t}k_{11} + \phi_{12}(\lambda_1)\,e^{\lambda_1 t}k_{21}, & \phi_{11}(\lambda_2)\,e^{\lambda_2 t}k_{12} + \phi_{12}(\lambda_2)\,e^{\lambda_2 t}k_{22} \\ \phi_{21}(\lambda_1)\,e^{\lambda_1 t}k_{11} + \phi_{22}(\lambda_1)\,e^{\lambda_1 t}k_{21}, & \phi_{21}(\lambda_2)\,e^{\lambda_2 t}k_{12} + \phi_{22}(\lambda_2)\,e^{\lambda_2 t}k_{22} \end{bmatrix}.$$

Hence

$$\phi(D)\,(kM) = \begin{bmatrix} \phi_{11}(\lambda_1)\,k_{11} + \phi_{12}(\lambda_1)\,k_{21}, & \phi_{11}(\lambda_2)\,k_{12} + \phi_{12}(\lambda_2)\,k_{22} \\ \phi_{21}(\lambda_1)\,k_{11} + \phi_{22}(\lambda_1)\,k_{21}, & \phi_{21}(\lambda_2)\,k_{12} + \phi_{22}(\lambda_2)\,k_{22} \end{bmatrix} M.$$

A generalisation of this formula will be used in § 6·4 to obtain the solution of a system of linear differential equations with constant coefficients.

(ii) *Differentiation of Bilinear Forms.* In applications to dynamics it is often necessary to differentiate bilinear or quadratic forms partially with respect to their variables. Let the bilinear form be (see § 1·15)

$$A(y, x) = yax = x'a'y', \qquad \text{......(1)}$$

in which there are n variables y and m variables x, and a is an $(\bar{n}, m)$ matrix.

Firstly, differentiate $A(y, x)$ with respect to one of the variables y, say y_1. Then

$$\frac{\partial A}{\partial y_1} = [1, 0, 0, \ldots, 0]\, ax = a_{11}x_1 + a_{12}x_2 + \ldots + a_{1m}x_m.$$

Hence $$\left\{\frac{\partial}{\partial y_i}\right\} A = ax \quad \text{and} \quad \left\lfloor\frac{\partial}{\partial y_j}\right\rfloor A = x'a'.$$

Similarly, $$\left\{\frac{\partial}{\partial x_i}\right\} A = a'y' \quad \text{and} \quad \left\lfloor\frac{\partial}{\partial x_j}\right\rfloor A = ya.$$

Again, suppose A to be the quadratic form

$$A(x, x) = x'ax, \qquad \text{......(2)}$$

in which the square matrix a is now assumed symmetrical (see § 1·15). Then

$$\left\{\frac{\partial}{\partial x_i}\right\} A = 2ax \quad \text{and} \quad \left\lfloor\frac{\partial}{\partial x_j}\right\rfloor A = 2x'a. \qquad \text{......(3)}$$

If the variables x, y and the coefficients a_{ij} in (1) are all functions of a parameter t, and if the differentiation is with respect to t, then

$$\frac{dA}{dt} = \frac{dy}{dt}ax + y\frac{da}{dt}x + ya\frac{dx}{dt},$$

while if A is the quadratic form (2)

$$\frac{dA}{dt} = \frac{dx'}{dt}ax + x'\frac{da}{dt}x + x'a\frac{dx}{dt} = 2\frac{dx'}{dt}ax + x'\frac{da}{dt}x.$$

2·9.* Change of the Independent Variables. The usual formulae of the differential calculus for the change from a set of n independent variables x to another set y are

$$\frac{\partial}{\partial x_i} = \sum_{j=1}^{n} \frac{\partial y_j}{\partial x_i}\frac{\partial}{\partial y_j},$$

in which $i = 1, 2, \ldots, n$. These relations can be expressed conveniently as

$$\left\{\frac{\partial}{\partial x_i}\right\} = a'\left\{\frac{\partial}{\partial y_i}\right\}, \qquad \text{......(1)}$$

in which a' is the transposed of the matrix

$$a \equiv \begin{bmatrix} \dfrac{\partial y_1}{\partial x_1} & \dfrac{\partial y_1}{\partial x_2} & \cdots & \dfrac{\partial y_1}{\partial x_n} \\ \dfrac{\partial y_2}{\partial x_1} & \dfrac{\partial y_2}{\partial x_2} & \cdots & \dfrac{\partial y_2}{\partial x_n} \\ \cdots & \cdots & \cdots & \cdots \\ \dfrac{\partial y_n}{\partial x_1} & \dfrac{\partial y_n}{\partial x_2} & \cdots & \dfrac{\partial y_n}{\partial x_n} \end{bmatrix}.$$

Similarly,

$$\left\{\frac{\partial}{\partial y_i}\right\} = b' \left\{\frac{\partial}{\partial x_i}\right\}, \qquad \text{......(2)}$$

where b is the matrix obtained when x and y are interchanged in a.

From (1) and (2) it follows immediately that $\left\{\frac{\partial}{\partial x_i}\right\} = a'b' \left\{\frac{\partial}{\partial x_i}\right\}$, so that $a'b' = I$; hence $ba = I = ab$. These yield the familiar relation $|a|\,|b| = 1$ between the Jacobians $|a| \equiv \frac{\partial(y_1, y_2, \ldots, y_n)}{\partial(x_1, x_2, \ldots, x_n)}$ and $|b| \equiv \frac{\partial(x_1, x_2, \ldots, x_n)}{\partial(y_1, y_2, \ldots, y_n)}$. It follows also that the adjoints A, B of a, b, respectively, have the properties

$$A = b\,|a|; \quad B = a\,|b|; \quad AB = I. \qquad \text{......(3)}$$

A further property is given by a theorem due to Jacobi, which states in effect that

$$\left\lfloor \frac{\partial}{\partial x_j} \right\rfloor A = 0 \quad \text{and} \quad \left\lfloor \frac{\partial}{\partial y_j} \right\rfloor B = 0. \qquad \text{......(4)}$$

For instance, if $n = 3$, and if A_{11}, A_{12}, A_{13} are the cofactors of the elements in the first row of a, then $\left\lfloor \frac{\partial}{\partial x_j} \right\rfloor \{A_{11}, A_{12}, A_{13}\}$ gives

$$\frac{\partial}{\partial x_1}\left(\frac{\partial y_2}{\partial x_2}\frac{\partial y_3}{\partial x_3} - \frac{\partial y_3}{\partial x_2}\frac{\partial y_2}{\partial x_3}\right) + \frac{\partial}{\partial x_2}\left(\frac{\partial y_2}{\partial x_3}\frac{\partial y_3}{\partial x_1} - \frac{\partial y_3}{\partial x_3}\frac{\partial y_2}{\partial x_1}\right) + \frac{\partial}{\partial x_3}\left(\frac{\partial y_2}{\partial x_1}\frac{\partial y_3}{\partial x_2} - \frac{\partial y_3}{\partial x_1}\frac{\partial y_2}{\partial x_2}\right),$$

and this vanishes identically. In the same way it can be shown that the other columns of A are annihilated by the operator $\left\lfloor \frac{\partial}{\partial x_j} \right\rfloor$. The truth of the general theorem can be verified on similar lines.

In the transformation of expressions which are quadratic in the differential operators, it is necessary to use in addition to the substitution (1) a substitution for the operational row $\left\lfloor \frac{\partial}{\partial x_j} \right\rfloor$. Now a direct transposition of equation (1) leads to the substitution

$$\left\lfloor \frac{\partial}{\partial x_j} \right\rfloor = \left\lfloor \frac{\partial}{\partial y_j} \right\rfloor a_0, \qquad \text{......(5)}$$

where the cipher suffix indicates the restriction that a is to be unaffected by the differentiations $\partial/\partial y_j$. This restriction, which is very inconvenient, can be removed by the following elegant modification of the substitution due to Smith.*

Let $\{B_{ri}\}$ denote the rth column of the adjoint matrix B, and suppose ϕ to be any scalar function of the variables. Then by (4) and (2)

$$\left\lfloor \frac{\partial}{\partial y_j} \right\rfloor \{B_{ri}\} \phi = B_{r1} \frac{\partial \phi}{\partial y_1} + B_{r2} \frac{\partial \phi}{\partial y_2} + \ldots + B_{rn} \frac{\partial \phi}{\partial y_n}$$

$$= \lfloor B_{rj} \rfloor \left\{ \frac{\partial}{\partial y_i} \right\} \phi$$

$$= \lfloor B_{rj} \rfloor b' \left\{ \frac{\partial}{\partial x_i} \right\} \phi.$$

Since $bB = |b| I$, and therefore also $B'b' = |b| I$, the last expression reduces to $|b| \frac{\partial \phi}{\partial x_r}$. Accordingly by (3)

$$\left\lfloor \frac{\partial}{\partial y_j} \right\rfloor |b| \{a_{ir}\} \phi = |b| \frac{\partial \phi}{\partial x_r},$$

so that

$$\left\lfloor \frac{\partial}{\partial y_j} \right\rfloor |b| a\phi = |b| \left\lfloor \frac{\partial}{\partial x_j} \right\rfloor \phi.$$

Since ϕ is arbitrary and $|a||b| = 1$ we obtain, as an equivalent to (5), Smith's transformation

$$\left\lfloor \frac{\partial}{\partial x_j} \right\rfloor = |a| \left\lfloor \frac{\partial}{\partial y_j} \right\rfloor |a|^{-1} a. \qquad \text{.....(6)}$$

EXAMPLES

(i) *Linear Partial Differential Equation of the First Order.* If the given differential equation is

$$u_1 \frac{\partial \phi}{\partial x_1} + u_2 \frac{\partial \phi}{\partial x_2} + \ldots + u_n \frac{\partial \phi}{\partial x_n} = \psi$$

* Ref. 4.

where $u_1, u_2, \ldots, u_n$ and ψ are given functions of the n variables x, the matrix equivalent is

$$\lfloor u_j \rfloor \left\{\frac{\partial}{\partial x_i}\right\} \phi = \psi.$$

When transformed to new variables y the equation becomes by (1)

$$ua' \left\{\frac{\partial}{\partial y_i}\right\} \phi = \psi.$$

(ii) *Transformation of the General Laplacian Operator* ∇^2. The general Laplacian operator in n variables is defined to be

$$\nabla^2 \equiv \frac{\partial^2}{\partial x_1^2} + \frac{\partial^2}{\partial x_2^2} + \ldots + \frac{\partial^2}{\partial x_n^2} \equiv \left\lfloor \frac{\partial}{\partial x_j} \right\rfloor \left\{\frac{\partial}{\partial x_i}\right\}.$$

After transformation to new variables y this operator becomes by (6) and (1)

$$\nabla^2 = |a| \left\lfloor \frac{\partial}{\partial y_j} \right\rfloor |a|^{-1} aa' \left\{\frac{\partial}{\partial y_i}\right\}.$$

In the important special case where $y_1 = \text{const.}$, $y_2 = \text{const.}$, etc. are orthogonal, the product aa' reduces to a diagonal matrix.

As an illustration let us consider the transformation of the usual three-dimensional operator $\nabla^2 \equiv \frac{\partial^2}{\partial x^2} + \frac{\partial^2}{\partial y^2} + \frac{\partial^2}{\partial z^2}$ to spherical polar coordinates r, θ, ϕ. The relations between the variables are here

$$x = r \sin\theta \cos\phi; \quad y = r \sin\theta \sin\phi; \quad z = r\cos\theta.$$

Assuming x, y, z and r, θ, ϕ to correspond, respectively, to x_1, x_2, x_3 and y_1, y_2, y_3, we have

$$b = \begin{bmatrix} \frac{\partial x}{\partial r} & \frac{\partial x}{\partial \theta} & \frac{\partial x}{\partial \phi} \\ \frac{\partial y}{\partial r} & \frac{\partial y}{\partial \theta} & \frac{\partial y}{\partial \phi} \\ \frac{\partial z}{\partial r} & \frac{\partial z}{\partial \theta} & \frac{\partial z}{\partial \phi} \end{bmatrix} = \begin{bmatrix} \sin\theta\cos\phi & r\cos\theta\cos\phi & -r\sin\theta\sin\phi \\ \sin\theta\sin\phi & r\cos\theta\sin\phi & r\sin\theta\cos\phi \\ \cos\theta & -r\sin\theta & 0 \end{bmatrix}.$$

This readily yields $|b| = |a|^{-1} = r^2 \sin\theta$, and

$$a = b^{-1} = \frac{1}{r^2\sin\theta} \begin{bmatrix} r^2\sin^2\theta\cos\phi & r^2\sin^2\theta\sin\phi & r^2\sin\theta\cos\theta \\ r^2\sin\theta\cos\theta\cos\phi & r\sin\theta\cos\theta\sin\phi & -r\sin^2\theta \\ -r\sin\phi & r\cos\phi & 0 \end{bmatrix},$$

so that

$$aa' = \frac{1}{r^2\sin^2\theta} \begin{bmatrix} r^2\sin^2\theta & 0 & 0 \\ 0 & \sin^2\theta & 0 \\ 0 & 0 & 1 \end{bmatrix}.$$

The transformed operator is accordingly

$$\nabla^2 = \frac{1}{r^2\sin\theta}\left[\frac{\partial}{\partial r}, \frac{\partial}{\partial \theta}, \frac{\partial}{\partial \phi}\right]\begin{bmatrix} r^2\sin\theta & 0 & 0 \\ 0 & \sin\theta & 0 \\ 0 & 0 & 1/\sin\theta \end{bmatrix}\left\{\frac{\partial}{\partial r}, \frac{\partial}{\partial \theta}, \frac{\partial}{\partial \phi}\right\}$$

$$= \frac{1}{r^2}\frac{\partial}{\partial r}\left(r^2\frac{\partial}{\partial r}\right) + \frac{1}{r^2\sin\theta}\frac{\partial}{\partial\theta}\left(\sin\theta\frac{\partial}{\partial\theta}\right) + \frac{1}{r^2\sin^2\theta}\frac{\partial^2}{\partial\phi^2}.$$

2·10. Integration of Matrices. When the elements of u are functions of a variable t, the integral of u with respect to t taken between the limits t_0 and t is defined to be the matrix which has for its (i, j)th element $\int_{t_0}^{t} u_{ij}dt$. The integrated matrix is often written $Q_{t_0}^{t}u$. When no uncertainty can arise as to the range of integration, $Q_{t_0}^{t}$ may be abbreviated to Q.

In defining the integrated matrix Qu we have tacitly assumed that the elements u_{ij} are continuous functions of a real variable t. The definition applies equally to a complex variable t, but in this case the variable must be viewed as represented in an Argand diagram (t-plane), and the integrations of the elements must be performed along a suitable path or curve connecting t_0 and t in that diagram. The path of integration can be chosen arbitrarily, apart from the restriction that it does not encounter certain "barriers" which can be constructed from a knowledge of the singularities of the elements. These barriers may be taken as the straight line continuations to infinity of the radii joining the point t_0 to the singularities.*

An upper bound to the elements of an integrated matrix can readily be found as follows. Suppose U to be the greatest modulus of all the elements in $u(t)$, so that for all points t of the range (or path) of integration, $|u_{ij}(t)| < U$. Then, if t is a real variable, by a known property of integrals

$$\int_{t_0}^{t} u_{ij}(t)\,dt \leqslant \int_{t_0}^{t} |u_{ij}(t)|\,dt \leqslant U(t-t_0).$$

Hence

$$Q_{t_0}^{t}u \leqslant U(t-t_0)\,[1],$$

where [1] is here used to denote the square matrix of order n having units for all its elements. More generally, if t is complex, let τ be a

* The singularities of any element are the points at which that element ceases to be an *analytic* function of t, i.e. at which the element fails to satisfy the conditions of being single-valued, continuous, and of having a unique derivative. The system of "barriers" referred to is spoken of as a "Mittag-Leffler star" (see for instance § 16·5 of Ref. 5).

current point on the path of integration, σ the arc of the curve up to the point t, and s the total arc from t_0 to t. In this case

$$\int_{t_0}^{t} u_{ij}(\tau)\,d\tau \leqslant \int_{t_0}^{t} |\, u_{ij}(\tau)\,d\tau\,| \leqslant U\int_0^s d\sigma \leqslant Us,$$

so that

$$Q_{t_0}^{t} u \leqslant Us\,[1].$$

2·11. The Matrizant. A brief discussion will now be given of series of a special type which are built up by means of repeated integrations of a given matrix $u(t)$. These series have their origin in the theory of systems of linear differential equations of the first order (see § 7·5), and to illustrate how they arise naturally, we shall begin by considering the single *scalar* linear differential equation

$$\frac{dy}{dt} = f(t)\,y.$$

For simplicity assume the variables to be real, and suppose that a solution is required such that $y = y_0$ at $t = t_0$. Direct integration with respect to t yields

$$y(t) = y_0 + \int_{t_0}^{t} f(\tau_1)\,y(\tau_1)\,d\tau_1, \qquad \text{......(1)}$$

in which τ_1 is a subsidiary variable. For the particular value $t = \tau_1$, this equation gives

$$y(\tau_1) = y_0 + \int_{t_0}^{\tau_1} f(\tau_2)\,y(\tau_2)\,d\tau_2, \qquad \text{......(2)}$$

where the subsidiary variable is now written τ_2. On substitution for $y(\tau_1)$ from (2) in (1) we obtain

$$y(t) = y_0 + y_0\int_{t_0}^{t} f(\tau_1)\,d\tau_1 + \int_{t_0}^{t} f(\tau_1)\int_{t_0}^{\tau_1} f(\tau_2)\,y(\tau_2)\,d\tau_2\,d\tau_1,$$

and continued applications of (2) lead to the result

$$y(t) = y_0\,\Omega_{t_0}^{t}(f),$$

where

$$\Omega_{t_0}^{t}(f) \equiv 1 + \int_{t_0}^{t} f(\tau_1)\,d\tau_1 + \int_{t_0}^{t} f(\tau_1)\int_{t_0}^{\tau_1} f(\tau_2)\,d\tau_2\,d\tau_1 + \int_{t_0}^{t} f(\tau_1)\int_{t_0}^{\tau_1} f(\tau_2)\int_{t_0}^{\tau_2} f(\tau_3)\,d\tau_3\,d\tau_2\,d\tau_1 + \dots.$$

In the notation of § 2·10, this series would be written more concisely as

$$\Omega_{t_0}^{t}(f) \equiv 1 + Q_{t_0}^{t} f + Q_{t_0}^{t} f Q_{t_0}^{t} f + \dots,$$

or even more simply as

$$\Omega(f) = 1 + Qf + QfQf + QfQfQf + \dots.$$

A correct interpretation of this series is important. The meaning of the first integral Qf is obvious. To obtain the next term $QfQf$ we first form the product fQf, and then, treating t in that product as a current variable, integrate between the limits t_0 and t. The succeeding terms $QfQfQf$, etc. are formed in a similar manner. The series formally satisfies the given differential equation and the assigned condition $y = y_0$ at $t = t_0$. However, before it can be accepted as a genuine solution, its convergence must be established. The proof is very simple. It is assumed that, for all values of t concerned, a positive number A can be chosen such that $|f(t)| \leqslant A$. Then $Qf \leqslant A(t-t_0)$, so that

$$QfQf \leqslant A^2 \int_{t_0}^{t} (t-t_0)\, dt \leqslant \frac{A^2(t-t_0)^2}{\underline{|2}},$$

and so on. Hence

$$\begin{aligned} \Omega(f) &\leqslant 1 + A(t-t_0) + \frac{A^2(t-t_0)^2}{\underline{|2}} + \dots \\ &\leqslant e^{A(t-t_0)}. \end{aligned}$$

This shows the series to be absolutely convergent.

Suppose now that we had commenced with a system of n simultaneous first-order linear differential equations instead of with a single equation. Such a system of equations can be written concisely as

$$\frac{dy}{dt} = u(t)\, y.$$

Here y denotes the column of n dependent variables, and u is a square matrix of order n having for elements given functions of t. Direct integration gives

$$y(t) = y(t_0) + \int_{t_0}^{t} u(\tau_1)\, y(\tau_1)\, d\tau_1,$$

where $y(t_0)$ is the column of values assigned at $t = t_0$. A process of repeated substitution and integration, precisely analogous to that already described, leads to the formal solution

$$y(t) = \Omega(u)\, y(t_0),$$

in which $\Omega(u)$ is the matrix series

$$\Omega(u) \equiv I_n + Qu + Qu\,Qu + Qu\,Qu\,Qu + \dots . \qquad \dots\dots(3)$$

This expression is called* the *matrizant* of u. Care should be taken to interpret the meaning of the series correctly. The first term is the unit

* The term is due to Baker; see p. 335 of Ref. 6.

matrix of order n, and the second term is the integral of u taken between the limits t_0 and t. To obtain $Qu\,Qu$, we multiply u and Qu in the order $u\,Qu$, treat t as current in the product matrix, and again integrate between the limits t_0 and t. The remaining terms are formed in succession in the same way.

The proof of the convergence of the matrizant follows the lines adopted for the scalar series. The variable t will be assumed to be complex, as no additional complication is thereby involved. The integrations are then supposed effected along paths in the Argand diagram (t-plane) which do not encounter any of the barriers referred to in § 2·10. In the region of the t-plane under consideration let U_{ij} be an upper bound for the modulus of the typical element $u_{ij}(t)$, so that $|u_{ij}| \leqslant U_{ij}$; further, let U be a positive number such that $U_{ij} \leqslant U$ for all the elements. Then, in the first place,

$$Qu \leqslant Us[1],$$

where s is the arc of integration from t_0 to t. Similarly,

$$Qu\,Qu \equiv \int_{t_0}^{t} u\,Qu\,dt \leqslant \int_0^s U[1]\,Us[1]\,ds,$$

or
$$Qu\,Qu \leqslant \frac{nU^2s^2}{\underline{|2}}[1].$$

In general, for m repeated integrations

$$Qu\,Qu \ldots Qu \leqslant \frac{n^{m-1}U^m s^m}{\underline{|m}}[1].$$

Hence
$$\Omega(u) \leqslant I + Us[1] + \frac{nU^2s^2}{\underline{|2}}[1] + \frac{n^2U^3s^3}{\underline{|3}}[1] + \ldots.$$

Each of the series comprising the elements of $\Omega(u)$ is accordingly less than

$$S = \frac{n-1+e^{nUs}}{n}.$$

It follows that the series in question are absolutely and uniformly convergent in the part of the t-plane under consideration.

By differentiation of the series (3) it follows that

$$\frac{d}{dt}\Omega(u) = u\Omega(u). \qquad \ldots\ldots(4)$$

Other properties of the matrizant are given in § 7·6.

EXAMPLE

The Matrizant of a Matrix of Constants. Suppose u to be a matrix of constants. Then

$$Qu = u\int_{t_0}^{t} dt = u(t-t_0); \quad Qu\,Qu = u^2\int_{t_0}^{t}(t-t_0)\,dt = \frac{u^2(t-t_0)^2}{\lfloor 2}, \text{ etc.}$$

Hence in this case

$$\Omega(u) = I + u(t-t_0) + \frac{u^2(t-t_0)^2}{\lfloor 2} + \text{etc.} = e^{u(t-t_0)}.$$

Equation (2·7·2) is a particular case of (4).

CHAPTER III

LAMBDA-MATRICES AND CANONICAL FORMS

3·1. Preliminary Remarks. Part I of the present Chapter gives an account of some of the properties of matrices which have for their elements polynomials of a parameter. The study of matrices of this type may be regarded as preparing the way for the applications to linear differential equations considered in Chapters V and VI. In Part II the important conception of the equivalence of matrices is explained, and a brief review is given of some of the canonical forms to which matrices can be reduced.

PART I. LAMBDA-MATRICES

3·2. Lambda-Matrices. A square matrix* $f(\lambda)$—or more briefly f—the elements $f_{ij}(\lambda)$ of which are rational integral functions of a scalar parameter λ, is referred to as a *λ-matrix*. If N is the highest degree in λ of any of the elements, then f is said to be of degree N. Such a matrix can evidently be expanded in the form

$$f = A_0\lambda^N + A_1\lambda^{N-1} + \ldots + A_{N-1}\lambda + A_N,$$

where A_0, A_1, etc. are matrices independent of λ. A λ-matrix is accordingly a polynomial in λ with matrix coefficients. The matrix A_0 will be spoken of as the *leading matrix coefficient*.

The rank of a λ-matrix is defined as follows: If at least one of the minor determinants of f of order r is not identically zero, whereas all the minors of order greater than r do vanish identically, then f is of rank r.

It should be noted that in the foregoing definition the stated conditions must be satisfied for a general value of λ. If particular values of λ are assigned, a λ-matrix having rank r in accordance with the definition may actually acquire a rank less than r in accordance with the definition of § 1·9.

* Rectangular λ-matrices will not be considered here.

EXAMPLES

(i) *Non-Singular λ-Matrices.*

Order 3, degree 1:

$$f = \begin{bmatrix} 2\lambda-2, & \lambda+2, & -3 \\ 3\lambda-1, & \lambda, & -1 \\ -1, & 4\lambda-3, & -\lambda+1 \end{bmatrix} = \begin{bmatrix} 2, & 1, & 0 \\ 3, & 1, & 0 \\ 0, & 4, & -1 \end{bmatrix}\lambda + \begin{bmatrix} -2, & 2, & -3 \\ -1, & 0, & -1 \\ -1, & -3, & 1 \end{bmatrix}.$$

Order 2, degree 2:

$$f = \begin{bmatrix} 7\lambda^2-34\lambda+102, & 3\lambda^2-6\lambda+28 \\ 3\lambda^2-18\lambda+50, & 2\lambda^2-2\lambda+15 \end{bmatrix}.$$

(ii) *Singular λ-Matrices.*

Order 3, degree 3, rank 2:

$$f = \begin{bmatrix} \lambda, & \lambda, & \lambda^2+1 \\ \lambda^2-1, & \lambda^2+1, & \lambda \\ \lambda^3, & \lambda^3+2\lambda, & 2\lambda^2+1 \end{bmatrix}.$$

Here $|f| = 0$, but the first minors do not all vanish identically.

Order 3, degree 2, rank 1:

$$f = \begin{bmatrix} \lambda^2, & 2\lambda^2, & 3\lambda^2 \\ \lambda, & 2\lambda, & 3\lambda \\ 1, & 2, & 3 \end{bmatrix}.$$

3·3. Multiplication and Division of Lambda-Matrices. Let f, g denote two λ-matrices of equal order and of degrees N and M, respectively. If their typical matrix coefficients are A_i, B_j, respectively, the product gf formed by the usual rules is

$$gf = \left(\sum_{j=0}^{M} B_j\lambda^{M-j}\right)\left(\sum_{i=0}^{N} A_i\lambda^{N-i}\right) = \sum_{k=0}^{M+N} C_k\lambda^{M+N-k},$$

where
$$C_0 = B_0A_0,$$
$$C_1 = B_0A_1 + B_1A_0,$$
and generally
$$C_k = \sum_{i+j=k} B_jA_i.$$

The product is therefore a λ-matrix of degree $N+M$ provided C_0 is not null. A similar conclusion is valid for the product fg.

The reciprocal matrix f^{-1}—the elements of which are obtained by division of the appropriate cofactors in f by $|f|$—is not, in general, a

λ-matrix. A special case of importance in applications to differential equations (see Chapter v) is where $|f|$ is independent of λ and not zero. The reciprocal f^{-1} then is a λ-matrix.

From the preceding remarks it is seen that normally a product such as $g^{-1}f$, where both f and g are λ-matrices, will not be a λ-matrix. We shall now show that, if B_0 is not singular, it is possible to express the result of this predivision in the form

$$g^{-1}f = \boldsymbol{Q} + g^{-1}\boldsymbol{R}, \qquad \text{......(1)}$$

where the quotient $\boldsymbol{Q}$ and the remainder $\boldsymbol{R}$ are λ-matrices, and the degree of $\boldsymbol{R}$ is lower than that of g. A convenient alternative to (1) is

$$f = g\boldsymbol{Q} + \boldsymbol{R}. \qquad \text{......(2)}$$

To establish (1) we note firstly that if the degree M of g exceeds N, a solution of (2) which satisfies all the stated conditions is $\boldsymbol{Q} = 0$ and $\boldsymbol{R} = f$. Dismissing this trivial case we proceed on the assumption that $M \leqslant N$. The identity to be satisfied is then of the form

$$\begin{aligned} A_0\lambda^N + A_1\lambda^{N-1} + \dots + A_{N-1}\lambda + A_N \\ = (B_0\lambda^M + B_1\lambda^{M-1} + \dots + B_M)(\boldsymbol{Q}_0\lambda^{N-M} + \boldsymbol{Q}_1\lambda^{N-M-1} + \dots + \boldsymbol{Q}_{N-M}) \\ + \boldsymbol{R}_0\lambda^{M-1} + \boldsymbol{R}_1\lambda^{M-2} + \dots + \boldsymbol{R}_{M-1}. \end{aligned}$$

Equating the coefficients of the successive powers of λ we thus require

$$\left.\begin{aligned} A_0 &= B_0\boldsymbol{Q}_0, \\ A_1 &= B_0\boldsymbol{Q}_1 + B_1\boldsymbol{Q}_0, \\ &\dots\dots\dots\dots \\ A_{N-M} &= B_0\boldsymbol{Q}_{N-M} + B_1\boldsymbol{Q}_{N-M-1} + \dots + B_{N-M}\boldsymbol{Q}_0, \end{aligned}\right\} \qquad \text{......(3)}$$

together with

$$\left.\begin{aligned} A_{N-M+1} &= B_1\boldsymbol{Q}_{N-M} + B_2\boldsymbol{Q}_{N-M-1} + \dots + B_{N-M+1}\boldsymbol{Q}_0 + \boldsymbol{R}_0, \\ A_{N-M+2} &= B_2\boldsymbol{Q}_{N-M} + B_3\boldsymbol{Q}_{N-M-1} + \dots + B_{N-M+2}\boldsymbol{Q}_0 + \boldsymbol{R}_1, \\ &\dots\dots\dots\dots\dots\dots\dots\dots \\ A_N &= B_M\boldsymbol{Q}_{N-M} + \boldsymbol{R}_{M-1}. \end{aligned}\right\} \qquad \text{......(4)}$$

Provided B_0 is not singular, the matrix coefficients $\boldsymbol{Q}_i$ of the quotient are uniquely determinable in succession by (3), and those $\boldsymbol{R}_j$ of the remainder are then given uniquely by (4).

In a similar way it can be proved that, if B_0 is not singular, a unique quotient Q and a unique remainder R of degree less than that of g can be found such that

$$f = Qg + R. \qquad \text{......(5)}$$

The operations expressed by equations (2) and (5) are sometimes referred to as *division on the left* and *division on the right*, respectively.*

Example

Division on the Left and on the Right. Suppose

$$f \equiv \begin{bmatrix} 2\lambda-2, & \lambda+2 \\ 3\lambda-1, & \lambda \end{bmatrix} \quad \text{and} \quad g \equiv \begin{bmatrix} \lambda^2+1, & \lambda \\ 3, & 1 \end{bmatrix}.$$

The degrees of f and g are respectively 1 and 2. The identities corresponding to (2) and (5), with f placed on the left of the equation in each case, are both trivial, since $M > N$. On the other hand, if f and g are interchanged, the appropriate identities are respectively

$$\begin{bmatrix} \lambda^2+1, & \lambda \\ 3, & 1 \end{bmatrix} = \begin{bmatrix} 2\lambda-2, & \lambda+2 \\ 3\lambda-1, & \lambda \end{bmatrix} \begin{bmatrix} -\lambda+7, & -1 \\ 3\lambda-22, & 3 \end{bmatrix} + \begin{bmatrix} 59, & -8 \\ 10, & 0 \end{bmatrix},$$

$$\begin{bmatrix} \lambda^2+1, & \lambda \\ 3, & 1 \end{bmatrix} = \begin{bmatrix} -\lambda+10, & \lambda-7 \\ 0, & 0 \end{bmatrix} \begin{bmatrix} 2\lambda-2, & \lambda+2 \\ 3\lambda-1, & \lambda \end{bmatrix} + \begin{bmatrix} 14, & -20 \\ 3, & 1 \end{bmatrix}.$$

3·4. Remainder Theorems for Lambda-Matrices. Let f be as defined in § 3·2, and suppose u to be a square matrix of the same order with elements independent of λ. Write

$$f_1(u) \equiv u^N A_0 + u^{N-1} A_1 + \dots + A_N,$$

so that

$$f(\lambda) - f_1(u) = \sum_{i=0}^{N-1} [(\lambda I)^{N-i} - u^{N-i}] A_i.$$

Now

$$(\lambda I)^{N-i} - u^{N-i} = (\lambda I - u)(I\lambda^{N-i-1} + u\lambda^{N-i-2} + \dots + u^{N-i-1}).$$

Hence $\lambda I - u$ is a factor (on the left) of $f(\lambda) - f_1(u)$, and we can therefore write

$$f(\lambda) - f_1(u) = (\lambda I - u)\, Q(\lambda),$$

where $Q(\lambda)$ is a certain λ-matrix. Accordingly $f_1(u)$ is the remainder when $f(\lambda)$ is divided on the left by $\lambda I - u$.

Similarly, it can be shown that when $f(\lambda)$ is divided on the right by $\lambda I - u$ the remainder is

$$f_2(u) \equiv A_0 u^N + A_1 u^{N-1} + \dots + A_N.$$

An important application of these remainder theorems is given in § 3·7.

* See § 2·03 of Ref. 7.

3·5. The Determinantal Equation and the Adjoint of a Lambda-Matrix. Let $\Delta(\lambda)$ denote the determinant of a λ-matrix $f(\lambda)$ of order m, so that

$$\Delta(\lambda) \equiv \begin{vmatrix} f_{11}(\lambda) & f_{12}(\lambda) & \dots & f_{1m}(\lambda) \\ f_{21}(\lambda) & f_{22}(\lambda) & \dots & f_{2m}(\lambda) \\ \dots & \dots & \dots & \dots \\ f_{m1}(\lambda) & f_{m2}(\lambda) & \dots & f_{mm}(\lambda) \end{vmatrix}.$$

Then $\Delta(\lambda) = 0$ is termed the *determinantal equation*, and its roots are denoted by $\lambda_1, \lambda_2, \dots, \lambda_n$, where n is the degree of $\Delta(\lambda)$ in λ. These roots are not necessarily all distinct, but it is convenient to specify them by n distinct symbols.

The adjoint $F(\lambda)$ of $f(\lambda)$ is itself a λ-matrix. It has the properties

$$f(\lambda)\,F(\lambda) = F(\lambda)f(\lambda) = I\Delta(\lambda),$$

and differentiation p times with respect to λ gives

$$\frac{d^p}{d\lambda^p}(fF) = \frac{d^p}{d\lambda^p}(Ff) = I\overset{(p)}{\Delta}(\lambda). \qquad \text{......(1)}$$

The products in (1) can if necessary be expanded in terms of $F(\lambda)$ and the derived adjoint matrices $\overset{(1)}{F}(\lambda)$, $\overset{(2)}{F}(\lambda)$, etc.

The following theorems are of importance:

(A) The matrix $f(\lambda_s)$ obtained by substitution of any root λ_s for λ in $f(\lambda)$ is necessarily singular. When λ_s is an unrepeated root, $f(\lambda_s)$ is necessarily simply degenerate.

(B) When $f(\lambda_s)$ has degeneracy q, at least q of the roots $\lambda_1, \lambda_2, \dots, \lambda_n$ are equal to λ_s.

(C) The matrix $f(\lambda_s)$ is not necessarily multiply degenerate when λ_s is a multiple root.

(D) When $f(\lambda_s)$ is simply degenerate, the adjoint $F(\lambda_s)$ is a matrix of unit rank and is expressible as a product of the form

$$F(\lambda_s) \equiv \{k_{1s}, k_{2s}, \dots, k_{ms}\}\,[\kappa_{1s}, \kappa_{2s}, \dots, \kappa_{ms}] \equiv \{k_{is}\}\,\lfloor \kappa_{is} \rfloor,$$

where k_{is}, κ_{is} are constants appropriate to the selected root, and at least one constant of each type is not zero. For conciseness the foregoing relations will often be written

$$F(\lambda_s) = k_s \kappa_s.$$

When this abbreviated notation is used the column matrix k_s and the row matrix κ_s will be understood to be appropriate to the root λ_s.

(E) When $f(\lambda_s)$ has degeneracy q, where $q > 1$, the adjoint matrix $F(\lambda)$ and its derivatives up to and including $\overset{(q-2)}{F}(\lambda)$ at least are all null for $\lambda = \lambda_s$.

Theorems (A) and (B) follow at once from the results given in example (i) of § 1·8. Since $\overset{(1)}{\Delta}(\lambda_s) \neq 0$* for an unrepeated root λ_s, and since $\overset{(1)}{\Delta}(\lambda)$ is linear and homogeneous in the first minors of $\Delta(\lambda)$, these minors cannot all be zero for $\lambda = \lambda_s$. Hence $f(\lambda_s)$ is simply degenerate (Theorem A). When $f(\lambda_s)$ has degeneracy q, so that all the $(q-1)$th minors of $f(\lambda)$ vanish when $\lambda = \lambda_s$, then all the derivatives of $\Delta(\lambda)$ up to and including $\overset{(q-1)}{\Delta}(\lambda)$ are zero for $\lambda = \lambda_s$. This implies that λ_s is at least a q-fold root of $\Delta(\lambda)$ (Theorem B).

With regard to Theorem (C) it is obvious that $\overset{(1)}{\Delta}(\lambda)$, for instance, can vanish for $\lambda = \lambda_s$ without all the first minors in $\Delta(\lambda)$ being zero

Theorem (D) is easily proved on reference to § 1·12. For if $f(\lambda_s)$ is simply degenerate, $F(\lambda_s)$ by definition cannot be null. Hence, since the product $f(\lambda_s)\,F(\lambda_s)$—which equals $\Delta(\lambda_s)\,I$—is null, $F(\lambda_s)$ must be a matrix of unit rank. It is therefore expressible as a product of the form stated (see example (iv) of § 1·9).

Lastly, Theorem (E) is obvious from the results given in example (ii) of § 1·8, which show at once that if $f(\lambda_s)$ has degeneracy q then $(\lambda-\lambda_s)^{q-1}$ at least is a factor of every first minor of $\Delta(\lambda)$, and is thus a factor of $F(\lambda)$. Clearly in this case the derived adjoint matrices up to and including $\overset{(q-2)}{F}(\lambda)$ at least are null when $\lambda = \lambda_s$.

EXAMPLES

(i) *Distinct Roots.* If

$$f(\lambda) = \begin{bmatrix} \lambda-1 & 0 & 0 \\ 0 & \lambda-2 & 0 \\ 0 & 0 & \lambda-3 \end{bmatrix},$$

then $\Delta(\lambda) = (\lambda-1)(\lambda-2)(\lambda-3)$, so that $\lambda_1 = 1$, $\lambda_2 = 2$, $\lambda_3 = 3$. Here $f(\lambda)$ is simply degenerate for each root (Theorem A). The adjoint, for a general value of λ, is

$$F(\lambda) = \begin{bmatrix} (\lambda-2)(\lambda-3) & 0 & 0 \\ 0 & (\lambda-3)(\lambda-1) & 0 \\ 0 & 0 & (\lambda-1)(\lambda-2) \end{bmatrix}.$$

* $\overset{(1)}{\Delta}(\lambda_s)$ is an abbreviation for $\left(\dfrac{d\Delta(\lambda)}{d\lambda}\right)_{\lambda=\lambda_s}$.

Hence
$$F(\lambda_1) = \begin{bmatrix} 2 & 0 & 0 \\ 0 & 0 & 0 \\ 0 & 0 & 0 \end{bmatrix} = \{2,0,0\}[1,0,0].$$

Similarly,
$$F(\lambda_2) = \{0,1,0\}[0,-1,0],$$
$$F(\lambda_3) = \{0,0,2\}[0,\ \ 0,1].$$

These results illustrate Theorem (D).

(ii) *Repeated Roots and Simple Degeneracy.* Let

$$f(\lambda) = \begin{bmatrix} \lambda-2 & -1 & 0 \\ 0 & \lambda-2 & 0 \\ 0 & 0 & \lambda-3 \end{bmatrix}.$$

Then $\Delta(\lambda) = (\lambda-2)^2(\lambda-3)$, giving $\lambda_1 = 2$, $\lambda_2 = 2$, $\lambda_3 = 3$, and for each root $f(\lambda)$ is simply degenerate. The adjoint is

$$F(\lambda) = \begin{bmatrix} (\lambda-2)(\lambda-3) & (\lambda-3) & 0 \\ 0 & (\lambda-3)(\lambda-2) & 0 \\ 0 & 0 & (\lambda-2)^2 \end{bmatrix},$$

so that
$$F(\lambda_1) = F(\lambda_2) = \begin{bmatrix} 0 & -1 & 0 \\ 0 & 0 & 0 \\ 0 & 0 & 0 \end{bmatrix} = \{1,0,0\}[0,-1,0].$$

The results exemplify Theorems (C) and (D).

(iii) *Repeated Roots and Multiple Degeneracy.* Suppose

$$f(\lambda) = \begin{bmatrix} \lambda-2 & 0 & 0 \\ 0 & \lambda-2 & 0 \\ 0 & 0 & \lambda-3 \end{bmatrix},$$

which again yields $\Delta(\lambda) = (\lambda-2)^2(\lambda-3)$ and $\lambda_1 = 2$, $\lambda_2 = 2$, $\lambda_3 = 3$. However, in this case $f(\lambda_2)$ has degeneracy 2 and the adjoint $F(\lambda_2)$ is obviously null. This exemplifies Theorems (B) and (E).

As a further simple illustration assume

$$f(\lambda) = \begin{bmatrix} 1 & 0 & 0 & 0 \\ 0 & \lambda & 0 & 0 \\ 0 & 0 & \lambda^2 & 0 \\ 0 & 0 & 0 & \lambda^3 \end{bmatrix}.$$

Then $\Delta(\lambda) = \lambda^6$, and $f(\lambda)$ has degeneracy 3 for the sextuple root $\lambda = 0$. With λ left general the adjoint is

$$F(\lambda) = \lambda^3 \begin{bmatrix} \lambda^3 & 0 & 0 & 0 \\ 0 & \lambda^2 & 0 & 0 \\ 0 & 0 & \lambda & 0 \\ 0 & 0 & 0 & 1 \end{bmatrix},$$

so that $F(\lambda)$, $\dfrac{dF}{d\lambda}$ and $\dfrac{d^2F}{d\lambda^2}$ are all null when $\lambda = 0$.

3·6. The Characteristic Matrix of a Square Matrix and the Latent Roots. If u is a given square matrix of order n with constant elements, then the matrix

$$f(\lambda) \equiv \lambda I - u \qquad \text{......(1)}$$

is called the *characteristic matrix* of u. This very simple and important type of λ-matrix has a natural origin in applications of linear substitutions relating to the conditions under which a set of transformed variables y is proportional to the original set of variables x. If the linear substitution is given by $y = ux$, and if this relation is to be satisfied by, say, $y = \lambda x$, where λ is a scalar factor of proportionality, then we require

$$(\lambda I - u)\,x \equiv f(\lambda)\,x = 0. \qquad \text{......(2)}$$

In order that this equation may be satisfied by a column x which is not null, it is necessary that

$$\Delta(\lambda) \equiv |\,\lambda I - u\,| = 0. \qquad \text{......(3)}$$

The determinant $\Delta(\lambda)$ in this special case is usually referred to as the *characteristic function* of the matrix u, and $\Delta(\lambda) = 0$ is the *characteristic equation*. Moreover, the n roots $\lambda_1, \lambda_2, \ldots, \lambda_n$ of $\Delta(\lambda) = 0$ are called the *latent roots** of u. Theorems (A) to (E) of § 3·5 are, of course, directly applicable to the characteristic matrix $f(\lambda)$ and its adjoint $F(\lambda)$.

Suppose now that the latent roots have been found, and let a solution of (2) appropriate to the typical root λ_s be denoted by

$$x(\lambda_s) \equiv \{x_{1s}, x_{2s}, \ldots, x_{ns}\}.$$

Such a column of quantities will for convenience be spoken of as a *modal column* appropriate to λ_s.

Now equation (2) requires that

$$(\lambda_s I - u)\,x(\lambda_s) \equiv f(\lambda_s)\,x(\lambda_s) = 0. \qquad \text{......(4)}$$

* Sometimes also called the characteristic numbers of u.

If $f(\lambda_s)$ has degeneracy q, this equation will have q linearly distinct solutions (see § 1·9).

Firstly, assume λ_s to be a simple root. Then by Theorems (A) and (D) of § 3·5 $f(\lambda_s)$ is simply degenerate and $F(\lambda_s)$ is a matrix product of the type $k_s\kappa_s$ in which neither the column k_s nor the row κ_s is null. But since $f(\lambda_s)\,F(\lambda_s) = F(\lambda_s)f(\lambda_s) = 0$, it follows that $f(\lambda_s)\,k_s\kappa_s = 0$ and $k_s\kappa_s f(\lambda_s) = 0$, and these obviously require that

$$f(\lambda_s)\,k_s = 0, \qquad \text{......(5)}$$

$$\kappa_s f(\lambda_s) = 0. \qquad \text{......(6)}$$

On comparison of (4) and (5) we see that the modal column $x(\lambda_s)$ can here be taken proportional to any non-zero column of the adjoint $F(\lambda_s)$.

Next suppose λ_s to be one of a set of s equal latent roots. Then if $f(\lambda_s)$ is simply degenerate the modal column can again be chosen proportional to a non-vanishing column k_s of the adjoint, and this will be the only solution corresponding to the whole set of s equal roots. If, on the other hand, $f(\lambda_s)$ has degeneracy $q>1$, the adjoint $F(\lambda_s)$ is null by Theorem (E). The q linearly distinct solutions must then be obtained from columns of such of the derived adjoint matrices as are not null. To illustrate the method suppose $f(\lambda_s)$ to have degeneracy s, in which case the derived adjoint matrices up to and including $\overset{(s-2)}{F}(\lambda)$ are all null for $\lambda = \lambda_s$. Putting $p = s-1$ in (3·5·1), expanding the first product, and substituting $\lambda = \lambda_s$, we have

$$f(\lambda_s)\,\overset{(s-1)}{F}(\lambda_s) = 0.$$

It can be shown that the matrix $\overset{(s-1)}{F}(\lambda_s)$ is in this case of rank s and therefore a matrix product of the type $\boldsymbol{k}_s\boldsymbol{\varkappa}_s$, where the s columns of the rectangular matrix $\boldsymbol{k}_s$ and the s rows of $\boldsymbol{\varkappa}_s$ are all distinct (see § 1·9). The s columns of $\boldsymbol{k}_s$ can be chosen as the required modal columns.

In the important special case where all the latent roots of u are distinct, there is a distinct modal column k_s and a corresponding row κ_s appropriate to each root. The n modal columns, taken for all the roots $\lambda_1, \lambda_2, \ldots, \lambda_n$, form a square matrix

$$k \equiv \begin{bmatrix} k_{11} & k_{12} & \cdots & k_{1n} \\ k_{21} & k_{22} & \cdots & k_{2n} \\ \cdots & \cdots & \cdots & \cdots \\ k_{n1} & k_{n2} & \cdots & k_{nn} \end{bmatrix}, \qquad \text{......(7)}$$

which will be termed the *modal matrix*. Since the modal columns can be taken in arbitrary multiples, the matrix k is indeterminate to the extent that it can always be postmultiplied by a non-singular diagonal matrix of constants. It can be shown without difficulty that k is necessarily non-singular.

In the same way the n rows κ_s can be arranged to form a square matrix*

$$\kappa \equiv \begin{bmatrix} \kappa_{11} & \kappa_{21} & \cdots & \kappa_{n1} \\ \kappa_{12} & \kappa_{22} & \cdots & \kappa_{n2} \\ \cdots & \cdots & \cdots & \cdots \\ \kappa_{1n} & \kappa_{2n} & \cdots & \kappa_{nn} \end{bmatrix}. \qquad \text{......(8)}$$

By means of the modal matrix the n matrix equations $(\lambda_s I - u)\, k_s = 0$, taken for all the latent roots, can be combined into the single matrix equation

$$uk = \begin{bmatrix} \lambda_1 k_{11} & \lambda_2 k_{12} & \cdots & \lambda_n k_{1n} \\ \lambda_1 k_{21} & \lambda_2 k_{22} & \cdots & \lambda_n k_{2n} \\ \cdots & \cdots & \cdots & \cdots \\ \lambda_1 k_{n1} & \lambda_2 k_{n2} & \cdots & \lambda_n k_{nn} \end{bmatrix}.$$

This can be written concisely as

$$uk = k\Lambda, \qquad \text{......(9)}$$

where

$$\Lambda \equiv \begin{bmatrix} \lambda_1 & 0 & \cdots & 0 \\ 0 & \lambda_2 & \cdots & 0 \\ \cdots & \cdots & \cdots & \cdots \\ 0 & 0 & \cdots & \lambda_n \end{bmatrix} \qquad \text{......(10)}$$

denotes the diagonal matrix of the latent roots. Since k is non-singular, equation (9) gives

$$u = k\Lambda k^{-1}. \qquad \text{......(11)}$$

When two square matrices, say u and v, are related by an equation of the type

$$u = pvp^{-1},$$

where p is a non-singular square matrix, they are said to be connected by a *collineatory transformation*. Equation (11) states that a square matrix which has all its latent roots distinct is reducible by a collineatory transformation to the diagonal matrix of its latent roots. This particular transformation is often very useful. For instance, when u is expressed in the form (11),

$$u^2 = k\Lambda k^{-1} k\Lambda k^{-1} = k\Lambda^2 k^{-1}.$$

* For convenience the second suffix of the elements k_{is} and κ_{is} indicates the particular latent root with which the elements are associated. With this convention the suffices in the square matrix κ appear transposed.

Similarly,

$$u^m = k\Lambda^m k^{-1} = k\begin{bmatrix} \lambda_1^m & 0 & \dots & 0 \\ 0 & \lambda_2^m & \dots & 0 \\ \dots & \dots & \dots & \dots \\ 0 & 0 & \dots & \lambda_n^m \end{bmatrix} k^{-1},$$

and more generally, if $P(u)$ is any polynomial of u,

$$P(u) = kP(\Lambda)\,k^{-1} = k\begin{bmatrix} P(\lambda_1) & 0 & \dots & 0 \\ 0 & P(\lambda_2) & \dots & 0 \\ \dots & \dots & \dots & \dots \\ 0 & 0 & \dots & P(\lambda_n) \end{bmatrix} k^{-1}. \quad \text{......(12)}$$

In particular, we may choose $P(\lambda)$ to be the characteristic function $\Delta(\lambda) \equiv |\lambda I - u|$. Then, since $\Delta(\lambda)$ vanishes when λ is any latent root, equation (12) reduces to $\Delta(u) = 0$. The matrix u thus has the important property that it satisfies its own characteristic equation. In § 3·7 it will be shown that this holds generally for any square matrix.

If there are s roots equal to λ_s, and if $f(\lambda_s)$ is simply degenerate, equation (4) yields only a single modal column corresponding to the whole set of s roots. It is not then possible to reduce u to the diagonal form. However, the reduction to the form (11) still holds good if for every set of s equal roots λ_s the degeneracy of $f(\lambda_s)$ is s, i.e. if $f(\lambda_s)$ has "full degeneracy" for every set of repeated roots. For in that case there are s distinct modal columns for each set of repeated roots—namely, those of the $(\overline{m}, s)$ matrix $\boldsymbol{k}_s$—and the modal matrix k is composed, as before, of the n modal columns taken for all the latent roots. Similarly, there is a matrix κ composed of the rows of the matrices $\boldsymbol{\varkappa}_s$.

Examples

(i) *Latent Roots all Distinct.* Suppose

$$u = \begin{bmatrix} 2 & -2 & 3 \\ 1 & 1 & 1 \\ 1 & 3 & -1 \end{bmatrix}.$$

Then the characteristic matrix is

$$f(\lambda) \equiv \lambda I - u = \begin{bmatrix} \lambda-2 & 2 & -3 \\ -1 & \lambda-1 & -1 \\ -1 & -3 & \lambda+1 \end{bmatrix}.$$

The characteristic equation is

$$\begin{vmatrix} \lambda-2 & 2 & -3 \\ -1 & \lambda-1 & -1 \\ -1 & -3 & \lambda+1 \end{vmatrix} = \lambda^3-2\lambda^2-5\lambda+6 = (\lambda-1)(\lambda+2)(\lambda-3) = 0.$$

The latent roots are therefore $\lambda_1 = 1, \lambda_2 = -2, \lambda_3 = 3$.

When $\lambda = \lambda_1$ in $f(\lambda)$, we have

$$f(\lambda_1) = \begin{bmatrix} -1 & 2 & -3 \\ -1 & 0 & -1 \\ -1 & -3 & 2 \end{bmatrix} \quad \text{with} \quad F(\lambda_1) = \begin{bmatrix} -3 & 5 & -2 \\ 3 & -5 & 2 \\ 3 & -5 & 2 \end{bmatrix}.$$

Hence we may choose $k_1 = \{-1, 1, 1\}$.

Similarly $k_2 = \{11, 1, -14\}$ and $k_3 = \{1, 1, 1\}$. The modal matrix may accordingly be taken as

$$k = \begin{bmatrix} -1 & 11 & 1 \\ 1 & 1 & 1 \\ 1 & -14 & 1 \end{bmatrix}, \quad \text{giving} \quad k^{-1} = \tfrac{1}{30}\begin{bmatrix} -15 & 25 & -10 \\ 0 & 2 & -2 \\ 15 & 3 & 12 \end{bmatrix}.$$

Equation (11) then is

$$\begin{bmatrix} 2 & -2 & 3 \\ 1 & 1 & 1 \\ 1 & 3 & -1 \end{bmatrix} = \tfrac{1}{30}\begin{bmatrix} -1 & 11 & 1 \\ 1 & 1 & 1 \\ 1 & -14 & 1 \end{bmatrix}\begin{bmatrix} 1 & 0 & 0 \\ 0 & -2 & 0 \\ 0 & 0 & 3 \end{bmatrix}\begin{bmatrix} -15 & 25 & -10 \\ 0 & 2 & -2 \\ 15 & 3 & 12 \end{bmatrix}.$$

(ii) *Repeated Latent Roots and Simple Degeneracy.* If

$$u = \begin{bmatrix} 2 & -2 & 3 \\ 10 & -4 & 5 \\ 5 & -4 & 6 \end{bmatrix},$$

then

$$\Delta(\lambda) = \begin{vmatrix} \lambda-2 & 2 & -3 \\ -10 & \lambda+4 & -5 \\ -5 & 4 & \lambda-6 \end{vmatrix} = \lambda^3-4\lambda^2+5\lambda-2 = (\lambda-1)^2(\lambda-2).$$

The latent roots are $\lambda_1 = 1, \lambda_2 = 1, \lambda_3 = 2$. For the repeated root,

$$f(\lambda_2) = \begin{bmatrix} -1 & 2 & -3 \\ -10 & 5 & -5 \\ -5 & 4 & -5 \end{bmatrix},$$

which is simply degenerate. The adjoint is

$$F(\lambda_2) = \begin{bmatrix} -5 & -2 & 5 \\ -25 & -10 & 25 \\ -15 & -6 & 15 \end{bmatrix},$$

and this yields the single modal column $\{1, 5, 3\}$ for the two roots λ_1, λ_2. In this case the matrix u is not reducible to the diagonal form.

(iii) *Repeated Latent Roots and Multiple Degeneracy.* Let

$$u = \begin{bmatrix} 7 & 4 & -1 \\ 4 & 7 & -1 \\ -4 & -4 & 4 \end{bmatrix},$$

so that

$$\Delta(\lambda) = \begin{vmatrix} \lambda-7 & -4 & 1 \\ -4 & \lambda-7 & 1 \\ 4 & 4 & \lambda-4 \end{vmatrix} = \lambda^3 - 18\lambda^2 + 81\lambda - 108 = (\lambda-3)^2(\lambda-12).$$

Denoting the latent roots as $\lambda_1 = \lambda_2 = 3$ and $\lambda_3 = 12$, we obtain

$$f(\lambda_2) = \begin{bmatrix} -4 & -4 & 1 \\ -4 & -4 & 1 \\ 4 & 4 & -1 \end{bmatrix},$$

which has degeneracy 2. The adjoint $F(\lambda_2)$ is null, but

$$\overset{(1)}{F}(\lambda_2) = \begin{bmatrix} -5 & 4 & -1 \\ 4 & -5 & -1 \\ -4 & -4 & -8 \end{bmatrix} = \begin{bmatrix} -5 & 4 \\ 4 & -5 \\ -4 & -4 \end{bmatrix} \begin{bmatrix} 1 & 0 & 1 \\ 0 & 1 & 1 \end{bmatrix}.$$

Hence we may choose the two modal columns corresponding to the double root to be $k_1 = \{5, -4, 4\}$ and $k_2 = \{-4, 5, 4\}$. The modal column appropriate to the simple root λ_3 is found to be $\{-1, -1, 1\}$, so that

$$k = \begin{bmatrix} 5 & -4 & -1 \\ -4 & 5 & -1 \\ 4 & 4 & 1 \end{bmatrix}, \quad \text{giving} \quad k^{-1} = \tfrac{1}{9}\begin{bmatrix} 1 & 0 & 1 \\ 0 & 1 & 1 \\ -4 & -4 & 1 \end{bmatrix}.$$

Accordingly,

$$\begin{bmatrix} 7 & 4 & -1 \\ 4 & 7 & -1 \\ -4 & -4 & 4 \end{bmatrix} = \tfrac{1}{9}\begin{bmatrix} 5 & -4 & -1 \\ -4 & 5 & -1 \\ 4 & 4 & 1 \end{bmatrix} \begin{bmatrix} 3 & 0 & 0 \\ 0 & 3 & 0 \\ 0 & 0 & 12 \end{bmatrix} \begin{bmatrix} 1 & 0 & 1 \\ 0 & 1 & 1 \\ -4 & -4 & 1 \end{bmatrix}.$$

(iv) *Latent Roots of Matrices Connected by a Collineatory Transformation.* If $u = pvp^{-1}$, then $\lambda I - u = p(\lambda I - v)p^{-1}$. Hence

$$|\lambda I - u| = |\lambda I - v|,$$

so that u and v have the same latent roots. When u is reducible to the diagonal form $k\Lambda k^{-1}$ we can write $v = p^{-1}k\Lambda k^{-1}p$. The modal matrix of v then is $p^{-1}k$.

(v) *Latent Roots and Modal Matrix of $P(u)$.* Applying the theorem of example (iv) to the transformation (12), we see that when u is reducible to the diagonal form, $P(\lambda_1)$, $P(\lambda_2)$, ..., $P(\lambda_n)$ are the latent roots of $P(u)$, and that u and $P(u)$ have the same modal matrix k.

The matrix $P(u)$ can have repeated latent roots even when $\lambda_1, \lambda_2, \ldots, \lambda_n$ are all distinct. In this case if $P(u)$ has s latent roots equal to λ_s, then its characteristic matrix will have degeneracy s for the root λ_s.

3·7. The Cayley-Hamilton Theorem. In § 3·6 it was incidentally shown that a square matrix whose latent roots are all distinct satisfies its own characteristic equation. The Cayley-Hamilton theorem states that this property holds good generally for any square matrix u. Thus

$$\Delta(u) = 0. \qquad \text{......(1)}$$

Since $\Delta(\lambda) \equiv (-1)^n (\lambda_1 - \lambda)(\lambda_2 - \lambda) \ldots (\lambda_n - \lambda)$, the theorem may also be written in the factorised form

$$(\lambda_1 I - u)(\lambda_2 I - u) \ldots (\lambda_n I - u) = 0,$$

or

$$f(\lambda_1) f(\lambda_2) \ldots f(\lambda_n) = 0.$$

The proposition follows immediately from the remainder theorems of § 3·4. For, if $F(\lambda)$ is the adjoint of the characteristic matrix, then

$$f(\lambda) F(\lambda) = \Delta(\lambda) I. \qquad \text{......(2)}$$

Now $\Delta(\lambda) I$ is a λ-matrix, and the last equation shows that it has zero remainder when divided on the left by $\lambda I - u$. Hence by the remainder theorem $\Delta(u) = 0$.

In the foregoing proof no restriction has been placed on the nature of the latent roots. The theorem thus holds for matrices with repeated latent roots. However, if there are any repeated roots for which $f(\lambda)$ is multiply degenerate, the theorem admits some extension. Suppose all the first minors in $f(\lambda)$ to have a common factor $\theta(\lambda)$. This will be a factor of $\Delta(\lambda)$ so that we may write $\Delta(\lambda) = \theta(\lambda) \Delta_\rho(\lambda)$, and $\theta(\lambda)$ will evidently also be a factor of all the elements of $F(\lambda)$, so that, say, $F(\lambda) = \theta(\lambda) F_\rho(\lambda)$. Hence in place of equation (2) we can now use

$$f(\lambda) F_\rho(\lambda) = \Delta_\rho(\lambda) I.$$

By the same argument as before it follows that

$$\Delta_\rho(u) = 0. \qquad \text{......(3)}$$

In the special case where $\theta(\lambda)$ is chosen to be the highest common factor of the first minors of $\Delta(\lambda)$, written with unity as the coefficient of the highest power of λ, then $\Delta_\rho(\lambda) \equiv \Delta(\lambda)/\theta(\lambda)$ is spoken of as the *reduced characteristic function*. It can be shown* that $\Delta_\rho(u)$ then is the vanishing polynomial of u of lowest possible degree.

* See, for instance, Chap. v of Ref. 2.

Examples

(i) *Latent Roots all Distinct* (see example (i) of § 3·6). If

$$u = \begin{bmatrix} 2 & -2 & 3 \\ 1 & 1 & 1 \\ 1 & 3 & -1 \end{bmatrix},$$

then $$\Delta(\lambda) = \lambda^3 - 2\lambda^2 - 5\lambda + 6 = (\lambda-1)(\lambda+2)(\lambda-3).$$

Hence $$u^3 - 2u^2 - 5u + 6I = 0.$$

The factorised form of the theorem is

$$(I-u)(-2I-u)(3I-u) = 0,$$

which yields the identity

$$\begin{bmatrix} 1 & -2 & 3 \\ 1 & 0 & 1 \\ 1 & 3 & -2 \end{bmatrix} \begin{bmatrix} 4 & -2 & 3 \\ 1 & 3 & 1 \\ 1 & 3 & 1 \end{bmatrix} \begin{bmatrix} -1 & -2 & 3 \\ 1 & -2 & 1 \\ 1 & 3 & -4 \end{bmatrix} = 0.$$

(ii) *Repeated Latent Roots and Simple Degeneracy* (see example (ii) of § 3·6). If

$$u = \begin{bmatrix} 2 & -2 & 3 \\ 10 & -4 & 5 \\ 5 & -4 & 6 \end{bmatrix},$$

then $$\Delta(\lambda) = \lambda^3 - 4\lambda^2 + 5\lambda - 2 = (\lambda-1)^2(\lambda-2).$$

Hence $$u^3 - 4u^2 + 5u - 2I = 0.$$

In this case $f(\lambda)$ is simply degenerate for the repeated root $\lambda = 1$, and the ordinary characteristic function is the vanishing polynomial of lowest degree.

(iii) *Repeated Latent Roots and Multiple Degeneracy* (see example (iii) of § 3·6). Let

$$u = \begin{bmatrix} 7 & 4 & -1 \\ 4 & 7 & -1 \\ -4 & -4 & 4 \end{bmatrix},$$

so that

$$\Delta(\lambda) = \begin{vmatrix} \lambda-7, & -4, & 1 \\ -4, & \lambda-7, & 1 \\ 4, & 4, & \lambda-4 \end{vmatrix} = \lambda^3 - 18\lambda^2 + 81\lambda - 108 = (\lambda-3)^2(\lambda-12).$$

The normal form of the theorem gives

$$u^3 - 18u^2 + 81u - 108I = 0.$$

However, $f(\lambda)$ has degeneracy 2 for the double latent root $\lambda = 3$. Extracting from $\Delta(\lambda)$ the common factor $(\lambda-3)$ of the first minors, we obtain the reduced characteristic function

$$\Delta_\rho(\lambda) = (\lambda-3)(\lambda-12) = \lambda^2-15\lambda+36.$$

Hence the reduced form of the Cayley-Hamilton theorem gives $u^2-15u+36I = 0$, or alternatively

$$(3I-u)(12I-u) = \begin{bmatrix} -4 & -4 & 1 \\ -4 & -4 & 1 \\ 4 & 4 & -1 \end{bmatrix} \begin{bmatrix} 5 & -4 & 1 \\ -4 & 5 & 1 \\ 4 & 4 & 8 \end{bmatrix} = 0.$$

(iv) *Computation of Positive Powers of a Matrix.* Since $\Delta(u) = 0$, we have $u\Delta(u) = 0$, $u^2\Delta(u) = 0$, and so on. By successive applications of these equations any positive integral power of u is linearly expressible in terms of the unit matrix and of the first $n-1$ powers of u, where n is the order of u.

For instance, suppose u to be as for example (i). Then

$$u^3 = 2u^2+5u-6I,$$
$$u^4 = 2u^3+5u^2-6u = 9u^2+4u-12I,$$
$$u^5 = 2u^4+5u^3-6u^2 = 22u^2+33u-54I,$$

and so on.

(v) *Computation of Negative Powers of a Matrix.* If u is not singular the method of example (iv) can also be used to evaluate the negative powers, and in particular, the reciprocal, of a matrix. Thus, if u is as for example (i),

$$6u^{-1} = -u^2+2u+5I,$$
$$36u^{-2} = -6u+12I+30u^{-1} = -5u^2+4u+37I,$$

and so on.

(vi) *Reduction of Polynomials.* Since any positive integral power of u is linearly expressible in terms of I and of the first $n-1$ powers of u, so also is any polynomial $P(u)$. Thus

$$P(u) = A_0u^{n-1}+A_1u^{n-2}+\dots+A_{n-2}u+A_{n-1}I,$$

where the coefficients A_i are scalars.

By a similar process the corresponding scalar polynomial $P(\lambda_r)$ is reducible to

$$P(\lambda_r) = A_0\lambda_r^{n-1}+A_1\lambda_r^{n-2}+\dots+A_{n-2}\lambda_r+A_{n-1}.$$

However, this expression would not be valid for a general value of λ.

The argument also applies for expressions $P(u)$ involving negative powers, provided that u is not singular.

(vii) *Linear Difference Equation Satisfied by the Powers of a Matrix.* If the characteristic equation is

$$\Delta(\lambda) = \lambda^n + p_1\lambda^{n-1} + \dots + p_n = 0,$$

then $u^s\Delta(u) = 0$. Hence if $Y(s) \equiv u^{s+n}$, the sequence of matrices $Y(s)$, $Y(s-1)$, etc. satisfies the difference equation

$$Y(s) + p_1 Y(s-1) + \dots + p_n Y(s-n) = 0.$$

This implies that a sequence formed by any corresponding elements taken from $Y(s)$, $Y(s-1)$, etc. also satisfies the difference equation. Let E_s denote the value in $Y(s)$ of the selected element. Now it is known from the theory of difference equations that

$$E_s = \sum_{i=1}^{n} e_i \lambda_i^s, \qquad \dots\dots(4)$$

where the coefficients e_i are constants and $\lambda_1, \lambda_2, \dots, \lambda_n$ are the roots of $\Delta(\lambda) = 0$, here assumed all distinct. Hence also

$$Y_s = \sum_{i=1}^{n} \mathbf{e}_i \lambda_i^s,$$

where the coefficients $\mathbf{e}_i$ are matrices of constants. The last result is a particular case of Sylvester's theorem (see § 3·9).

3·8. The Adjoint and Derived Adjoints of the Characteristic Matrix. Simple expressions for the adjoint and derived adjoints of the characteristic matrix will now be obtained.

Firstly, it will be convenient to construct certain identities involving merely scalars. Write the expanded form of $\Delta(\lambda) \equiv | f(\lambda) |$ as

$$\Delta(\lambda) = \lambda^n + p_1\lambda^{n-1} + p_2\lambda^{n-2} + \dots + p_{n-1}\lambda + p_n = \prod_{j=1}^{n} (\lambda - \lambda_j). \qquad \dots\dots(1)$$

Then identically

$$\frac{\Delta(\lambda) - \Delta(x)}{\lambda - x} = \lambda^{n-1} + (x + p_1)\lambda^{n-2} + \dots + (x^{n-1} + p_1 x^{n-2} + \dots + p_{n-2}x + p_{n-1})$$
$$= \phi(\lambda, x), \text{ say.} \qquad \dots\dots(2).$$

If $x = \lambda_r$, where λ_r is a typical simple root of $\Delta(\lambda) = 0$, the preceding formula gives

$$\phi(\lambda, \lambda_r) = \Delta(\lambda)/(\lambda - \lambda_r) = \prod_{j \neq r} (\lambda - \lambda_j). \qquad \dots\dots(3)$$

Next, suppose $\Delta(\lambda) = 0$ to have two equal roots, say $\lambda_1 = \lambda_2$. On differentiation of (2) with respect to x, we obtain

$$\frac{\partial}{\partial x}\phi(\lambda, x) = \frac{\Delta(\lambda) - \Delta(x)}{(\lambda - x)^2} - \frac{\overset{(1)}{\Delta}(x)}{(\lambda - x)}.$$

Hence when $x = \lambda_2 = \lambda_1$,

$$\frac{\partial}{\partial \lambda_2}\phi(\lambda, \lambda_2) = \frac{\Delta(\lambda)}{(\lambda - \lambda_2)^2} = (\lambda - \lambda_3)(\lambda - \lambda_4)\dots(\lambda - \lambda_n), \quad \dots(4)$$

since $\Delta(\lambda_2)$ and $\overset{(1)}{\Delta}(\lambda_2)$ both vanish.

More generally, suppose there to be a set of s equal roots $\lambda_1, \lambda_2, \dots, \lambda_s$, so that $\overset{(i)}{\Delta}(\lambda_s) = 0$ for $i = 0, 1, \dots, s-1$. Differentiating (2) p times with respect to x, where $p \leqslant s-1$, and substituting $x = \lambda_s$, we obtain

$$\frac{1}{\underline{|p}}\frac{\partial^p}{\partial \lambda_s^p}\phi(\lambda, \lambda_s) = \frac{\Delta(\lambda)}{(\lambda - \lambda_s)^{p+1}} = (\lambda - \lambda_s)^{s-p-1}(\lambda - \lambda_{s+1})\dots(\lambda - \lambda_n). \quad \dots\dots(5)$$

The formulae (3) and (5) are scalar identities which are true for all values of λ, and when λ is replaced by the square matrix u corresponding matrix identities are derived. Thus (2) and (3) yield

$$\phi(u, \lambda_r) = u^{n-1} + (\lambda_r + p_1)u^{n-2} + \dots + (\lambda_r^{n-1} + p_1\lambda_r^{n-2} + \dots + p_{n-1})I$$
$$= \prod_{j \neq r}(u - \lambda_j I),$$

while, if there is a set of s equal latent roots, equation (5) gives

$$\frac{1}{\underline{|p}}\frac{\partial^p}{\partial \lambda_s^p}\phi(u, \lambda_s) = (u - \lambda_s I)^{s-p-1}(u - \lambda_{s+1}I)\dots(u - \lambda_n I), \quad \dots\dots(6)$$

in which $0 \leqslant p \leqslant s-1$.

Now replace λ by u and x by λI in the identity (2). Then, since $\Delta(u) = 0$, we obtain

$$(\lambda I - u)\phi(u, \lambda) = \Delta(\lambda) I.$$

This shows that $\phi(u, \lambda)$ is the adjoint matrix of $f(\lambda)$, for general values of λ. Hence

$$F(\lambda) = \phi(u, \lambda) = u^{n-1} + (\lambda + p_1)u^{n-2} + \dots$$
$$+ (\lambda^{n-1} + p_1\lambda^{n-2} + \dots + p_{n-1})I. \quad \dots\dots(7)$$

In particular, when $\lambda = \lambda_r$, we have

$$F(\lambda_r) = \phi(u, \lambda_r) = (-1)^{n-1}\prod_{j \neq r}(\lambda_j I - u). \quad \dots\dots(8)$$

More generally, if there are s equal roots we obtain from equation (6)

$$\frac{1}{\underline{|p}}\left(\frac{\partial^p F}{\partial \lambda^p}\right)_{\lambda=\lambda_s} = (u-\lambda_s I)^{s-p-1}(u-\lambda_{s+1}I)\dots(u-\lambda_n I),$$

for $0 \leqslant p \leqslant s-1$.

The results may be summarised as follows: If $F(\lambda)$ is the adjoint of the characteristic matrix $f(\lambda)$, then for any latent root λ_r

$$F(\lambda_r) = (-1)^{n-1}\prod_{j\neq r} f(\lambda_j), \qquad \text{......(9)}$$

while for any set of s equal roots $\lambda_1, \lambda_2, \dots, \lambda_s$, and for $0 \leqslant p \leqslant s-1$,

$$\frac{1}{\underline{|p}}\overset{(p)}{F}(\lambda_s) = (-1)^{n-p-1}[f(\lambda_s)]^{s-p-1}f(\lambda_{s+1})f(\lambda_{s+2})\dots f(\lambda_n), \qquad \text{......(10)}$$

in which $\overset{(p)}{F}(\lambda_s)$ is, as usual, an abbreviation for $\left(\frac{d^pF}{d\lambda^p}\right)_{\lambda=\lambda_s}$.

Examples

(i) *Latent Roots all Distinct* (see example (i) of § 3·6). Suppose

$$u = \begin{bmatrix} 2 & -2 & 3 \\ 1 & 1 & 1 \\ 1 & 3 & -1 \end{bmatrix}, \quad \text{with} \quad \lambda_1 = 1, \lambda_2 = -2, \lambda_3 = 3,$$

and

$$f(\lambda) = \begin{bmatrix} \lambda-2, & 2, & -3 \\ -1, & \lambda-1, & -1 \\ -1, & -3, & \lambda+1 \end{bmatrix}.$$

Then, by (9),

$$F(1) = (-1)^2 f(-2)f(3) = \begin{bmatrix} -4 & 2 & -3 \\ -1 & -3 & -1 \\ -1 & -3 & -1 \end{bmatrix}\begin{bmatrix} 1 & 2 & -3 \\ -1 & 2 & -1 \\ -1 & -3 & 4 \end{bmatrix}$$

$$= \begin{bmatrix} -3 & 5 & -2 \\ 3 & -5 & 2 \\ 3 & -5 & 2 \end{bmatrix}.$$

Similarly,
$$F(-2) = f(1)f(3) = \begin{bmatrix} 0 & 11 & -11 \\ 0 & 1 & -1 \\ 0 & -14 & 14 \end{bmatrix},$$

$$F(3) = f(1)f(-2) = \begin{bmatrix} 5 & 1 & 4 \\ 5 & 1 & 4 \\ 5 & 1 & 4 \end{bmatrix}.$$

(ii) *Repeated Latent Roots and Simple Degeneracy* (see example (ii) of § 3·6). Let

$$u = \begin{bmatrix} 2 & -2 & 3 \\ 10 & -4 & 5 \\ 5 & -4 & 6 \end{bmatrix}, \quad \text{so that} \quad \lambda_1 = 1,\ \lambda_2 = 1,\ \lambda_3 = 2,$$

and

$$f(\lambda) = \begin{bmatrix} \lambda-2, & 2, & -3 \\ -10, & \lambda+4, & -5 \\ -5, & 4, & \lambda-6 \end{bmatrix}.$$

By (9),

$$F(1) = (-1)^2 f(1) f(2) = \begin{bmatrix} -5 & -2 & 5 \\ -25 & -10 & 25 \\ -15 & -6 & 15 \end{bmatrix},$$

while (10) gives, for $n = 3$, $s = 2$, $p = 1$,

$$\frac{1}{\lfloor 1}\overset{(1)}{F}(1) = (-1) f(2) = \begin{bmatrix} 0 & -2 & 3 \\ 10 & -6 & 5 \\ 5 & -4 & 4 \end{bmatrix}.$$

We may note that $F(1)$ and $\overset{(1)}{F}(1)$ here jointly contain only two linearly independent columns. Thus $F(1)$ is of unit rank and yields only one distinct column, proportional to $\{1, 5, 3\}$. Again, the first column of $\overset{(1)}{F}(1)$ is proportional to $\{0, 2, 1\}$, and the property to be particularly noted is that the remaining two columns of $\overset{(1)}{F}(1)$ are linear combinations of $\{1, 5, 3\}$ and $\{0, 2, 1\}$. Thus

$$\{-2, -6, -4\} = -2\{1, 5, 3\} + 2\{0, 2, 1\},$$
$$\{3, 5, 4\} = 3\{1, 5, 3\} - 5\{0, 2, 1\}.$$

The general rule is that if λ_s is a latent root of multiplicity s, then s (and only s) linearly distinct non-vanishing columns will be included amongst the s matrices $F(\lambda_s)$, $\overset{(1)}{F}(\lambda_s)$, ..., $\overset{(s-1)}{F}(\lambda_s)$.

(iii) *Repeated Latent Roots and Multiple Degeneracy* (see example (iii) of § 3·6). Take

$$u = \begin{bmatrix} 7 & 4 & -1 \\ 4 & 7 & -1 \\ -4 & -4 & 4 \end{bmatrix}, \quad \text{with} \quad \lambda_1 = 3,\ \lambda_2 = 3,\ \lambda_3 = 12,$$

and

$$f(\lambda) = \begin{bmatrix} \lambda-7, & -4, & 1 \\ -4, & \lambda-7, & 1 \\ 4, & 4, & \lambda-4 \end{bmatrix}.$$

Then

$$F(3) = (-1)^2 f(3) f(12) = 0; \quad \overset{(1)}{F}(3) = (-1) f(12) = \begin{bmatrix} -5 & 4 & -1 \\ 4 & -5 & -1 \\ -4 & -4 & -8 \end{bmatrix}.$$

Here the adjoint $F(3)$ corresponding to the double root is null, while $\overset{(1)}{F}(3)$ has only two linearly independent columns.

(iv) *Matrices* $\overset{(p)}{F}(\lambda)$ *and* $f(\lambda)$ *Permutable.* Since in the case of the characteristic matrix $\dfrac{df}{d\lambda} = I$ while $\dfrac{d^p f}{d\lambda^p} = 0$ if $p > 1$, equation (3·5·1) gives

$$f(\lambda)\overset{(p)}{F}(\lambda) + p\overset{(p-1)}{F}(\lambda) = \overset{(p)}{F}(\lambda) f(\lambda) + p\overset{(p-1)}{F}(\lambda) = I\overset{(p)}{\Delta}(\lambda).$$

Hence

$$f(\lambda)\overset{(p)}{F}(\lambda) = \overset{(p)}{F}(\lambda) f(\lambda).$$

(v) *Properties of Modal Columns.* In (3·5·1) put $p = 1$ and $\lambda = \lambda_r$, where λ_r is an unrepeated latent root. Then

$$f(\lambda_r)\overset{(1)}{F}(\lambda_r) + F(\lambda_r) = \overset{(1)}{\Delta}(\lambda_r) I,$$

which, on premultiplication by $F(\lambda_r)$ and use of the relation

$$F(\lambda_r) f(\lambda_r) = 0,$$

yields

$$F^2(\lambda_r) = \overset{(1)}{\Delta}(\lambda_r) F(\lambda_r).$$

In the notation of § 3·5

$$F^2(\lambda_r) = k_r \kappa_r k_r \kappa_r = \kappa_r k_r F(\lambda_r),$$

since $\kappa_r k_r$ is a scalar. Hence

$$\kappa_r k_r = \overset{(1)}{\Delta}(\lambda_r).$$

Again, from (9) it is obvious that $F(\lambda_p) F(\lambda_q) = 0$ provided that λ_p and λ_q do not belong to a common set of equal roots. In particular, if λ_p and λ_q are both simple roots

$$\kappa_p k_q = 0. \qquad \text{......(11)}$$

These results show that when the latent roots are all distinct the modal matrix k and the matrix κ of § 3·6 are connected by the relation $\kappa k = d$, or

$$k^{-1} = d^{-1}\kappa, \qquad \text{......(12)}$$

where d is the diagonal matrix of the quantities $\overset{(1)}{\Delta}(\lambda_r)$. If u, and therefore every adjoint $F(\lambda_r)$, is symmetrical we can choose $\kappa_r = k'_r$, so that $\kappa = k'$. In this particular case $k'k = d$, and the modal matrix thus has a generalised orthogonal property (compare § 1·17).

(vi) *Partial Fractions Formulae.* In the present example use is made of the familiar method of resolution of a rational fraction into partial fractions.

For simplicity all the latent roots of u are assumed distinct. Then since $(\lambda I - u) F(\lambda) = \Delta(\lambda) I$, we may write

$$(\lambda I - u)^{-1} = \frac{F(\lambda)}{\Delta(\lambda)} = \frac{F(\lambda)}{(\lambda - \lambda_1)(\lambda - \lambda_2) \dots (\lambda - \lambda_n)}.$$

Now every element of the matrix $F(\lambda)/\Delta(\lambda)$ is a rational fraction, which can be expressed by the methods of ordinary algebra in simple partial fractions. Hence, noting that the elements of $F(\lambda)$ are of degree $n-1$ at most in λ while $\Delta(\lambda)$ has degree n, we may assume for the complete matrix an expansion of the form

$$\frac{F(\lambda)}{\Delta(\lambda)} = \frac{A_1}{\lambda - \lambda_1} + \frac{A_2}{\lambda - \lambda_2} + \dots + \frac{A_n}{\lambda - \lambda_n}.$$

The coefficients A_1, A_2, etc. are determinable by the usual methods. Thus to find A_r, multiply the equations throughout by $(\lambda - \lambda_r)$ and then substitute $\lambda = \lambda_r$. This gives

$$A_r = F(\lambda_r)/\overset{(1)}{\Delta}(\lambda_r).$$

Hence we derive the identity

$$(\lambda I - u)^{-1} = \sum_{r=1}^{n} \frac{F(\lambda_r)}{\overset{(1)}{\Delta}(\lambda_r)(\lambda - \lambda_r)}. \qquad \text{......(13)}$$

This result is also deducible from Sylvester's theorem (see § 3·9). The matrix $(\lambda I - u)^{-1}$ is usually called the *resolvent* of u.

3·9. Sylvester's Theorem. This important theorem states that if the n latent roots of u are all distinct and $P(u)$ is any polynomial of u, then

$$P(u) = \sum_{r=1}^{n} P(\lambda_r) Z_0(\lambda_r), \qquad \text{......(1)}$$

where $$Z_0(\lambda_r) \equiv F(\lambda_r)/\overset{(1)}{\Delta}(\lambda_r) \equiv \prod_{j \neq r} (\lambda_j I - u) / \prod_{j \neq r} (\lambda_j - \lambda_r),$$

and $\overset{(1)}{\Delta}(\lambda_r)$ denotes $\left(\frac{d\Delta}{d\lambda}\right)_{\lambda=\lambda_r}$.

The matrices Z which appear as coefficients of the scalar polynomials $P(\lambda_r)$ are here given a cipher suffix in order to distinguish them from the more general coefficients which are introduced when repeated latent roots occur (see § 3·10).

To prove the theorem* we note firstly (see example (vi) of § 3·7) that $P(u)$ and $P(\lambda_r)$ are reducible to the similar forms

$$P(u) = A_0 u^{n-1} + A_1 u^{n-2} + \dots + A_{n-1} I,$$
$$P(\lambda_r) = A_0 \lambda_r^{n-1} + A_1 \lambda_r^{n-2} + \dots + A_{n-1}.$$

Now use formula (2·3·5), with λ_r substituted for a_r. This formula becomes applicable since $P(u)$ is here expressed as an equivalent polynomial of degree not exceeding $n-1$, and it immediately gives (1). The theorem is also valid if $P(u)$ contains negative integral powers of u, provided that u is not singular.†

The matrices Z_0 have the properties

$$Z_0(\lambda_r)\, Z_0(\lambda_p) = 0 \quad \text{if } r \neq p, \qquad \dots\dots(2)$$

and‡

$$Z_0^m(\lambda_r) = Z_0(\lambda_r), \qquad \dots\dots(3)$$

where m is any positive integer. Also

$$\sum_{r=1}^{n} Z_0(\lambda_r) = I. \qquad \dots\dots(4)$$

Equations (2) and (3) are obtained from results given in example (v) of § 3·8, while (4) follows on substitution of I for $P(u)$ in (1).

Examples

(i) *Third Order Matrix.* Suppose

$$u = \begin{bmatrix} 2 & -2 & 3 \\ 1 & 1 & 1 \\ 1 & 3 & -1 \end{bmatrix}, \quad \text{with} \quad \lambda_1 = 1,\ \lambda_2 = -2,\ \lambda_3 = 3.$$

From the results of example (i), § 3·8,

$$F(1) = \{-1, 1, 1\}[3, -5, 2]; \qquad \overset{(1)}{\Delta}(1) = -6;$$
$$F(-2) = \{11, 1, -14\}[0, 1, -1]; \quad \overset{(1)}{\Delta}(-2) = 15;$$
$$F(3) = \{1, 1, 1\}[5, 1, 4]; \qquad \overset{(1)}{\Delta}(3) = 10.$$

Hence

$$30P(u) = P(1)\{-1, 1, 1\}[-15, 25, -10] + P(-2)\{11, 1, -14\}[0, 2, -2] + P(3)\{1, 1, 1\}[15, 3, 12], \qquad \dots\dots(5)$$

* An alternative proof can be based on the identity (3·6·12) and the relation (3·8·12).

† Note that if u is not singular no latent root can be zero.

‡ On account of the property (3) the matrices Z_0 are said to be *idempotent*.

which can be written

$$P(u) = \frac{1}{30}\begin{bmatrix} -1 & 11 & 1 \\ 1 & 1 & 1 \\ 1 & -14 & 1 \end{bmatrix}\begin{bmatrix} P(1) & 0 & 0 \\ 0 & P(-2) & 0 \\ 0 & 0 & P(3) \end{bmatrix}\begin{bmatrix} -15 & 25 & -10 \\ 0 & 2 & -2 \\ 15 & 3 & 12 \end{bmatrix}.$$

This should be compared with the last equation in example (i) of § 3·6.

If $P(u) = u^m$, and m is large, equation (5) gives the approximation

$$u^m = \tfrac{1}{30}\{1, 1, 1\}[15, 3, 12]\,3^m.$$

Here λ_3 ($=3$) is the *dominant* latent root, namely, the root having the largest modulus. A similar approximation for a high power of a matrix holds good in all cases where there is a single dominant root.

(ii) *Fourth Order Matrix.* Suppose

$$u = \begin{bmatrix} 7 & 4 & 3 & 2 \\ -3 & -2 & -5 & 2 \\ -6 & -4 & 0 & -4 \\ -6 & -4 & 1 & -5 \end{bmatrix}.$$

In this case $\Delta(\lambda) = |\lambda I - u| = (\lambda^2-1)(\lambda^2-4)$, giving the latent roots $\lambda_1 = 1$; $\lambda_2 = -1$; $\lambda_3 = 2$; $\lambda_4 = -2$. The corresponding adjoint matrices and values of $\overset{(1)}{\Delta}(\lambda)$ are

$$F(1) = -6\{-1, -1, 2, 2\}[1, 0, 2, -1]; \qquad \overset{(1)}{\Delta}(1) = -6;$$
$$F(-1) = 6\{1, 1, -2, -3\}[0, 0, 1, -1]; \qquad \overset{(1)}{\Delta}(-1) = 6;$$
$$F(2) = 12\{1, 0, -1, -1\}[3, 1, 2, 0]; \qquad \overset{(1)}{\Delta}(2) = 12;$$
$$F(-2) = -12\{-1, 1, 1, 1\}[1, 1, 1, 0]; \qquad \overset{(1)}{\Delta}(-2) = -12.$$

Hence

$$\begin{aligned} P(u) = {} & \{-1, -1, 2, 2\}[1, 0, 2, -1]\,P(1) \\ & + \{1, 1, -2, -3\}[0, 0, 1, -1]\,P(-1) \\ & + \{1, 0, -1, -1\}[3, 1, 2, 0]\,P(2) \\ & + \{-1, 1, 1, 1\}[1, 1, 1, 0]\,P(-2). \end{aligned}$$

If $P = u^m$, and m is large, the value of u^m is approximately given by

$$2^{-m}u^m = \{1, 0, -1, -1\}[3, 1, 2, 0] + (-1)^m\{-1, 1, 1, 1\}[1, 1, 1, 0].$$

In this case the two latent roots λ_3 and λ_4 have equal moduli and are dominant. The two corresponding matrices must be retained in the approximate expression for u^m.

(iii) *Summation of Infinite Power Series* (see § 2·4). Conditions for the convergence of an infinite power series in u can be found by the use of Sylvester's theorem. Thus if

$$S(u) = \sum_{i=0}^{\infty} \alpha_i u^i,$$

where the coefficients α are all scalar, it is seen at once that the series will converge to a finite limit provided that all the corresponding scalar series $S(\lambda_r)$ are convergent.

As a simple example of the summation of a series by Sylvester's theorem we may assume

$$u = \begin{bmatrix} 0·15 & -0·01 \\ -0·25 & 0·15 \end{bmatrix},$$

and
$$S(u) = I + u + u^2 + u^3 + \dots.$$

Here $f(\lambda) = \begin{bmatrix} \lambda - 0·15 & 0·01 \\ 0·25 & \lambda - 0·15 \end{bmatrix}$; $F(\lambda) = \begin{bmatrix} \lambda - 0·15 & -0·01 \\ -0·25 & \lambda - 0·15 \end{bmatrix}$,

and
$$\Delta(\lambda) = \lambda^2 - 0·3\lambda + 0·02.$$

The latent roots are $\lambda_1 = 0·1$ and $\lambda_2 = 0·2$, and these give

$$F(0·1) = \begin{bmatrix} -0·05 & -0·01 \\ -0·25 & -0·05 \end{bmatrix}; \quad F(0·2) = \begin{bmatrix} 0·05 & -0·01 \\ -0·25 & 0·05 \end{bmatrix},$$

with
$$\overset{(1)}{\Delta}(0·1) = -0·1 \quad \text{and} \quad \overset{(1)}{\Delta}(0·2) = 0·1.$$

Again
$$S(0·1) = \frac{1}{1-0·1} = \frac{10}{9}; \quad S(0·2) = \frac{1}{1-0·2} = \frac{10}{8}.$$

Hence finally (see also example, § 2·4)

$$S(u) = \begin{bmatrix} -0·05 & -0·01 \\ -0·25 & -0·05 \end{bmatrix} \frac{S(0·1)}{(-0·1)} + \begin{bmatrix} 0·05 & -0·01 \\ -0·25 & 0·05 \end{bmatrix} \frac{S(0·2)}{0·1}$$
$$= \tfrac{1}{72} \begin{bmatrix} 85 & -1 \\ -25 & 85 \end{bmatrix}.$$

(iv) *Fractional Powers of a Matrix* (see § 2·2). Sylvester's theorem has only been proved for rational integral functions $P(u)$, but it suggests that if, for example, u is a second order square matrix, then

$$[\lambda_1^{\frac{1}{2}} Z_0(\lambda_1) + \lambda_2^{\frac{1}{2}} Z_0(\lambda_2)]^2 = u. \qquad \dots\dots(6)$$

Now by direct multiplication

$$[\lambda_1^{\frac{1}{2}} Z_0(\lambda_1) + \lambda_2^{\frac{1}{2}} Z_0(\lambda_2)]^2$$
$$= \lambda_1 Z_0^2(\lambda_1) + \lambda_1^{\frac{1}{2}} \lambda_2^{\frac{1}{2}} [Z_0(\lambda_1) Z_0(\lambda_2) + Z_0(\lambda_2) Z_0(\lambda_1)] + \lambda_2 Z_0^2(\lambda_2).$$

On account of the properties (2) and (3) the right-hand side of this equation reduces to $\lambda_1 Z_0(\lambda_1) + \lambda_2 Z_0(\lambda_2)$, i.e. to u. Hence (6) is verified.

Evidently by the foregoing method four possible square roots of u can be obtained, namely $\pm \lambda_1^{\frac{1}{2}} Z_0(\lambda_1) \pm \lambda_2^{\frac{1}{2}} Z_0(\lambda_2)$, the signs being here associated in all possible combinations. More generally, if u is a square matrix of order n with distinct latent roots, then $\Sigma\, \lambda_r^{1/m} Z_0(\lambda_r)$ is an mth root of u, and since $\lambda_r^{1/m}$ has m distinct values and there are n latent roots, it will be possible to construct m^n such roots. The foregoing argument would, of course, require modification if some of the latent roots were repeated.

(v) *The Matrices Z_0 as Selective Operators.* Suppose we have to deal with sets of objects which are drawn from n mutually exclusive classes $1, 2, \ldots, n-1, n$. Let U represent a set, and U_r the subset of U which belongs to the class r. Then obviously

$$U = \sum_{r=1}^{n} U_r. \qquad \text{......(7)}$$

Again, let S_r denote the operation of selecting from U all the objects belonging to the class r; then

$$S_r U = U_r. \qquad \text{......(8)}$$

Accordingly,

$$U = \sum_{r=1}^{n} S_r U,$$

or

$$\sum_{r=1}^{n} S_r = 1. \qquad \text{......(9)}$$

Next, apply S_r to both sides of (8). Now clearly $S_r U_r = U_r$, since U_r consists entirely of members of the class r. Hence

$$S_r^2 U = U_r = S_r U,$$

so that

$$S_r^2 = S_r. \qquad \text{......(10)}$$

Further, $S_p U_r = 0$ if $p \neq r$, for U_r contains no members of the class p. Apply S_p to equation (8). The result is $S_p S_r U = 0$, or

$$S_p S_r = 0 \quad (p \neq r). \qquad \text{......(11)}$$

Operators which possess the properties given by (9), (10) and (11) are said to form a *spectral set*.* From the present standpoint the interest lies in the fact that the matrices $Z_0(\lambda_r)$ can be regarded as selective operators of this type.

* For a simple account see, for instance, Chapter XII of Ref. 8.

3·10.* Confluent Form of Sylvester's Theorem. The theorem given in § 3·9 requires modification if two or more of the latent roots are equal. Before discussing this modification we shall give an alternative proof of the theorem for the case of distinct roots.

Since $P(\lambda)I - P(u)$ is exactly divisible by $\lambda I - u$, it follows that

$$P(u) = P(\lambda)I + Q_1(\lambda)(\lambda I - u),$$

where $Q_1(\lambda)$ is a polynomial of λ and u which need not be determined. This equation may be written

$$P(u)(\lambda I - u)^{-1} = \frac{P(\lambda)F(\lambda)}{\Delta(\lambda)} + Q_1(\lambda).$$

The first matrix on the right can be resolved into partial fractions, as illustrated in example (vi) of § 3·8. Since the degree in λ of $P(\lambda)F(\lambda)$ will in general exceed that of $\Delta(\lambda)$, the correct form to assume in the present case is

$$\frac{P(\lambda)F(\lambda)}{\Delta(\lambda)} = \sum_{r=1}^{n} \frac{A_r}{(\lambda - \lambda_r)} + Q_2(\lambda),$$

and the typical coefficient A_r is given by

$$A_r = P(\lambda_r)F(\lambda_r)/\overset{(1)}{\Delta}(\lambda_r).$$

It follows that

$$P(u)(\lambda I - u)^{-1} = \sum_{r=1}^{n} \frac{P(\lambda_r)F(\lambda_r)}{\overset{(1)}{\Delta}(\lambda_r)(\lambda - \lambda_r)} + Q_1(\lambda) + Q_2(\lambda).$$

Evidently, since $P(u)$ is independent of λ, the quotient $Q_1 + Q_2$ must be zero. Hence

$$P(u)(\lambda I - u)^{-1} = \sum_{r=1}^{n} \frac{P(\lambda_r)F(\lambda_r)}{\overset{(1)}{\Delta}(\lambda_r)(\lambda - \lambda_r)}. \qquad \text{......(1)}$$

Similarly, replacing $P(u)$ by $uP(u)$ we have

$$uP(u)(\lambda I - u)^{-1} = \sum_{r=1}^{n} \frac{\lambda_r P(\lambda_r)F(\lambda_r)}{\overset{(1)}{\Delta}(\lambda_r)(\lambda - \lambda_r)}. \qquad \text{......(2)}$$

Multiply (1) throughout by λ and subtract (2). Then

$$P(u) = \sum_{r=1}^{n} \frac{P(\lambda_r)F(\lambda_r)}{\overset{(1)}{\Delta}(\lambda_r)}, \qquad \text{......(3)}$$

which is the theorem of § 3·9.

To extend the theorem to cover the case of multiple latent roots, it is only necessary to modify appropriately the expansion involving the partial fractions, and it will be enough to consider those terms which arise from a typical set of s equal roots $\lambda_1, \lambda_2, \ldots, \lambda_s$. Thus assume

$$\frac{P(\lambda)\,F(\lambda)}{\Delta(\lambda)} = \sum_{i=1}^{s} \frac{A_i}{(\lambda-\lambda_s)^i} + R,$$

where R includes the quotient and the partial fractions arising from the remaining roots. Then by the usual methods of algebra

$$A_i = \frac{1}{\lfloor s-i}\left[\frac{d^{s-i}}{d\lambda^{s-i}}\frac{P(\lambda)\,F(\lambda)}{\Delta_s(\lambda)}\right]_{\lambda=\lambda_s},$$

where $$\Delta_s(\lambda) \equiv (\lambda-\lambda_{s+1})(\lambda-\lambda_{s+2})\ldots(\lambda-\lambda_n).$$

The terms in the right-hand side of the equation corresponding to (1), due to the set of equal roots considered, are then

$$\sum_{i=1}^{s} \frac{A_i}{(\lambda-\lambda_s)^i}.$$

Similarly, those contributed to the equation corresponding to (2) are

$$\sum_{i=1}^{s} \frac{B_i}{(\lambda-\lambda_s)^i},$$

where $$B_i = \frac{1}{\lfloor s-i}\left[\frac{d^{s-i}}{d\lambda^{s-i}}\frac{\lambda P(\lambda)\,F(\lambda)}{\Delta_s(\lambda)}\right]_{\lambda=\lambda_s}$$

Hence the terms contributed to the equation corresponding to (3) are

$$\sum_{i=1}^{s} \frac{\lambda A_i - B_i}{(\lambda-\lambda_s)^i}.$$

But $B_i = \lambda_s A_i + A_{i+1}$, so that the last summation reduces to

$$\sum_{i=1}^{s} \frac{(\lambda-\lambda_s)\,A_i - A_{i+1}}{(\lambda-\lambda_s)^i} = A_1.$$

The term in the expression for $P(u)$ due to the roots λ_s is accordingly

$$\frac{1}{\lfloor s-1}\left[\frac{d^{s-1}}{d\lambda^{s-1}}\frac{P(\lambda)\,F(\lambda)}{\Delta_s(\lambda)}\right]_{\lambda=\lambda_s}$$

The *confluent form of Sylvester's theorem* may thus be stated as

$$P(u) = \Sigma\, T(\lambda_s) = \Sigma \frac{1}{\lfloor s-1}\left[\frac{d^{s-1}}{d\lambda^{s-1}}\frac{P(\lambda)\,F(\lambda)}{\Delta_s(\lambda)}\right]_{\lambda=\lambda_s}, \quad \ldots\ldots(4)$$

where $\Delta_s(\lambda) \equiv (\lambda-\lambda_{s+1})(\lambda-\lambda_{s+2})\ldots(\lambda-\lambda_n)$, and the summation is taken for all distinct values λ_s of the latent roots.

The typical term $T(\lambda_s)$ is expressible in the expanded form

$$T(\lambda_s) = P(\lambda_s)\, Z_{s-1}(\lambda_s) + \frac{\overset{(1)}{P}(\lambda_s)}{\underline{|1}}\, Z_{s-2}(\lambda_s) + \ldots + \frac{\overset{(s-1)}{P}(\lambda_s)}{\underline{|s-1}}\, Z_0(\lambda_s),$$

in which
$$Z_i(\lambda_s) \equiv \frac{1}{\underline{|i}} \left[\frac{d^i}{d\lambda^i} \frac{F(\lambda)}{\Delta_s(\lambda)} \right]_{\lambda=\lambda_s},$$

for $i=0, 1, \ldots, s-1$, and $\overset{(i)}{P}(\lambda_s)$ denotes $\left(\dfrac{d^i P(\lambda)}{d\lambda^i}\right)_{\lambda=\lambda_s}$

An alternative is

$$T(\lambda_s) = \frac{Y(\lambda_s)\, \overset{(s-1)}{F}(\lambda_s)}{\underline{|s-1}} + \frac{\overset{(1)}{Y}(\lambda_s)\, \overset{(s-2)}{F}(\lambda_s)}{\underline{|1}\,\underline{|s-2}} + \frac{\overset{(2)}{Y}(\lambda_s)\, \overset{(s-3)}{F}(\lambda_s)}{\underline{|2}\,\underline{|s-3}} + \ldots + \frac{\overset{(s-1)}{Y}(\lambda_s)\, F(\lambda_s)}{\underline{|s-1}},$$

where $Y(\lambda) \equiv P(\lambda)/\Delta_s(\lambda)$ and $\overset{(i)}{Y}(\lambda_s)$, $\overset{(i)}{F}(\lambda_s)$ have meanings similar to $\overset{(i)}{P}(\lambda_s)$. On application of (3·8·10) the last equation may be written also as

$$T(\lambda_s) = (-1)^{n-s} \Bigg[Y(\lambda_s) I - \overset{(1)}{Y}(\lambda_s) f(\lambda_s) + \frac{\overset{(2)}{Y}(\lambda_s) f^2(\lambda_s)}{\underline{|2}} + \ldots$$
$$+ (-1)^{s-1} \frac{\overset{(s-1)}{Y}(\lambda_s)\, f^{s-1}(\lambda_s)}{\underline{|s-1}} \Bigg] f(\lambda_{s+1}) f(\lambda_{s+2}) \ldots f(\lambda_n).$$

If $f(\lambda_s)$ has degeneracy q, then the highest power of $f(\lambda_s)$ present in the foregoing expression will be $s-q$. This also follows from a consideration of the reduced characteristic function (see § 3·7).

EXAMPLES

(i) *Matrix with Latent Roots all Equal.* If the latent roots of u are all equal, so that $s = n$, then $\Delta_s = 1$, $Y = P(\lambda_s)$. Hence

$$P(u) = P(\lambda_n) - \overset{(1)}{P}(\lambda_n)\,(\lambda_n I - u) + \frac{\overset{(2)}{P}(\lambda_n)}{\underline{|2}} (\lambda_n I - u)^2 + \ldots$$
$$+ (-1)^{n-1} \frac{\overset{(n-1)}{P}(\lambda_n)}{\underline{|n-1}} (\lambda_n I - u)^{n-1}.$$

In particular

$$u^m = \lambda_n^m I - m\lambda_n^{m-1} (\lambda_n I - u) + \frac{m(m-1)}{\underline{|2}} \lambda_n^{m-2} (\lambda_n I - u)^2 + \ldots$$
$$+ (-1)^{n-1} \frac{m(m-1)\ldots(m-n+2)}{\underline{|n-1}} \lambda_n^{m-n+1} (\lambda_n I - u)^{n-1}.$$

This result is verified immediately by the identity

$$u^m \equiv [\lambda_n I - (\lambda_n I - u)]^m$$

and an application of the Cayley-Hamilton theorem, which in this case gives $(\lambda_n I - u)^n = 0$.

(ii) *Sylvester's Expansion for Matrix u itself.* If $P(u) = u$, then

$$u = \Sigma_1 \lambda_r Z_0(\lambda_r) + \Sigma_2 [\lambda_s Z_{s-1}(\lambda_s) + Z_{s-2}(\lambda_s)], \qquad \text{......(5)}$$

where the first summation includes all unrepeated roots and the second is taken for all distinct values λ_s of repeated roots.

In the particular case where the characteristic matrix has full degeneracy s for every set of s repeated roots, the derived adjoint matrices up to, and including, $\overset{(s-2)}{F}(\lambda_s)$ are all null (see Theorem (E) of § 3·5). Hence $Z_{s-2}(\lambda_s) = 0$ and $Z_{s-1}(\lambda_s) = \overset{(s-1)}{F}(\lambda_s) / \underline{|s-1}\, \Delta_s(\lambda_s)$. Equation (5) may then be written in the notation of § 3·6

$$u = \Sigma \frac{\lambda_s \boldsymbol{k}_s \boldsymbol{\kappa}_s}{\underline{|s-1}\, \Delta_s(\lambda_s)},$$

where the summation is for all *distinct* values of the latent roots. A comparison of this with the equation $u = k\Lambda k^{-1}$, in which the modal matrix k is now composed of all the columns of the matrices $\boldsymbol{k}_s$ (see § 3·6), readily yields the identity

$$\kappa k = d \quad \text{or} \quad k^{-1} = d^{-1}\kappa, \qquad \text{......(6)}$$

where d is a diagonal matrix which is expressible in partitioned form as

$$\begin{bmatrix} \underline{|a-1}\, \Delta_a(\lambda_a) I_a & 0 & \cdots & 0 \\ 0 & \underline{|b-1}\, \Delta_b(\lambda_b) I_b & \cdots & 0 \\ \cdots\cdots & \cdots\cdots & \cdots\cdots & \cdots\cdots \\ 0 & 0 & \cdots & \underline{|p-1}\, \Delta_p(\lambda_p) I_p \end{bmatrix},$$

and $a, b, \ldots, p$ are the multiplicities in the sets of equal roots. Equation (6) is a generalisation of (3·8·12).

(iii) *Two Equal Latent Roots and Simple Degeneracy.* If

$$u = \begin{bmatrix} 2 & -2 & 3 \\ 10 & -4 & 5 \\ 5 & -4 & 6 \end{bmatrix},$$

then $\Delta(\lambda) = (\lambda-1)^2(\lambda-2)$ and $\lambda_1 = \lambda_2 = 1$, $\lambda_3 = 2$. The confluent form of Sylvester's theorem gives

$$P(u) = \left[\frac{d}{d\lambda}\cdot\frac{P(\lambda)\,F(\lambda)}{(\lambda-2)}\right]_{\lambda=1} + \frac{P(2)\,F(2)}{(2-1)^2}$$
$$= P(1)\,[-\overset{(1)}{F}(1) - F(1)] - \overset{(1)}{P}(1)\,F(1) + P(2)\,F(2).$$

Using results obtained in example (ii) of § 3·8, we obtain

$$P(u) = \begin{bmatrix} 5 & 4 & -8 \\ 15 & 16 & -30 \\ 10 & 10 & -19 \end{bmatrix} P(1) + \begin{bmatrix} 5 & 2 & -5 \\ 25 & 10 & -25 \\ 15 & 6 & -15 \end{bmatrix} \overset{(1)}{P}(1)$$
$$+ \begin{bmatrix} -4 & -4 & 8 \\ -15 & -15 & 30 \\ -10 & -10 & 20 \end{bmatrix} P(2).$$

(iv) *Two Equal Latent Roots and Multiple Degeneracy.* If

$$u = \begin{bmatrix} 7 & 4 & -1 \\ 4 & 7 & -1 \\ -4 & -4 & 4 \end{bmatrix},$$

the latent roots are $\lambda_1 = \lambda_2 = 3$, $\lambda_3 = 12$, and on reference to the results of example (iii) of § 3·8, we find

$$9P(u) = \begin{bmatrix} 5 & -4 & 1 \\ -4 & 5 & 1 \\ 4 & 4 & 8 \end{bmatrix} P(3) + \begin{bmatrix} 4 & 4 & -1 \\ 4 & 4 & -1 \\ -4 & -4 & 1 \end{bmatrix} P(12).$$

The term involving $\overset{(1)}{P}(3)$ is absent since $F(3) = 0$.

PART II. CANONICAL FORMS

3·11. Elementary Operations on Matrices. An important conception relating to matrices is that of equivalence, but before defining equivalent matrices we shall consider certain elementary operations which are, in fact, extensively used in the reduction of determinants. These operations are:

(I) Interchange of two rows, or of two columns.

(II) Addition to a row of a multiple of another row, or addition to a column of a multiple of another column.

(III) Multiplication of a row or of a column by a non-vanishing constant.

It will now be shown that the performance of any one of these operations upon a square matrix u is equivalent to premultiplication or postmultiplication of u by a non-singular matrix.

Operation of Type I. Let J denote the unit matrix I with the ith and the jth rows interchanged. Then it is easy to see that the product Ju is the matrix obtained when the ith and jth rows of u are interchanged. Similarly uJ is the matrix obtained from u by interchange of the ith and jth columns.

The determinant of J is -1.

EXAMPLES

(i)
$$\begin{bmatrix} 1 & 0 & 0 \\ 0 & 0 & 1 \\ 0 & 1 & 0 \end{bmatrix}\begin{bmatrix} u_{11} & u_{12} & u_{13} \\ u_{21} & u_{22} & u_{23} \\ u_{31} & u_{32} & u_{33} \end{bmatrix} = \begin{bmatrix} u_{11} & u_{12} & u_{13} \\ u_{31} & u_{32} & u_{33} \\ u_{21} & u_{22} & u_{23} \end{bmatrix}.$$

(ii)
$$\begin{bmatrix} u_{11} & u_{12} & u_{13} \\ u_{21} & u_{22} & u_{23} \\ u_{31} & u_{32} & u_{33} \end{bmatrix}\begin{bmatrix} 1 & 0 & 0 \\ 0 & 0 & 1 \\ 0 & 1 & 0 \end{bmatrix} = \begin{bmatrix} u_{11} & u_{13} & u_{12} \\ u_{21} & u_{23} & u_{22} \\ u_{31} & u_{33} & u_{32} \end{bmatrix}.$$

Operation of Type II. Let L be the unit matrix modified by the introduction of the element l in the ith row and jth column, where i and j are unequal. Then Lu is the matrix obtained from u by addition of l times the jth row to the ith row. Similarly uL is the matrix obtained from u by addition of l times the ith column to the jth column.

The determinant of L is 1.

EXAMPLES

(i)
$$\begin{bmatrix} 1 & 0 & 0 \\ 0 & 1 & 0 \\ 0 & l & 1 \end{bmatrix}\begin{bmatrix} u_{11} & u_{12} & u_{13} \\ u_{21} & u_{22} & u_{23} \\ u_{31} & u_{32} & u_{33} \end{bmatrix} = \begin{bmatrix} u_{11}, & u_{12}, & u_{13} \\ u_{21}, & u_{22}, & u_{23} \\ u_{31}+lu_{21}, & u_{32}+lu_{22}, & u_{33}+lu_{23} \end{bmatrix}.$$

(ii)
$$\begin{bmatrix} u_{11} & u_{12} & u_{13} \\ u_{21} & u_{22} & u_{23} \\ u_{31} & u_{32} & u_{33} \end{bmatrix}\begin{bmatrix} 1 & 0 & 0 \\ 0 & 1 & 0 \\ 0 & l & 1 \end{bmatrix} = \begin{bmatrix} u_{11}, & u_{12}+lu_{13}, & u_{13} \\ u_{21}, & u_{22}+lu_{23}, & u_{23} \\ u_{31}, & u_{32}+lu_{33}, & u_{33} \end{bmatrix}.$$

Operation of Type III. Let H denote the unit matrix with h $(\neq 0)$ substituted for unity in the ith element in the principal diagonal. Then Hu equals the matrix u with its ith row of elements multiplied by h, and uH is the matrix u with its ith column multiplied by h.

The determinant of H is h.

EXAMPLES

(i)
$$\begin{bmatrix}1 & 0 & 0\\ 0 & 1 & 0\\ 0 & 0 & h\end{bmatrix}\begin{bmatrix}u_{11} & u_{12} & u_{13}\\ u_{21} & u_{22} & u_{23}\\ u_{31} & u_{32} & u_{33}\end{bmatrix} = \begin{bmatrix}u_{11}, & u_{12}, & u_{13}\\ u_{21}, & u_{22}, & u_{23}\\ hu_{31}, & hu_{32}, & hu_{33}\end{bmatrix}.$$

(ii)
$$\begin{bmatrix}u_{11} & u_{12} & u_{13}\\ u_{21} & u_{22} & u_{23}\\ u_{31} & u_{32} & u_{33}\end{bmatrix}\begin{bmatrix}1 & 0 & 0\\ 0 & 1 & 0\\ 0 & 0 & h\end{bmatrix} = \begin{bmatrix}u_{11} & u_{12} & hu_{13}\\ u_{21} & u_{22} & hu_{23}\\ u_{31} & u_{32} & hu_{33}\end{bmatrix}.$$

3·12. Equivalent Matrices. Two matrices are said to be *equivalent* if one can be derived from the other by any finite number of operations of the types specified in § 3·11. It is easy to see that the relation of equivalence is of a reciprocal character, since the inverse of any one of the three elementary operations is another operation of the same type.

Since any elementary operation upon a matrix u is equivalent to premultiplication or postmultiplication of u by a non-singular matrix, it follows that any matrix v which is equivalent to u must be related to u by the equation

$$v = PuQ, \qquad \text{......(1)}$$

where P and Q are non-singular matrices. It can in fact be proved* that if $v = PuQ$, where P and Q are non-singular, then u and v are necessarily equivalent in accordance with the definition already given.

It is important to note that, since the rank of a matrix is clearly unaltered by any of the elementary operations referred to in § 3·11, equivalent matrices have the same rank.

3·13. A Canonical Form for Square Matrices of Rank r. We shall now show that any square matrix u of rank r is equivalent to a *canonical* (or standard type) matrix C, whose elements are all zero with the exception of r units occupying the first r places in the principal diagonal.

If the matrix is null, its rank is zero, and it is already in the canonical form. If it is not null, then bring a non-zero element to the top place in the principal diagonal by operations of type I (§ 3·11), and reduce this element to unity by an operation of type III. Next reduce all the other elements of the top row and of the first column to zero, by operations of type II. If the elements lying below the first row are now not

* See, for instance, Ref. 1.

all zero, bring a finite element to the second place in the principal diagonal: this can always be done without alteration of the first row or column. The element in question can then be reduced to unity, and the remaining elements of the second row and second column reduced to zero, in the same way as before. It is evident that in this way u can always be reduced to a matrix C of the canonical form, with say t consecutive units in the principal diagonal. Since C is equivalent to u, and thus has the same rank as u (see § 3·12), it follows that $t = r$.

Examples

(i) *Non-Singular Matrix.* In the trivial case where u is non-singular the canonical matrix C is simply I. Hence in equation (3·12·1) we may choose $P = u^{-1}$ and $Q = I$.

(ii) *Singular Matrix.* Suppose

$$u = \begin{bmatrix} 1 & 2 & -3 \\ -1 & 2 & -1 \\ -1 & -3 & 4 \end{bmatrix},$$

which is of rank 2. Then by combined operations of type II

$$\begin{bmatrix} 1 & 0 & 0 \\ 1 & 1 & 0 \\ 0 & -1 & 1 \end{bmatrix}\begin{bmatrix} 1 & 2 & -3 \\ -1 & 2 & -1 \\ -1 & -3 & 4 \end{bmatrix}\begin{bmatrix} 1 & 1 & 0 \\ 0 & 1 & \frac{3}{2} \\ 0 & 1 & 1 \end{bmatrix} = \begin{bmatrix} 1 & 0 & 0 \\ 0 & 0 & 2 \\ 0 & 0 & -\frac{5}{2} \end{bmatrix},$$

and by further elementary operations

$$\begin{bmatrix} 1 & 0 & 0 \\ 0 & 0 & 1 \\ 0 & 1 & \frac{4}{5} \end{bmatrix}\begin{bmatrix} 1 & 0 & 0 \\ 0 & 0 & 2 \\ 0 & 0 & -\frac{5}{2} \end{bmatrix}\begin{bmatrix} 1 & 0 & 0 \\ 0 & 0 & 1 \\ 0 & -\frac{2}{5} & 0 \end{bmatrix} = \begin{bmatrix} 1 & 0 & 0 \\ 0 & 1 & 0 \\ 0 & 0 & 0 \end{bmatrix}.$$

These relations yield

$$C \equiv \begin{bmatrix} 1 & 0 & 0 \\ 0 & 1 & 0 \\ 0 & 0 & 0 \end{bmatrix} = \begin{bmatrix} 1 & 0 & 0 \\ 0 & -1 & 1 \\ 1 & \frac{1}{5} & \frac{4}{5} \end{bmatrix}\begin{bmatrix} 1 & 2 & -3 \\ -1 & 2 & -1 \\ -1 & -3 & 4 \end{bmatrix}\begin{bmatrix} 1 & 0 & 1 \\ 0 & -\frac{3}{5} & 1 \\ 0 & -\frac{2}{5} & 1 \end{bmatrix}.$$

3·14. Equivalent Lambda-Matrices. Two λ-matrices are equivalent when one can be derived from the other by means of the three elementary operations defined in § 3·11. The elements of the matrix multipliers which correspond to the performance of an operation may depend on λ, but the restriction is introduced that the determinants of these matrices must be independent of λ, and as before they must not vanish. If $g(\lambda)$ is equivalent to $f(\lambda)$, then

$$g(\lambda) = Pf(\lambda)\,Q, \qquad \text{......(1)}$$

where P and Q are, in general, λ-matrices whose determinants are non-vanishing constants. Conversely, if g and f are related as in equation (1), and if $|P|$ and $|Q|$ are non-vanishing constants, then g and f are equivalent in accordance with the definition given.

Clearly if g is equivalent to f, and h is equivalent to g, then h is also equivalent to f: this is a transitive property of equivalence.

3·15. Smith's Canonical Form for Lambda-Matrices. A canonical diagonal form for λ-matrices was established by Smith in accordance with the following theorem.*

Any λ-matrix $f(\lambda)$ of order n and rank r can be reduced to an equivalent diagonal form

$$E(\lambda) \equiv \begin{bmatrix} E_1(\lambda), & 0, & \dots, & 0, & \dots, 0 \\ 0, & E_2(\lambda), & \dots, & 0, & \dots, 0 \\ \dots & \dots & \dots & \dots & \dots \\ 0, & 0, & \dots, & E_r(\lambda), & \dots, 0 \\ \dots & \dots & \dots & \dots & \dots \\ 0, & 0, & \dots, & 0, & \dots, 0 \end{bmatrix} \qquad \dots\dots(1)$$

containing r isolated elements $E_p(\lambda)$ in the principal diagonal. The elements $E_p(\lambda)$ are such that

$$\left.\begin{aligned} E_1(\lambda) &= \delta_1, \\ E_2(\lambda) &= \delta_1\delta_2, \\ &\dots\dots\dots\dots \\ E_p(\lambda) &= \delta_1\delta_2\dots\delta_p, \\ &\dots\dots\dots\dots \end{aligned}\right\} \qquad \dots\dots(2)$$

where each δ is a unit or a polynomial in λ having unity for the coefficient of its highest power. Moreover, this diagonal form is unique, and is the same for all matrices equivalent to the given λ-matrix. The elements $E_p(\lambda)$ are called the *invariant factors* of $f(\lambda)$. Each invariant factor is a factor of all those that follow it in the sequence.

Let $D_p(\lambda)$ be the greatest common divisor† of the p-rowed minor determinants of $f(\lambda)$, written with unity as coefficient of the highest power of λ, and adopt the convention that $D_0(\lambda) = 1$. Then it can be shown that the invariant factors are given by

$$E_p(\lambda) = D_p(\lambda)/D_{p-1}(\lambda). \qquad \dots\dots(3)$$

Suppose that $$E_p(\lambda) = (\lambda-\lambda_1)^{e_{p1}}(\lambda-\lambda_2)^{e_{p2}}\dots. \qquad \dots\dots(4)$$

* For proofs see Refs. 1 or 2.

† The criterion of divisibility is absence of a remainder. A zero determinant is therefore divisible by any polynomial of λ.

Then those factors $(\lambda-\lambda_q)^{e_{pq}}$ which are not mere constants are called the *elementary divisors* of $f(\lambda)$, while $(\lambda-\lambda_1)$, $(\lambda-\lambda_2)$, etc. are the *linear factors*. The invariant factors, elementary divisors, and linear factors are invariant: that is to say, they are the same for all matrices equivalent to $f(\lambda)$.

Since $E = PfQ$, it follows that $|E| = |P||f||Q|$. But $|P|$ and $|Q|$ are independent of λ. Hence the roots of $\Delta(\lambda) \equiv |f(\lambda)| = 0$ are the same as the roots of $|E| = 0$, and these are the numbers λ_1, λ_2, etc. which appear in (4). The total multiplicity of the root λ_s is $\sum_p e_{ps}$. If the roots are not repeated, it follows that $E_p(\lambda) = 1$ when $p \neq n$, while $E_n(\lambda)$ is $\Delta(\lambda)$ divided by the coefficient of the highest power of λ in $\Delta(\lambda)$.

Example

Canonical Form for a Lambda-Matrix. It can readily be verified that if

$$f(\lambda) = \begin{bmatrix} \lambda & -\lambda & 0 \\ 0 & \lambda^2-\lambda & \lambda^3-\lambda^2 \\ \lambda^3 & -\lambda^2 & \lambda^4-\lambda \end{bmatrix},$$

then

$$f(\lambda) = P(\lambda) \begin{bmatrix} \lambda & 0 & 0 \\ 0 & \lambda(\lambda-1) & 0 \\ 0 & 0 & \lambda(\lambda^2-1) \end{bmatrix} Q(\lambda),$$

where

$$P(\lambda) = \begin{bmatrix} 1 & 0 & 0 \\ 0 & 1 & 0 \\ \lambda^2 & \lambda & 1 \end{bmatrix} \quad \text{and} \quad Q(\lambda) = \begin{bmatrix} 1 & -1 & 0 \\ 0 & 1 & \lambda \\ 0 & 0 & 1 \end{bmatrix}.$$

Here $|P(\lambda)| = 1$ and $|Q(\lambda)| = 1$, so that both determinants are non-vanishing constants. Smith's canonical form for $f(\lambda)$ is

$$E(\lambda) = \begin{bmatrix} \lambda & 0 & 0 \\ 0 & \lambda(\lambda-1) & 0 \\ 0 & 0 & \lambda(\lambda^2-1) \end{bmatrix} \equiv \begin{bmatrix} E_1(\lambda) & 0 & 0 \\ 0 & E_2(\lambda) & 0 \\ 0 & 0 & E_3(\lambda) \end{bmatrix}.$$

Note that here $\Delta(\lambda) = \lambda^3(\lambda-1)^2(\lambda+1)$, that the 2-rowed minor determinants of $f(\lambda)$ have $\lambda^2(\lambda-1)$ as a common factor, and that the 1-rowed minor determinants have λ as a common factor. Hence (see (3)), $D_3(\lambda) = \lambda^3(\lambda-1)(\lambda^2-1)$; $D_2(\lambda) = \lambda^2(\lambda-1)$; $D_1(\lambda) = \lambda$; and the invariant factors are

$$E_3(\lambda) = \lambda(\lambda^2-1); \quad E_2(\lambda) = \lambda(\lambda-1); \quad E_1(\lambda) = \lambda.$$

The linear factors are λ, $\lambda-1$ and $\lambda+1$.

3·16. Collineatory Transformation of a Numerical Matrix to a Canonical Form. It has been shown in § 3·13 that an arbitrary numerical* matrix u is equivalent to a certain diagonal matrix C, so that

$$u = pCq,$$

where p and q are not singular. When the matrices p and q in this equation are reciprocal, then u and C have the same latent roots (see § 3·6, example (iv)). Moreover, if $P(u)$ is any polynomial of u, then

$$P(C) = p^{-1}P(u)p,$$

as illustrated by equation (3·6·12). The usefulness of this collineatory transformation is so obvious, and the ease of calculation with a diagonal matrix so great, that it is natural to consider whether such a transformation can always be effected.

It has already been shown in § 3·6 that the collineatory transformation is always possible when the latent roots of u are all distinct. However, the transformation is not always possible when there are repeated latent roots. It is found† that the general canonical form is a diagonal matrix with a certain number of unit elements added in the superdiagonal.‡ The rule is that the unit element is certainly absent if the place in the superdiagonal is adjacent horizontally and vertically to distinct elements in the principal diagonal, but that it may be present when the adjacent elements in the principal diagonal are the same. Thus the typical canonical form is

$$\left[\begin{array}{ccc|cc|ccc|c} \alpha & 1 & & & & & & & \\ & \alpha & 1 & & & & & & \\ & & \alpha & & & & & & \\ \hline & & & \alpha & 1 & & & & \\ & & & & \alpha & & & & \\ \hline & & & & & \beta & 1 & & \\ & & & & & & \beta & 1 & \\ & & & & & & & \beta & \\ \hline & & & & & & & & \ddots \end{array}\right], \qquad \text{......(1)}$$

* The adjective "numerical" is added in order to emphasise that a λ-matrix is not intended.

† For a full discussion of this question see Chap. VI of Ref. 2.

‡ The elements immediately to the right of the principal diagonal are the *superdiagonal* elements.

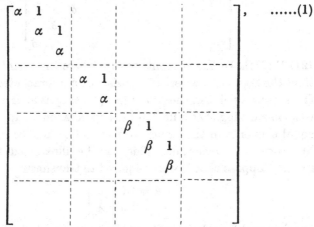

where the elements in the principal diagonal are, of course, the latent roots. It is clear that the canonical matrix can be partitioned into a diagonal matrix of square submatrices whose rows and columns do not overlap, and which are all of the *simple classical type*

$$\begin{bmatrix} \alpha & 1 & & \\ & \alpha & 1 & \\ & & \alpha & 1 \\ & & & \alpha \end{bmatrix}.$$

The order of the simple classical submatrix can range from 1 (the elementary case) to n, the order of the matrix itself.

The exact type of the canonical matrix is conveniently specified by means of its *Segre characteristic*. This consists of a set of integers which are the orders of the classical submatrices; those integers which correspond to submatrices containing the same latent root are bracketed together. For example, the Segre characteristic of

$$\begin{bmatrix} \alpha & 1 & & & & & & & & \\ & \alpha & & & & & & & & \\ & & \alpha & & & & & & & \\ & & & \beta & 1 & & & & & \\ & & & & \beta & 1 & & & & \\ & & & & & \beta & & & & \\ & & & & & & \gamma & & & \\ & & & & & & & \delta & 1 & \\ & & & & & & & & \delta & \\ & & & & & & & & & \delta \end{bmatrix}$$

is [(21) 31 (21)]. In the elementary case where the latent roots are all distinct the Segre characteristic consists of n unbracketed units.

The essential point in the proof that (1) really is an irreducible form consists in showing that a unit in the superdiagonal adjacent to a pair of equal elements in the principal diagonal cannot be removed by a collineatory transformation. This can be illustrated by a simple example. Suppose that it were required to transform

$$A = \begin{bmatrix} \lambda_1 & 1 \\ 0 & \lambda_2 \end{bmatrix}$$

to the diagonal form by a collineatory transformation. Let

$$p = \begin{bmatrix} a & b \\ c & d \end{bmatrix},$$

so that
$$p^{-1} = \delta^{-1}\begin{bmatrix} d & -b \\ -c & a \end{bmatrix},$$
where $\delta = ad - bc \neq 0$. Then
$$p^{-1}Ap = \delta^{-1}\begin{bmatrix} d(\lambda_1 a + c) - \lambda_2 bc, & d(\lambda_1 b + d) - \lambda_2 bd \\ -c(\lambda_1 a + c) + \lambda_2 ac, & -c(\lambda_1 b + d) + \lambda_2 ad \end{bmatrix}.$$
If both the non-diagonal elements are to vanish,
$$(\lambda_1 - \lambda_2)\, bd + d^2 = 0,$$
and
$$(\lambda_1 - \lambda_2)\, ac + c^2 = 0.$$
When $\lambda_1 \neq \lambda_2$ these equations can be satisfied without violation of the condition $\delta \neq 0$, but when $\lambda_1 = \lambda_2$ the only solution is $d = c = 0$, which is not permissible, for then $\delta = 0$.

CHAPTER IV

MISCELLANEOUS NUMERICAL METHODS

4·1. Range of the Subjects Treated. This chapter is divided into three sections. Part I deals with the computation of determinants, reciprocal matrices, and adjoint matrices, and with the numerical solution of systems of linear algebraic equations. Part II is concerned with the limiting forms of high powers of matrices and with the approximate calculation of latent roots. In this connection iterative methods are developed which are applied in Chapter X to the solution of dynamical problems. The approximate solution of algebraic equations of general degree and the computation of Sturm's functions and test functions for stability are briefly considered in Part III.

Part I. Determinants, Reciprocal and Adjoint Matrices, and Systems of Linear Algebraic Equations

4·2. Preliminary Remarks. Before describing the actual methods of computation we shall deal briefly with their underlying principles.

A system of n linearly independent linear equations in n variables x may be represented by

$$ax = h,$$

where a is a given non-singular square matrix of order n, and h is a column of n constants. The formal solution of the equations is

$$x = a^{-1}h.$$

If a^{-1} can be determined, the values of x corresponding to any set of constants h can be found immediately. Now $a^{-1} = A/|a|$, where A is the adjoint of a, and it is therefore possible to obtain a^{-1} by a computation of the determinant $|a|$ and of its first minors. Unless the matrix a is of low order this process is very laborious, and it is usually preferable to build up the reciprocal matrix in stages by elementary operations on the columns or the rows of a.

One possible procedure is as follows. Postmultiply the equation $a^{-1}a = I$ by a non-singular matrix M_1 to give $a^{-1}(aM_1) = IM_1$. The postmultiplication of a by M_1 represents some operation performed on the columns of a, while the postmultiplication of I by M_1 represents the same operation on the columns of the unit matrix. A succession of s operations of this kind would give

$$a^{-1}\sigma = M_1 M_2 \dots M_s,$$

where $\sigma = aM_1 M_2 \dots M_s$. Hence if σ is such that its reciprocal can be found easily, then a^{-1} is calculable by the formula $a^{-1} = M_1 M_2 \dots M_s \sigma^{-1}$. Again, if the determinant $|\sigma|$ can be evaluated easily, and if each of the multipliers M has unit determinant, then

$$|a| = |\sigma|.$$

An alternative is to apply to the equation $aa^{-1} = I$ a succession of premultipliers, representing operations performed on rows. The choice of rows, or of columns, for the operations is in fact decided by convenience only.

The derivation of suitable types of matrix σ and of the appropriate multipliers will now be considered.

4·3. Triangular and Related Matrices. A matrix having only zero elements either to the right or to the left of its principal diagonal will be referred to as a *triangular* matrix. If the matrix is non-singular, none of its principal diagonal elements can be zero. Simple examples of non-singular triangular matrices are

$$\begin{bmatrix} 2 & 0 & 0 \\ 3 & 3 & 0 \\ 4 & 5 & 1 \end{bmatrix}, \qquad \begin{bmatrix} 4 & 5 & 1 \\ 0 & 3 & 3 \\ 0 & 0 & 2 \end{bmatrix}.$$

A non-singular matrix can always be reduced to a matrix of triangular form by operations on columns only, or on rows only, in an indefinitely large number of ways. One possible scheme of reduction, by operations on columns only, will be described.

Firstly, choosing any convenient non-zero element of the top row,* say the ith, we reduce to zero all the other elements of the top row by subtracting suitable multiples of the ith column from the other columns in turn, and leave the ith column unchanged. Next, choosing any convenient non-zero element of the second row of the new matrix, say the jth (where $j \neq i$), and leaving both the ith and the jth columns unchanged, we can in the same way annul all the elements of the second row other than the ith and the jth. By a continuation of this

* Each row of a non-singular matrix must contain at least one non-zero element.

process we are ultimately led to a matrix which, if not immediately triangular, is a triangular matrix with its columns deranged.

The operations described can conveniently be represented by a sequence of postmultiplications of the given matrix a. To avoid unwieldy general formulae the process will be illustrated with reference to a matrix of order four.

Dealing firstly with the simplest case where the chosen element of the first row is a_{11} (shown "starred" below), we have

$$\begin{bmatrix} a_{11}^* & a_{12} & a_{13} & a_{14} \\ a_{21} & a_{22} & a_{23} & a_{24} \\ a_{31} & a_{32} & a_{33} & a_{34} \\ a_{41} & a_{42} & a_{43} & a_{44} \end{bmatrix} \begin{bmatrix} 1 & -a_{12}/a_{11} & -a_{13}/a_{11} & -a_{14}/a_{11} \\ 0 & 1 & 0 & 0 \\ 0 & 0 & 1 & 0 \\ 0 & 0 & 0 & 1 \end{bmatrix} = \begin{bmatrix} a_{11} & 0 & 0 & 0 \\ a_{21} & b_{11} & b_{12} & b_{13} \\ a_{31} & b_{21} & b_{22} & b_{23} \\ a_{41} & b_{31} & b_{32} & b_{33} \end{bmatrix}, \qquad \text{......(1)}$$

where b_{ij} are new elements which need not here be specified. The postmultiplier in this equation is the unit matrix I_4 augmented by a first row of elements derived from the first row of a in the manner shown. Next, if b_{11} (shown starred below) is the chosen element of the new second row,

$$\begin{bmatrix} a_{11} & 0 & 0 & 0 \\ a_{21} & b_{11}^* & b_{12} & b_{13} \\ a_{31} & b_{21} & b_{22} & b_{23} \\ a_{41} & b_{31} & b_{32} & b_{33} \end{bmatrix} \begin{bmatrix} 1 & 0 & 0 & 0 \\ 0 & 1 & -b_{12}/b_{11} & -b_{13}/b_{11} \\ 0 & 0 & 1 & 0 \\ 0 & 0 & 0 & 1 \end{bmatrix} = \begin{bmatrix} a_{11} & 0 & 0 & 0 \\ a_{21} & b_{11} & 0 & 0 \\ a_{31} & b_{21} & c_{11} & c_{12} \\ a_{41} & b_{31} & c_{21} & c_{22} \end{bmatrix}, \qquad \text{......(2)}$$

where c_{ij} are new elements. Lastly, if c_{11} (shown starred below) is the chosen element of the third row,

$$\begin{bmatrix} a_{11} & 0 & 0 & 0 \\ a_{21} & b_{11} & 0 & 0 \\ a_{31} & b_{21} & c_{11}^* & c_{12} \\ a_{41} & b_{31} & c_{21} & c_{22} \end{bmatrix} \begin{bmatrix} 1 & 0 & 0 & 0 \\ 0 & 1 & 0 & 0 \\ 0 & 0 & 1 & -c_{12}/c_{11} \\ 0 & 0 & 0 & 1 \end{bmatrix} = \begin{bmatrix} a_{11} & 0 & 0 & 0 \\ a_{21} & b_{11} & 0 & 0 \\ a_{31} & b_{21} & c_{11} & 0 \\ a_{41} & b_{31} & c_{21} & d_{11} \end{bmatrix}. \qquad \text{......(3)}$$

Denoting the postmultipliers in equations (1), (2), (3), respectively, by M_1, M_2, M_3, and the final triangular matrix by τ, we obtain

$$aM_1M_2M_3 = \tau.$$

Since each of the multipliers M has unit determinant, it follows that

$$|a| = |\tau| = a_{11}b_{11}c_{11}d_{11}.$$

It is easy to show that the product $M_1M_2M_3$ is itself a triangular matrix. Thus

$$M_1M_2 = \begin{bmatrix} 1 & -a_{12}/a_{11} & -a_{13}/a_{11} & -a_{14}/a_{11} \\ 0 & 1 & 0 & 0 \\ 0 & 0 & 1 & 0 \\ 0 & 0 & 0 & 1 \end{bmatrix} \begin{bmatrix} 1 & 0 & 0 & 0 \\ 0 & 1 & -b_{12}/b_{11} & -b_{13}/b_{11} \\ 0 & 0 & 1 & 0 \\ 0 & 0 & 0 & 1 \end{bmatrix}$$

$$= \begin{bmatrix} 1 & -a_{12}/a_{11} & \alpha_1 & \alpha_2 \\ 0 & 1 & -b_{12}/b_{11} & -b_{13}/b_{11} \\ 0 & 0 & 1 & 0 \\ 0 & 0 & 0 & 1 \end{bmatrix},$$

where α_1, α_2 are new elements. Hence

$$M_1M_2M_3 = \begin{bmatrix} 1 & -a_{12}/a_{11} & \alpha_1 & \alpha_2 \\ 0 & 1 & -b_{12}/b_{11} & -b_{13}/b_{11} \\ 0 & 0 & 1 & 0 \\ 0 & 0 & 0 & 1 \end{bmatrix} \begin{bmatrix} 1 & 0 & 0 & 0 \\ 0 & 1 & 0 & 0 \\ 0 & 0 & 1 & -c_{12}/c_{11} \\ 0 & 0 & 0 & 1 \end{bmatrix}$$

$$= \begin{bmatrix} 1 & -a_{12}/a_{11} & \alpha_1 & \beta_1 \\ 0 & 1 & -b_{12}/b_{11} & \beta_2 \\ 0 & 0 & 1 & -c_{12}/c_{11} \\ 0 & 0 & 0 & 1 \end{bmatrix},$$

in which β_1, β_2 are new elements. The product $M_1M_2M_3$ in this case is triangular and "opposite-handed" to the triangular matrix τ.

The particular method of reduction given would fail if any one of the starred elements were zero. A more general treatment is possible in which the rows are taken in any order and the starred elements are not necessarily in the principal diagonal. In this case the final matrix will, in general, not be triangular, but will be a triangular matrix with its rows in some way interchanged, and possibly with its columns also in some way interchanged. To illustrate the rule for the construction of the multipliers, suppose a_{32} to be the first starred element; then

$$\begin{bmatrix} a_{11} & a_{12} & a_{13} & a_{14} \\ a_{21} & a_{22} & a_{23} & a_{24} \\ a_{31} & a_{32}^* & a_{33} & a_{34} \\ a_{41} & a_{42} & a_{43} & a_{44} \end{bmatrix} \begin{bmatrix} 1 & 0 & 0 & 0 \\ -a_{31}/a_{32} & 1 & -a_{33}/a_{32} & -a_{34}/a_{32} \\ 0 & 0 & 1 & 0 \\ 0 & 0 & 0 & 1 \end{bmatrix} = \begin{bmatrix} b_{11} & a_{12} & b_{12} & b_{13} \\ b_{21} & a_{22} & b_{22} & b_{23} \\ 0 & a_{32} & 0 & 0 \\ b_{31} & a_{42} & b_{32} & b_{33} \end{bmatrix}.$$

The multiplier is here constructed similarly to that in (1), except that its second row now contains the elements derived from the third row of a. More generally, if the starred element in a lies in the ith row and jth column, the elements derived from the ith row of a are entered in the jth row of the multiplier. The process of reduction will be clear from the numerical examples which follow.

EXAMPLES

(i) *Rows and Columns taken in Consecutive Order.* If

$$a = \begin{bmatrix} 1 & 4 & 1 & 3 \\ 0 & -1 & 3 & -1 \\ 3 & 1 & 0 & 2 \\ 1 & -2 & 5 & 1 \end{bmatrix},$$

then the scheme of reduction, with the starred elements in the principal diagonal, is

$$\begin{bmatrix} 1^* & 4 & 1 & 3 \\ 0 & -1 & 3 & -1 \\ 3 & 1 & 0 & 2 \\ 1 & -2 & 5 & 1 \end{bmatrix} \begin{bmatrix} 1 & -4 & -1 & -3 \\ 0 & 1 & 0 & 0 \\ 0 & 0 & 1 & 0 \\ 0 & 0 & 0 & 1 \end{bmatrix} = \begin{bmatrix} 1 & 0 & 0 & 0 \\ 0 & -1 & 3 & -1 \\ 3 & -11 & -3 & -7 \\ 1 & -6 & 4 & -2 \end{bmatrix},$$

$$\begin{bmatrix} 1 & 0 & 0 & 0 \\ 0 & -1^* & 3 & -1 \\ 3 & -11 & -3 & -7 \\ 1 & -6 & 4 & -2 \end{bmatrix} \begin{bmatrix} 1 & 0 & 0 & 0 \\ 0 & 1 & 3 & -1 \\ 0 & 0 & 1 & 0 \\ 0 & 0 & 0 & 1 \end{bmatrix} = \begin{bmatrix} 1 & 0 & 0 & 0 \\ 0 & -1 & 0 & 0 \\ 3 & -11 & -36 & 4 \\ 1 & -6 & -14 & 4 \end{bmatrix},$$

$$\begin{bmatrix} 1 & 0 & 0 & 0 \\ 0 & -1 & 0 & 0 \\ 3 & -11 & -36^* & 4 \\ 1 & -6 & -14 & 4 \end{bmatrix} \begin{bmatrix} 1 & 0 & 0 & 0 \\ 0 & 1 & 0 & 0 \\ 0 & 0 & 1 & \frac{1}{9} \\ 0 & 0 & 0 & 1 \end{bmatrix} = \begin{bmatrix} 1 & 0 & 0 & 0 \\ 0 & -1 & 0 & 0 \\ 3 & -11 & -36 & 0 \\ 1 & -6 & -14 & \frac{22}{9} \end{bmatrix}.$$

Hence $$|a| = 1 \times (-1) \times (-36) \times (\tfrac{22}{9}) = 88.$$

The product of the multipliers works out as

$$M_1 M_2 M_3 = \begin{bmatrix} 1 & -4 & -13 & -\frac{4}{9} \\ 0 & 1 & 3 & -\frac{2}{3} \\ 0 & 0 & 1 & \frac{1}{9} \\ 0 & 0 & 0 & 1 \end{bmatrix}.$$

(ii) *Rows, but not Columns, taken in Consecutive Order.* With a as for example (i), another possible scheme of reduction is

$$\begin{bmatrix} 1 & 4 & 1^* & 3 \\ 0 & -1 & 3 & -1 \\ 3 & 1 & 0 & 2 \\ 1 & -2 & 5 & 1 \end{bmatrix} \begin{bmatrix} 1 & 0 & 0 & 0 \\ 0 & 1 & 0 & 0 \\ -1 & -4 & 1 & -3 \\ 0 & 0 & 0 & 1 \end{bmatrix} = \begin{bmatrix} 0 & 0 & 1 & 0 \\ -3 & -13 & 3 & -10 \\ 3 & 1 & 0 & 2 \\ -4 & -22 & 5 & -14 \end{bmatrix},$$

$$\begin{bmatrix} 0 & 0 & 1 & 0 \\ -3^* & -13 & 3 & -10 \\ 3 & 1 & 0 & 2 \\ -4 & -22 & 5 & -14 \end{bmatrix} \begin{bmatrix} 1 & -\frac{13}{3} & 0 & -\frac{10}{3} \\ 0 & 1 & 0 & 0 \\ 0 & 0 & 1 & 0 \\ 0 & 0 & 0 & 1 \end{bmatrix} = \begin{bmatrix} 0 & 0 & 1 & 0 \\ -3 & 0 & 3 & 0 \\ 3 & -12 & 0 & -8 \\ -4 & -\frac{14}{3} & 5 & -\frac{2}{3} \end{bmatrix},$$

$$\begin{bmatrix} 0 & 0 & 1 & 0 \\ -3 & 0 & 3 & 0 \\ 3 & -12 & 0 & -8^* \\ -4 & -\frac{14}{3} & 5 & -\frac{2}{3} \end{bmatrix} \begin{bmatrix} 1 & 0 & 0 & 0 \\ 0 & 1 & 0 & 0 \\ 0 & 0 & 1 & 0 \\ 0 & -\frac{3}{2} & 0 & 1 \end{bmatrix} = \begin{bmatrix} 0 & 0 & 1 & 0 \\ -3 & 0 & 3 & 0 \\ 3 & 0 & 0 & -8 \\ -4 & -\frac{11}{3} & 5 & -\frac{2}{3} \end{bmatrix}.$$

By three interchanges of its columns the final matrix can be brought to the triangular matrix

$$\begin{bmatrix} 1 & 0 & 0 & 0 \\ 3 & -3 & 0 & 0 \\ 0 & 3 & -8 & 0 \\ 5 & -4 & -\frac{2}{3} & -\frac{11}{3} \end{bmatrix}$$

Hence $|a| = 1 \times (-3) \times (-8) \times (-\frac{11}{3}) \times (-1)^3 = 88$, as before. The product of the multipliers in the present case is

$$M_1 M_2 M_3 = \begin{bmatrix} 0 & 1 & \frac{-0}{3} & \frac{2}{3} \\ 0 & 0 & 0 & 1 \\ 1 & -1 & \frac{1}{3} & -\frac{1}{6} \\ 0 & 0 & 1 & -\frac{3}{2} \end{bmatrix}.$$

This matrix has unit determinant. It can be represented as a triangular matrix by a rearrangement of its rows, but not by a rearrangement of its columns.

(iii) *Columns, but not Rows, taken in Consecutive Order.* With a again as for example (i), an illustrative sequence of operations is

$$\begin{bmatrix} 1 & 4 & 1 & 3 \\ 0 & -1 & 3 & -1 \\ 3 & 1 & 0 & 2 \\ 1^* & -2 & 5 & 1 \end{bmatrix} \begin{bmatrix} 1 & 2 & -5 & -1 \\ 0 & 1 & 0 & 0 \\ 0 & 0 & 1 & 0 \\ 0 & 0 & 0 & 1 \end{bmatrix} = \begin{bmatrix} 1 & 6 & -4 & 2 \\ 0 & -1 & 3 & -1 \\ 3 & 7 & -15 & -1 \\ 1 & 0 & 0 & 0 \end{bmatrix},$$

$$\begin{bmatrix}1 & 6 & -4 & 2\\ 0 & -1^* & 3 & -1\\ 3 & 7 & -15 & -1\\ 1 & 0 & 0 & 0\end{bmatrix}\begin{bmatrix}1 & 0 & 0 & 0\\ 0 & 1 & 3 & -1\\ 0 & 0 & 1 & 0\\ 0 & 0 & 0 & 1\end{bmatrix}=\begin{bmatrix}1 & 6 & 14 & -4\\ 0 & -1 & 0 & 0\\ 3 & 7 & 6 & -8\\ 1 & 0 & 0 & 0\end{bmatrix},$$

$$\begin{bmatrix}1 & 6 & 14^* & -4\\ 0 & -1 & 0 & 0\\ 3 & 7 & 6 & -8\\ 1 & 0 & 0 & 0\end{bmatrix}\begin{bmatrix}1 & 0 & 0 & 0\\ 0 & 1 & 0 & 0\\ 0 & 0 & 1 & \frac{2}{7}\\ 0 & 0 & 0 & 1\end{bmatrix}=\begin{bmatrix}1 & 6 & 14 & 0\\ 0 & -1 & 0 & 0\\ 3 & 7 & 6 & -\frac{44}{7}\\ 1 & 0 & 0 & 0\end{bmatrix}.$$

If the rows of the final matrix are taken in the order 4, 2, 1, 3—corresponding to the row order chosen for the starred elements—the result is the triangular matrix

$$\begin{bmatrix}1 & 0 & 0 & 0\\ 0 & -1 & 0 & 0\\ 1 & 6 & 14 & 0\\ 3 & 7 & 6 & -\frac{44}{7}\end{bmatrix}.$$

From this it is seen that $|a| = 1\times(-1)\times 14\times(-\frac{44}{7})\times(-1)^4 = 88.$

4·4. Reduction of Triangular and Related Matrices to Diagonal Form. The reduction of a triangular matrix to the diagonal form is a simple matter. If the given non-singular matrix is

$$\tau = \begin{bmatrix}\tau_{11} & 0 & 0 & 0\\ \tau_{21} & \tau_{22} & 0 & 0\\ \tau_{31} & \tau_{32} & \tau_{33} & 0\\ \tau_{41} & \tau_{42} & \tau_{43} & \tau_{44}\end{bmatrix}, \qquad \text{......(1)}$$

then by operations on columns only

$$\begin{bmatrix}\tau_{11} & 0 & 0 & 0\\ \tau_{21} & \tau_{22} & 0 & 0\\ \tau_{31} & \tau_{32} & \tau_{33} & 0\\ \tau_{41} & \tau_{42} & \tau_{43} & \tau_{44}^*\end{bmatrix}\begin{bmatrix}1 & 0 & 0 & 0\\ 0 & 1 & 0 & 0\\ 0 & 0 & 1 & 0\\ -\frac{\tau_{41}}{\tau_{44}} & -\frac{\tau_{42}}{\tau_{44}} & -\frac{\tau_{43}}{\tau_{44}} & 1\end{bmatrix}=\begin{bmatrix}\tau_{11} & 0 & 0 & 0\\ \tau_{21} & \tau_{22} & 0 & 0\\ \tau_{31} & \tau_{32} & \tau_{33} & 0\\ 0 & 0 & 0 & \tau_{44}\end{bmatrix}.$$

The result of the postmultiplication is that all the elements of the last row with the exception of τ_{44} are annulled, but the matrix is otherwise unaltered. Clearly, by a succession of such operations τ is reducible to the diagonal form

$$\delta = \begin{bmatrix}\tau_{11} & 0 & 0 & 0\\ 0 & \tau_{22} & 0 & 0\\ 0 & 0 & \tau_{33} & 0\\ 0 & 0 & 0 & \tau_{44}\end{bmatrix}. \qquad \text{......(2)}$$

If the given matrix is a triangular matrix with its rows interchanged, a similar method may be adopted. For instance, if

$$\sigma = \begin{bmatrix} \sigma_{11} & \sigma_{12} & \sigma_{13} & 0 \\ \sigma_{21} & \sigma_{22} & 0 & 0 \\ \sigma_{31} & \sigma_{32} & \sigma_{33} & \sigma_{34} \\ \sigma_{41} & 0 & 0 & 0 \end{bmatrix},$$

then by three successive postmultiplications this matrix is reducible to

$$\phi = \begin{bmatrix} 0 & 0 & \sigma_{13} & 0 \\ 0 & \sigma_{22} & 0 & 0 \\ 0 & 0 & 0 & \sigma_{34} \\ \sigma_{41} & 0 & 0 & 0 \end{bmatrix}.$$

4·5. Reciprocals of Triangular and Related Matrices. The reciprocal of the triangular matrix τ (see (4·4·1)) can be obtained as follows. Consider the system of linear equations

$$\begin{aligned} y_1 \qquad\qquad\qquad\qquad\qquad &= h_1, \\ \frac{\tau_{21}}{\tau_{11}}y_1 + \quad y_2 \qquad\qquad\qquad &= h_2, \\ \frac{\tau_{31}}{\tau_{11}}y_1 + \frac{\tau_{32}}{\tau_{22}}y_2 + \quad y_3 \qquad &= h_3, \\ \frac{\tau_{41}}{\tau_{11}}y_1 + \frac{\tau_{42}}{\tau_{22}}y_2 + \frac{\tau_{43}}{\tau_{33}}y_3 + y_4 &= h_4, \end{aligned}$$

in which h_1, h_2, h_3, h_4 are arbitrary. These equations are equivalent to $\tau\delta^{-1}y = h$ (where δ is the diagonal matrix (4·4·2)) and their solution can be written

$$y = \omega h,$$

where $\omega^{-1} = \tau\delta^{-1}$ or $\tau^{-1} \equiv \delta^{-1}\omega$. Hence τ^{-1} can be obtained by multiplication of the successive rows of ω by $1/\tau_{11}$, $1/\tau_{22}$, $1/\tau_{33}$, $1/\tau_{44}$, respectively.

Now the elements in the first column of ω are the values of y when $h = \{1, 0, 0, 0\}$: similarly, the elements in the second column are the values of y when $h = \{0, 1, 0, 0\}$; and so on. But when $h = \{1, 0, 0, 0\}$

$$\begin{aligned} y_1 &= 1, \\ y_2 &= \left[-\frac{\tau_{21}}{\tau_{11}}\right] \times 1, \\ y_3 &= \left[-\frac{\tau_{31}}{\tau_{11}}, -\frac{\tau_{32}}{\tau_{22}}\right]\{y_1, y_2\}, \\ y_4 &= \left[-\frac{\tau_{41}}{\tau_{11}}, -\frac{\tau_{42}}{\tau_{22}}, -\frac{\tau_{43}}{\tau_{33}}\right]\{y_1, y_2, y_3\}. \end{aligned}$$

This suggests the following scheme of computation. The scalar multipliers by which the rows of ω have to be multiplied to give τ^{-1} are written on the extreme right.

—	—	—	—	1	0	0	0	$(1/\tau_{11})$
$-\frac{\tau_{21}}{\tau_{11}}$,	—	—	—	ω_{21}	1	0	0	$(1/\tau_{22})$
$-\frac{\tau_{31}}{\tau_{11}}$,	$-\frac{\tau_{32}}{\tau_{22}}$,	—	—	ω_{31}	ω_{32}	1	0	$(1/\tau_{33})$
$-\frac{\tau_{41}}{\tau_{11}}$,	$-\frac{\tau_{42}}{\tau_{22}}$,	$-\frac{\tau_{43}}{\tau_{33}}$,	—	ω_{41}	ω_{42}	ω_{43}	1	$(1/\tau_{44})$
X				Y (matrix ω)				

Rules: (i) To form the left-hand array (X) enter blanks in, and to the right of, the principal diagonal. Derive the remaining elements from τ as shown.

(ii) Commence the right-hand array (Y) by entering units in the principal diagonal, and ciphers to the right of that diagonal.

(iii) Calculate the remaining elements of ω in succession by the following method. To obtain ω_{ij} postmultiply the row of (X) level with ω_{ij} by the part-column of (Y) standing above ω_{ij}. Blank elements of (X) are to be disregarded.

(iv) To find τ^{-1}, multiply the rows of ω respectively by the scalar factors on the extreme right.

Column No. 1 of (Y), for instance, is completed as below:

$$\omega_{21} = \left[-\frac{\tau_{21}}{\tau_{11}}\right] \times \{1\},$$

$$\omega_{31} = \left[-\frac{\tau_{31}}{\tau_{11}}, -\frac{\tau_{32}}{\tau_{22}}\right] \times \{1, \omega_{21}\},$$

$$\omega_{41} = \left[-\frac{\tau_{41}}{\tau_{11}}, -\frac{\tau_{42}}{\tau_{22}}, -\frac{\tau_{43}}{\tau_{33}}\right] \times \{1, \omega_{21}, \omega_{31}\}.$$

A similar procedure is possible if the given matrix, say σ, is derived from a triangular matrix τ by interchange of its rows, but the construction of the arrays (X) and (Y) in this case requires care. The simplest and safest method is to convert σ first to the triangular form by actual interchange of the rows. Suppose for example

$$\sigma = \begin{bmatrix} \sigma_{11} & \sigma_{12} & \sigma_{13}^* & 0 \\ \sigma_{21} & \sigma_{22}^* & 0 & 0 \\ \sigma_{31} & \sigma_{32} & \sigma_{33} & \sigma_{34}^* \\ \sigma_{41}^* & 0 & 0 & 0 \end{bmatrix}.$$

The stars indicate the last non-zero elements of the rows. If the rows are taken in the order 4, 2, 1, 3 to bring the starred elements into the principal diagonal, the resulting matrix is triangular. This process can be represented by

$$\begin{bmatrix} 0 & 0 & 0 & 1 \\ 0 & 1 & 0 & 0 \\ 1 & 0 & 0 & 0 \\ 0 & 0 & 1 & 0 \end{bmatrix} \begin{bmatrix} \sigma_{11} & \sigma_{12} & \sigma_{13}^* & 0 \\ \sigma_{21} & \sigma_{22}^* & 0 & 0 \\ \sigma_{31} & \sigma_{32} & \sigma_{33} & \sigma_{34}^* \\ \sigma_{41}^* & 0 & 0 & 0 \end{bmatrix} = \begin{bmatrix} \sigma_{41}^* & 0 & 0 & 0 \\ \sigma_{21} & \sigma_{22}^* & 0 & 0 \\ \sigma_{11} & \sigma_{12} & \sigma_{13}^* & 0 \\ \sigma_{31} & \sigma_{32} & \sigma_{33} & \sigma_{34}^* \end{bmatrix}.$$

For brevity denote this as $N\sigma = \tau$. ......(1)

Then the reciprocal of σ is given by

$$\sigma^{-1} = \tau^{-1}N. \qquad \text{......(2)}$$

The appropriate multiplier N in any given case is found by the following rule: N is the transposed of σ with its starred elements replaced by units and all other elements replaced by ciphers.

If σ is derived from a triangular matrix τ by interchange of its columns, then the equations corresponding to (1) and (2) are $\sigma M = \tau$ and $\sigma^{-1} = M\tau^{-1}$. In this case the first non-zero elements of the columns of σ are starred, and the multiplier M is then derived as before.

Examples

(i) *Reciprocal of Triangular Matrix.* Suppose (see example (i) of § 4·3)

$$\tau = \begin{bmatrix} 1 & 0 & 0 & 0 \\ 0 & -1 & 0 & 0 \\ 3 & -11 & -36 & 0 \\ 1 & -6 & -14 & \frac{22}{9} \end{bmatrix}.$$

The first step is to prepare the following scheme in accordance with rules (i) and (ii).

—	—	—	—	1	0	0	0	(1)
0	—	—	—		1	0	0	(−1)
−3	−11	—	—			1	0	$(-\frac{1}{36})$
−1	−6	$-\frac{7}{18}$	—				1	$(\frac{9}{22})$

X $\qquad$ Y (matrix ω)

To complete the first column of ω, we calculate in succession

$$\omega_{21} = [0]\{1\} = 0,$$
$$\omega_{31} = [-3, -11]\{1, 0\} = -3,$$
$$\omega_{41} = [-1, -6, -\tfrac{7}{18}]\{1, 0, -3\} = \tfrac{1}{6}.$$

Similarly, $\omega_{32} = [-3, -11]\{0, 1\} = -11$, and so on. The results are entered in (Y) as they are obtained, and give

$$\omega = \begin{bmatrix} 1 & 0 & 0 & 0 \\ 0 & 1 & 0 & 0 \\ -3 & -11 & 1 & 0 \\ \frac{1}{6} & -\frac{31}{18} & -\frac{7}{18} & 1 \end{bmatrix}.$$

Hence

$$\tau^{-1} = \begin{bmatrix} 1 & 0 & 0 & 0 \\ 0 & -1 & 0 & 0 \\ \frac{1}{12} & \frac{11}{36} & -\frac{1}{36} & 0 \\ \frac{3}{44} & -\frac{31}{44} & -\frac{7}{44} & \frac{9}{22} \end{bmatrix}.$$

(ii) *Reciprocal of Triangular Matrix with Rows Interchanged.* Suppose

$$\sigma = \begin{bmatrix} 3 & -11 & -36^* & 0 \\ 0 & -1^* & 0 & 0 \\ 1 & -6 & -14 & \frac{22}{9}^* \\ 1^* & 0 & 0 & 0 \end{bmatrix}.$$

Taking the rows in the order 4, 2, 1, 3 to bring the starred elements into the principal diagonal, we have the triangular matrix already considered in example (i). The appropriate multiplier N is here

$$\begin{bmatrix} 0 & 0 & 0 & 1 \\ 0 & 1 & 0 & 0 \\ 1 & 0 & 0 & 0 \\ 0 & 0 & 1 & 0 \end{bmatrix}.$$

Hence, by the result of example (i),

$$\sigma^{-1} = \begin{bmatrix} 1 & 0 & 0 & 0 \\ 0 & -1 & 0 & 0 \\ \frac{1}{12} & \frac{11}{36} & -\frac{1}{36} & 0 \\ \frac{3}{44} & -\frac{31}{44} & -\frac{7}{44} & \frac{9}{22} \end{bmatrix} \begin{bmatrix} 0 & 0 & 0 & 1 \\ 0 & 1 & 0 & 0 \\ 1 & 0 & 0 & 0 \\ 0 & 0 & 1 & 0 \end{bmatrix}$$

$$= \begin{bmatrix} 0 & 0 & 0 & 1 \\ 0 & -1 & 0 & 0 \\ -\frac{1}{36} & \frac{11}{36} & 0 & \frac{1}{12} \\ -\frac{7}{44} & -\frac{31}{44} & \frac{9}{22} & \frac{3}{44} \end{bmatrix}.$$

4·6. Computation of Determinants. The abridged schemes of computation of determinants now to be described are based on the theorems given in §§ 4·2 and 4·3. The treatment is applicable generally, but for brevity the given determinant will be assumed to be of order 4. Thus $\Delta \equiv |a|$, where

$$a = \begin{bmatrix} a_{11} & a_{12} & a_{13} & a_{14} \\ a_{21} & a_{22} & a_{23} & a_{24} \\ a_{31} & a_{32} & a_{33} & a_{34} \\ a_{41} & a_{42} & a_{43} & a_{44} \end{bmatrix}.$$

The process consists effectively in the reduction of the matrix a to a triangular form. Many methods of computation of determinants are of course based on the same principle, but the reductions will here be expressed by means of matrices.

A convenient abridged scheme of computation, which corresponds to the sequence of postmultiplications represented by equations (4·3·1, 4·3·2, 4·3·3), is as follows:

$$\begin{array}{cccccc|cccc}
 & a_{11}^* & a_{12} & a_{13} & a_{14} & & -\frac{a_{12}}{a_{11}} & -\frac{a_{13}}{a_{11}} & -\frac{a_{14}}{a_{11}} & \\
\text{(A)} & a_{21} & a_{22} & a_{23} & a_{24} & & 1 & 0 & 0 & \text{(A}'\text{)} \\
 & a_{31} & a_{32} & a_{33} & a_{34} & & 0 & 1 & 0 & \\
 & a_{41} & a_{42} & a_{43} & a_{44} & & 0 & 0 & 1 & \\
\hline
 & & b_{11}^* & b_{12} & b_{13} & & -\frac{b_{12}}{b_{11}} & -\frac{b_{13}}{b_{11}} & & \\
\text{(B)} & & b_{21} & b_{22} & b_{23} & & 1 & 0 & \text{(B}'\text{)} & \\
 & & b_{31} & b_{32} & b_{33} & & 0 & 1 & & \\
\hline
 & \text{(C)} & & c_{11}^* & c_{12} & & -\frac{c_{12}}{c_{11}} & \text{(C}'\text{)} & & \\
 & & & c_{21} & c_{22} & & 1 & & & \\
\hline
 & & \text{(D)} & & d_{11}^* & & & & & \\
\hline\hline
\end{array}$$

The value of $\Delta \equiv |a|$ is given by the product $a_{11}b_{11}c_{11}d_{11}$.

The top left-hand array (A) is the array of the elements in the given determinant Δ. The leading element a_{11} (shown "starred") is divided into the remaining first row elements, and the results with their signs changed form the first row of the top right-hand array (A′); this array is completed by the elements of the unit matrix I_3. To obtain the array (B), the starred row is omitted from (A), and the remaining rows are postmultiplied by (A′), in accordance with (4·3·1). The array (C) is derived from (B), and (D) from (C), in a similar manner.

In the preceding arrangement the starred element on the left is in each case the leading element of the corresponding array. However, by slightly generalising the scheme, we can remove this restriction and star any convenient element in any left-hand array. The rule in this connection is as follows:

Rule: If the chosen starred element in any left-hand array (say B) lies in the ith row and jth column of that array, then the corresponding right-hand array (B′) is constructed from the ith row of (B) and from a unit matrix as previously, except that the elements derived from (B) must be entered as the jth *row* of (B′). The value of Δ is then the product of the starred elements multiplied by the factor $(-1)^N$, where N is the total number of interchanges required to bring the starred elements into the leading places of their arrays.

Example

Evaluate the determinant

$$\Delta = \begin{vmatrix} 559·2 & 0 & 24·0 & 0 & 174 \\ 0 & 601·3 & 0 & 117·2 & 0 \\ 0·21 & 1·899 & -1 & 1·862 & 0·21 \\ 1·258 & 0·005 & 0·669 & -1 & 0·580 \\ 0·008 & 0·832 & 0·008 & 0·670 & -1 \end{vmatrix}$$

559·2	0	24·0	0	174	1	0	0	0
0	601·3	0	117·2	0	0	1	0	0
0·21	1·899	−1*	1·862	0·21	0·21	1·899	1·862	0·21
1·258	0·005	0·669	−1	0·580	0	0	1	0
0·008	0·832	0·008	0·670	−1	0	0	0	1
	564·24	45·576	44·688	179·04	1	0	0	
	0	601·3	117·2	0	0	1	0	
	1·39849	1·27543	0·245678	0·72049	0	0	1	
	0·00968	0·847192	0·684896	−0·99832*	0·009696	0·848618	0·686049	
		565·976	197·513	167·518	1	0		
		0	601·3	117·2*	0	1		
		1·40548	1·88685	0·739969	0	−5·13055		
			565·976	−661·946	1·35868			
			1·40548*	−1·90960	1			
				107·034*				

Hence

$$\Delta = (-1)\times(-0·99832)\times 117·2\times 1·40548\times 107·034\times(-1)^{14}$$
$$= 17601·3.$$

4·7. Computation of Reciprocal Matrices. Three different methods will be described. They will be dealt with under separate headings, and will be referred to as (i) the method of postmultipliers, (ii) the method of submatrices, and (iii) the method of direct operation on rows.

In the first two methods definite rules for procedure must be learnt. In the third the operations are extremely simple, and are left to a large extent to the judgment of the computer.

4·8. Reciprocation by the Method of Postmultipliers. It has been shown in § 4·3 that by the application of $n-1$ successive post-multipliers, $M_1, M_2, \ldots, M_{n-1}$, each of unit determinant, a given non-singular matrix a of order n can be reduced to a matrix σ, which if not immediately triangular is a triangular matrix with its rows interchanged. The reciprocal of σ can be found readily by the methods of § 4·5, and then

$$a^{-1} = (M_1 M_2 \ldots M_{n-1})\, \sigma^{-1}.$$

The construction of σ and of the product $M_1 M_2 \ldots M_{n-1}$ can be effected simultaneously by a slight extension of the scheme already described in § 4·6 for the computation of determinants. To avoid unnecessary complication the starred elements will be restricted to lie in the first columns of their arrays, although they need not necessarily lie in the first rows. A representative abridged scheme of computation, appropriate to a matrix of order 4, is as follows.

	a_{11}	a_{12}	a_{13}	a_{14}	$-\dfrac{a_{32}}{a_{31}}$	$-\dfrac{a_{33}}{a_{31}}$	$-\dfrac{a_{34}}{a_{31}}$	
(A)	a_{21}	a_{22}	a_{23}	a_{24}	1	0	0	(A′)
	a_{31}^*	a_{32}	a_{33}	a_{34}	0	1	0	
	a_{41}	a_{42}	a_{43}	a_{44}	0	0	1	
		—	—	—	α_1	α_2		
		b_{11}	b_{12}	b_{13}	$-\dfrac{b_{32}}{b_{31}}$	$-\dfrac{b_{33}}{b_{31}}$		
(B)		b_{21}	b_{22}	b_{23}	1	0		(B′)
		b_{31}^*	b_{32}	b_{33}	0	1		
			—	—	β_1			
			—	—	β_2			
(C)			c_{11}^*	c_{12}	$-\dfrac{c_{12}}{c_{11}}$			(C′)
			c_{21}	c_{22}	1			
		(D)		d_{11}^*				

Rules: (i) Proceed as for the computation of determinants (see § 4·6), but introduce blank upper rows on the left as necessary, so as to preserve a constant number of rows. Temporarily leave empty rows on the right opposite the blank rows on the left.

(ii) After calculation of d_{11} fill the empty rows on the right in succession. To find $[\alpha_1, \alpha_2]$ premultiply the two part-columns of (B′) already determined by the top row of (A′). Similarly, to find $\{\beta_1, \beta_2\}$ premultiply the part-column of (C′) already determined by the two top rows of (B′).†

(iii) To form the matrix σ, set down in succession the first columns of the left-hand arrays (disregarding places filled by blanks), and always write ciphers to the right of each starred element.

(iv) To form the product $(M_1 M_2 M_3)$, write down the successive first columns of the right-hand arrays, and precede them by the column $\{1, 0, 0, 0\}$.

In the particular case taken the matrices σ and $M_1 M_2 M_3$ formed in accordance with the rules given would be

$$\sigma = \begin{bmatrix} a_{11} & b_{11} & c_{11}^* & 0 \\ a_{21} & b_{21} & c_{21} & d_{11}^* \\ a_{31}^* & 0 & 0 & 0 \\ a_{41} & b_{31}^* & 0 & 0 \end{bmatrix}; \quad M_1 M_2 M_3 = \begin{bmatrix} 1 & -a_{32}/a_{31} & \alpha_1 & \beta_1 \\ 0 & 1 & -b_{32}/b_{31} & \beta_2 \\ 0 & 0 & 1 & -c_{12}/c_{11} \\ 0 & 0 & 0 & 1 \end{bmatrix}.$$

Moreover, $$|a| = a_{31} b_{31} c_{11} d_{11} (-1)^4.$$

The reciprocal σ^{-1} can now be found by the methods given in § 4·5, and then
$$a^{-1} = (M_1 M_2 M_3)\, \sigma^{-1}.$$

If required the adjoint of a can also be deduced by use of the relation $A = a^{-1} | a |$. It may be noted that, since σ^{-1} necessarily contains a column proportional to $\{0, 0, 0, 1\}$, one column of a^{-1}—and therefore of the adjoint A—will be proportional to the last column of $M_1 M_2 M_3$. Accordingly, the adjoint has a column proportional to the last right-hand array of the abridged scheme.

Example

Suppose $$a = \begin{bmatrix} 1 & 4 & 1 & 3 \\ 0 & -1 & 3 & -1 \\ 3 & 1 & 0 & 2 \\ 1 & -2 & 5 & 1 \end{bmatrix}.$$

† More generally, to complete any right-hand array (K'), premultiply the submatrix of K' already determined by the appropriate submatrix taken from the top of the completed preceding right-hand array.

The abridged scheme which follows should be compared with the calculations given *in extenso* in example (iii) of § 4·3.

$$
\begin{array}{c|cccc|ccc|c}
 & 1 & 4 & 1 & 3 & 2 & -5 & -1 & \\
(A) & 0 & -1 & 3 & -1 & 1 & 0 & 0 & (A') \\
 & 3 & 1 & 0 & 2 & 0 & 1 & 0 & \\
 & 1^* & -2 & 5 & 1 & 0 & 0 & 1 & \\
\hline
 & & — & — & — & 1 & -3 & & \\
\hdashline
(B) & & 6 & -4 & 2 & 3 & -1 & & (B') \\
 & & -1^* & 3 & -1 & 1 & 0 & & \\
 & & 7 & -15 & -1 & 0 & 1 & & \\
\hline
 & & & — & — & -\frac{19}{7} & & & \\
 & & & — & — & -\frac{1}{7} & & & \\
\hdashline
(C) & & & 14^* & -4 & \frac{2}{7} & & & (C') \\
 & & & 6 & -8 & 1 & & & \\
\hline
(D) & & & & -\frac{44}{7}^* & & & & \\
\end{array}
$$

Hence

$$
\sigma = \begin{bmatrix} 1 & 6 & 14^* & 0 \\ 0 & -1^* & 0 & 0 \\ 3 & 7 & 6 & -\frac{44}{7}^* \\ 1^* & 0 & 0 & 0 \end{bmatrix} \quad \text{and} \quad M_1 M_2 M_3 = \begin{bmatrix} 1 & 2 & 1 & -\frac{19}{7} \\ 0 & 1 & 3 & -\frac{1}{7} \\ 0 & 0 & 1 & \frac{2}{7} \\ 0 & 0 & 0 & 1 \end{bmatrix}.
$$

Also $$|a| = 1 \times (-1) \times 14 \times (-\tfrac{44}{7}) \times (-1)^4 = 88.$$

Note that the array (B′) is completed by

$$
[2 \;\; -5 \;\; -1] \begin{bmatrix} 3 & -1 \\ 1 & 0 \\ 0 & 1 \end{bmatrix} = [1, -3],
$$

and that (C′) is then completed by

$$
\begin{bmatrix} 1 & -3 \\ 3 & -1 \end{bmatrix} \begin{bmatrix} \frac{2}{7} \\ 1 \end{bmatrix} = \begin{bmatrix} -\frac{19}{7} \\ -\frac{1}{7} \end{bmatrix}.
$$

To find σ^{-1} the method already illustrated in example (ii) of § 4·5 can be used. Thus

$$
N = \begin{bmatrix} 0 & 0 & 0 & 1 \\ 0 & 1 & 0 & 0 \\ 1 & 0 & 0 & 0 \\ 0 & 0 & 1 & 0 \end{bmatrix} \quad \text{and} \quad N\sigma = \tau = \begin{bmatrix} 1^* & 0 & 0 & 0 \\ 0 & -1^* & 0 & 0 \\ 1 & 6 & 14^* & 0 \\ 3 & 7 & 6 & -\frac{44}{7}^* \end{bmatrix}.
$$

The scheme for the computation of ω is here

—	—	—	—	1	0	0	0	(1)
0	—	—	—	0	1	0	0	(−1)
−1	6	—	—	−1	6	1	0	($\frac{1}{14}$)
−3	7	$-\frac{3}{7}$	—	$-\frac{18}{7}$	$\frac{31}{7}$	$-\frac{3}{7}$	1	($-\frac{7}{44}$)
	X				Y (matrix ω)			

Hence
$$\sigma^{-1} = \begin{bmatrix} 1 & 0 & 0 & 0 \\ 0 & -1 & 0 & 0 \\ -\frac{1}{14} & \frac{3}{7} & \frac{1}{14} & 0 \\ \frac{9}{22} & -\frac{31}{44} & \frac{3}{44} & -\frac{7}{44} \end{bmatrix} \begin{bmatrix} 0 & 0 & 0 & 1 \\ 0 & 1 & 0 & 0 \\ 1 & 0 & 0 & 0 \\ 0 & 0 & 1 & 0 \end{bmatrix}.$$

Finally,
$$a^{-1} = (M_1 M_2 M_3)\,\sigma^{-1}$$
$$= \begin{bmatrix} 1 & 2 & 1 & -\frac{19}{7} \\ 0 & 1 & 3 & -\frac{1}{7} \\ 0 & 0 & 1 & \frac{2}{7} \\ 0 & 0 & 0 & 1 \end{bmatrix} \begin{bmatrix} 0 & 0 & 0 & 1 \\ 0 & -1 & 0 & 0 \\ \frac{1}{14} & \frac{3}{7} & 0 & -\frac{1}{14} \\ \frac{3}{44} & -\frac{31}{44} & -\frac{7}{44} & \frac{9}{22} \end{bmatrix}$$
$$= \frac{1}{44}\begin{bmatrix} -5 & 15 & 19 & -8 \\ 9 & 17 & 1 & -12 \\ 4 & 10 & -2 & 2 \\ 3 & -31 & -7 & 18 \end{bmatrix}.$$

If required, the adjoint of a can now be deduced by use of the relation $A = a^{-1}|a|$, and the known value $|a| = 88$. One column of the adjoint is proportional to $\{19, 1, -2, -7\}$, or to column (C′) in the abridged scheme.

4·9. Reciprocation by the Method of Submatrices. This is an independent method based on the properties of partitioned matrices.

Suppose α to denote a non-singular square matrix of order m. Then α can be partitioned as below into four submatrices:

$$\alpha = \begin{bmatrix} \alpha_{11}(\bar{r}, r), & \alpha_{12}(\bar{r}, s) \\ \alpha_{21}(\bar{s}, r), & \alpha_{22}(\bar{s}, s) \end{bmatrix}.$$

The orders of the submatrices are as indicated, and $r+s=m$.

Let the reciprocal $\beta \equiv \alpha^{-1}$ be correspondingly partitioned; thus

$$\beta = \begin{bmatrix} \beta_{11}(\bar{r}, r), & \beta_{12}(\bar{r}, s) \\ \beta_{21}(\bar{s}, r), & \beta_{22}(\bar{s}, s) \end{bmatrix}.$$

Then, since $\beta\alpha = I_m$, we have the four relations

$$\begin{aligned}\beta_{11}\alpha_{11}+\beta_{12}\alpha_{21} &= I_r,\\ \beta_{11}\alpha_{12}+\beta_{12}\alpha_{22} &= 0,\\ \beta_{21}\alpha_{11}+\beta_{22}\alpha_{21} &= 0,\\ \beta_{21}\alpha_{12}+\beta_{22}\alpha_{22} &= I_s.\end{aligned}$$

These equations may now (in general) be solved to give the submatrices of β explicitly. The results may be expressed as follows. Let

$$X \equiv \alpha_{11}^{-1}\alpha_{12}; \quad Y \equiv \alpha_{21}\alpha_{11}^{-1}; \quad \theta \equiv \alpha_{22} - Y\alpha_{12} = \alpha_{22} - \alpha_{21}X.$$

Then

$$\begin{aligned}\beta_{11} &= \alpha_{11}^{-1} + X\theta^{-1}Y,\\ \beta_{12} &= -X\theta^{-1},\\ \beta_{21} &= -\theta^{-1}Y,\\ \beta_{22} &= \theta^{-1}.\end{aligned}$$

These formulae serve to determine β provided the reciprocals α_{11}^{-1} and θ^{-1} exist. If α_{11}^{-1} is known, the quantities X, Y, θ can be calculated, and β can then be deduced. An arrangement of the numerical work which is convenient and self-explanatory is suggested below:

	α_{21}	α_{22}
$X = \alpha_{11}^{-1}\alpha_{12}$	α_{11}^{-1}	α_{12}
θ^{-1}	$Y = \alpha_{21}\alpha_{11}^{-1}$	$\theta = \alpha_{22} - Y\alpha_{12}$

$$\alpha^{-1} = \left[\begin{array}{c|c} \alpha_{11}^{-1} + X\theta^{-1}Y & -X\theta^{-1} \\ \hline -\theta^{-1}Y & \theta^{-1} \end{array}\right].$$

In the simplest application $r = m-1$ and $s = 1$: then α_{21} is a single row, α_{12} is a single column, α_{22} is a single element, and θ is a scalar. Moreover, since in this case θ^{-1} is the last diagonal element of α^{-1}, namely $|\alpha_{11}|/|\alpha|$, we have the relation

$$\theta = |\alpha|/|\alpha_{11}|. \qquad \text{......(1)}$$

The general procedure in dealing with a given numerical matrix will now be explained. Suppose the matrix to be

$$a = \begin{bmatrix} a_{11} & a_{12} & \dots & a_{1n} \\ a_{21} & a_{22} & \dots & a_{2n} \\ \dots & \dots & \dots & \dots \\ a_{n1} & a_{n2} & \dots & a_{nn} \end{bmatrix}.$$

Then a sequence of matrices S_1, S_2, S_3, etc., progressively including more and more of the principal diagonal elements of a, can be formed as follows:

$$S_1 \equiv [a_{11}],$$

$$S_2 \equiv \begin{bmatrix} a_{11} & a_{12} \\ a_{21} & a_{22} \end{bmatrix},$$

$$S_3 \equiv \begin{bmatrix} a_{11} & a_{12} & a_{13} \\ a_{21} & a_{22} & a_{23} \\ a_{31} & a_{32} & a_{33} \end{bmatrix} = \left[\begin{array}{cc|c} \multicolumn{2}{c|}{S_2} & \begin{matrix} a_{13} \\ a_{23} \end{matrix} \\ \hline a_{31} & a_{32} & a_{33} \end{array}\right],$$

$$S_4 \equiv \begin{bmatrix} a_{11} & a_{12} & a_{13} & a_{14} \\ a_{21} & a_{22} & a_{23} & a_{24} \\ a_{31} & a_{32} & a_{33} & a_{34} \\ a_{41} & a_{42} & a_{43} & a_{44} \end{bmatrix} = \left[\begin{array}{ccc|c} \multicolumn{3}{c|}{S_3} & \begin{matrix} a_{14} \\ a_{24} \\ a_{34} \end{matrix} \\ \hline a_{41} & a_{42} & a_{43} & a_{44} \end{array}\right],$$

and so on.

Commencing with the second of these, we can write down the reciprocal of S_2 immediately; thus

$$S_2^{-1} = \frac{1}{a_{11}a_{22} - a_{12}a_{21}} \begin{bmatrix} a_{22} & -a_{12} \\ -a_{21} & a_{11} \end{bmatrix}.$$

Using the value of S_2^{-1}, we can now apply to S_3 the scheme of computation just described, and so obtain S_3^{-1}. Next, using S_3^{-1}, and applying the scheme to S_4, we derive S_4^{-1}. Proceeding in this way, we finally obtain S_n^{-1} or a^{-1}.

In the exceptional case where a singular matrix, say S_i, is encountered (as indicated by $\theta = 0$), the simplest procedure is to transfer the ith row of a to last place, and then to continue as normally. This rearrangement of the rows will be equivalent to premultiplication of a by N, where N is the unit matrix with its ith row transferred to last place. The final reciprocal matrix as computed will then have to be *postmultiplied* by N to give a^{-1} (compare (4·5·2)). More generally, if several rows of a are transferred in the foregoing way during the computations, the appropriate multiplier N can readily be constructed.

If θ_i denotes the value of θ obtained in the construction of S_i^{-1}, then by (1)

$$\theta_i = |S_i|/|S_{i-1}|.$$

Hence

$$|a| = \theta_n \theta_{n-1} \dots \theta_2 \theta_1.$$

Examples

(i) *Normal Case* (*Matrices S all Non-singular*). Suppose the given matrix to be (see example, § 4·8)

$$a = \begin{bmatrix} 1 & 4 & 1 & 3 \\ 0 & -1 & 3 & -1 \\ 3 & 1 & 0 & 2 \\ 1 & -2 & 5 & 1 \end{bmatrix}.$$

Then

$$S_2 = \begin{bmatrix} 1 & 4 \\ 0 & -1 \end{bmatrix} = S_2^{-1}.$$

The scheme for the computation of S_3^{-1} is as follows:

$$\begin{array}{c|c|cc|c|c}
\cline{3-5}
 & \multicolumn{1}{c}{} & 3 & 1 & 0 & \\
\cline{2-5}
X & 13 & 1 & 4 & 1 & \\
 & -3 & 0 & -1 & 3 & \\
\cline{2-5}
\theta^{-1} & -\frac{1}{36} & 3 & 11 & -36 & \theta \\
\cline{2-5}
\multicolumn{6}{c}{Y}
\end{array} \quad ; \quad X\theta^{-1}Y = \begin{bmatrix} -\frac{13}{12} & -\frac{143}{36} \\ \frac{1}{4} & \frac{11}{12} \end{bmatrix}.$$

Hence

$$S_3^{-1} = \left[\begin{array}{cc:c} -\frac{1}{12} & \frac{1}{36} & \frac{13}{36} \\ \frac{1}{4} & -\frac{1}{12} & -\frac{1}{12} \\ \hdashline \frac{1}{12} & \frac{11}{36} & -\frac{1}{36} \end{array}\right].$$

Proceeding to S_4^{-1}, we have

$$\begin{array}{c|c|ccc|c|c}
\cline{3-6}
 & \multicolumn{1}{c}{} & 1 & -2 & 5 & 1 & \\
\cline{2-6}
 & \frac{4}{9} & -\frac{1}{12} & \frac{1}{36} & \frac{13}{36} & 3 & \\
X & \frac{2}{3} & \frac{1}{4} & -\frac{1}{12} & -\frac{1}{12} & -1 & \\
 & -\frac{1}{9} & \frac{1}{12} & \frac{11}{36} & -\frac{1}{36} & 2 & \\
\cline{2-6}
\theta^{-1} & \frac{9}{22} & -\frac{1}{6} & \frac{31}{18} & \frac{7}{18} & \frac{22}{9} & \theta \\
\cline{2-6}
\multicolumn{7}{c}{Y}
\end{array} \quad ; \quad X\theta^{-1}Y = \begin{bmatrix} -\frac{1}{33} & \frac{31}{99} & \frac{7}{99} \\ -\frac{1}{22} & \frac{31}{66} & \frac{7}{66} \\ \frac{1}{132} & -\frac{31}{396} & -\frac{7}{396} \end{bmatrix}.$$

Hence (compare result of example in § 4·8)

$$S_4^{-1} \equiv a^{-1} = \left[\begin{array}{ccc:c} -\frac{5}{44} & \frac{15}{44} & \frac{19}{44} & -\frac{2}{11} \\ \frac{9}{44} & \frac{17}{44} & \frac{1}{44} & -\frac{3}{11} \\ \frac{4}{44} & \frac{10}{44} & -\frac{2}{44} & \frac{1}{22} \\ \hdashline \frac{3}{44} & -\frac{31}{44} & -\frac{7}{44} & \frac{9}{22} \end{array}\right]$$

$$= \tfrac{1}{44}\begin{bmatrix} -5 & 15 & 19 & -8 \\ 9 & 17 & 1 & -12 \\ 4 & 10 & -2 & 2 \\ 3 & -31 & -7 & 18 \end{bmatrix}.$$

The determinant $|a|$ is here given by

$$\theta_4\theta_3\,|S_2| = \tfrac{22}{9} \times (-36) \times (-1) = 88.$$

(ii) *Case of Singular Matrix S*. Suppose

$$a = \begin{bmatrix} 1 & 1 & 2 & 1 \\ 1 & 2 & 4 & 3 \\ 1 & 0 & 0 & 2 \\ 3 & -1 & 1 & 5 \end{bmatrix}.$$

Here $$S_2 = \begin{bmatrix} 1 & 1 \\ 1 & 2 \end{bmatrix} \quad \text{and} \quad S_2^{-1}\begin{bmatrix} 2 & -1 \\ -1 & 1 \end{bmatrix}.$$

Proceeding to S_3, we find that $\theta = 0$: thus

1	0	0
2	−1	2
−1	1	4
2	−1	$\theta = 0$

This result indicates that S_3 is singular. Transferring the third row of a to last place, and continuing as normally, we have

		3	−1	1	
X	0	2	−1	2	
	2	−1	1	4	
θ^{-1}	$\frac{1}{3}$	7	−4	3	θ
		Y			

$$;\quad X\theta^{-1}Y = \begin{bmatrix} 0 & 0 \\ \frac{14}{3} & -\frac{8}{3} \end{bmatrix}.$$

Lastly,

		1	0	0	2	
	−1	2	−1	0	1	
X	$-\frac{14}{3}$	$\frac{11}{3}$	$-\frac{5}{3}$	$-\frac{2}{3}$	3	
	$\frac{10}{3}$	$-\frac{7}{3}$	$\frac{4}{3}$	$\frac{1}{3}$	5	
θ^{-1}	$\frac{1}{3}$	2	−1	0	3	θ
		Y				

$$;\quad X\theta^{-1}Y = \begin{bmatrix} -\frac{2}{3} & \frac{1}{3} & 0 \\ -\frac{28}{9} & \frac{14}{9} & 0 \\ \frac{20}{9} & -\frac{10}{9} & 0 \end{bmatrix}.$$

The required reciprocal matrix is accordingly

$$a^{-1} = \begin{bmatrix} \frac{4}{3} & -\frac{2}{3} & 0 & \frac{1}{3} \\ \frac{5}{9} & -\frac{1}{9} & -\frac{2}{3} & \frac{14}{9} \\ -\frac{1}{9} & \frac{2}{9} & \frac{1}{3} & -\frac{10}{9} \\ -\frac{2}{3} & \frac{1}{3} & 0 & \frac{1}{3} \end{bmatrix} \begin{bmatrix} 1 & 0 & 0 & 0 \\ 0 & 1 & 0 & 0 \\ 0 & 0 & 0 & 1 \\ 0 & 0 & 1 & 0 \end{bmatrix}$$

$$= \frac{1}{9}\begin{bmatrix} 12 & -6 & 3 & 0 \\ 5 & -1 & 14 & -6 \\ -1 & 2 & -10 & 3 \\ -6 & 3 & 3 & 0 \end{bmatrix}.$$

(iii) *Specimen Calculation for Sixth Order Matrix.* In the preceding examples the successive steps have been explained in some detail. To illustrate the compactness of the method, we shall now compute the reciprocal of a symmetrical sixth order matrix, omitting all unessential steps. The given matrix is

$$a = \begin{bmatrix} 10{\cdot}472 & 0{\cdot}506 & 0 & -3{\cdot}935 & -0{\cdot}521 & 0 \\ 0{\cdot}506 & 11{\cdot}016 & 5{\cdot}000 & -0{\cdot}521 & -1{\cdot}046 & 3{\cdot}750 \\ 0 & 5{\cdot}000 & 26{\cdot}000 & 0 & 0 & -1{\cdot}050 \\ -3{\cdot}935 & -0{\cdot}521 & 0 & 6{\cdot}322 & 0{\cdot}536 & 0{\cdot}355 \\ -0{\cdot}521 & -1{\cdot}046 & 0 & 0{\cdot}536 & 2{\cdot}737 & 0 \\ 0 & 3{\cdot}750 & -1{\cdot}050 & 0{\cdot}355 & 0 & 3{\cdot}881 \end{bmatrix},$$

and the computations begin with S_2^{-1} bordered (see next page).

To illustrate the accuracy, the product of the given matrix a and the computed reciprocal is given below:

$$aa^{-1} =$$

$$\begin{bmatrix} 0{\cdot}999995, & 0{\cdot}000000, & 0{\cdot}000000, & 0{\cdot}000003, & 0{\cdot}000000, & 0{\cdot}000000 \\ 0{\cdot}000000, & 1{\cdot}000003, & -0{\cdot}000001, & 0{\cdot}000001, & 0{\cdot}000001, & -0{\cdot}000002 \\ 0{\cdot}000000, & 0{\cdot}000003, & 0{\cdot}999999, & 0{\cdot}000001, & 0{\cdot}000002, & -0{\cdot}000006 \\ 0{\cdot}000002, & 0{\cdot}000000, & 0{\cdot}000000, & 0{\cdot}999996, & 0{\cdot}000000, & 0{\cdot}000000 \\ 0{\cdot}000000, & 0{\cdot}000000, & 0{\cdot}000000, & 0{\cdot}000000, & 1{\cdot}000000, & 0{\cdot}000000 \\ 0{\cdot}000000, & 0{\cdot}000001, & 0{\cdot}000000, & 0{\cdot}000000, & 0{\cdot}000000, & 0{\cdot}999998 \end{bmatrix}.$$

	0	5·000	26·000
−0·0219802	0·0957052,	−0·00439604	0
0·454895	−0·00439604,	0·0909790	5·000
0·0421487	−0·0219802,	0·454895	23·7255

	−3·935	−0·521	0	6·322
−0·374170	0·0957256,	−0·00481747	0·000926437	−3·935
−0·0329874	−0·00481747,	0·0997008	−0·0191732	−0·521
0·00634371	0·000926437,	−0·0191732	0·0421487	0
0·206934	−0·374170,	−0·0329874,	0·00634371	4·83245

	−0·521	−1·046	0	0·536	2·737
−0·0210980	0·124697,	−0·00226331,	0·000435253	0·0774285	−0·521
−0·0996846	−0·00226331,	0·0999260,	−0·0192165	0·00682621	−1·046
0·0191701	0·000435253,	−0·0192165,	0·0421570	−0·00131273.	0
0·0634362	0·0774285,	0·00682621,	−0·00131273	0·206934	0·536
0·386438	−0·0210980,	−0·0996846,	0·0191701,	0·0634362	2·58774

	0	3·750	−1·050	0·355	0	3·881
0·0215710	0·124869,	−0·00145057,	0·000278958,	0·0769113	0·00815307	0
0·411631	−0·00145057,	0·103766,	−0·0199550,	0·00438253	0·0385219	3·750
−0·119544	0·000278958,	−0·0199550,	0·0422990,	−0·000842790	−0·00740806	−1·050
0·0913330	0·0769113,	0·00438252,	−0·000842790,	0·208489	−0·0245142	0·355
0·143533	0·00815307,	0·0385219,	−0·00740806,	−0·0245142	0·386438	0
0·458833	0·0215710,	0·411631,	−0·119544,	0·0913330,	0·143533	2·17944

$a^{-1}=$	0·125082,	0·00262354,	−0·000904232,	0·0778153,	0·00957369	−0·00989749
	0·00262354,	0·181511,	−0·0425333,	0·0216326,	0·0656310	−0·188870
	−0·000904232,	−0·0425333,	0·0488561,	−0·00585247,	−0·0152809	0·0548507
	0·0778153,	0·0216326,	−0·00585247,	0·212316,	−0·0184992	−0·0419066
	0·00957369,	0·0656310,	−0·0152809,	−0·0184992,	0·395891	−0·0658577
	−0·00989749,	−0·188870,	0·0548507,	−0·0419066,	−0·0658577	0·458833

4·10. Reciprocation by Direct Operations on Rows. This method offers the great advantage that no formulae have to be memorised. The underlying principle is the reduction of the given matrix a to a triangular form and then to the unit matrix by a succession of simple operations on rows. These operations are simultaneously performed on the unit matrix, which is ultimately transformed into the required reciprocal. The process is thus essentially the same as that given in § 4·8, but with premultipliers used instead of postmultipliers. However, the actual scheme of numerical calculation is considerably different.

A simple example will make the procedure clear. Suppose the given matrix to be

$$a = \begin{bmatrix} 36 & 16 & 4 \\ 15 & 9 & 3 \\ 6 & 4 & 2 \end{bmatrix}.$$

Then the required reciprocal is such that

$$\begin{matrix} (r_1) \\ (r_2) \\ (r_3) \end{matrix} \begin{bmatrix} 36 & 16 & 4 \\ 15 & 9 & 3 \\ 6 & 4 & 2 \end{bmatrix} a^{-1} = \begin{bmatrix} 1 & 0 & 0 \\ 0 & 1 & 0 \\ 0 & 0 & 1 \end{bmatrix}.$$

The symbols r_i are used in this and in subsequent equations merely to identify the rows. Now operate on the left-hand and right-hand rows of the equation in the manner indicated below:

$$\begin{array}{cc} \text{Operation} & \text{New Row No.} \\ \frac{1}{4}r_1 & r_4 \\ \frac{1}{3}r_2 & r_5 \\ \frac{1}{2}r_3 & r_6 \end{array} \qquad \begin{bmatrix} 9 & 4 & 1 \\ 5 & 3 & 1 \\ 3 & 2 & 1 \end{bmatrix} a^{-1} = \begin{bmatrix} \frac{1}{4} & 0 & 0 \\ 0 & \frac{1}{3} & 0 \\ 0 & 0 & \frac{1}{2} \end{bmatrix}$$

$$\begin{array}{cc} r_4 - r_6 & r_7 \\ r_5 - r_6 & r_8 \\ \text{—} & r_6 \end{array} \qquad \begin{bmatrix} 6 & 2 & 0 \\ 2 & 1 & 0 \\ 3 & 2 & 1 \end{bmatrix} a^{-1} = \begin{bmatrix} \frac{1}{4} & 0 & -\frac{1}{2} \\ 0 & \frac{1}{3} & -\frac{1}{2} \\ 0 & 0 & \frac{1}{2} \end{bmatrix}$$

The last column of the matrix on the left now contains only one non-zero element (namely the unit), and one further operation completes the reduction to a triangular form. Thus

$$\begin{array}{cc} \text{Operation} & \text{New Row No.} \\ \frac{1}{2}r_7 - r_8 & r_9 \\ & r_8 \\ & r_6 \end{array} \qquad \begin{bmatrix} 1 & 0 & 0 \\ 2 & 1 & 0 \\ 3 & 2 & 1 \end{bmatrix} a^{-1} = \begin{bmatrix} \frac{1}{8} & -\frac{1}{3} & \frac{1}{4} \\ 0 & \frac{1}{3} & -\frac{1}{2} \\ 0 & 0 & \frac{1}{2} \end{bmatrix}$$

The foregoing reduction to triangular form can conveniently be arranged as below:

Operation	Row No.	Left Array			Right Array		
—	r_1	36	16	4	1	0	0
—	r_2	15	9	3	0	1	0
—	r_3	6	4	2	0	0	1
$\frac{1}{4}r_1$	r_4	9	4	1	$\frac{1}{4}$	0	0
$\frac{1}{3}r_2$	r_5	5	3	1	0	$\frac{1}{3}$	0
$\frac{1}{2}r_3$	r_6	3	2	1	0	0	$\frac{1}{2}$
r_4-r_6	r_7	6	2	0	$\frac{1}{4}$	0	$-\frac{1}{2}$
r_5-r_6	r_8	2	1	0	0	$\frac{1}{3}$	$-\frac{1}{2}$
$\frac{1}{2}r_7-r_8$	r_9	1	0	0	$\frac{1}{8}$	$-\frac{1}{3}$	$\frac{1}{4}$
—	r_9	1	0	0	$\frac{1}{8}$	$-\frac{1}{3}$	$\frac{1}{4}$
—	r_8	2	1	0	0	$\frac{1}{3}$	$-\frac{1}{2}$
—	r_6	3	2	1	0	0	$\frac{1}{2}$

The value of $|a|$, if required, is now given by

$$|a| = (\tfrac{1}{8}\times\tfrac{1}{3}\times\tfrac{1}{2})^{-1} = 48.$$

The operations on rows can be continued to effect the reduction of a to the unit matrix. Thus

Operation	Row No.	Left Array			Right Array		
r_8-2r_9	r_{10}	0	1	0	$-\frac{1}{4}$	1	-1
r_6-2r_9	r_{11}	0	2	1	$-\frac{3}{8}$	1	$-\frac{1}{4}$
$r_{11}-2r_{10}$	r_{12}	0	0	1	$\frac{1}{8}$	-1	$\frac{7}{4}$
—	r_9	1	0	0	$\frac{1}{8}$	$-\frac{1}{3}$	$\frac{1}{4}$
—	r_{10}	0	1	0	$-\frac{1}{4}$	1	-1
—	r_{12}	0	0	1	$\frac{1}{8}$	-1	$\frac{7}{4}$

The final right-hand array gives the required reciprocal.

4·11. Improvement of the Accuracy of an Approximate Reciprocal Matrix. If an approximate reciprocal matrix has been obtained, and greater accuracy is required, the following method is usually effective. Let a be the given matrix, ρ the approximate reciprocal, and $\rho+\delta\rho$ the exact reciprocal. Then $a(\rho+\delta\rho) = I$, or

$$a\,\delta\rho = I - a\rho.$$

Premultiplication by ρ gives, approximately, $\delta\rho = \rho(I-a\rho)$. Hence the next approximation to the reciprocal is

$$\rho+\delta\rho = \rho(2I-a\rho).$$

The conditions for the convergence of this method of approximation are readily obtained. If $\rho = \rho_0$ and $\rho + \delta\rho = \rho_1$, the last equation gives

$$I - a\rho_1 = I - a\rho_0(2I - a\rho_0) = (I - a\rho_0)^2.$$

Similarly, for a second approximation,

$$I - a\rho_2 = (I - a\rho_1)^2 = (I - a\rho_0)^4,$$

and generally
$$I - a\rho_r = (I - a\rho_0)^{2^r}.$$

Hence provided that ρ_0 is such that the latent roots of $I - a\rho_0$ all have moduli less than unity, $I - a\rho_r$ will tend rapidly to zero as r increases, i.e. ρ_r will tend rapidly to a^{-1}. In particular, this condition is satisfied if ρ_0 is a fairly accurate reciprocal of a, so that the elements of $I - a\rho_0$ are all small. In practice, if a very accurate reciprocal of a matrix is required, it is probable that labour will be saved and the possibility of error reduced by a preliminary calculation of an approximate reciprocal and a subsequent application of one or two corrections according to the above scheme.

EXAMPLE

If $a = \begin{bmatrix} 7 & 3 \\ 2 & 1 \end{bmatrix}$, then the exact reciprocal is

$$a^{-1} = \begin{bmatrix} 1 & -3 \\ -2 & 7 \end{bmatrix}.$$

Suppose a known approximate reciprocal to be

$$\rho = \begin{bmatrix} 0\cdot998 & -3\cdot005 \\ -1\cdot994 & 7\cdot013 \end{bmatrix}.$$

Then $2I - a\rho = \begin{bmatrix} 2 & 0 \\ 0 & 2 \end{bmatrix} - \begin{bmatrix} 1\cdot004 & 0\cdot004 \\ 0\cdot002 & 1\cdot003 \end{bmatrix} = \begin{bmatrix} 0\cdot996 & -0\cdot004 \\ -0\cdot002 & 0\cdot997 \end{bmatrix}.$

Hence the next approximation to a^{-1} is

$$\rho(2I - a\rho) = \begin{bmatrix} 0\cdot998 & -3\cdot005 \\ -1\cdot994 & 7\cdot013 \end{bmatrix} \begin{bmatrix} 0\cdot996 & -0\cdot004 \\ -0\cdot002 & 0\cdot997 \end{bmatrix}$$
$$= \begin{bmatrix} 1\cdot000018 & -2\cdot999977 \\ -2\cdot000050 & 6\cdot999937 \end{bmatrix}.$$

4·12. Computation of the Adjoint of a Singular Matrix. The adjoint A of a non-singular matrix a can be obtained by the first or the third of the methods of reciprocation already described, and use of the relation $A = |a| a^{-1}$. Attention will here be restricted to the computation of the adjoint of a singular matrix. It will be remembered (see §§ 1·10, 1·12) that the adjoint of a simply degenerate matrix has proportional rows and proportional columns, while the adjoint of a multiply degenerate matrix is null. For many purposes it is sufficient to obtain a column of numbers proportional to a non-vanishing column of the adjoint, if such a column exists.

A numerical example will sufficiently explain the procedure, which is an adaptation of the method of § 4·10. Suppose the given singular matrix to be

$$a = \begin{bmatrix} 2 & 4 & 6 \\ 3 & 9 & 15 \\ 4 & 16 & 28 \end{bmatrix}.$$

Let any convenient element, say a_{23} ($\equiv 15$) be given a small increment ϵ. Then

$$\begin{vmatrix} 2 & 4 & 6 \\ 3 & 9 & 15+\epsilon \\ 4 & 16 & 28 \end{vmatrix} = \begin{vmatrix} 2 & 4 & 6 \\ 3 & 9 & 15 \\ 4 & 16 & 28 \end{vmatrix} + \begin{vmatrix} 2 & 4 & 0 \\ 3 & 9 & \epsilon \\ 4 & 16 & 0 \end{vmatrix} = \epsilon A_{23},$$

where A_{23} denotes the cofactor of a_{23}. Hence, if the increment of the adjoint due to ϵ is ϵB,

$$\begin{bmatrix} 2 & 4 & 6 \\ 3 & 9 & 15+\epsilon \\ 4 & 16 & 28 \end{bmatrix} (A+\epsilon B) = \epsilon A_{23} \begin{bmatrix} 1 & 0 & 0 \\ 0 & 1 & 0 \\ 0 & 0 & 1 \end{bmatrix}. \qquad \text{......(1)}$$

Direct operations on the rows of this equation are now performed, as in § 4·10, until a row is obtained on the left containing only one non-zero element which is proportional to ϵ. Thus

Operation	Row No.	Left Array			Right Array		
	r_1	2	4	6	1	0	0
	r_2	3	9	$15+\epsilon$	0	1	0
	r_3	4	16	28	0	0	1
$\frac{1}{2}r_1$	r_4	1	2	3	$\frac{1}{2}$	0	0
$\frac{1}{3}r_2$	r_5	1	3	$5+\epsilon/3$	0	$\frac{1}{3}$	0
$\frac{1}{4}r_3$	r_6	1	4	7	0	0	$\frac{1}{4}$
$r_5 - r_4$	r_7	0	1	$2+\epsilon/3$	$-\frac{1}{2}$	$\frac{1}{3}$	0
$r_6 - r_4$	r_8	0	2	4	$-\frac{1}{2}$	0	$\frac{1}{4}$
$\frac{1}{2}r_8$	r_9	0	1	2	$-\frac{1}{4}$	0	$\frac{1}{8}$
$r_7 - r_9$	r_{10}	0	0	$\epsilon/3$	$-\frac{1}{4}$	$\frac{1}{3}$	$-\frac{1}{8}$

The last row obtained yields on division by $\epsilon/3$ (see (1)),

$$[0, 0, 1](A+\epsilon B) = A_{23}[-\tfrac{3}{4}, 1, -\tfrac{3}{8}].$$

Hence when ϵ is made zero

$$[0, 0, 1]A = [A_{13}, A_{23}, A_{33}] = A_{23}[-\tfrac{3}{4}, 1, -\tfrac{3}{8}].$$

Since A has unit rank, it follows that each row is proportional to $[-\tfrac{3}{4}, 1, -\tfrac{3}{8}]$.

In practice it is not necessary to introduce the increment ϵ explicitly. The operations are performed simply on the matrix a, and are continued until a *null* row is obtained on the left. The corresponding row on the right will then be proportional to the rows of the adjoint.

Next, if the transposed of a is taken for the initial left-hand array, and the process is repeated, a row will be derived on the right which, when transposed, is proportional to the columns of A. The adjoint is then determined apart from a scalar multiplier. If required, this multiplier can be obtained by evaluation of a non-vanishing first minor of a: the calculation is usually simple, since by the preceding operations some of the first minors will already have been reduced to a triangular form.

A caution should be added. If during the evaluation of the adjoint it is found possible to derive on the left two or more null rows from a, corresponding to linearly independent rows on the right, then a is multiply degenerate and the adjoint is null. Accordingly, when one null row has been obtained it is always advisable to ascertain whether another can be derived independently.

The scheme of computation described can also be viewed as follows. Suppose $r_1, r_2, \ldots, r_n$ to be the rows of the given matrix a. Since a is degenerate, there will be one or more relations of the type $\sum_{i=1}^{n} A_i r_i = 0$ between the rows, where the multipliers A_i are scalars; the number of such relations will equal the degeneracy of a. When a is simply degenerate, the row $[A_1, A_2, \ldots, A_n]$ is proportional to the rows of the adjoint of a, and in the general case the relation $\sum_{i=1}^{n} A_i r_i = 0$ is equivalent to $[A_1, A_2, \ldots, A_n]\,a = 0$. Evidently the process described, in which various multiples of the rows of a are added until the sum total vanishes, amounts simply to a determination of the coefficients A_i.

Examples

(i) *Simply Degenerate Matrix.* The complete calculations for the matrix already considered are summarised below:

Matrix as Given						Matrix Transposed					
Left Array			Right Array			Left Array			Right Array		
2	4	6	1	0	0	2	3	4	1	0	0
3	9	15	0	1	0	4	9	16	0	1	0
4	16	28	0	0	1	6	15	28	0	0	1
1	2	3	$\frac{1}{2}$	0	0	1	$\frac{3}{2}$	2	$\frac{1}{2}$	0	0
1	3	5	0	$\frac{1}{3}$	0	1	$\frac{9}{4}$	4	0	$\frac{1}{4}$	0
1	4	7	0	0	$\frac{1}{4}$	1	$\frac{5}{2}$	$\frac{14}{3}$	0	0	$\frac{1}{6}$
0	1	2	$-\frac{1}{2}$	$\frac{1}{3}$	0	0	$\frac{3}{4}$	2	$-\frac{1}{2}$	$\frac{1}{4}$	0
0	2	4	$-\frac{1}{2}$	0	$\frac{1}{4}$	0	1	$\frac{8}{3}$	$-\frac{1}{2}$	0	$\frac{1}{6}$
0	1	2	$-\frac{1}{4}$	0	$\frac{1}{8}$	0	1	$\frac{8}{3}$	$-\frac{2}{3}$	$\frac{1}{3}$	0
0	0	0	$-\frac{1}{4}$	$\frac{1}{3}$	$-\frac{1}{8}$	0	0	0	$\frac{1}{6}$	$-\frac{1}{3}$	$\frac{1}{6}$

Hence $A = c\{\frac{1}{6}, -\frac{1}{3}, \frac{1}{6}\}[-\frac{1}{4}, \frac{1}{3}, -\frac{1}{8}]$, where c is an unknown constant. Also the cofactor of a_{33} is, in the fourth and seventh rows of the above table, reduced to triangular form. Hence

$$A_{33} = (1\times 1)\div(\tfrac{1}{2}\times\tfrac{1}{3}) = 6.$$

Alternatively, from the operations on the transposed matrix,

$$A_{33} = (1\times\tfrac{3}{4})\div(\tfrac{1}{2}\times\tfrac{1}{4}) = 6.$$

Hence $c = -6\times 48$ and

$$A = \{1, -2, 1\}[12, -16, 6].$$

Note that the row and the column determined have the properties

$$[-\tfrac{1}{4}, \tfrac{1}{3}, -\tfrac{1}{8}]a = a\{\tfrac{1}{6}, -\tfrac{1}{3}, \tfrac{1}{6}\} = 0.$$

(ii) *Multiply Degenerate Matrix.* If

$$a = \begin{bmatrix} 2 & -1 & 3 & 5 \\ 3 & 2 & -1 & 4 \\ 1 & -4 & 7 & 6 \\ -4 & -5 & 5 & -3 \end{bmatrix}$$

the computations are as follows:

Matrix as Given								Matrix Transposed							
Left Array				Right Array				Left Array				Right Array			
2	-1	3	5	1	0	0	0	2	3	1	-4	1	0	0	0
3	2	-1	4	0	1	0	0	-1	2	-4	-5	0	1	0	0
1	-4	7	6	0	0	1	0	3	-1	7	5	0	0	1	0
-4	-5	5	-3	0	0	0	1	5	4	6	-3	0	0	0	1
1	$-\frac{1}{2}$	$\frac{3}{2}$	$\frac{5}{2}$	$\frac{1}{2}$	0	0	0	1	$\frac{3}{2}$	$\frac{1}{2}$	-2	$\frac{1}{2}$	0	0	0
1	$\frac{2}{3}$	$-\frac{1}{3}$	$\frac{4}{3}$	0	$\frac{1}{3}$	0	0	1	-2	4	5	0	-1	0	0
1	-4	7	6	0	0	1	0	1	$-\frac{1}{3}$	$\frac{7}{3}$	$\frac{5}{3}$	0	0	$\frac{1}{3}$	0
1	$\frac{5}{4}$	$-\frac{5}{4}$	$\frac{3}{4}$	0	0	0	$-\frac{1}{4}$	1	$\frac{4}{5}$	$\frac{6}{5}$	$-\frac{3}{5}$	0	0	0	$\frac{1}{5}$
0	$\frac{7}{6}$	$-\frac{11}{6}$	$-\frac{7}{6}$	$-\frac{1}{2}$	$\frac{1}{3}$	0	0	0	$-\frac{7}{2}$	$\frac{7}{2}$	7	$-\frac{1}{2}$	-1	0	0
0	$-\frac{7}{2}$	$\frac{11}{2}$	$\frac{7}{2}$	$-\frac{1}{2}$	0	1	0	0	$-\frac{11}{6}$	$\frac{11}{6}$	$\frac{11}{3}$	$-\frac{1}{2}$	0	$\frac{1}{3}$	0
0	$\frac{7}{4}$	$-\frac{11}{4}$	$-\frac{7}{4}$	$-\frac{1}{2}$	0	0	$-\frac{1}{4}$	0	$-\frac{7}{10}$	$\frac{7}{10}$	$\frac{7}{5}$	$-\frac{1}{2}$	0	0	$\frac{1}{5}$
0	1	$-\frac{11}{7}$	-1	$-\frac{3}{7}$	$\frac{2}{7}$	0	0	0	1	-1	2	$\frac{1}{7}$	$\frac{2}{7}$	0	0
0	1	$-\frac{11}{7}$	-1	$\frac{1}{7}$	0	$-\frac{2}{7}$	0	0	1	-1	2	$\frac{3}{11}$	0	$-\frac{2}{11}$	0
0	1	$-\frac{11}{7}$	-1	$-\frac{2}{7}$	0	0	$-\frac{1}{7}$	0	1	-1	2	$\frac{5}{7}$	0	0	$-\frac{2}{7}$
0	0	0	0	$\frac{4}{7}$	$-\frac{2}{7}$	$-\frac{2}{7}$	0	0	0	0	0	$\frac{10}{77}$	$-\frac{2}{7}$	$-\frac{2}{11}$	0
0	0	0	0	$\frac{1}{7}$	$-\frac{2}{7}$	0	$-\frac{1}{7}$	0	0	0	0	$\frac{4}{7}$	$-\frac{2}{7}$	0	$-\frac{2}{7}$

The process yields two null rows from a, corresponding to linearly independent rows on the right, so that the adjoint A is null. The following four relations, derived from the rows corresponding to the null rows of the left-hand arrays, may be noted:

$$[2, \ -1, \ -1, \ 0]a = 0,$$
$$[1, \ -2, \ 0, \ -1]a = 0,$$
$$a\{5, \ -11, \ -7, \ 0\} = 0,$$
$$a\{2, \ -1, \ 0, \ -1\} = 0.$$

4·13. Numerical Solution of Simultaneous Linear Algebraic Equations.

(*a*) *Preliminary Remarks on the General Nature of the Solution.* Before dealing with methods of solution we shall briefly consider the general nature of the solution.

Suppose the given system of m equations in m unknowns x to be $ax = h$. Then, if a is non-singular, the equations have a unique solution. On the other hand, if a is singular, the m equations are either incompatible or not linearly independent. To make this clear, let us suppose a to be of rank r. Then a is equivalent to a canonical matrix C which has units in the first r places of the principal diagonal and zero elements elsewhere (see § 3·13). We may therefore substitute pCq for a

(where p and q are non-singular matrices) and write the given equations as

$$pCqx = h.$$

Hence if $y = qx$ and $p^{-1}h = l$, we obtain the equivalent set of equations

$$Cy = l. \qquad \text{......(1)}$$

From the form of the canonical matrix C it is seen at once that the m scalar equations contained in (1) are incompatible unless

$$l_{r+1} = l_{r+2} = \ldots = l_m = 0,$$

and that when these conditions happen to be satisfied the solution is $y_i = l_i$ for $i \leqslant r$, with $y_{r+1}, y_{r+2}, \ldots, y_m$ arbitrary. The corresponding values of x are in this case given by

$$x = q^{-1}\{l_1, l_2, \ldots, l_r, y_{r+1}, y_{r+2}, \ldots, y_m\}.$$

The most general solution of the original equations in x is accordingly the sum of a unique column and of arbitrary multiples of $m-r$ other columns. For instance, if the equations are

$$\begin{bmatrix} 2 & 2 & 5 & 3 \\ 6 & 1 & 5 & 4 \\ 4 & -1 & 0 & 1 \\ 2 & 0 & 1 & 1 \end{bmatrix} \begin{bmatrix} x_1 \\ x_2 \\ x_3 \\ x_4 \end{bmatrix} = \begin{bmatrix} 5 \\ 5 \\ 0 \\ 1 \end{bmatrix}, \qquad \text{......(2)}$$

the matrix a is of rank 2, and the most general solution is found to be

$$x = \{\tfrac{1}{2}, 2, 0, 0\} + \alpha\{-\tfrac{1}{2}, -2, 1, 0\} + \beta\{-\tfrac{1}{2}, -1, 0, 1\},$$

where α and β are arbitrary constants.

When a is non-singular and the solution of a system of equations $ax = h$ is required with the numbers h left general, the methods of reciprocation given in §§ 4·8–4·10 can be used immediately to give the required solution $x = a^{-1}h$. In the discussion which follows it is assumed that the numbers h have assigned values.

(b) *Solution by Method of Postmultiplication.* In this method operations are performed on columns, and it is necessary to assume the equations to be given in the form $xa = h$, where x and h now denote row matrices, and a is a given square matrix. When postmultiplied by the $n-1$ matrices M_i (see § 4·8), this equation becomes

$$x(aM_1M_2 \ldots M_{n-1}) \equiv x\sigma = h(M_1M_2 \ldots M_{n-1}).$$

Hence
$$x = h(M_1M_2 \ldots M_{n-1})\,\sigma^{-1}.$$

The following representative scheme of computation for a system of four equations is an obvious modification of the scheme given in § 4·8. The essential difference is that, whereas formerly a computation of the matrix product $M_1 M_2 \dots M_{n-1}$ was necessary, now only the row $hM_1 M_2 \dots M_{n-1}$ need be obtained. The calculation of this row is effected simultaneously with the reduction of a to σ.

h_1	h_2	h_3	h_4	—	—	—
a_{11}	a_{12}	a_{13}	a_{14}	$-\frac{a_{32}}{a_{31}}$	$-\frac{a_{33}}{a_{31}}$	$-\frac{a_{34}}{a_{31}}$
a_{21}	a_{22}	a_{23}	a_{24}	1	0	0
a_{31}^*	a_{32}	a_{33}	a_{34}	0	1	0
a_{41}	a_{42}	a_{43}	a_{44}	0	0	1
	h_2'	h_3'	h_4'	—	—	
	b_{11}	b_{12}	b_{13}	$-\frac{b_{32}}{b_{31}}$	$-\frac{b_{33}}{b_{31}}$	
	b_{21}	b_{22}	b_{23}	1	0	
	b_{31}^*	b_{32}	b_{33}	0	1	
		h_3''	h_4''	—		
		c_{11}^*	c_{12}	$-\frac{c_{12}}{c_{11}}$		
		c_{21}	c_{22}	1		
			h_4'''			
			d_{11}^*			

These calculations give the result

$$x\sigma \equiv x\begin{bmatrix} a_{11} & b_{11} & c_{11}^* & 0 \\ a_{21} & b_{21} & c_{21} & d_{11}^* \\ a_{31}^* & 0 & 0 & 0 \\ a_{41} & b_{31}^* & 0 & 0 \end{bmatrix} = [h_1, h_2', h_3'', h_4'''].$$

Commencing with the last column of σ and working backwards to the first, we can now derive the values of x_2, x_1, x_4 and x_3 in succession. Thus

$$x_2 d_{11} = h_4''' \quad \text{or} \quad x_2 = h_4'''/d_{11},$$

$$x_1 c_{11} + x_2 c_{21} = h_3'' \quad \text{or} \quad x_1 = \frac{1}{c_{11}}(h_3'' - c_{21}x_2),$$

and so on.

If the entire first column of any left-hand array—for instance $\{h_3'', c_{11}, c_{21}\}$—is null, the given equations are not linearly independent. On the other hand, if such a column is null with the exception of its first element (for instance, if $h_3'' \neq 0$ but $c_{11} = c_{21} = 0$), then the equations are incompatible.

Examples

(i) *Equations Linearly Independent.* As a simple illustration suppose the given equations to be

$$\begin{aligned} x_1 \qquad + 3x_3 + \; x_4 &= 2,\\ 4x_1 - \; x_2 + \; x_3 - 2x_4 &= 1,\\ x_1 + 3x_2 \qquad + 5x_4 &= -1,\\ 3x_1 - \; x_2 + 2x_3 + \; x_4 &= 3, \end{aligned}$$

or

$$x\begin{bmatrix} 1 & 4 & 1 & 3 \\ 0 & -1 & 3 & -1 \\ 3 & 1 & 0 & 2 \\ 1 & -2 & 5 & 1 \end{bmatrix} = [2, 1, -1, 3].$$

The calculations which follow should be compared with those in the example to § 4·8:

2	1	-1	3	—	—	—
1	4	1	3	2	-5	-1
0	-1	3	-1	1	0	0
3	1	0	2	0	1	0
1*	-2	5	1	0	0	1
	5.	-11	1	—	—	
	6	-4	2	3	-1	
	-1*	3	-1	1	0	
	7	-15	-1	0	1	
		4	-4	—		
		14*	-4	$\frac{2}{7}$		
		6	-8	1		
			$-\frac{20}{7}$			
			$-\frac{44}{7}$*			

Hence

$$[x_1, x_2, x_3, x_4]\begin{bmatrix} 1 & 6 & 14^* & 0 \\ 0 & -1^* & 0 & 0 \\ 3 & 7 & 6 & -\frac{44}{7}^* \\ 1^* & 0 & 0 & 0 \end{bmatrix} = [2, 5, 4, -\tfrac{20}{7}].$$

Accordingly

$$\begin{aligned} x_3(-\tfrac{44}{7}) &= -\tfrac{20}{7}, &\text{or}\quad x_3 &= \tfrac{5}{11};\\ x_1(14) &= 4 - 6x_3, &\text{or}\quad x_1 &= \tfrac{1}{11};\\ x_2(-1) &= 5 - 6x_1 - 7x_3, &\text{or}\quad x_2 &= -\tfrac{14}{11},\\ x_4 &= 2 - x_1 - 3x_3, &\text{or}\quad x_4 &= \tfrac{6}{11}. \end{aligned}$$

(ii) *Equations not Linearly Independent.* As an illustration we may take the set of equations (2), which can be expressed as

$$x\begin{bmatrix} 2 & 6 & 4 & 2 \\ 2 & 1 & -1 & 0 \\ 5 & 5 & 0 & 1 \\ 3 & 4 & 1 & 1 \end{bmatrix} = [5, 5, 0, 1].$$

Then

$$\begin{array}{rrrr|rrr}
5 & 5 & 0 & 1 & - & - & - \\
\hline
2^* & 6 & 4 & 2 & -3 & -2 & -1 \\
2 & 1 & -1 & 0 & 1 & 0 & 0 \\
5 & 5 & 0 & 1 & 0 & 1 & 0 \\
3 & 4 & 1 & 1 & 0 & 0 & 1 \\
\hline
 & -10 & -10 & -4 & - & - & \\
\hline
 & -5^* & -5 & -2 & -1 & -\frac{2}{5} & \\
 & -10 & -10 & -4 & 1 & 0 & \\
 & -5 & -5 & -2 & 0 & 1 & \\
\hline
 & & 0 & 0 & & & \\
\hline
 & & 0 & 0 & & & \\
 & & 0 & 0 & & & \\
\hline\hline
\end{array}$$

Hence $$[x_1, x_2, x_3, x_4]\begin{bmatrix} 2^* & 0 & 0 & 0 \\ 2 & -5^* & 0 & 0 \\ 5 & -10 & 0 & 0 \\ 3 & -5 & 0 & 0 \end{bmatrix} = [5, -10, 0, 0],$$

so that two unknowns are arbitrary (say $x_3 = \alpha$ and $x_4 = \beta$), while

$$x_2 = 2-2\alpha-\beta, \quad \text{and} \quad 2x_1 = 1-\alpha-\beta.$$

The solution is accordingly

$$x = [\tfrac{1}{2}, 2, 0, 0]+\alpha[-\tfrac{1}{2}, -2, 1, 0]+\beta[-\tfrac{1}{2}, -1, 0, 1].$$

(iii) *Equations Incompatible.* Suppose

$$[x_1, x_2, x_3]\begin{bmatrix} 1 & 2 & 1 \\ 1 & 3 & 4 \\ -2 & -1 & 7 \end{bmatrix} = [4, 1, 3].$$

Then

$$\begin{array}{rrr|rr}
4 & 1 & 3 & - & - \\
\hline
1^* & 2 & 1 & -2 & -1 \\
1 & 3 & 4 & 1 & 0 \\
-2 & -1 & 7 & 0 & 1 \\
\hline
 & -7 & -1 & - & \\
\hline
 & 1^* & 3 & -3 & \\
 & 3 & 9 & 1 & \\
\hline
 & & 20 & & \\
\hline
 & & 0^* & & \\
\hline\hline
\end{array}$$

The given equations are thus equivalent to

$$[x_1, x_2, x_3]\begin{bmatrix} 1^* & 0 & 0 \\ 1 & 1^* & 0 \\ -2 & 3 & 0^* \end{bmatrix} = [4, -7, 20],$$

and they are therefore incompatible.

(c) *Solution by Direct Operations on Rows.* For the application of this method, which is a variant of that described in § 4·10, the equations are taken in the form $ax = h$. The initial right-hand array can be taken to be the column h, and only single elements are written below on the right. If during the process a null row can be formed on the left corresponding to a non-zero element on the right, the equations are incompatible; however, if this element also is zero, the equations are consistent but not linearly independent. If q null rows with corresponding zero elements on the right can be formed, the general solution will contain q arbitrary constants.

EXAMPLES

(iv) *Equations Linearly Independent.* The equations are assumed to be those given in example (i), and the computations follow the scheme of § 4·10.

Operation	Row No.	Left Array				Column
	r_1	1	0	3	1	2
	r_2	4	−1	1	−2	1
	r_3	1	3	0	5	−1
	r_4	3	−1	2	1	3
$r_2 - 4r_1$	r_5	0	−1	−11	−6	−7
$r_3 - r_1$	r_6	0	3	−3	4	−3
$r_4 - 3r_1$	r_7	0	−1	−7	−2	−3
$3r_5 + r_6$	r_8	0	0	−36	−14	−24
$r_7 - r_5$	r_9	0	0	4	4	4
$9r_9 + r_8$	r_{10}	0	0	0	22	12
$\frac{1}{22}r_{10}$	r_{11}	0	0	0	1	$\frac{6}{11}$
$\frac{1}{4}r_9 - r_{11}$	r_{12}	0	0	1	0	$\frac{5}{11}$
$r_7 + 2r_{11}$	r_{13}	0	−1	−7	0	$-\frac{21}{11}$
$r_1 - r_{11}$	r_{14}	1	0	3	0	$\frac{16}{11}$
$r_{13} + 7r_{12}$	r_{15}	0	−1	0	0	$\frac{14}{11}$
$r_{14} - 3r_{12}$	r_{16}	1	0	0	0	$\frac{1}{11}$
r_{16}	—	1	0	0	0	$\frac{1}{11}$
$-r_{15}$	—	0	1	0	0	$-\frac{14}{11}$
r_{12}	—	0	0	1	0	$\frac{5}{11}$
r_{11}	—	0	0	0	1	$\frac{6}{11}$

Hence $\{x_1, x_2, x_3, x_4\} = \{\frac{1}{11}, -\frac{14}{11}, \frac{5}{11}, \frac{6}{11}\}$.

(v) *Equations not Linearly Independent.* If the equations are those of example (ii), the scheme of computation is

Operation	Row No.	Left Array				Column
	r_1	2	2	5	3	5
	r_2	6	1	5	4	5
	r_3	4	−1	0	1	0
	r_4	2	0	1	1	1
r_2-3r_1	r_5	0	−5	−10	−5	−10
r_3-2r_1	r_6	0	−5	−10	−5	−10
r_4-r_1	r_7	0	−2	−4	−2	−4
r_6-r_5	r_8	0	0	0	0	0
$5r_7-2r_5$	r_9	0	0	0	0	0

Since two null rows are derived, only two of the original equations are linearly independent, and these may be chosen to correspond to r_1 and r_5. Thus

$$\begin{bmatrix} 2 & 2 & 5 & 3 \\ 0 & 1 & 2 & 1 \end{bmatrix} \{x_1, x_2, x_3, x_4\} = \begin{bmatrix} 5 \\ 2 \end{bmatrix}.$$

Two of the unknowns may be assigned arbitrarily, say $x_3 = \alpha$ and $x_4 = \beta$. The equations can then be written

$$\begin{bmatrix} 2 & 2 \\ 0 & 1 \end{bmatrix} \begin{bmatrix} x_1 \\ x_2 \end{bmatrix} + \alpha \begin{bmatrix} 5 \\ 2 \end{bmatrix} + \beta \begin{bmatrix} 3 \\ 1 \end{bmatrix} = \begin{bmatrix} 5 \\ 2 \end{bmatrix},$$

and the calculations are continued as previously.

Operation	Row No.	Left Array		Column
	r_{10}	2	2	$5-5\alpha-3\beta$
	r_{11}	0	1	$2-2\alpha-\beta$
$\frac{1}{2}r_{10}-r_{11}$	r_{12}	1	0	$\frac{1}{2}-\frac{1}{2}\alpha-\frac{1}{2}\beta$
r_{11}	—	0	1	$2-2\alpha-\beta$

Hence $\{x_1, x_2\} = \{\frac{1}{2}, 2\} - \alpha\{\frac{1}{2}, 2\} - \beta\{\frac{1}{2}, 1\}$,

or $\{x_1, x_2, x_3, x_4\} = \{\frac{1}{2}, 2, 0, 0\} + \alpha\{-\frac{1}{2}, -2, 1, 0\} + \beta\{-\frac{1}{2}, -1, 0, 1\}$.

(vi) *Equations Incompatible.* Taking the same equations as those in example (iii), we obtain

Operation	Row No.	Left Array			Column
	r_1	1	1	−2	4
	r_2	2	3	−1	1
	r_3	1	4	7	3
r_2-2r_1	r_4	0	1	3	−7
r_3-r_1	r_5	0	3	9	−1
r_5-3r_4	r_6	0	0	0	20

Row r_6 shows that the equations are incompatible.

(*d*) *Iterative Methods of Solution.* In certain cases it may be possible to apply successfully an iterative method of solution. Denote the equations as $ax = h$, and let $a = v + w$, where v is chosen to be a non-singular matrix which readily admits inversion. The exact solution then is

$$x = a^{-1}h = (I + v^{-1}w)^{-1}v^{-1}h.$$

If $f = -v^{-1}w$ and $g = v^{-1}h$, the solution may be developed as

$$x = (I + f + f^2 + f^3 + \dots)g,$$

provided that this matrix series converges. The conditions for convergence are satisfied if the moduli of the latent roots of f—namely, the roots of the equation $|\lambda I - f| = 0$—are all less than unity. This equation can be written alternatively as

$$|v\lambda + w| = 0.$$

When the foregoing conditions for convergence are satisfied, the successive approximations to x, say $x(0)$, $x(1)$, $x(2)$, etc. may be taken as

$$x(0) = g,$$
$$x(1) = (I + f)g \qquad = g + fx(0),$$
$$x(2) = (I + f + f^2)g = g + fx(1),$$

and generally
$$x(r+1) = g + fx(r).$$

If an approximate solution is already known, this will naturally be used as the first approximation in place of $x(0) = g$; the computations are thereby shortened.

Two methods of solution based on this principle have been developed. A tabular method due to Morris is effectively the same as that described, with the matrix v chosen to have zero elements to the right of its principal diagonal and the same elements as a elsewhere; thus w has zero elements in and to the left of its principal diagonal, and the same elements as a elsewhere. Since v is triangular, its reciprocal can readily be found as in § 4·5. Morris has applied his method successfully to the solution of sets of linear equations such as arise in certain structural problems.* In these problems the matrix a is symmetrical, and has its origin in a quadratic form which has all its discriminants positive. It may be noted that it is possible to convert any system of linearly independent algebraic equations $ax = h$ to a system of the foregoing type by premultiplication by the matrix a' (see §§ 1·13, 1·16).

* Ref. 9. For such problems the method is always convergent.

In the second method the matrix v is chosen to be diagonal or more simply the unit matrix. In certain types of statical problem the equations arise naturally in the form $x = g + fx$, and provided the latent roots of f are all less than unity, the process described, with $v = I$, provides the solution. An illustration is given in § 10·9.

EXAMPLE

(vii) *Morris' Type of Solution.* Let the equations be

$$\begin{bmatrix} 25 & 2 & 1 \\ 2 & 10 & 1 \\ 1 & 1 & 4 \end{bmatrix} \begin{bmatrix} x_1 \\ x_2 \\ x_3 \end{bmatrix} = \begin{bmatrix} 69 \\ 63 \\ 43 \end{bmatrix},$$

for which the exact solution is $x = \{2, 5, 9\}$. Then

$$v = \begin{bmatrix} 25 & 0 & 0 \\ 2 & 10 & 0 \\ 1 & 1 & 4 \end{bmatrix}; \quad v^{-1} = \begin{bmatrix} 0\cdot04, & 0, & 0 \\ -0\cdot008, & 0\cdot1, & 0 \\ -0\cdot008, & -0\cdot025, & 0\cdot25 \end{bmatrix};$$

$$f = -v^{-1}w = \begin{bmatrix} 0, & -0\cdot08, & -0\cdot04 \\ 0, & 0\cdot016, & -0\cdot092 \\ 0, & 0\cdot016, & 0\cdot033 \end{bmatrix}; \quad g = v^{-1}h = \begin{bmatrix} 2\cdot760 \\ 5\cdot748 \\ 8\cdot623 \end{bmatrix}.$$

Hence the iterative process yields

$x(0) = g$	$fx(0)$	$x(1)$	$fx(1)$	$x(2)$	$fx(2)$	$x(3)$	$fx(3)$	$x(4)$	$fx(4)$	$x(5)$
2·760	—	—	—	—	—	—	—	—	−0·7600	2·0000
5·748	−0·7013	5·0467	−0·7472	5·0008	−0·7481	4·9999	−0·7480	5·0000	−0·7480	5·0000
8·623	0·3765	8·9995	0·3777	9·0007	0·3770	9·0000	0·3770	9·0000	0·3770	9·0000

It may be noted that until the final step is reached it is unnecessary to compute the leading element in $x(r)$, since this is always multiplied by a cipher in the iteration.

PART II. HIGH POWERS OF A MATRIX AND THE LATENT ROOTS

4·14. Preliminary Summary of Sylvester's Theorem. A high power of a matrix in general approximates to a relatively simple form, and the dominant latent roots—namely, the roots of greatest modulus—can be found from this limiting form. The method of calculation* is based on simple applications of Sylvester's theorem. For convenience of reference the simple and the confluent forms of this theorem will first be restated from §§ 3·9 and 3·10.

* Refs. 10, 11.

Simple Form of Sylvester's Theorem. If $P(u)$ denotes any polynomial of a square matrix u of order n, and if all the latent roots $\lambda_1, \lambda_2, \ldots, \lambda_n$ of u are distinct, then

$$P(u) = \sum_{r=1}^{n} P(\lambda_r)\, Z_0(\lambda_r),$$

where $Z_0(\lambda_r)$ denotes the square matrix

$$F(\lambda_r)/\overset{(1)}{\Delta}(\lambda_r) \equiv k_r \kappa_r/\overset{(1)}{\Delta}(\lambda_r) \equiv \prod_{j \neq r} (\lambda_j I - u) / \prod_{j \neq r} (\lambda_j - \lambda_r),$$

and $\overset{(1)}{\Delta}(\lambda_r)$ is an abbreviation for $\left(\dfrac{d\Delta}{d\lambda}\right)_{\lambda=\lambda_r}$.

Confluent Form of Sylvester's Theorem. If the latent roots are not all distinct, the terms contributed to $P(u)$ by a typical set of s equal roots $\lambda_1, \lambda_2, \ldots, \lambda_s$ are given by

$$T(\lambda_s) = \frac{1}{\lfloor s-1}\left[\frac{d^{s-1}}{d\lambda^{s-1}} \frac{P(\lambda)\, F(\lambda)}{\Delta_s(\lambda)}\right]_{\lambda=\lambda_s},$$

where $\Delta_s(\lambda) \equiv (\lambda - \lambda_{s+1})(\lambda - \lambda_{s+2}) \ldots (\lambda - \lambda_n)$. An alternative is

$$T(\lambda_s) = P(\lambda_s)\, Z_{s-1}(\lambda_s) + \frac{\overset{(1)}{P}(\lambda_s)}{\lfloor 1} Z_{s-2}(\lambda_s) + \ldots + \frac{\overset{(s-1)}{P}(\lambda_s)}{\lfloor s-1} Z_0(\lambda_s), \quad \ldots\ldots(1)$$

in which

$$Z_i(\lambda_s) \equiv \frac{1}{\lfloor i}\left[\frac{d^i}{d\lambda^i} \frac{F(\lambda)}{\Delta_s(\lambda)}\right]_{\lambda=\lambda_s},$$

for $i = 0, 1, \ldots, s-1$, and

$$\overset{(i)}{P}(\lambda_s) \equiv \left(\frac{d^i P(\lambda)}{d\lambda^i}\right)_{\lambda=\lambda_s}.$$

Any matrix coefficient of the type $Z_0(\lambda_s)$ either has unit rank or is null. It is of unit rank, and therefore expressible as a matrix product $k_s \kappa_s$, when the characteristic matrix $f(\lambda) \equiv \lambda I - u$ is simply degenerate for $\lambda = \lambda_s$. In other cases it is null.

4·15. Evaluation of the Dominant Latent Roots from the Limiting Form of a High Power of a Matrix. The limiting form of a high power of a matrix depends upon the number and the nature of the dominant latent roots (see also examples (i) and (ii) of § 3·9). Some of the possible cases are discussed below.

(i) *Single Dominant Real Root.* If there is a single dominant root, say λ_1, the general form of Sylvester's theorem shows that when $P(u) = u^m$ and m is large

$$\lambda_1^{-m} u^m \to Z_0(\lambda_1). \quad \ldots\ldots(1)$$

It will be convenient to express this as

$$u^m \doteqdot \lambda_1^m Z_0(\lambda_1),$$

where $\doteqdot$ denotes approximate equality. If E_m is an element occupying a particular position in u^m, it follows that

$$E_{m+1}/E_m \to \lambda_1. \qquad \text{......(2)}$$

(ii) *Repeated Dominant Real Root.* Next, let the roots of greatest modulus consist of a set of s equal roots $\lambda_1, \lambda_2, \ldots, \lambda_s$. Choose

$$P(u) = u^m P_0(u),$$

where $P_0(u)$ is a polynomial of u independent of m. When m is large

$$\lambda_s^{-m} u^m P_0(u) \to \frac{1}{\lfloor s-1} \lambda_s^{-m} \left[\frac{d^{s-1}}{d\lambda^{s-1}} \cdot \frac{\lambda^m P_0(\lambda)\, F(\lambda)}{\Delta_s(\lambda)}\right]_{\lambda=\lambda_s}. \qquad \text{......(3)}$$

If $P_0(u) = (\lambda_s I - u)^s$, all the terms in the expanded form of the expression on the right of (3) vanish, so that

$$\lambda_s^{-m} u^m (\lambda_s I - u)^s \to 0.$$

Accordingly, if again E_m is an element having a selected position in u^m, then for m large λ_s will be given approximately by the equation

$$E_m \lambda_s^s - s E_{m+1} \lambda_s^{s-1} + \frac{s(s-1)}{\lfloor 2} E_{m+2} \lambda_s^{s-2} + \ldots + (-1)^s E_{m+s} = 0. \qquad \text{......(4)}$$

Only one of the roots of this equation is the true value of λ_s. The true root is readily determinable, since it will be common to (4) and the similar equations which can be obtained from elements having other selected positions in u^m.

When the characteristic matrix has multiple degeneracy for $\lambda = \lambda_s$, equation (4) can be replaced by one of lower degree. Suppose the degeneracy to be $q(>1)$: then $F(\lambda)$ and the derived adjoint matrices up to and including $\overset{(q-2)}{F}(\lambda)$ are all null for $\lambda = \lambda_s$ (see Theorem (E) of § 3·5). Accordingly, if in (3) we choose $P_0(u) = (\lambda_s I - u)^{s-q+1}$, the terms in the expanded form of the expression on the right again vanish, and the final equation for λ_s is

$$E_m \lambda_s^{s-q+1} - (s-q+1) E_{m+1} \lambda_s^{s-q} + \frac{(s-q+1)(s-q)}{\lfloor 2} E_{m+2} \lambda_s^{s-q-1} + \ldots$$
$$+ (-1)^{s-q+1} E_{m+s-q+1} = 0.$$

In particular, when $q = s$, the root λ_s is given by

$$E_m \lambda_s - E_{m+1} = 0.$$

(iii) *Dominant Roots a Conjugate Complex Pair.* Suppose the roots of greatest modulus to be $\lambda_1 = \mu + i\omega$ and $\lambda_2 = \mu - i\omega$. Then if as before $P_0(u)$ is a polynomial independent of m,

$$u^m P_0(u) \doteqdot \lambda_1^m P_0(\lambda_1)\, Z_0(\lambda_1) + \lambda_2^m P_0(\lambda_2)\, Z_0(\lambda_2). \qquad \text{......(5)}$$

When $$P_0(u) = (\lambda_1 I - u)(\lambda_2 I - u),$$

(5) gives $$u^m(\lambda_1 I - u)(\lambda_2 I - u) \doteqdot 0,$$

or, approximately,

$$(\mu^2 + \omega^2)\, E_m - 2\mu E_{m+1} + E_{m+2} = 0. \qquad \text{......(6)}$$

The values of μ and ω can be found from (6) and a companion equation which is either the next equation in the same sequence or the corresponding equation for some other element of the matrix.

Other possible cases arise when the dominant roots embrace several distinct roots or several distinct sets of equal roots. When required, the formulae appropriate to such cases can be constructed by the methods already exemplified. The nature of the dominant root or roots is in general disclosed by the manner in which E_{m+1}/E_m varies with m. If this ratio quickly converges to a constant, then the dominant root is real and equal to this constant. If the ratio does not change sign but is very slowly convergent, there are two or more equal or nearly equal dominant roots. On the other hand, if the ratio does show changes of sign when m is large, then the dominant roots are complex.

Simplified methods for obtaining the latent roots are given in §4·17.

Examples

(i) *Single Dominant Real Root.* Suppose

$$u = \begin{bmatrix} 2 & 4 & 6 \\ 3 & 9 & 15 \\ 4 & 16 & 36 \end{bmatrix}.$$

Here, by direct multiplication,

$$u^4 = \begin{bmatrix} 72220 & 264188 & 557820 \\ 173289 & 633957 & 1338633 \\ 388448 & 1421248 & 3001248 \end{bmatrix},$$

$$u^5 = \begin{bmatrix} 3168284 & 11591692 & 24477660 \\ 7602981 & 27816897 & 58739877 \\ 17045632 & 62364992 & 131694336 \end{bmatrix}.$$

The ratios of the elements of u^5 to the corresponding elements of u^4 are

$$\begin{matrix} 43{\cdot}8699, & 43{\cdot}8767, & 43{\cdot}8809, \\ 43{\cdot}8746, & 43{\cdot}8782, & 43{\cdot}8805, \\ 43{\cdot}8814, & 43{\cdot}8804, & 43{\cdot}8799. \end{matrix}$$

By (2) these ratios, whose mean is 43·8781, should give an approximate value for the dominant root. The actual latent roots are 43·8800, 2·71747, and 0·402541, so that although the powers of u employed are very low, the approximation to the dominant root is quite good. The separation of this root from the adjacent root is, however, unusually great in the present example. Generally, the approximation is not so rapid.

(ii) *Once Repeated Dominant Real Root.* The typical equation giving the latent root in this case is (see (4))

$$E_m \lambda^2 - 2E_{m+1}\lambda + E_{m+2} = 0.$$

Suppose
$$u = \begin{bmatrix} 4 & 1 & 2 \\ 2 & 4 & -3 \\ 3 & 1 & 3 \end{bmatrix},$$

of which the latent roots are 5, 5, 1. It is found that

$$u^8 = \begin{bmatrix} 952149 & 625000 & 63476 \\ -463868 & -234375 & -161132 \\ 952148 & 625000 & 63477 \end{bmatrix},$$

$$u^9 = \begin{bmatrix} 5249024 & 3515625 & 219726 \\ -2807618 & -1562500 & -708007 \\ 5249023 & 3515625 & 219727 \end{bmatrix},$$

$$u^{10} = \begin{bmatrix} 28686524 & 19531250 & 610351 \\ -16479493 & -9765625 & -3051757 \\ 28686523 & 19531250 & 610352 \end{bmatrix}.$$

The leading diagonal elements give

$$952149\lambda^2 - 2(5249024)\lambda + 28686524 = 0,$$

so that
$$\lambda = 5{\cdot}000008 \quad \text{or} \quad 6{\cdot}025628.$$

Similarly the last diagonal elements yield

$$\lambda = 4{\cdot}999966 \quad \text{or} \quad 1{\cdot}923078.$$

Hence $\lambda = 5$.

(iii) *Unrepeated Conjugate Complex Dominant Roots.* Let

$$u = \begin{bmatrix} 26 & -54 & 4 \\ 13 & -28 & 3 \\ 26 & -56 & 5 \end{bmatrix},$$

of which the latent roots can be shown to be 1 and $1 \pm 5i$. Here

$$u^8 = \begin{bmatrix} -2280668 & 4933944 & -372606 \\ -1184118 & 2645932 & -277695 \\ -2372214 & 5117340 & -372911 \end{bmatrix},$$

$$u^9 = \begin{bmatrix} -4843852 & 5871576 & 3816130 \\ -3610022 & 5407196 & 1812849 \\ -4847830 & 5697052 & 3998609 \end{bmatrix},$$

$$u^{10} = \begin{bmatrix} 49609716 & -116539400 & 17319970 \\ 23567050 & -57979844 & 10845745 \\ 51981930 & -121656740 & 17692881 \end{bmatrix}.$$

If E is the first element in the top row, equation (6) becomes

$$-2280668(\mu^2+\omega^2)+2(4843852)\,\mu+49609716 = 0,$$

while if E is the second element in the top row

$$4933944(\mu^2+\omega^2)-2(5871576)\,\mu-116539400 = 0.$$

These equations yield $\mu^2+\omega^2 = 25{\cdot}999977$, and $2\mu = 1{\cdot}999979$. Hence $\mu = 1$ and $\omega = 5$ with an error of less than 1 in 10^4.

4·16. Evaluation of the Matrix Coefficients Z for the Dominant Roots. It is sometimes required to calculate not only the latent roots but also the corresponding matrix coefficients Z in Sylvester's expansion for $P(u)$. In dynamical applications, for instance, these coefficients effectively determine the modes of the constituent motions. It will now be shown how the matrices Z appropriate to the dominant root or roots can be found, once these roots have been determined.

(i) *Single Dominant Real Root.* If λ_1 is the dominant root, then by (4·15·1)

$$\lambda_1^{-m}u^m \to Z_0(\lambda_1).$$

(ii) *Repeated Dominant Real Root.* Suppose firstly that the dominant roots consist of two equal roots λ_1, λ_2. In this case, when $P(u) = u^m$ and m is large, (4·14·1) gives

$$\lambda_2^{-m+1}u^m \to \lambda_2 Z_1(\lambda_2)+mZ_0(\lambda_2). \qquad \text{......(1)}$$

The next relation of the sequence is

$$\lambda_2^{-m} u^{m+1} \to \lambda_2 Z_1(\lambda_2) + (m+1)\, Z_0(\lambda_2),$$

so that

$$-\lambda_2^{-m} u^m (\lambda_2 I - u) \to Z_0(\lambda_2), \qquad \text{......(2)}$$

and

$$\lambda_2^{-m-1} u^m [\lambda_2 I + m(\lambda_2 I - u)] \to Z_1(\lambda_2). \qquad \text{......(3)}$$

When $\lambda_2 I - u$ is doubly degenerate, $Z_0(\lambda_2)$ is null, and then (1) gives at once

$$\lambda_2^{-m} u^m \to Z_1(\lambda_2).$$

The derivation of the appropriate formulae is rather more troublesome if the roots of greatest modulus consist of s equal roots $\lambda_1, \lambda_2, \ldots, \lambda_s$. In this case, taking $P(u) = u^m P_0(u)$ and assuming m large, we obtain by (4·14·1)

$$u^m P_0(u) \doteqdot \lambda_s^m P_0(\lambda_s)\, Z_{s-1}(\lambda_s) + \frac{1}{\underline{|1}} \left(\frac{d}{d\lambda} \lambda^m P_0(\lambda) \right)_{\lambda=\lambda_s} Z_{s-2}(\lambda_s) + \ldots$$
$$+ \frac{1}{\underline{|s-1}} \left(\frac{d^{s-1}}{d\lambda^{s-1}} \lambda^m P_0(\lambda) \right)_{\lambda=\lambda_s} Z_0(\lambda_s).$$

To isolate any particular matrix coefficient, say $Z_i(\lambda_s)$, it is necessary to choose $P_0(u)$ in such a way that $\dfrac{d^p}{d\lambda^p}(\lambda^m P_0(\lambda))$ vanishes when $\lambda = \lambda_s$ and $p \neq s - i - 1$, but does not vanish when $\lambda = \lambda_s$ and $p = s - i - 1$. It can be verified that the appropriate polynomial $P_0(u, i)$ in this case is given by

$$(\lambda_s I - u)^{s-i-1} \Bigg[\lambda_s^i I + m \lambda_s^{i-1} (\lambda_s I - u) + \frac{m(m+1)}{\underline{|2}} \lambda_s^{i-2} (\lambda_s I - u)^2 + \ldots$$
$$+ \frac{m(m+1)\ldots(m+i-1)}{\underline{|i}} (\lambda_s I - u)^i \Bigg],$$

where $i \leqslant s - 1$. The matrix $Z_i(\lambda_s)$ is then given by the formula

$$(-1)^{s-i-1} \lambda_s^{-m-i} u^m P_0(u, i) \to Z_i(\lambda_s).$$

As for the once repeated dominant root, the formulae can be simplified if $\lambda_s I - u$ is multiply degenerate.

(iii) *Dominant Roots a Conjugate Complex Pair*. In this case, when $P_0(u) = \lambda_2 I - u$ in (4·15·5),

$$u^m(\lambda_2 I - u)/\lambda_1^m(\lambda_2 - \lambda_1) \to Z_0(\lambda_1). \qquad \text{......(4)}$$

In general $Z_0(\lambda_1)$ is complex, and $Z_0(\lambda_2)$ is its conjugate.

EXAMPLES

(i) *Single Dominant Real Root.* Assuming u to be as for example (i) of § 4·15, and applying (4·15·1) with $m = 5$ and $\lambda_1 = 43{\cdot}8781$ (the mean computed dominant root), we obtain the approximation

$$Z_0(\lambda_1) = \lambda_1^{-5}u^5$$
$$= \begin{bmatrix} 0{\cdot}0194799 & 0{\cdot}0712705 & 0{\cdot}150499 \\ 0{\cdot}0467463 & 0{\cdot}171030 & 0{\cdot}361157 \\ 0{\cdot}104804 & 0{\cdot}383446 & 0{\cdot}809711 \end{bmatrix}.$$

The value of $Z_0(\lambda_1)$ computed directly from $F(\lambda_1)/\overset{(1)}{\Delta}(\lambda_1)$ for the true root is

$$\begin{bmatrix} 0{\cdot}0194753 & 0{\cdot}0712545 & 0{\cdot}150466 \\ 0{\cdot}0467356 & 0{\cdot}170992 & 0{\cdot}361078 \\ 0{\cdot}104781 & 0{\cdot}383361 & 0{\cdot}809532 \end{bmatrix}.$$

(ii) *Once Repeated Dominant Real Root.* Suppose u to be the matrix in example (ii) of § 4·15, and apply formulae (2) and (3), with $m = 8$ and $\lambda_2 = 5$. The approximate results work out as

$$Z_0(\lambda_2) = \begin{bmatrix} 1{\cdot}249994 & 1{\cdot}000000 & -0{\cdot}249994 \\ -1{\cdot}249992 & -1{\cdot}000000 & 0{\cdot}249992 \\ 1{\cdot}250004 & 1{\cdot}000000 & -0{\cdot}250004 \end{bmatrix},$$

and
$$Z_1(\lambda_2) = \begin{bmatrix} 0{\cdot}437511 & 0 & 0{\cdot}562489 \\ 0{\cdot}812485 & 1{\cdot}0 & -0{\cdot}812485 \\ 0{\cdot}437492 & 0 & 0{\cdot}562508 \end{bmatrix},$$

while the accurate values are

$$Z_0(\lambda_2) = \begin{bmatrix} 1{\cdot}25 & 1{\cdot}0 & -0{\cdot}25 \\ -1{\cdot}25 & -1{\cdot}0 & 0{\cdot}25 \\ 1{\cdot}25 & 1{\cdot}0 & -0{\cdot}25 \end{bmatrix},$$

and
$$Z_1(\lambda_2) = \begin{bmatrix} 0{\cdot}4375 & 0 & 0{\cdot}5625 \\ 0{\cdot}8125 & 1{\cdot}0 & -0{\cdot}8125 \\ 0{\cdot}4375 & 0 & 0{\cdot}5625 \end{bmatrix}.$$

4·17. Simplified Iterative Methods. It has been shown in §§ 4·15 and 4·16 that the dominant latent roots of a matrix u and the corresponding matrix coefficients Z can be evaluated by raising u to a high power. The same result can be achieved, and usually with less labour, by repeated premultiplications of an arbitrary column by u, or alternatively by repeated postmultiplications of an arbitrary row by u.

To illustrate the method, consider firstly the simple case where there is a single real dominant root λ_1. Let $x(0)$ be an arbitrary column of n elements. Then by (4·15·1)

$$u^m x(0) \doteqdot \lambda_1^m Z_0(\lambda_1)\, x(0).$$

This may be written as

$$u^m x(0) \doteqdot \lambda_1^m k_1 \kappa_1 x(0)/\overset{(1)}{\Delta}(\lambda_1),$$

or, if Φ denotes the scalar factor $\kappa_1 x(0)/\overset{(1)}{\Delta}(\lambda_1)$, as

$$u^m x(0) \doteqdot \lambda_1^m \Phi k_1.$$

Accordingly, unless $x(0)$ happens to be so chosen that Φ vanishes, λ_1 is given by the ratio of corresponding elements in $u^{m+1}x(0)$ and $u^m x(0)$. The formula (4·15·2) is therefore still applicable if E_m represents one of the elements of $u^m x(0)$. Moreover, the modal column k_1 is proportional to $u^m x(0)$.

Again, if $y(0)$ denotes an arbitrary row of n elements, then

$$y(0)\, u^m \doteqdot \lambda_1^m \Psi \kappa_1,$$

where $$\Psi = y(0)\, k_1/\overset{(1)}{\Delta}(\lambda_1).$$

Hence continued postmultiplication of $y(0)$ by u yields a row, the elements of which are proportional to the corresponding elements of κ_1. Also λ_1 can be derived as before.

A further case of importance is where the dominant roots are conjugate complex and unrepeated. Equation (4·15·6) is clearly still applicable if E_m is interpreted to mean an element of the column $C_m \equiv u^m x(0)$, and in this case by (4·16·4)

$$Z_0(\lambda_1)\, x(0) \equiv k_1 \kappa_1 x(0)/\overset{(1)}{\Delta}(\lambda_1) \doteqdot (\lambda_2 C_m - C_{m+1})/\lambda_1^m(\lambda_2 - \lambda_1).$$

The modal column k_1 is thus proportional to

$$\lambda_2 C_m - C_{m+1}.$$

Similarly, if E_m is an element of the row $R_m \equiv y(0)\, u^m$, the row κ_1 is proportional to

$$\lambda_2 R_m - R_{m+1}.$$

Another application of the iterative method, which may be mentioned here, is to the construction of the characteristic equation of a matrix u. If this equation is written

$$\lambda^n + p_1 \lambda^{n-1} + p_2 \lambda^{n-2} + \ldots + p_n = 0,$$

then by the Cayley-Hamilton theorem

$$u^n + p_1 u^{n-1} + p_2 u^{n-2} + \ldots + p_n I = 0.$$

Postmultiplying this equation by an arbitrary column $x(0)$ and writing $ux(0) = x(1)$, $u^2x(0) = ux(1) = x(2)$, etc., we obtain

$$x(n) + p_1 x(n-1) + p_2 x(n-2) + \ldots + p_n x(0) = 0.$$

This yields n simultaneous scalar equations for the n unknown coefficients p.

Examples

(i) *Determination of the Dominant Latent Root.* Consider the matrix u of example (i) of §4·15. If $\{0, 0, 1\}$ is chosen as an arbitrary column, the first premultiplication gives

$$\begin{bmatrix} 2 & 4 & 6 \\ 3 & 9 & 15 \\ 4 & 16 & 36 \end{bmatrix} \begin{bmatrix} 0 \\ 0 \\ 1 \end{bmatrix} = \begin{bmatrix} 6 \\ 15 \\ 36 \end{bmatrix}.$$

It is convenient, as each new column is obtained, to extract a scalar factor so as to reduce a certain element (say the last) to unity. The iterative process can then be tabulated as follows:

Initial column	Iteration number									
	1		2		3		4		5	
0	6	0·1667	8·00	0·1846	8·15	0·1859	8·16	0·1860	8·16	0·1860
0	15	0·4167	19·25	0·4443	19·55	0·4458	19·57	0·4460	19·57	0·4460
1	36	1·0	43·33	1·0	43·85	1·0	43·88	1·0	43·88	1·0

The fifth iteration repeats the fourth. Hence

$$\begin{bmatrix} 2 & 4 & 6 \\ 3 & 9 & 15 \\ 4 & 16 & 36 \end{bmatrix} \begin{bmatrix} 0·1860 \\ 0·4460 \\ 1·0 \end{bmatrix} = 43·88 \begin{bmatrix} 0·1860 \\ 0·4460 \\ 1·0 \end{bmatrix},$$

and 43·88 is therefore the dominant latent root, and $\{0·1860, 0·4460, 1·0\}$ is the associated modal column.

(ii) *Construction of the Characteristic Equation.* The characteristic equation of the matrix used in example (i) can be found by the direct iteration

$$u\{0, 0, 1\} = \{6, 15, 36\},$$

$$u\{6, 15, 36\} = \{288, 693, 1560\},$$

$$u\{288, 693, 1560\} = \{12708, 30501, 68400\}.$$

Hence
$$\begin{bmatrix}12708\\30501\\68400\end{bmatrix}+p_1\begin{bmatrix}288\\693\\1560\end{bmatrix}+p_2\begin{bmatrix}6\\15\\36\end{bmatrix}+p_3\begin{bmatrix}0\\0\\1\end{bmatrix}=0,$$

or
$$\begin{bmatrix}288, & 6, & 0\\693, & 15, & 0\\1560, & 36, & 1\end{bmatrix}\begin{bmatrix}p_1\\p_2\\p_3\end{bmatrix}=\begin{bmatrix}-12708\\-30501\\-68400\end{bmatrix}.$$

The solution, which is readily found by any of the methods of § 4·13, is

$$\{p_1, p_2, p_3\} = \{-47, 138, -48\}.$$

The characteristic equation of u is therefore

$$\lambda^3-47\lambda^2+138\lambda-48 = 0.$$

4·18. Computation of the Non-Dominant Latent Roots. When the dominant latent root or roots have been found, the remaining roots can be obtained successively in the order of their moduli by an extension of the methods already explained. The essence of the extension consists in the construction of a modified matrix v which possesses all the latent roots of u except the dominant roots.

For simplicity, suppose there to be a single dominant root λ_1, and assume that λ_1 and the corresponding row κ_1 have already been calculated. Then, if λ_s is any other root, we have by (3·8·11)

$$\kappa_1 k_s = 0. \qquad \text{......(1)}$$

Let κ_{r1} be any non-zero element of κ_1, and denote as w the square matrix which has κ_1/κ_{r1} for its rth row and its remaining $n-1$ rows all null. Then in view of (1) we can write $k_s = (I-w)k_s$. But $(\lambda_s I-u)k_s = 0$: hence also

$$(\lambda_s I-v)k_s = 0, \qquad \text{......(2)}$$

where
$$v \equiv u(I-w) \equiv u-\frac{1}{\kappa_{r1}}\{u_{ir}\}\kappa_1. \qquad \text{......(3)}$$

From (2) it is clear that $\lambda_2, \lambda_3, \ldots, \lambda_n$ are latent roots of v, and that $k_2, k_3, \ldots, k_n$ are the corresponding modal columns* of v. The remaining latent root of v is $\lambda = 0$: this follows from the fact that the rth column of v is null, as is obvious from (3). When v has been obtained, the dominant latent root of this matrix (which is necessarily the subdominant latent root of u) can be calculated by the methods of §§ 4·15

* It is also easily shown that the sth row of the matrix κ appropriate to v is given by $\dfrac{\kappa_1}{\kappa_{r1}}-\dfrac{\kappa_s}{\kappa_{rs}}$.

or 4·17. Evidently all the latent roots of u can be found successively by a continuation of the process.

An alternative procedure in the calculation of the subdominant root is to omit the rth column and the rth row of v, and to apply the iterative method to the matrix of reduced order so derived. When this treatment is adopted the typical modal column k_s will be determined apart from the element k_{rs}, which will have to be calculated independently from (1).

If there are two distinct dominant roots λ_1 and λ_2, and if κ_{r1}, κ_{q2} $(r \neq q)$ are, respectively, non-zero elements of κ_1 and κ_2, the matrix w must be constructed to have κ_1/κ_{r1} and κ_2/κ_{q2} for its rth and qth rows respectively, and to have its remaining $n-2$ rows all null. The matrix $v = u(I-w)$ in this case has its rth and qth columns null, and has therefore two zero latent roots. The remaining latent roots are $\lambda_3, \lambda_4, \ldots, \lambda_n$, and the corresponding modal columns are $k_3, k_4, \ldots, k_n$. The method of extension to cases of three or more distinct dominant roots will be obvious.

It may be noted that in the special case of a complex pair of dominant roots λ_1, λ_2, since κ_1 is proportional to the row $\lambda_2 R_m - R_{m+1}$, and similarly κ_2 is proportional to the row $\lambda_1 R_m - R_{m+1}$ (see § 4·17), the two equations corresponding to (1) yield $R_m k_s = 0$ and $R_{m+1} k_s = 0$. Hence the elements E can be used directly in place of the elements κ in the construction of the matrix v.

Example

Computation of the Subdominant Latent Root. The method will be used to determine the subdominant latent root of the matrix u of example (i) of § 4·17.

Firstly, repeated postmultiplication of an arbitrary row by u yields

$$[0{\cdot}1294, 0{\cdot}4736, 1{\cdot}0]\begin{bmatrix}2 & 4 & 6\\ 3 & 9 & 15\\ 4 & 16 & 36\end{bmatrix} = 43{\cdot}88\,[0{\cdot}1294, 0{\cdot}4736, 1{\cdot}0].$$

Hence $\kappa_1 = [0{\cdot}1294, 0{\cdot}4736, 1{\cdot}0]$, and choosing $r = 1$ in (3), we obtain

$$v = \begin{bmatrix}2 & 4 & 6\\ 3 & 9 & 15\\ 4 & 16 & 36\end{bmatrix} - \begin{bmatrix}2\\ 3\\ 4\end{bmatrix}[1, 3{\cdot}660, 7{\cdot}728]$$

$$= \begin{bmatrix}0, & -3{\cdot}320, & -9{\cdot}456\\ 0, & -1{\cdot}980, & -8{\cdot}184\\ 0, & 1{\cdot}360, & 5{\cdot}088\end{bmatrix}.$$

The results of the iterative process, with v as premultiplier and $\{0, 0, 1\}$ as an arbitrary column, are given below. It should be noted particularly that, except in the last step, the top elements of the columns need not be calculated.

Initial column	Iteration number					
	1		2		3	
0 0 1	— −8·184 5·088	— −1·608 1·0	— −5·000 2·901	— −1·724 1·0	— −4·770 2·743	— −1·739 1·0

Iteration number					
4		5		6	
— −4·741 2·723	— −1·741 1·0	— −4·737 2·720	— −1·742 1·0	−3·673 −4·735 2·719	−1·351 −1·742 1·0

Since the sixth iteration repeats the fifth, the top element is computed, and the value 2·719 is deduced for the latent root. The corresponding modal column is proportional to $\{-1{\cdot}351, -1{\cdot}742, 1{\cdot}0\}$.

4·19. Upper Bounds to the Powers of a Matrix. Formulae giving upper bounds to the powers of a matrix are sometimes required, and we shall now obtain some simple formulae of this type. For definiteness, suppose u to be a square matrix of order 3 with real or complex elements, and write the typical element of u^s as

$$u_{ij}(s) \equiv U_{ij}(s)\,\alpha,$$

where $U_{ij}(s)$ is the essentially positive modulus, and α has unit modulus. The value of α in each case will not be required, so that the suffices which should strictly be associated with this symbol are omitted. Further, denote as r_i, c_j the sum of the moduli of the elements in the ith row and the jth column, respectively, of u, and let R_i, C_j be, respectively, the greatest modulus in the ith row and jth column.

Then

$$u_{11}(2) = U_{11}(1)\,\alpha\, U_{11}(1)\,\alpha + U_{12}(1)\,\alpha\, U_{21}(1)\,\alpha + U_{13}(1)\,\alpha\, U_{31}(1)\,\alpha,$$

so that $$U_{11}(2) \leqslant U_{11}^2(1) + U_{12}(1)\,U_{21}(1) + U_{13}(1)\,U_{31}(1).$$

Hence $U_{11}(2) \leqslant r_1 C_1$ and also $\leqslant R_1 c_1$. The remaining elements of u^2 may be treated on a similar basis. It follows that the moduli of the elements of u^2 are all less than the corresponding elements of both the matrices

$$A \equiv \begin{bmatrix} r_1C_1 & r_1C_2 & r_1C_3 \\ r_2C_1 & r_2C_2 & r_2C_3 \\ r_3C_1 & r_3C_2 & r_3C_3 \end{bmatrix} \quad \text{and} \quad B \equiv \begin{bmatrix} R_1c_1 & R_1c_2 & R_1c_3 \\ R_2c_1 & R_2c_2 & R_2c_3 \\ R_3c_1 & R_3c_2 & R_3c_3 \end{bmatrix}.$$

Proceeding next to u^3, and using A, we have

$$u^3 \equiv \begin{bmatrix} r_1C_1\beta & r_1C_2\beta & r_1C_3\beta \\ r_2C_1\beta & r_2C_2\beta & r_2C_3\beta \\ r_3C_1\beta & r_3C_2\beta & r_3C_3\beta \end{bmatrix} \begin{bmatrix} U_{11}(1)\alpha & U_{12}(1)\alpha & U_{13}(1)\alpha \\ U_{21}(1)\alpha & U_{22}(1)\alpha & U_{23}(1)\alpha \\ U_{31}(1)\alpha & U_{32}(1)\alpha & U_{33}(1)\alpha \end{bmatrix},$$

where β represents a quantity of modulus not exceeding unity. Hence, for instance, $U_{11}(3) \leqslant (C_1+C_2+C_3)\,r_1C_1$, and $U_{21}(3) \leqslant (C_1+C_2+C_3)\,r_2C_1$. This shows that the moduli of the elements of u^3 are dominated by the corresponding elements of $(C_1+C_2+C_3)\,A$, or briefly u^3 is dominated by $(C_1+C_2+C_3)\,A$.

Again, using B, we obtain

$$u^3 = \begin{bmatrix} U_{11}(1)\alpha & U_{12}(1)\alpha & U_{13}(1)\alpha \\ U_{21}(1)\alpha & U_{22}(1)\alpha & U_{23}(1)\alpha \\ U_{31}(1)\alpha & U_{32}(1)\alpha & U_{33}(1)\alpha \end{bmatrix} \begin{bmatrix} R_1c_1\beta & R_1c_2\beta & R_1c_3\beta \\ R_2c_1\beta & R_2c_2\beta & R_2c_3\beta \\ R_3c_1\beta & R_3c_2\beta & R_3c_3\beta \end{bmatrix}.$$

Consequently $U_{11}(3) \leqslant (R_1+R_2+R_3)\,R_1c_1$ etc., so that u^3 is also dominated by $(R_1+R_2+R_3)\,B$.

The argument can be extended to show that u^s is dominated by the two matrices

$$A_s = (C_1+C_2+C_3)^{s-2}\,A,$$
$$B_s = (R_1+R_2+R_3)^{s-2}\,B.$$

The corresponding theorem for a square matrix u of general order n is as follows: Let r_i, c_j be the sum of the moduli in the ith row and the jth column of u, respectively, and R_i, C_j be the greatest modulus in the ith row, and in the jth column, respectively. Then the moduli of the elements of u^s will not exceed the corresponding elements in the two matrices

$$A_s \equiv (C_1+C_2+\ldots+C_n)^{s-2}\{r_1, r_2, \ldots, r_n\}\,[C_1, C_2, \ldots, C_n],$$
$$B_s \equiv (R_1+R_2+\ldots+R_n)^{s-2}\{R_1, R_2, \ldots, R_n\}\,[c_1, c_2, \ldots, c_n],$$

where $s \geqslant 2$. Clearly, for the upper bound to the modulus of any specific element of u^s the smaller of the two corresponding values given by A_s and B_s can be selected.

From these results another upper bound can be derived which, though considerably cruder, is often sufficient in the discussion of the convergence of a matrix series. Thus, if U is the greatest modulus in u, then $r_i < nU$ and $c_j < nU$. Hence u^s is *a fortiori* dominated by

$$n^{s-2}U^{s-2}\{nU, nU, \ldots, nU\}[U, U, \ldots, U].$$

An upper bound to the modulus of any element of u^s is therefore $n^{s-1}U^s$.

Examples

Take
$$u = \begin{bmatrix} 1 & 0 & -2 \\ 3 & -3 & 1 \\ 0 & 2 & -1 \end{bmatrix}.$$

Then
$$r_1 = 3, \quad c_1 = 4, \quad R_1 = 2, \quad C_1 = 3.$$
$$r_2 = 7, \quad c_2 = 5, \quad R_2 = 3, \quad C_2 = 3.$$
$$r_3 = 3, \quad c_3 = 4, \quad R_3 = 2, \quad C_3 = 2.$$

Hence
$$A_6 = 8^4\{3, 7, 3\}[3, 3, 2] = \begin{bmatrix} 36864 & 36864 & 24576 \\ 86016 & 86016 & 57344 \\ 36864 & 36864 & 24576 \end{bmatrix},$$

$$B_6 = 7^4\{2, 3, 2\}[4, 5, 4] = \begin{bmatrix} 19208 & 24010 & 19208 \\ 28812 & 36015 & 28812 \\ 19208 & 24010 & 19208 \end{bmatrix}.$$

In this case B_6 offers the smaller bounds to the moduli of the elements in u^6. In actual fact

$$u^6 = \begin{bmatrix} 553, & -948, & 600 \\ -2322, & 4123, & -2670 \\ 1422, & -2496, & 1627 \end{bmatrix}.$$

It may be noted that the bounds given by B_6 are much lower than the value 177,147 ($= 3^5 3^6$) obtained from $n^{s-1}U^s$.

PART III. ALGEBRAIC EQUATIONS OF GENERAL DEGREE

4·20. Solution of Algebraic Equations and Adaptation of Aitken's Formulae. It is easy to verify that the equation

$$\lambda^n + p_1\lambda^{n-1} + \dots + p_{n-1}\lambda + p_n = 0 \qquad \text{......(1)}$$

is the characteristic equation of the matrix

$$u = \begin{bmatrix} 0, & 1, & 0, & \dots, & 0, & 0 \\ 0, & 0, & 1, & \dots, & 0, & 0 \\ \dots & \dots & \dots & \dots & \dots & \dots \\ 0, & 0, & 0, & \dots, & 0, & 1 \\ -p_n, & -p_{n-1}, & -p_{n-2}, & \dots, & -p_2, & -p_1 \end{bmatrix}.$$

The methods of Part II can be used to obtain the latent roots of u, and therefore the roots of (1).

It has been shown in example (vii) of § 3·7 that, for any square matrix u whose characteristic equation is (1), the elements E having a specified position in the matrices $Y(s) = u^{s+n}$ satisfy the difference equation

$$E(s) + p_1 E(s-1) + \dots + p_n E(s-n) = 0. \qquad \text{......(2)}$$

Consider firstly the simple product $ux(0)$, where $x(0)$ is a column of arbitrary elements $x_1, x_2, \dots, x_n$. This yields

$$ux(0) = \{x_2, x_3, \dots, x_n, x_{n+1}\},$$

where $$-x_{n+1} = p_1 x_n + p_2 x_{n-1} + \dots + p_{n-1}x_2 + p_n x_1. \qquad \text{......(3)}$$

The value of x_{n+1} given by (3) is precisely the value of $E(s)$ given by the difference equation (2) if

$$E(s-1) = x_n,\ E(s-2) = x_{n-1},\ \dots,\ E(s-n) = x_1.$$

A second premultiplication by u leads to $\{x_3, x_4, \dots, x_{n+2}\}$, where x_{n+2} is obtained from $x_{n+1}, x_n, \dots, x_2$ in the same way that x_{n+1} is obtained from $x_n, x_{n-1}, \dots, x_1$. Hence repeated premultiplication of $x(0)$ by u generates a single sequence of elements x which are the successive terms in that solution of the difference equation which begins with the set of numbers $x_1, x_2, \dots, x_n$. The theory given in Part II shows that the dominant root or roots of (1) can be obtained from the members of this sequence. When the dominant root λ_1 is real its value is the limit of x_{p+1}/x_p when p is large. This is the method of solution invented by Daniel Bernoulli in 1728.

Bernoulli's method of solution of algebraic equations in its original form is only applicable when the dominant root is real, but an extension has been developed by Aitken in which this restriction is removed.* Aitken gives a number of useful formulae, and, as these are all consequences of the fact that x_s is given by an expression of the form (3), they can be immediately adapted to the case where the sequence of elements results from repeated premultiplication of an arbitrary column by any square matrix u. The most useful of these formulae can be written

$$\lambda_1\lambda_2\ldots\lambda_p = D_{s+1}/D_s, \qquad \ldots\ldots(4)$$

where

$$D_s \equiv \begin{vmatrix} E_s & E_{s+1} & \cdots & E_{s+p-1} \\ E_{s+1} & E_{s+2} & \cdots & E_{s+p} \\ \cdots & \cdots & \cdots & \cdots \\ E_{s+p-1} & E_{s+p} & \cdots & E_{s+2p-2} \end{vmatrix},$$

and E_r represents the value of an element in some arbitrary fixed position in u^r, $u^r x$, or yu^r. It is assumed that the latent roots are arranged in the descending order of their moduli, and that s is so large that the contributions to the elements E due to the roots following λ_p are negligible. Aitken gives a scheme based on this formula for successive approximations to the values of λ_1, $\lambda_1\lambda_2$, $\lambda_1\lambda_2\lambda_3$, etc. (see also § 10·11).

The formula (4) can be proved as follows. By equation (3·7·4) the expression for the general element is approximately

$$E_r = \sum_{q=1}^{p} e_q\lambda_q^r.$$

Accordingly

$$D_s = \begin{vmatrix} \lambda_1^s & \lambda_2^s & \cdots & \lambda_p^s \\ \lambda_1^{s+1} & \lambda_2^{s+1} & \cdots & \lambda_p^{s+1} \\ \cdots & \cdots & \cdots & \cdots \\ \lambda_1^{s+p-1} & \lambda_2^{s+p-1} & \cdots & \lambda_p^{s+p-1} \end{vmatrix} \times \begin{vmatrix} e_1 & \lambda_1 e_1 & \cdots & \lambda_1^{p-1}e_1 \\ e_2 & \lambda_2 e_2 & \cdots & \lambda_2^{p-1}e_2 \\ \cdots & \cdots & \cdots & \cdots \\ e_p & \lambda_p e_p & \cdots & \lambda_p^{p-1}e_p \end{vmatrix}$$

and

$$D_{s+1} = \begin{vmatrix} \lambda_1^{s+1} & \lambda_2^{s+1} & \cdots & \lambda_p^{s+1} \\ \lambda_1^{s+2} & \lambda_2^{s+2} & \cdots & \lambda_p^{s+2} \\ \cdots & \cdots & \cdots & \cdots \\ \lambda_1^{s+p} & \lambda_2^{s+p} & \cdots & \lambda_p^{s+p} \end{vmatrix} \times \begin{vmatrix} e_1 & \lambda_1 e_1 & \cdots & \lambda_1^{p-1}e_1 \\ e_2 & \lambda_2 e_2 & \cdots & \lambda_2^{p-1}e_2 \\ \cdots & \cdots & \cdots & \cdots \\ e_p & \lambda_p e_p & \cdots & \lambda_p^{p-1}e_p \end{vmatrix}.$$

Hence

$$D_{s+1}/D_s = \lambda_1\lambda_2\ldots\lambda_p.$$

* Ref. 12.

When $p = 2$ and λ_1 and λ_2 are conjugate complexes $\mu \pm i\omega$, the formula becomes

$$\lambda_1\lambda_2 = \mu^2+\omega^2 = \frac{\begin{vmatrix} E_{s+1} & E_{s+2} \\ E_{s+2} & E_{s+3} \end{vmatrix}}{\begin{vmatrix} E_s & E_{s+1} \\ E_{s+1} & E_{s+2} \end{vmatrix}}. \qquad \text{......(5)}$$

This can also be deduced from (4·15·6) and a consecutive equation obtained by substituting $m+1$ for m. Another useful formula, which is deducible by a similar method, is

$$\lambda_1+\lambda_2 = \frac{\begin{vmatrix} E_s & E_{s+2} \\ E_{s+1} & E_{s+3} \end{vmatrix}}{\begin{vmatrix} E_s & E_{s+1} \\ E_{s+1} & E_{s+2} \end{vmatrix}}. \qquad \text{......(6)}$$

When λ_1 and λ_2 are conjugate complex quantities, this equation yields a value for 2μ which agrees with that given by equation (4·15·6) and the consecutive equation.

4·21. General Remarks on Iterative Methods. The latent roots of a matrix are normally calculated by expansion of the characteristic function and solution of the resulting algebraic equation by the usual methods. However, when the order of the matrix is high, this process is excessively laborious, and the methods given in preceding sections become advantageous, more especially when only a few of the roots with greatest moduli are required.* Repeated multiplications by a matrix are readily performed with any ordinary calculating machine, and a machine could no doubt be devised to perform most of the necessary operations automatically.

The rapidity of the convergence in the matrix method depends acutely on the separation of the moduli of the latent roots. This can obviously be increased by using some positive power of the matrix in place of the matrix itself, but this does not always result in a saving of labour. For example, when the original matrix is of a simple form, the simplicity will probably not be wholly preserved in its powers. The modal columns derived from a power of the matrix are the same as those appropriate to the matrix itself, and the latent roots are the corresponding powers of the original latent roots (see example (v),

* The latent roots with least moduli can be obtained by use of the reciprocal of the matrix in place of the matrix itself.

§ 3·6). The ambiguities which arise in extracting the true value λ_s of a latent root from a power may be resolved by substitution of the known elements of the corresponding modal column k_s in one of the n scalar equations represented by $(\lambda_s I - u) k_s = 0$.

When the chief object is to obtain the latent roots, and little interest attaches to the modal columns, it is preferable to use the process of repeated postmultiplication of a row matrix rather than repeated premultiplication of a column, for the first process leads to the evaluation of the matrix κ required in the determination of the subdominant latent root (see § 4·18).

It may be well to point out that in general Aitken's formula (4·20·4) will fail to give the values of more than a very few of the latent roots unless extremely high accuracy is preserved in the calculations. On the other hand, equation (4·20·4) is of considerable value when there is a set of latent roots with nearly equal moduli which are considerably larger than the moduli of the other roots. The values of a pair of such roots with nearly equal moduli can be obtained as the roots of a quadratic by means of equations (4·20·5) and (4·20·6).

4·22. Situation of the Roots of an Algebraic Equation. For the sake of completeness a brief reference may be made to certain further methods of computation which, though not directly relevant to the subject of matrices, are of considerable value in investigations on stability. The first method is based on the process described by Routh* as *cross-multiplication*. This process is most simply defined as follows: Let $a_0, a_1, a_2, \ldots, a_n$ and $b_0, b_1, b_2, \ldots, b_n$ be any two sets of $n+1$ numbers, some of which may be zero. Then by cross-multiplication we shall mean the derivation of a set of n numbers $c_0, c_1, \ldots, c_{n-1}$ such that

$$[c_0, c_1, \ldots, c_{n-1}] = b_0[a_1, a_2, \ldots, a_n] - a_0[b_1, b_2, \ldots, b_n].$$

A convenient scheme of entry is represented by

$$\begin{array}{rc|ccccc} \textit{Row No. } 1 & a_0 & a_1 & a_2 & a_3 & a_4 & \ldots \\ 2 & b_0 & b_1 & b_2 & b_3 & b_4 & \ldots \\ 3 & c_0 & c_1 & c_2 & c_3 & \ldots & \ldots \end{array}$$

Any member of row 3 is here derived by cross-multiplication by b_0 and a_0 of the two numbers situated above and one step to the right. For instance,

$$c_3 = b_0 a_4 - a_0 b_4.$$

* See p. 226 of Ref. 13.

(*a*) *Number and Situation of the Real Roots of an Equation* (*Sturm's Functions*). Sturm's theorem may be stated as follows:† Let $f(x) = 0$ be the given algebraic equation, and denote by $f_1(x)$ the first derived function of $f(x)$. Denote by $f_2(x), f_3(x), \ldots, f_n(x)$ the remainders, with their signs changed, which occur in the process of finding the highest common factor of $f(x)$ and $f_1(x)$. Let $N(x)$ be the number of changes of sign in the sequence $f(x), f_1(x), \ldots, f_n(x)$: then the number of real roots of $f(x) = 0$ between $x = a$ and $x = b$ is $N(a) - N(b)$.

The presence of equal roots is indicated by the vanishing of one or more of the auxiliary functions. If $f_r(x)$ is the last of these functions not to vanish identically, then the difference between the number of changes of sign when a and b are substituted in $f, f_1, \ldots, f_r$ is equal to the number of real roots between a and b, each multiple root counting only once.

A convenient scheme for the calculation of the coefficients of the auxiliary functions is as follows: Write

$$f(x) = p_0 x^n + p_1 x^{n-1} + \ldots + p_{n-1} x + p_n,$$
$$f_1(x) = p_0(1)\, x^{n-1} + p_1(1)\, x^{n-2} + \ldots + p_{n-1}(1),$$
$$f_2(x) = p_0(2)\, x^{n-2} + p_1(2)\, x^{n-3} + \ldots + p_{n-2}(2),$$

and so on. Enter the coefficients of $f(x)$ and of $f_1(x)$ respectively as rows 1 and 2, and derive in succession by cross-multiplication row 3 from rows 1 and 2, and row 4 from rows 2 and 3. Then the numbers in row 4 will differ by a constant positive factor from the coefficients of $f_2(x)$. The presence of this positive factor is immaterial, since in forming Sturm's functions positive numerical multipliers common to any row can be introduced or suppressed at convenience. The process is then continued to give the coefficients of $f_3(x)$, by the adoption of the coefficients of $f_1(x)$ and $f_2(x)$ as new initial rows. A representative scheme of computation is set out below:

Row No.					
1	p_0	p_1	p_2	p_3	...
2	$p_0(1)$	$p_1(1)$	$p_2(1)$	$p_3(1)$	...
*3	×	×	×	×	...
4	$p_0(2)$	$p_1(2)$	$p_2(2)$	$p_3(2)$	...
*5	×	×	×	×	...
6	$p_0(3)$	$p_1(3)$	$p_2(3)$	$p_3(3)$	...
etc.	etc.	etc.			

† § 96 of Ref. 14.

Each starred row is constructed by cross-multiplication of the two immediately preceding unstarred rows. Each unstarred row (subsequent to row 2) is formed by cross-multiplication of its two immediate predecessors. To reduce labour any convenient positive multiplier may be extracted from any row: the resulting row is then entered immediately below, and the original row is of course completely ignored. The process will be clear from the appended example.

Example

Assume the equation to be†

$$f(x) = x^4 - 2x^3 - 3x^2 + 10x - 4 = 0,$$

so that

$$f_1(x) = 4x^3 - 6x^2 - 6x + 10,$$

or, on suppression of the positive multiplier 2,

$$f_1(x) = 2x^3 - 3x^2 - 3x + 5.$$

The scheme is as follows:

f	1	−2	−3	10	−4
f_1	2	−3	−3	5	
*	−1	−3	15	−8	
f_2	9	−27	11		
*	27	−49	45		
f_3	−288	−108			
f_3	−8	−3			
*	243	−88			
f_4	−1433				

Hence Sturm's functions may be taken as $f(x)$ and $f_1(x)$ together with

$$f_2 = 9x^2 - 27x + 11; \quad f_3 = -8x - 3; \quad f_4 = -1433.$$

The signs are as tabulated below:

x	$f(x)$	$f_1(x)$	$f_2(x)$	$f_3(x)$	$f_4(x)$	$N(x)$
∞	+	+	+	−	−	1
1	+	+	−	−	−	1
0	−	+	+	−	−	2
−2	−	−	+	+	−	2
−3	+	−	+	+	−	3
$-\infty$	+	−	+	+	−	3

There are thus two real roots, one situated between 0 and 1, and the other between −2 and −3.

† Example selected from § 96 of Ref. 14.

(*b*) *Number of Roots whose Real Parts are Positive.* The rules as given by Routh* in this connection may be stated as follows. Let the given equation be

$$p_0 x^n + p_1 x^{n-1} + \dots + p_{n-1} x + p_n = 0,$$

and assume p_0 to be made positive. A sequence of *test functions* R_i may now be derived as follows. Commence with $R_0 = p_0$, and derive R_1 from R_0, R_2 from R_1, and so on, by writing the lower element for the upper element in the columns of the schedule

p_0	p_1	p_2	p_3	etc.
p_1	$p_2 - \dfrac{p_0 p_3}{p_1}$	p_3	$p_4 - \dfrac{p_0 p_5}{p_1}$	etc.

A cipher is to be written for any letter when the suffix exceeds the degree of the equation. For instance, in the case of a biquadratic, p_5, p_6, etc. are zero, and the test functions as constructed on the foregoing basis are

$$R_0 = p_0; \quad R_1 = p_1; \quad R_2 = p_2 - \frac{p_0 p_3}{p_1};$$

$$R_3 = p_3 - \frac{p_1 p_4}{\left(p_2 - \dfrac{p_0 p_3}{p_1}\right)} = \frac{p_1 p_2 p_3 - p_0 p_3^2 - p_1^2 p_4}{p_1 p_2 - p_0 p_3}; \quad R_4 = p_4.$$

For an equation of general degree the number of roots with their real parts positive equals the number of changes of sign in the sequence of test functions R_i.

Routh's test functions can be expressed conveniently in terms of the sequence of *test determinants*

$$T_0 = p_0; \quad T_1 = p_1; \quad T_2 = \begin{vmatrix} p_1 & p_0 \\ p_3 & p_2 \end{vmatrix};$$

$$T_3 = \begin{vmatrix} p_1 & p_0 & 0 \\ p_3 & p_2 & p_1 \\ p_5 & p_4 & p_3 \end{vmatrix}; \qquad T_4 = \begin{vmatrix} p_1 & p_0 & 0 & 0 \\ p_3 & p_2 & p_1 & p_0 \\ p_5 & p_4 & p_3 & p_2 \\ p_7 & p_6 & p_5 & p_4 \end{vmatrix},$$

and so on. As before, a cipher is to be substituted for any coefficient with a suffix exceeding n. It can be shown† that Routh's substitution,

* §§ 297–301 of Ref. 13. † Ref. 15.

as given above, converts the determinant T_r of the sequence into T_{r+1}/p_1. Hence

$$R_0 = T_0;\quad R_1 = T_1;\quad R_2 = T_2/T_1;\quad R_3 = T_3/T_2,\quad \text{etc.}$$

The penultimate test function is T_{n-1}/T_{n-2}, and the final one is T_n/T_{n-1}, which reduces to p_n. An equivalent set, which offers the advantage that no member becomes infinite when its immediate predecessor vanishes, is

$$R_0' = T_0;\quad R_1' = T_1;\quad R_2' = T_2T_1;\quad R_3' = T_3T_2;\quad \ldots;$$
$$R_{n-1}' = T_{n-1}T_{n-2};\quad R_n' = p_n.$$

The necessary and sufficient condition for the real parts of all the roots to be negative is that all the test functions, or all the test determinants, shall be positive. This is the condition for stability of a linear dynamical system.

It may be noted that the vanishing of T_{n-1} indicates either a pair of purely imaginary roots or else a pair of equal and opposite real roots.

(c) *Situation of the Latent Roots of a Matrix.* The following theorems* relating to the situation of the latent roots of a matrix u with real or complex elements may be noted. The typical latent root is denoted as $\lambda \equiv \mu + i\omega$.

(i) If u is Hermitian (see § 1·17), and in particular if u is real and symmetrical, the latent roots are all real.

(ii) If u is general, μ lies between the greatest and least latent root of the Hermitian matrix $\frac{1}{2}(u+\bar{u}')$, and ω lies between the greatest and least latent root of the Hermitian matrix $\frac{1}{2}i(u-\bar{u}')$.

* For proofs and historical references, see Chap. VIII of Ref. 2.

CHAPTER V

LINEAR ORDINARY DIFFERENTIAL EQUATIONS WITH CONSTANT COEFFICIENTS

Part I. General Properties

5·1. Systems of Simultaneous Differential Equations. An important field of application of the theory of matrices will now be considered, namely, the solution of systems of linear ordinary differential equations. Such systems arise in many branches of mathematical physics, but the majority of the illustrative problems discussed in later Chapters will relate to dynamics. Hence it will be convenient to regard the independent variable as representing time, and to denote it by t.

The most general system of linear ordinary differential equations with constant coefficients consists of a set of m equations connecting the m dependent variables $x_1, x_2, \ldots, x_m$ with t. These equations may be expressed as

$$\begin{aligned} U_1 &\equiv f_{11}(D)\,x_1 + f_{12}(D)\,x_2 + \ldots + f_{1m}(D)\,x_m - \xi_1(t) = 0, \\ U_2 &\equiv f_{21}(D)\,x_1 + f_{22}(D)\,x_2 + \ldots + f_{2m}(D)\,x_m - \xi_2(t) = 0, \\ &\cdots\cdots\cdots\cdots\cdots\cdots\cdots\cdots\cdots\cdots\cdots\cdots \\ U_m &\equiv f_{m1}(D)\,x_1 + f_{m2}(D)\,x_2 + \ldots + f_{mm}(D)\,x_m - \xi_m(t) = 0. \end{aligned}$$

Here D denotes the differential operator d/dt; $f_{ij}(D)$ is a polynomial of D having constant coefficients; while $\xi_1(t)$, etc. are given functions of t. The set of equations is expressible as the single matrix equation

$$\{U_i\} \equiv [f_{ij}(D)]\{x_i\} - \{\xi(t_i)\} = 0, \qquad \ldots\ldots(1)$$

or, when no ambiguity can result, even more concisely as

$$U \equiv fx - \xi = 0. \qquad \ldots\ldots(2)$$

It may be referred to simply as the system f.

If the elements of the matrix $f(D)$ are polynomials of degree N at most in D, the system is said to be of order N. It is sometimes convenient to express equations (2) in the alternative form

$$(A_0 D^N + A_1 D^{N-1} + \ldots + A_{N-1} D + A_N)\,x = \xi. \qquad \ldots\ldots(3)$$

Each of the coefficients A_0, A_1, etc. is a square matrix of order m having constant elements. In particular cases one or more of these elements may be zero, but one element at least of A_0 must not be zero.

The system f is said to be *homogeneous* when the functions $\xi_i(t)$ are all absent: if one or more of these functions is present, the system is *non-homogeneous*. The general solution of a non-homogeneous system is the sum of the *complementary function*, namely, the general solution of the homogeneous system obtained by omission of all the functions $\xi_i(t)$, and of any *particular integral* of the complete system.

The theory is intimately bound up with the properties of the matrix $f(D)$, which are obviously closely analogous to those of λ-matrices (see Chapter III). The usual process of construction of the complementary function practically amounts to a substitution of λ for the operator D in $f(D)$. The method, which is well known, is to assume a constituent of the complementary function to be $x = e^{\lambda t}k$, where $e^{\lambda t}$ is a scalar multiplier, and k is a column of constants to be found. Then λ and k must be such that

$$f(D)\,e^{\lambda t}k = e^{\lambda t}f(\lambda)\,k = 0.$$

The condition of consistency requires the index λ to be a root of the algebraic equation

$$\Delta(\lambda) \equiv |\,f(\lambda)\,| = 0, \qquad \text{......(4)}$$

while the column k_r of constants corresponding to any given root λ_r is determined to an arbitrary common multiplier by $f(\lambda_r)\,k_r = 0$. This process yields the complete complementary function in the standard simple case where all the roots of $\Delta(\lambda) = 0$ are distinct.

Both the purely operational determinant $|\,f(D)\,|$ and the corresponding algebraic determinant $|\,f(\lambda)\,|$ are sometimes referred to as the "characteristic determinant". Although in most cases this can cause no confusion, it is preferable to adopt terms which distinguish between operators and algebraic quantities. It is also common practice to speak of equation (4)—sometimes also written in the quite meaningless form $|\,f(D)\,| = 0$—as the "characteristic equation". However, this term already has an established meaning in relation to λ-matrices of the special simple type $\lambda I - u$. The following definitions will be adopted here:

$f(D)$ .... the D-matrix.

$f(\lambda)$ .... the λ-matrix.

$\Delta(D) \equiv |\,f(D)\,|$ ... the D-determinant.

$\Delta(\lambda) \equiv |\,f(\lambda)\,|$ ... the λ-determinant.

$\Delta(\lambda) = 0$ .... the determinantal equation.

The degree in λ of $\Delta(\lambda)$ is assumed to be n, and the roots of the determinantal equation are denoted by $\lambda_1, \lambda_2, \ldots, \lambda_n$. A distinct symbol is therefore used to represent each root, even when multiple roots occur.

5·2. Equivalent Systems. From a given system f of linear differential equations it is possible to form new systems by operations with polynomials of D on one or more of the equations U_1, U_2, etc., and addition. New complete systems g may thus be derived which will have the form

$$gy-\eta=\mu(D)(fy-\xi)=0. \qquad \text{......(1)}$$

The square matrix $\mu(D)$ here applied as a premultiplier to the system f is of order m, and its elements are given polynomials of D. The determinant $|\mu(D)|$ will be assumed not to vanish, but it may be constant in the sense that it is independent of D.

In (1) the dependent variables are denoted by y whereas in (5·1·2) they are denoted by x. This distinction must usually be made since the derived system g normally contains greater generality than the parent system f. Clearly, every solution of f satisfies g, but every solution of g does not necessarily satisfy f. However, the important case is where both conditions are satisfied, so that every solution of f satisfies g, and conversely. The two systems are then *equivalent*, and the variables x and y can legitimately be identified. The condition for equivalence is that the determinant $|\mu(D)|$ shall be independent of D and not zero. In this particular case the elements of the reciprocal matrix $\mu^{-1}(D)$ are clearly also all polynomials of D; and since $fy-\xi=\mu^{-1}(D)(gy-\eta)$, it follows that every solution of g then satisfies f, as required. The condition $|\mu(D)|=\text{const.}$ is necessary and sufficient for the equivalence of the two systems. It is evident that the D-determinants of equivalent systems can only differ by constant multipliers.

Example

Suppose f to be the system

$$\left.\begin{aligned}(D^2+1)x_1+2D^3x_2&=0,\\ Dx_1+(2D^2-1)x_2&=\cos t\end{aligned}\right\}. \qquad \text{......(2)}$$

Then if $\mu(D)$ is chosen to be $\begin{bmatrix}1, & -D\\ -D, & D^2+1\end{bmatrix}$, the derived system g is

$$\begin{bmatrix}1, & -D\\ -D, & D^2+1\end{bmatrix}\begin{bmatrix}D^2+1, & 2D^3\\ D, & 2D^2-1\end{bmatrix}\begin{bmatrix}y_1\\ y_2\end{bmatrix}=\begin{bmatrix}1, & -D\\ -D, & D^2+1\end{bmatrix}\begin{bmatrix}0\\ \cos t\end{bmatrix},$$

or

$$\left.\begin{aligned}y_1+Dy_2&=\sin t,\\ (D^2-1)y_2&=0.\end{aligned}\right\} \qquad \text{......(3)}$$

Since $|\mu(D)| = 1$ (a non-zero constant), the two systems are equivalent. The solution of (3) is here obvious by inspection. Evidently

$$y_2 = c_1 e^t + c_2 e^{-t}$$

and hence

$$y_1 = \sin t - c_1 e^t + c_2 e^{-t},$$

where c_1, c_2 are arbitrary constants. Thus the solution of (2) is also

$$\left.\begin{aligned} x_1 &= \sin t - c_1 e^t + c_2 e^{-t}, \\ x_2 &= c_1 e^t + c_2 e^{-t}. \end{aligned}\right\} \qquad \text{......(4)}$$

The important point to note is that equations (2) can be derived from (3) purely by operations with polynomials of D, just as (3) were derived from (2) by such operations. To obtain f from g the premultiplier to be used would be

$$\mu^{-1}(D) \equiv \begin{bmatrix} 1, & -D \\ -D, & D^2+1 \end{bmatrix}^{-1} = \begin{bmatrix} D^2+1, & D \\ D, & 1 \end{bmatrix}.$$

When $|\mu(D)|$ is not a constant, then although g can be derived from f by steps involving only differentiations, yet f cannot be constructed from g without operations of integration and the attendant introduction of further arbitrary constants.

5·3. Transformation of the Dependent Variables. In the preceding section attention was restricted to the use of matrices $\mu(D)$ as premultipliers to the equations. However, D-matrices are applied as postmultipliers when a direct transformation of the dependent variables x to a new set, say X, is effected such that

$$x = \mu(D)\, X. \qquad \text{......(1)}$$

The following systems are then evidently the same:

$$f(D)\, x - \xi = 0, \qquad \text{......(2)}$$

$$f(D)\, \mu(D)\, X - \xi = 0. \qquad \text{......(3)}$$

If the general solution of (3) in the variables X can be found, then the corresponding values of x are given uniquely by (1), even when $|\mu(D)|$ is not a constant. On the other hand, if the solution of (2) in the variables x is known, the most general values of X will not usually be deducible from (1) without operations of integration and the introduction of further arbitrary constants. The exceptional case, again, is when the determinant $|\mu(D)|$ is a constant other than zero: the values of X are then determinable uniquely by means of the inverse relation $X = \mu^{-1}(D)\, x$.

EXAMPLE

If in (5·2·2) we substitute

$$\begin{bmatrix} x_1 \\ x_2 \end{bmatrix} = \begin{bmatrix} 2D, & 2D^2-1 \\ -1, & -D \end{bmatrix} \begin{bmatrix} X_1 \\ X_2 \end{bmatrix},$$

the system is converted into

$$\begin{bmatrix} D^2+1, & 2D^3 \\ D, & 2D^2-1 \end{bmatrix} \begin{bmatrix} 2D, & 2D^2-1 \\ -1, & -D \end{bmatrix} \begin{bmatrix} X_1 \\ X_2 \end{bmatrix} = \begin{bmatrix} 0 \\ \cos t \end{bmatrix},$$

or

$$2DX_1+(D^2-1)X_2=0,$$
$$X_1=\cos t.$$

Hence $X_2=-\sin t+c_1e^t+c_2e^{-t}$, so that

$$\begin{bmatrix} x_1 \\ x_2 \end{bmatrix} = \begin{bmatrix} 2D, & 2D^2-1 \\ -1, & -D \end{bmatrix} \begin{bmatrix} \cos t \\ -\sin t+c_1e^t+c_2e^{-t} \end{bmatrix}$$
$$= \begin{bmatrix} \sin t+c_1e^t+c_2e^{-t} \\ -c_1e^t+c_2e^{-t} \end{bmatrix}.$$

This solution is effectively the same as (5·2·4). Conversely, if the values of x just obtained are supposed known, then the corresponding values X would be given uniquely by

$$\begin{bmatrix} X_1 \\ X_2 \end{bmatrix} = \begin{bmatrix} 2D, & 2D^2-1 \\ -1, & -D \end{bmatrix}^{-1} \begin{bmatrix} x_1 \\ x_2 \end{bmatrix} = \begin{bmatrix} D, & 2D^2-1 \\ -1, & -2D \end{bmatrix} \begin{bmatrix} x_1 \\ x_2 \end{bmatrix}.$$

5·4. Triangular Systems and a Fundamental Theorem.

A system, say $hx-\xi=0$, is commonly called a "diagonal" system when $h(D)$ has the special triangular arrangement

$$h(D) \equiv \begin{bmatrix} h_{11}(D) & h_{12}(D) & \dots & h_{1m}(D) \\ 0 & h_{22}(D) & \dots & h_{2m}(D) \\ \dots & \dots & \dots & \dots \\ 0 & \dots & 0 & h_{mm}(D) \end{bmatrix}.$$

It is preferable to describe such a system as *triangular*, since with matrices the term "diagonal" implies that only the principal diagonal elements are present.

The first equation of the system here involves x_1 and (in general) all the remaining variables, the second equation involves all the variables with the exception of x_1, and so on, until finally the last equation involves x_m only. In this case the variables are said to be in the *diagonal order* $x_1, x_2, \ldots, x_m$.

It can be shown* that any system of linear differential equations with constant coefficients is reducible to an equivalent triangular system in which the dependent variables have any assigned diagonal order (e.g. that adopted above). The number of arbitrary constants involved in the solution of any triangular system is clearly the sum of the degrees in D of the diagonal coefficients, and it therefore equals the degree in D of the product $h_{11}h_{22}\ldots h_{mm}$. This product, however, is equal to the D-determinant of the triangular system, and is thus merely a constant multiple of the D-determinant of any equivalent system. The fundamental theorem follows, that *the number of arbitrary constants entering into the solution of any linear system with constant coefficients equals the degree in D of the determinant* $\Delta(D) \equiv |f(D)|$.

EXAMPLES

(i) *A Given System Expressed as an Equivalent Triangular System.* Suppose the system f to be

$$\left.\begin{aligned} (D+1)x_1 + D^2x_2 + (D+1)x_3 &= 0, \\ (D-1)x_1 + Dx_2 + (D-1)x_3 &= 0, \\ x_1 + x_2 + Dx_3 &= 0. \end{aligned}\right\} \qquad \ldots\ldots(1)$$

Use as premultiplier

$$\mu(D) = \begin{bmatrix} -1, & -D, & D^2 \\ 0, & -1, & (D-1) \\ 1, & (-D^2+D+1), & (D^3-2D^2-D) \end{bmatrix},$$

the constant determinant of which is -1. Then the result is

$$\begin{bmatrix} -1, & -D^2, & (D^3-D^2-1) \\ 0, & -1, & (D^2-2D+1) \\ 0, & 0, & (D^4-3D^3+D^2+D) \end{bmatrix} \begin{bmatrix} x_1 \\ x_2 \\ x_3 \end{bmatrix} = 0, \qquad \ldots\ldots(2)$$

or

$$\begin{aligned} -x_1 - D^2x_2 + (D^3-D^2-1)x_3 &= 0, \\ -x_2 + (D^2-2D+1)x_3 &= 0, \\ (D^4-3D^3+D^2+D)x_3 &= 0. \end{aligned}$$

* A formal proof is given in § 6·5 of Ref. 5.

Triangular systems of the type (2), in which all the principal diagonal coefficients with the exception of the last are independent of D, are of special simplicity, and may be referred to as *simple triangular systems*.

(ii) *The Fundamental Theorem*. The number of arbitrary constants entering into the general solution of (2) is evidently four (the degree of the last coefficient). Hence four arbitrary constants appear in the general solution of (1). This equals the degree in D of $|f(D)|$.

5·5. Conversion of a System of General Order into a First-Order System.

If the system $fx-\xi=0$ is not already of first order, suppose it to be expressed in the form (see (5·1·3))

$$(A_0D^N+A_1D^{N-1}+\dots+A_{N-1}D+A_N)x=\xi, \qquad \text{......(1)}$$

in which the coefficients A are square matrices of order m with constant elements, and $N>1$. One element at least of A_0 is assumed not to be zero.

Case I. Leading Coefficient Not Singular. Suppose firstly that $|A_0|\neq 0$, so that the reciprocal matrix A_0^{-1} exists. In this case, equation (1) can be premultiplied by A_0^{-1} to give

$$D^Nx=-a_Nx-a_{N-1}Dx-\dots-a_1D^{N-1}x+\zeta, \qquad \text{......(2)}$$

where $$a_i=A_0^{-1}A_i,$$

and $$\zeta=A_0^{-1}\xi.$$

Now write $\quad X_1=x,\quad X_2=Dx,\quad \dots,\quad X_N=D^{N-1}x,$

so that $$DX_1=X_2,\quad DX_2=X_3,\quad \dots,\quad DX_{N-1}=X_N, \qquad \text{......(3)}$$

while equation (2) becomes

$$DX_N=-a_NX_1-a_{N-1}X_2-\dots-a_1X_N+\zeta. \qquad \text{......(4)}$$

If the N sets $X_1, X_2, \dots, X_m$—each containing m variables—are rega-ded as a single set of mN new variables y defined by

$$y=\{X_i\}=\{x_1,\dots,x_m,\ Dx_1,\dots,Dx_m,\ \dots,\ D^{N-1}x_1,\dots,D^{N-1}x_m\},$$

the mN scalar equations satisfied by the variables y are represented by (3) and (4). This first-order system will be denoted by

$$Dy=uy+\eta,$$

where $$\eta=\{0,0,\dots,0,\zeta\},$$

and where u is a square matrix of order mN which can be partitioned as below into square matrices of order m:

$$u = \begin{bmatrix} 0, & I_m, & 0, & \dots, & 0, & 0 \\ 0, & 0, & I_m, & \dots, & 0, & 0 \\ \dots & \dots & \dots & \dots & \dots & \dots \\ 0, & 0, & 0, & \dots, & 0, & I_m \\ -a_N, & -a_{N-1}, & -a_{N-2}, & \dots, & -a_2, & -a_1 \end{bmatrix}.$$

A system of differential equations of general order can usually be converted into a first-order system in an indefinitely large number of different ways. The special method given above has the advantage of simplicity.

Case II. Leading Coefficient Singular. Next, let A_0 be of rank r. Then (see § 3·13) non-singular square matrices P, Q, of order m, with constant elements, can be found such that

$$\alpha_0 = PA_0Q,$$

where α_0 has zero elements with the exception of r units occupying the first r places in the principal diagonal. Now replace the m variables x in equation (1) by a set $\boldsymbol{x}$ such that $x = Q\boldsymbol{x}$, and premultiply the resulting equation by P. Then

$$(\alpha_0 D^N + PA_1QD^{N-1} + \dots + PA_NQ)\boldsymbol{x} = P\xi,$$

or say $$\alpha_0 D^N\boldsymbol{x} = -a_N\boldsymbol{x} - a_{N-1}D\boldsymbol{x} - \dots - a_1 D^{N-1}\boldsymbol{x} + \zeta. \quad \dots\dots(5)$$

The system of scalar equations represented by (5) resembles (2) except that the left-hand side of the ith equation is equal to $D^N\boldsymbol{x}_i$ when $i \leqslant r$, but is zero when $i > r$. Hence if a new set of dependent variables y is introduced such that

$$\{y\} = \{\boldsymbol{x}_1, \dots, \boldsymbol{x}_m, D\boldsymbol{x}_1, \dots, D\boldsymbol{x}_m, \dots, D^{N-1}\boldsymbol{x}_1, \dots, D^{N-1}\boldsymbol{x}_m\},$$

the equations (5) will be reducible to a system of mN equations of first order similar to (3) and (4), except that the last $m-r$ equations will consist purely of linear algebraic relations connecting the mN variables y. These $m-r$ linear relations can be used to eliminate $m-r$ of the variables y from the remaining equations of the system. The final system will then consist of only $mN-m+r$ independent equations.

In practice, when A_0 is singular the conversion to a first-order system can usually be effected by a judicious manipulation of the equations, without actual determination of the matrices P and Q.

Examples

(i) *A Single Equation of Order n.* Let the given single equation be

$$D^n x + p_1 D^{n-1} x + \dots + p_n x = \xi.$$

Then if $\{y_1, y_2, \dots, y_n\} \equiv \{x, Dx, \dots, D^{n-1}x\}$,

the equivalent system of first order is

$$D\begin{bmatrix} y_1 \\ y_2 \\ \dots \\ \dots \\ y_n \end{bmatrix} = \begin{bmatrix} 0, & 1, & 0, & \dots, & 0, & 0 \\ 0, & 0, & 1, & \dots, & 0, & 0 \\ \dots & \dots & \dots & \dots & \dots & \dots \\ 0, & 0, & 0, & \dots, & 0, & 1 \\ -p_n, & -p_{n-1}, & -p_{n-2}, & \dots, & -p_2, & -p_1 \end{bmatrix}\begin{bmatrix} y_1 \\ y_2 \\ \dots \\ \dots \\ y_n \end{bmatrix} + \begin{bmatrix} 0 \\ 0 \\ \dots \\ \dots \\ \xi(t) \end{bmatrix}.$$

(ii) *Linear Dynamical Equations of Lagrangian Type.* The Lagrangian equations for the small free motions of a dynamical system in m generalised coordinates $q_1, q_2, \dots, q_m$ are of the type

$$A\ddot{q} + B\dot{q} + Cq = 0,$$

where $|A| \neq 0$. They can be replaced by the system of $2m$ first-order equations represented by

$$Dy = uy,$$

where $y = \{q, \dot{q}\}$

and

$$u = \begin{bmatrix} 0, & I_m \\ -A^{-1}C, & -A^{-1}B \end{bmatrix}.$$

(iii) *The Matrix A_0 Singular.* Consider the system

$$\begin{bmatrix} 3 & 2 & 1 \\ 2 & 1 & -1 \\ 1 & 1 & 2 \end{bmatrix} D^2 x + \begin{bmatrix} 0 & 0 & 1 \\ 0 & 1 & 0 \\ 1 & 0 & 0 \end{bmatrix} Dx + \begin{bmatrix} 1 & 2 & 1 \\ 0 & -1 & 2 \\ 2 & 1 & 0 \end{bmatrix} x = 0.$$

Here A_0 is of rank 2. Subtraction of the sum of the second and third equations from the first yields

$$-Dx_1 - Dx_2 + Dx_3 - x_1 + 2x_2 - x_3 = 0, \qquad \dots\dots(6)$$

which can be used to express D^2x_3 in terms of D^2x_1 and D^2x_2 and lower derivatives. By this means, the first two equations may be replaced by

$$\left.\begin{aligned} D^2x_1 - Dx_1 + 3Dx_2 - Dx_3 - x_2 + 2x_3 &= 0, \\ 3D^2x_2 + 5Dx_1 - 14Dx_2 + 6Dx_3 + x_1 + 6x_2 - 7x_3 &= 0. \end{aligned}\right\} \qquad \dots\dots(7)$$

Hence, if $\{y_1, y_2, y_3, y_4, y_5, y_6\} = \{x_1, x_2, x_3, Dx_1, Dx_2, Dx_3\}$, ...(8)

then

$$Dy_1 = y_4,$$
$$Dy_2 = y_5,$$
$$Dy_3 = y_6,$$

and from (7) and (8)

$$Dy_4 = y_4 - 3y_5 + y_6 + y_2 - 2y_3,$$
$$Dy_5 = -\tfrac{5}{3}y_4 + \tfrac{14}{3}y_5 - 2y_6 - \tfrac{1}{3}y_1 - 2y_2 + \tfrac{7}{3}y_3.$$

Also (6) is equivalent to

$$0 = y_4 + y_5 - y_6 + y_1 - 2y_2 + y_3.$$

These equations can be used as they stand, or reduced to a system of five equations by elimination of one variable, say y_6. The final system would then be

$$D\begin{bmatrix} y_1 \\ y_2 \\ y_3 \\ y_4 \\ y_5 \end{bmatrix} = \begin{bmatrix} 0 & 0 & 0 & 1 & 0 \\ 0 & 0 & 0 & 0 & 1 \\ 1 & -2 & 1 & 1 & 1 \\ 1 & -1 & -1 & 2 & -2 \\ -\tfrac{7}{3} & 2 & \tfrac{1}{3} & -\tfrac{11}{3} & \tfrac{8}{3} \end{bmatrix} \begin{bmatrix} y_1 \\ y_2 \\ y_3 \\ y_4 \\ y_5 \end{bmatrix}.$$

5·6. The Adjoint and Derived Adjoint Matrices. For convenience of reference a number of important definitions and theorems relating to λ-matrices will now be recalled.

If

$$f(\lambda) = \begin{bmatrix} f_{11}(\lambda), f_{12}(\lambda), \ldots, f_{1m}(\lambda) \\ f_{21}(\lambda), f_{22}(\lambda), \ldots, f_{2m}(\lambda) \\ \cdots\cdots\cdots\cdots\cdots\cdots \\ f_{m1}(\lambda), f_{m2}(\lambda), \ldots, f_{mm}(\lambda) \end{bmatrix},$$

the adjoint of f is defined to be

$$F(\lambda) = \begin{bmatrix} F_{11}(\lambda), F_{21}(\lambda), \ldots, F_{m1}(\lambda) \\ F_{12}(\lambda), F_{22}(\lambda), \ldots, F_{m2}(\lambda) \\ \cdots\cdots\cdots\cdots\cdots\cdots \\ F_{1m}(\lambda), F_{2m}(\lambda), \ldots, F_{mm}(\lambda) \end{bmatrix},$$

where F_{ij} is the cofactor of f_{ij} in the determinant $\Delta(\lambda) \equiv |f(\lambda)|$. Thus F is the transposed of the matrix of the cofactors of f. The two matrices f, F have the property

$$f(\lambda)\,F(\lambda) = F(\lambda)f(\lambda) = \Delta(\lambda)\,I_m.$$

The pth derived adjoint matrix $\frac{d^pF}{d\lambda^p}$ is denoted by $\overset{(p)}{F}(\lambda)$, and for brevity the matrix $\left(\frac{d^pF}{d\lambda^p}\right)_{\lambda=\lambda_r}$, where λ_r is any root of $\Delta(\lambda)=0$, is represented by $\overset{(p)}{F}(\lambda_r)$.

Theorems (A) to (E) of § 3·5 may be summarised as follows:

(*a*) *Simple Roots.* If λ_r is a simple root of the determinantal equation, $f(\lambda_r)$ is simply degenerate and $F(\lambda_r)$ is expressible as a matrix product

$$F(\lambda_r) \equiv \{k_{ir}\} \lfloor \kappa_{ir} \rfloor \equiv k_r \kappa_r,$$

in which k_{ir}, κ_{ir} are constants appropriate to the chosen root and at least one constant of each type is not zero.

(*b*) *Multiple Roots.* If λ_s is a multiple root and if $f(\lambda_s)$ is simply degenerate, then the adjoint can be represented as for case (*a*) by $F(\lambda_s) \equiv k_s \kappa_s$. On the other hand, if $f(\lambda_s)$ has degeneracy $q > 1$, the adjoint and the derived adjoints up to and including $\overset{(q-2)}{F}(\lambda_s)$ at least are all null. The root has multiplicity q at least when $f(\lambda_s)$ has degeneracy q.

The columns k_r and the rows κ_r have the properties (compare (3·6·5), (3·6·6))

$$f(\lambda_r)\, k_r = 0. \qquad \text{......(1)}$$

$$\kappa_r f(\lambda_r) = 0. \qquad \text{......(2)}$$

EXAMPLES

(i) *Simple Roots.* Assume

$$f(\lambda) = \begin{bmatrix} \lambda+2, & \lambda^3+2\lambda^2+2 \\ \lambda^2, & \lambda^4+2 \end{bmatrix}.$$

Then

$$\Delta(\lambda) = -2(\lambda-2)(\lambda+1) \quad \text{and} \quad F(\lambda) = \begin{bmatrix} \lambda^4+2, & -\lambda^3-2\lambda^2-2 \\ -\lambda^2, & \lambda+2 \end{bmatrix}.$$

The roots, say $\lambda_1 = 2$ and $\lambda_2 = -1$, are here both simple. They give

$$F(\lambda_1) = \begin{bmatrix} 18 & -18 \\ -4 & 4 \end{bmatrix} = \begin{bmatrix} 9 \\ -2 \end{bmatrix} [2, -2]$$

and

$$F(\lambda_2) = \begin{bmatrix} 3 & -3 \\ -1 & 1 \end{bmatrix} = \begin{bmatrix} 3 \\ -1 \end{bmatrix} [1, -1].$$

(ii) *Multiple Roots.* Suppose

$$f(\lambda) = \begin{bmatrix} 11\lambda^2+14\lambda+15, & 8\lambda^2+14\lambda+14, & 5\lambda^2+11\lambda+10 \\ 11\lambda^2+10\lambda+11, & 8\lambda^2+10\lambda+10, & 5\lambda^2+8\lambda+7 \\ 7\lambda^2+2\lambda+7, & 5\lambda^2+3\lambda+6, & 3\lambda^2+3\lambda+4 \end{bmatrix}.$$

In this case $\Delta(\lambda) = \lambda(\lambda-1)^2(\lambda+1)^2$, giving $\lambda_1 = 0$, $\lambda_2 = \lambda_3 = 1$, and $\lambda_4 = \lambda_5 = -1$. Substitution of the roots in $f(\lambda)$ gives

$$f(\lambda_1) = \begin{bmatrix} 15 & 14 & 10 \\ 11 & 10 & 7 \\ 7 & 6 & 4 \end{bmatrix}; \quad f(\lambda_3) = \begin{bmatrix} 40 & 36 & 26 \\ 32 & 28 & 20 \\ 16 & 14 & 10 \end{bmatrix}; \quad f(\lambda_5) = \begin{bmatrix} 12 & 8 & 4 \\ 12 & 8 & 4 \\ 12 & 8 & 4 \end{bmatrix}.$$

The first two of these are simply degenerate, while $f(\lambda_5)$ has degeneracy 2. The corresponding adjoints are

$$F(\lambda_1) = \begin{bmatrix} -2 & 4 & -2 \\ 5 & -10 & 5 \\ -4 & 8 & -4 \end{bmatrix} = \begin{bmatrix} -2 \\ 5 \\ -4 \end{bmatrix} [1, -2, 1],$$

$$F(\lambda_3) = \begin{bmatrix} 0 & 4 & -8 \\ 0 & -16 & 32 \\ 0 & 16 & -32 \end{bmatrix} = \begin{bmatrix} 4 \\ -16 \\ 16 \end{bmatrix} [0, 1, -2],$$

$$F(\lambda_5) = 0.$$

5·7. Construction of the Constituent Solutions. It will now be shown how the constituent solutions of a system of homogeneous differential equations $f(D)x = 0$ can be obtained. For clarity the constituents appropriate to simple roots and to multiple roots of the determinantal equation are considered under separate headings.

(*a*) *Simple Roots.* Let λ_r denote any simple root of $\Delta(\lambda) = 0$. Then since

$$f(D)\, e^{\lambda_r t} F(\lambda_r) = e^{\lambda_r t} f(\lambda_r)\, F(\lambda_r) = 0,$$

it follows that every column of the matrix $e^{\lambda_r t} F(\lambda_r)$ satisfies the given differential equations. But by § 5·6 the columns of $F(\lambda_r)$ are all proportional to k_r. Hence the constituent solution corresponding to the given simple root can be chosen to be an arbitrary multiple of $e^{\lambda_r t} k_r$.

(*b*) *Multiple Roots.* Let λ_s be a multiple root, and for simplicity suppose, firstly, λ_s to be only once repeated. Consider the two matrices

$$W_0(t, \lambda_s) \equiv e^{\lambda_s t} F(\lambda_s),$$

$$W_1(t, \lambda_s) \equiv \left[\frac{\partial}{\partial \lambda} e^{\lambda t} F(\lambda)\right]_{\lambda=\lambda_s} = e^{\lambda_s t}[\overset{(1)}{F}(\lambda_s) + tF(\lambda_s)].$$

The first matrix $W_0(t, \lambda_s)$ is similar to that already considered and all its columns thus satisfy $f(D)x = 0$: this conclusion remains true, but is nugatory, when $F(\lambda_s)$ happens to be null. We shall now show that all the columns of $W_1(t, \lambda_s)$ also satisfy the differential equations. By the usual properties of differential operators

$$\begin{aligned} f(D)\,W_1(t,\lambda) &= f\left(\frac{\partial}{\partial t}\right)\left[\frac{\partial}{\partial \lambda}e^{\lambda t}F(\lambda)\right] \\ &= \frac{\partial}{\partial \lambda}f\left(\frac{\partial}{\partial t}\right)[e^{\lambda t}F(\lambda)] \\ &= \frac{\partial}{\partial \lambda}[e^{\lambda t}f(\lambda)\,F(\lambda)] \\ &= e^{\lambda t}\left(\frac{\partial}{\partial \lambda}+t\right)\Delta(\lambda)\,I. \qquad \text{......(1)} \end{aligned}$$

Since λ_s is assumed to be a double root of $\Delta(\lambda)=0$, both $\Delta(\lambda)$ and $\dfrac{\partial\Delta(\lambda)}{\partial\lambda}$ vanish when $\lambda=\lambda_s$. Hence

$$f(D)\,W_1(t,\lambda_s)=0.$$

It can be proved by exactly the same method that, if λ_s is an s-fold root of $\Delta(\lambda)=0$, then every column of every member of the family of matrices $W_0(t,\lambda_s), W_1(t,\lambda_s), \ldots, W_{s-1}(t,\lambda_s)$ satisfies the differential equations, where the typical matrix of the family is defined to be*

$$W_p(t,\lambda_s)\equiv\left[\frac{\partial^p}{\partial\lambda^p}e^{\lambda t}F(\lambda)\right]_{\lambda=\lambda_s}=e^{\lambda_s t}\left(\frac{\partial}{\partial\lambda_s}+t\right)^p F(\lambda_s).$$

It will be convenient to write $W_p(t,\lambda_s)\equiv e^{\lambda_s t}U_p(t,\lambda_s)$, so that

$$U_p(t,\lambda_s)=\overset{(p)}{F}(\lambda_s)+pt\,\overset{(p-1)}{F}(\lambda_s)+\frac{p(p-1)}{\underline{|2}}t^2\,\overset{(p-2)}{F}(\lambda_s)+\ldots+t^pF(\lambda_s).$$

Since there are s matrices $U_0, U_1, \ldots, U_{s-1}$ corresponding to the s-fold root, and since each matrix has m columns, the total number of constituent solutions (including possibly some null solutions) obtained in this way would be ms. However, corresponding to any s-fold root, only s linearly independent solutions are to be expected. It follows that the ms columns concerned are necessarily connected by $s(m-1)$ linear relations.

In view of the theorems summarised under heading (*b*) of § 5·6 it is seen that when $f(\lambda_s)$ has degeneracy q the matrices $W_p(t,\lambda_s)$ and $U_p(t,\lambda_s)$ are null for values of p up to and including $q-2$ at least.

* More generally, if $\quad W_p(t,\lambda_s)=\left[\dfrac{\partial^p}{\partial\lambda^p}e^{\lambda(t-\tau)}F(\lambda)\right]_{\lambda=\lambda_s},$

where τ is an arbitrary constant, then every member of the family $W_0, W_1, \ldots, W_{s-1}$ also satisfies $f(D)\,x=0$.

EXAMPLES

(i) *Simple Case of a Diagonal System.* Suppose $f(D)\,x=0$, where

$$f(\lambda)=\begin{bmatrix}(\lambda-\alpha), & 0, & 0\\ 0, & (\lambda-\alpha)(\lambda-\beta), & 0\\ 0, & 0, & (\lambda-\alpha)(\lambda-\beta)(\lambda-\gamma)\end{bmatrix}$$

$$=(\lambda-\alpha)\begin{bmatrix}1, & 0, & 0\\ 0, & (\lambda-\beta), & 0\\ 0, & 0, & (\lambda-\beta)(\lambda-\gamma)\end{bmatrix},$$

and $\alpha\neq\beta\neq\gamma$. Here $\Delta(\lambda)=(\lambda-\alpha)^3(\lambda-\beta)^2(\lambda-\gamma)$, and

$$F(\lambda)=(\lambda-\alpha)^2(\lambda-\beta)\begin{bmatrix}(\lambda-\beta)(\lambda-\gamma), & 0, & 0\\ 0, & (\lambda-\gamma), & 0\\ 0, & 0, & 1\end{bmatrix}.$$

For the simple root $\lambda=\gamma$ the matrix $W_0(t,\gamma)=e^{\gamma t}F(\gamma)$ contains a non-zero column proportional to $\{0,0,1\}$. The one constituent corresponding to γ may thus be taken as $e^{\gamma t}\{0,0,1\}$.

The degeneracy of $f(\lambda)$ is 2 for the root β and 3 for the root α, and for each of these roots $F(\lambda)$ is null. Now

$$\overset{(1)}{F}(\beta)=(\beta-\alpha)^2\begin{bmatrix}0, & 0, & 0\\ 0, & (\beta-\gamma), & 0\\ 0, & 0, & 1\end{bmatrix},$$

and this contains two non-zero columns, which lead to the two distinct constituents $e^{\beta t}\{0,1,0\}$ and $e^{\beta t}\{0,0,1\}$.

Again, $\overset{(1)}{F}(\alpha)=0$, but

$$\overset{(2)}{F}(\alpha)=2(\alpha-\beta)\begin{bmatrix}(\alpha-\beta)(\alpha-\gamma), & 0, & 0\\ 0, & (\alpha-\gamma), & 0\\ 0, & 0, & 1\end{bmatrix},$$

which yields the three independent constituents $e^{\alpha t}\{1,0,0\}$, $e^{\alpha t}\{0,1,0\}$ and $e^{\alpha t}\{0,0,1\}$.

Next consider the modifications when $\gamma=\beta$. In this case

$$f(\lambda)=(\lambda-\alpha)\begin{bmatrix}1, & 0, & 0\\ 0, & (\lambda-\beta), & 0\\ 0, & 0, & (\lambda-\beta)^2\end{bmatrix},$$

and $\Delta(\lambda) = (\lambda-\alpha)^3(\lambda-\beta)^3$, which yields two triple roots α and β. As before, the matrix $f(\lambda)$ has degeneracy 2 for the root β and 3 for the root α, while $F(\lambda)$ is null for each root. The three constituents corresponding to $\lambda = \alpha$ have the same forms as before. However,

$$\overset{(1)}{F}(\beta) = (\beta-\alpha)^2 \begin{bmatrix} 0 & 0 & 0 \\ 0 & 0 & 0 \\ 0 & 0 & 1 \end{bmatrix},$$

which yields only one non-vanishing column. To obtain the full number of constituents corresponding to the triple root β it is necessary to proceed to

$$\overset{(2)}{F}(\beta) = 2(\beta-\alpha) \begin{bmatrix} 0 & 0 & 0 \\ 0 & (\beta-\alpha) & 0 \\ 0 & 0 & 2 \end{bmatrix}.$$

The three independent constituents required are now given by the third column of the matrix $e^{\beta t}\overset{(1)}{F}(\beta)$ and the third and second columns of $e^{\beta t}[\overset{(2)}{F}(\beta) + 2t\overset{(1)}{F}(\beta)]$. They may be taken as $e^{\beta t}\{0, 0, 1\}$, $e^{\beta t}\{0, 0, t\}$ and $e^{\beta t}\{0, 1, 0\}$.

Lastly, suppose $\alpha = \beta = \gamma$, so that

$$f(\lambda) = (\lambda-\alpha) \begin{bmatrix} 1, & 0, & 0 \\ 0, & \lambda-\alpha, & 0 \\ 0, & 0, & (\lambda-\alpha)^2 \end{bmatrix},$$

with $\Delta(\lambda) = (\lambda-\alpha)^6$. Here α is a sextuple root, and $f(\alpha)$ has degeneracy 3. Then

$$F(\lambda) = (\lambda-\alpha)^3 \begin{bmatrix} (\lambda-\alpha)^2, & 0, & 0 \\ 0, & \lambda-\alpha, & 0 \\ 0, & 0, & 1 \end{bmatrix}.$$

The derived adjoint matrices which are not null are

$$\overset{(3)}{F}(\alpha) = \begin{bmatrix} 0 & 0 & 0 \\ 0 & 0 & 0 \\ 0 & 0 & 6 \end{bmatrix}; \quad \overset{(4)}{F}(\alpha) = \begin{bmatrix} 0 & 0 & 0 \\ 0 & 24 & 0 \\ 0 & 0 & 0 \end{bmatrix}; \quad \overset{(5)}{F}(\alpha) = \begin{bmatrix} 120 & 0 & 0 \\ 0 & 0 & 0 \\ 0 & 0 & 0 \end{bmatrix};$$

while the matrices whose non-vanishing columns yield possible constituents are

$$e^{\alpha t}\overset{(3)}{F}(\alpha), \quad e^{\alpha t}[\overset{(4)}{F}(\alpha) + 4t\overset{(3)}{F}(\alpha)], \quad \text{and} \quad e^{\alpha t}[\overset{(5)}{F}(\alpha) + 5t\overset{(4)}{F}(\alpha) + 10t^2\overset{(3)}{F}(\alpha)].$$

The six independent constituents may clearly be chosen as the six columns of the matrix

$$e^{at}\begin{bmatrix} 0 & 0 & 0 & 1 & 0 & 0 \\ 0 & 1 & 0 & 0 & t & 0 \\ 1 & 0 & t & 0 & 0 & t^2 \end{bmatrix}.$$

(ii) *A more General System.* Suppose

$$f(\lambda) = \begin{bmatrix} \lambda^3-\lambda^2-2\lambda, & \lambda^3-\lambda^2+\lambda, & -\lambda \\ \lambda^3-3\lambda, & \lambda^3, & -\lambda \\ \lambda^3+2\lambda^2-\lambda, & -\lambda^2, & \lambda^3+\lambda^2-\lambda \end{bmatrix}.$$

Here $\Delta(\lambda) = -2\lambda^3(\lambda-1)^2(\lambda+1)$, giving the roots $\lambda = 0$ (triple), $\lambda = 1$ (double), $\lambda = -1$ (single). In this case

$$F(\lambda) = \lambda^2(\lambda-1)\begin{bmatrix} \lambda(\lambda+1)^2, & -(\lambda^3+\lambda^2+1), & 1 \\ -(\lambda^2-1)(\lambda+2), & (\lambda^3+\lambda^2-2\lambda-1), & -1 \\ -\lambda(\lambda+1)(\lambda+3), & (\lambda^3+3\lambda^2+1), & -3 \end{bmatrix}.$$

This is null for $\lambda = 0$ and $\lambda = 1$, and for $\lambda = -1$ it reduces to

$$F(-1) = -2\begin{bmatrix} 0 & -1 & 1 \\ 0 & 1 & -1 \\ 0 & 3 & -3 \end{bmatrix}.$$

The constituent corresponding to the single root $\lambda = -1$ may thus be taken as $e^{-t}\{1, -1, -3\}$.

Since $F(\lambda)$ is null for the double root $\lambda = 1$, the two constituents corresponding to this root will be given by two of the columns of $e^t\overset{(1)}{F}(1)$. By direct differentiation of the expression for $F(\lambda)$, it is seen that

$$\overset{(1)}{F}(1) = \begin{bmatrix} 4 & -3 & 1 \\ 0 & -1 & -1 \\ -8 & 5 & -3 \end{bmatrix}.$$

The three columns of this matrix are not linearly independent: the sum of the first two yields the third.

Lastly, it is evident that $\overset{(1)}{F}(0) = F(0) = 0$, and that

$$\overset{(2)}{F}(0) = -2\begin{bmatrix} 0 & -1 & 1 \\ 2 & -1 & -1 \\ 0 & 1 & -3 \end{bmatrix}.$$

The three independent constituents corresponding to the triple root $\lambda = 0$ may therefore be taken proportional to the three columns of this matrix.

It should be noted that the arithmetical work involved in the evaluation of the derived adjoint matrices for the multiple roots is often greatly simplified by the extraction, as a scalar multiplier, of any factors common to the elements of $F(\lambda)$ (e.g. the extraction of $\lambda^2(\lambda-1)$ as a common factor in this example).

5·8. Numerical Evaluation of the Constituent Solutions. As explained in § 5·7 the constituent solutions appropriate to any given root λ_s of $\Delta(\lambda) = 0$ are found from the independent columns of the family of matrices $W_p(t, \lambda_s)$, which are linear in $F(\lambda_s)$ and its derivatives. A possible method of computation of the constituents is to construct $F(\lambda)$ from the cofactors in $f(\lambda)$, and to obtain the necessary derived adjoint matrices by direct differentiation with λ kept general. The value $\lambda = \lambda_s$ is then inserted. However, in practice, the construction of $F(\lambda)$ for a general value of λ is extremely laborious when the order of $f(\lambda)$ is even moderately large. A simpler treatment, which is actually an extension of the methods given in § 4·12, will now be described.

When λ_s is a simple root, the corresponding constituent can be taken proportional to any non-zero column of the matrix $W_0(t, \lambda_s)$. A column proportional to the columns of $F(\lambda_s)$ can be calculated at once by the method of § 4·12.

When λ_s is a double root, the constituent solutions are to be chosen proportional to any two linearly independent columns of W_0 and W_1. More generally, the required columns can be chosen from the matrices

$$aW_0 \equiv e^{\lambda_s t}[aF(\lambda_s)],$$

and

$$aW_1 + bW_0 \equiv e^{\lambda_s t}[a\overset{(1)}{F}(\lambda_s) + bF(\lambda_s) + taF(\lambda_s)],$$

in which a and b are arbitrary scalars. But since $f(\lambda_s)\,F(\lambda_s) = 0$ and $\dfrac{\partial}{\partial \lambda_s} f(\lambda_s)\,F(\lambda_s) = 0$ when λ_s is a double root, we have also

$$f(\lambda_s)\,[aF(\lambda_s)] = 0,$$

$$f(\lambda_s)\,[a\overset{(1)}{F}(\lambda_s) + bF(\lambda_s)] + \overset{(1)}{f}(\lambda_s)\,[aF(\lambda_s)] = 0.$$

Hence, if ϕ denotes a column of $aF(\lambda_s)$, and ϕ_1 the corresponding column of $a\overset{(1)}{F}(\lambda_s) + bF(\lambda_s)$,

$$f(\lambda_s)\,\phi = 0, \qquad \text{......(1)}$$

$$f(\lambda_s)\,\phi_1 + \overset{(1)}{f}(\lambda_s)\,\phi = 0. \qquad \text{......(2)}$$

Equation (1) when treated by the method of § 4·12 yields ϕ, and if the column so determined is substituted in (2), ϕ_1 can be found by operation on the rows of $f(\lambda_s)$ and $\overset{(1)}{f}(\lambda_s)\phi$ in a way similar to that described in § 4·12. If $f(\lambda_s)$ is doubly degenerate $F(\lambda_s)$ is null, and the equation to be solved then is

$$f(\lambda_s)\,\phi_1 = 0.$$

This yields two linearly independent solutions.

When λ_s is an s-fold root, an extension of the foregoing process yields successive columns which may be used directly to construct the constituent solutions.

Examples

(i) *First-Order System.* As a very simple first example, suppose

$$f(D)\,x \equiv \begin{bmatrix} D+1, & -4 \\ 1, & D-3 \end{bmatrix} \begin{bmatrix} x_1 \\ x_2 \end{bmatrix} = 0.$$

The determinantal equation $\Delta(\lambda) = 0$ has a double root $\lambda = 1$, and

$$f(1) = \begin{bmatrix} 2, & -4 \\ 1, & -2 \end{bmatrix}, \qquad \overset{(1)}{f}(1) = \begin{bmatrix} 1, 0 \\ 0, 1 \end{bmatrix}.$$

The method of § 4·12 shows at once that the columns of the adjoint $F(1)$ are proportional to $\{2, 1\}$, and we may therefore choose $\phi = \{2, 1\}$. Then

$$\begin{bmatrix} 2, & -4 \\ 1, & -2 \end{bmatrix} \phi_1 + \begin{bmatrix} 1, 0 \\ 0, 1 \end{bmatrix} \begin{bmatrix} 2 \\ 1 \end{bmatrix} = 0.$$

Since $f(1)$ is singular, the equations contained here are not linearly independent (see § 4·13). Writing $\phi_1 \equiv \{\alpha, \beta\}$, we obtain the single equation $\alpha - 2\beta + 1 = 0$, so that

$$\phi_1 = \{2\beta - 1, \beta\} = \{2, 1\}\beta - \{1, 0\}.$$

The column multiplied by β is an arbitrary multiple of ϕ and can be disregarded. Thus ϕ_1 can be taken to be $\{1, 0\}$, and the constituents can be chosen as $e^t\{2, 1\}$ and $e^t\{1, 0\} + te^t\{2, 1\}$.

The first columns of the actual adjoint $F(1)$ and of $\overset{(1)}{F}(1)$ are $\{-2, -1\}$ and $\{1, 0\}$.

(ii) *Second-Order System.* Next consider the system for which

$$f(\lambda) = \begin{bmatrix} 2\lambda^2 - 4\lambda + 1, & \lambda^2 - \lambda + 1, & 2\lambda - 1 \\ \lambda^2 - 3\lambda + 1, & \lambda^2 - 2\lambda + 1, & 2\lambda - 1 \\ \lambda^2 - 6\lambda + 3, & 2\lambda^2 - 6\lambda + 3, & 5\lambda - 3 \end{bmatrix}.$$

Here $\Delta(\lambda) = \lambda^2(\lambda-1)^3$, and $f(\lambda)$ is doubly degenerate for the double root $\lambda = 0$ and simply degenerate for the triple root $\lambda = 1$.

The method of § 4·12 at once yields the two constituent solutions corresponding to $\lambda = 0$. The rows of the transposed of $f(0)$ are combined as follows:

r_1	1	1	3	1	0	0
r_2	1	1	3	0	1	0
r_3	-1	-1	-3	0	0	1
r_1+r_3	0	0	0	1	0	1
r_2+r_3	0	0	0	0	1	1

Hence the two constituents may be chosen as $\{1, 0, 1\}$ and $\{0, 1, 1\}$.

For the triple root $\lambda = 1$ we have

$$f(1) = \begin{bmatrix} -1 & 1 & 1 \\ -1 & 0 & 1 \\ -2 & -1 & 2 \end{bmatrix}, \quad \overset{(1)}{f}(1) = \begin{bmatrix} 0 & 1 & 2 \\ -1 & 0 & 2 \\ -4 & -2 & 5 \end{bmatrix}, \quad \overset{(2)}{f}(1) = \begin{bmatrix} 4 & 2 & 0 \\ 2 & 2 & 0 \\ 2 & 4 & 0 \end{bmatrix}.$$

A column of $F(1)$ is found by the previous method to be $\phi = \{1, 0, 1\}$. The equation determining ϕ_1 is

$$f(1)\,\phi_1 + \overset{(1)}{f}(1)\,\phi = f(1)\,\phi_1 + \{2, 1, 1\} = 0; \text{ or, if } \phi_1 \equiv \{\alpha, \beta, \gamma\},$$

$$\begin{bmatrix} -1 & 1 & 1 \\ -1 & 0 & 1 \\ -2 & -1 & 2 \end{bmatrix}\begin{bmatrix} \alpha \\ \beta \\ \gamma \end{bmatrix} = \begin{bmatrix} -2 \\ -1 \\ -1 \end{bmatrix}.$$

If we reject the last scalar equation represented here, and regard γ as arbitrary, we can write

$$\begin{bmatrix} -1 & 1 \\ -1 & 0 \end{bmatrix}\begin{bmatrix} \alpha \\ \beta \end{bmatrix} + \gamma\begin{bmatrix} 1 \\ 1 \end{bmatrix} = \begin{bmatrix} -2 \\ -1 \end{bmatrix},$$

the solution of which is readily found to be

$$\begin{bmatrix} \alpha \\ \beta \end{bmatrix} = \gamma\begin{bmatrix} 1 \\ 0 \end{bmatrix} + \begin{bmatrix} 1 \\ -1 \end{bmatrix}.$$

Hence $$\phi_1 \equiv \{\alpha, \beta, \gamma\} = \gamma\{1, 0, 1\} + \{1, -1, 0\}.$$

Here, again, an arbitrary multiple of ϕ is present.

To determine the next independent column $\phi_2 \equiv \{\xi, \eta, \zeta\}$ we can assign γ at convenience; but for illustrative purposes the value will be kept general. The equation for ϕ_2 is then

$$f(1)\phi_2 + 2\overset{(1)}{f}(1)\phi_1 + \overset{(2)}{f}(1)\phi = 0,$$

or
$$\begin{bmatrix} -1 & 1 & 1 \\ -1 & 0 & 1 \\ -2 & -1 & 2 \end{bmatrix}\begin{bmatrix} \xi \\ \eta \\ \zeta \end{bmatrix} = \gamma\begin{bmatrix} -4 \\ -2 \\ -2 \end{bmatrix} + \begin{bmatrix} 2 \\ 2 \\ 4 \end{bmatrix} + \begin{bmatrix} -4 \\ -2 \\ -2 \end{bmatrix} = \gamma\begin{bmatrix} -4 \\ -2 \\ -2 \end{bmatrix} + \begin{bmatrix} -2 \\ 0 \\ 2 \end{bmatrix}.$$

Rejecting the last scalar equation, and keeping ζ general, we write

$$\begin{bmatrix} -1 & 1 \\ -1 & 0 \end{bmatrix}\begin{bmatrix} \xi \\ \eta \end{bmatrix} + \zeta\begin{bmatrix} 1 \\ 1 \end{bmatrix} = \gamma\begin{bmatrix} -4 \\ -2 \end{bmatrix} + \begin{bmatrix} -2 \\ 0 \end{bmatrix},$$

which gives
$$\begin{bmatrix} \xi \\ \eta \end{bmatrix} = \zeta\begin{bmatrix} 1 \\ 0 \end{bmatrix} + \gamma\begin{bmatrix} 2 \\ -2 \end{bmatrix} + \begin{bmatrix} 0 \\ -2 \end{bmatrix},$$

or
$$\phi_2 \equiv \{\xi, \eta, \zeta\} = \zeta\{1, 0, 1\} + 2\gamma\{1, -1, 0\} + \{0, -2, 0\}.$$

In this solution arbitrary multiples of ϕ and ϕ_1 are present, but it is legitimate to choose $\gamma = \zeta = 0$, and then $\phi_2 = \{0, -2, 0\}$. Hence the constituent solutions may be chosen to be

$$e^t\{1, 0, 1\}, \quad e^t\{1, -1, 0\} + te^t\{1, 0, 1\},$$

and
$$e^t\{0, -2, 0\} + 2te^t\{1, -1, 0\} + t^2e^t\{1, 0, 1\}.$$

Actually the first column of $F(1)$ is $\{1, 0, 1\}$; while the first columns of $\overset{(1)}{F}(1)$ and $\overset{(2)}{F}(1)$ are, respectively,

$$\{4, -1, 3\} = 3\{1, 0, 1\} + \{1, -1, 0\},$$

and
$$\{8, -8, 2\} = 2\{1, 0, 1\} + 6\{1, -1, 0\} + \{0, -2, 0\}.$$

5·9. Expansions in Partial Fractions. We shall next consider certain identities which will be used extensively later.

Suppose $G(\lambda)$ to denote an arbitrary $(\bar{q}, m)$ λ-matrix, and let $\Delta(\lambda)$, $F(\lambda)$ be, respectively, the λ-determinant and the adjoint matrix of $f(\lambda)$. Then each element of the matrix product $G(\lambda)F(\lambda)/\Delta(\lambda)$ can be developed in partial fractions by the usual methods of algebra, and a similar type of expansion can therefore be assumed for the complete matrix product. The precise form of the expansion depends on the nature of the roots of $\Delta(\lambda) = 0$.

Case I. Roots of $\Delta(\lambda) = 0$ *all Distinct.* In this case the correct form to assume is

$$\frac{G(\lambda)F(\lambda)}{\Delta(\lambda)} = Q(\lambda) + \sum_{r=1}^{n} \frac{A_r}{\lambda - \lambda_r}. \qquad \text{......(1)}$$

Here $Q(\lambda)$, if present, is a λ-matrix, while A_1, A_2, etc. are matrices of constants which are determinable by the usual rules. Thus to find A_r, multiply (1) throughout by $\lambda-\lambda_r$ and put $\lambda=\lambda_r$. This gives

$$A_r = G(\lambda_r)\,F(\lambda_r)/\overset{(1)}{\Delta}(\lambda_r).$$

Hence $$\frac{G(\lambda)\,F(\lambda)}{\Delta(\lambda)} = G(\lambda)f^{-1}(\lambda) = Q(\lambda)+\sum_{r=1}^{n}\frac{G(\lambda_r)\,F(\lambda_r)}{\overset{(1)}{\Delta}(\lambda_r)\,(\lambda-\lambda_r)}. \quad ......(2)$$

If the product $G(\lambda)\,F(\lambda)$ is of degree less than n (the degree of $\Delta(\lambda)$), then $Q(\lambda)=0$.

As a particular case assume $G(\lambda)=I_m$. Then (2) gives

$$\frac{F(\lambda)}{\Delta(\lambda)} = f^{-1}(\lambda) = Q(\lambda)+\sum_{r=1}^{n}\frac{F(\lambda_r)}{\overset{(1)}{\Delta}(\lambda_r)\,(\lambda-\lambda_r)}. \quad ......(3)$$

When the degree of $F(\lambda)$ does not exceed that of $\Delta(\lambda)$, the quotient in (3) will be a matrix of constants (or ciphers), say $Q=C$. To determine C put $\lambda=0$ in (3): then

$$\frac{F(0)}{\Delta(0)} = C-\sum_{r=1}^{n}\frac{F(\lambda_r)}{\lambda_r\overset{(1)}{\Delta}(\lambda_r)},$$

and substitution for C in (3) yields

$$\frac{F(\lambda)}{\Delta(\lambda)} = \frac{F(0)}{\Delta(0)}+\sum_{r=1}^{n}\frac{F(\lambda_r)\,\lambda}{\overset{(1)}{\Delta}(\lambda_r)\,(\lambda-\lambda_r)}. \quad ......(4)$$

This is the matrix equivalent of a well-known identity due to Heaviside.*

Case II. Roots of $\Delta(\lambda)=0$ *Repeated.* When repeated roots occur, a set of s roots equal to λ_s will give rise in (1) to a set of s terms of the form

$$\frac{B_1}{\lambda-\lambda_s}+\frac{B_2}{(\lambda-\lambda_s)^2}+\dots+\frac{B_s}{(\lambda-\lambda_s)^s}. \quad ......(5)$$

Writing $\Delta(\lambda)\equiv(\lambda-\lambda_s)^s\,\Delta_s(\lambda)$, we obtain in the usual way

$$B_i = \frac{1}{\underline{|s-i}}\left[\frac{\partial^{s-i}}{\partial\lambda^{s-i}}\frac{G(\lambda)\,F(\lambda)}{\Delta_s(\lambda)}\right]_{\lambda=\lambda_s}.$$

The preceding formulae are all identities in λ, and following the usual operational methods for scalar differential equations, we may replace λ by the differential operator $D\equiv\dfrac{d}{dt}$, and so derive corresponding "operational" identities. Such operational identities are particularly convenient in the solution of systems of differential equations with given initial conditions (see, for instance, § 6·9).

* For remarks on Heaviside's method of solution see § 6·9.

Examples

(i) *Operational Formulae for the Case of Distinct Roots.* In (2) substitute $\lambda = D$, and postmultiply both sides by $f(D)x(t) - \xi(t)$, where $x(t)$ and $\xi(t)$ are columns of any sets of m quantities dependent on t. Then, using the property $F(\lambda_r)f(\lambda_r) = 0$, we can write the resulting identity as

$$G(D)x(t) = Q(D)\{f(D)x(t) - \xi(t)\} + \sum_{r=1}^{n} A_r \frac{[f(D) - f(\lambda_r)]}{D - \lambda_r} x(t) + \frac{G(D)F(D)}{\Delta(D)} \xi(t) - \sum_{r=1}^{n} \frac{A_r}{D - \lambda_r} \xi(t). \quad \text{......(6)}$$

Since $f(D) - f(\lambda_r)$ is divisible by $D - \lambda_r$, the first summation in (6) is rational and integral in the operator D.

If $x(t)$ is now assumed to satisfy the differential equations

$$f(D)x(t) - \xi(t) = 0, \quad \text{......(7)}$$

then (6) reduces to

$$G(D)x(t) = \sum_{r=1}^{n} A_r \frac{[f(D) - f(\lambda_r)]}{D - \lambda_r} x(t) + \frac{G(D)F(D)}{\Delta(D)} \xi(t) - \sum_{r=1}^{n} \frac{A_r}{D - \lambda_r} \xi(t). \quad \text{......(8)}.$$

It should be noted that (8) holds good for arbitrary columns $x(t)$ and $\xi(t)$ only when $Q = 0$ in (2). If Q does not vanish, then (8) is only true in conjunction with (7). The formula will be used in § 6·5 to obtain a special form of solution of a system of differential equations.

(ii) *Operational Formulae for Case of Repeated Roots.* The modifications to (6) and (8) when repeated roots occur will next be considered briefly. The terms contributed to the right-hand sides of these equations due to a set of s roots equal to λ_s are clearly

$$\sum_{i=1}^{s} \frac{B_i}{(D - \lambda_s)^i} f(D)x(t) - \sum_{i=1}^{s} \frac{B_i}{(D - \lambda_s)^i} \xi(t). \quad \text{......(9)}$$

It is possible to substitute for the first summation in (9) another summation, each term of which is rational and integral in D. Thus it can be shown that

$$\sum_{i=1}^{s} \frac{B_i}{(D - \lambda_s)^i} f(D) \equiv \sum_{i=1}^{s} B_i \vartheta_i(D, \lambda_s), \quad \text{......(10)}$$

in which

$$(D - \lambda_s)^i \vartheta_i(D, \lambda_s) = f(D) - f(\lambda_s) - (D - \lambda_s)\overset{(1)}{f}(\lambda_s) - \dots - \frac{(D - \lambda_s)^{i-1}}{\lfloor i-1} \overset{(i-1)}{f}(\lambda_s). \quad \text{......(11)}$$

By expansion of $f(D) \equiv f(\lambda_s + \overline{D-\lambda_s})$ in a Taylor's series it is at once seen that $\vartheta_i(D, \lambda_s)$ is rational and integral in D. To establish (10) it is only necessary to collect together the separate powers of $(D-\lambda_s)^{-1}$ and to show that the total coefficients of these terms vanish: use should be made of the relation $F(\lambda)f(\lambda)/\Delta_s(\lambda) = (\lambda-\lambda_s)^s I$. The actual proof can be left to the reader.

The terms contributed to the right-hand sides of (6) and (8) may thus be taken as

$$\sum_{i=1}^{s} B_i \vartheta_i(D, \lambda_s)\, x(t) - \sum_{i=1}^{s} \frac{B_i}{(D-\lambda_s)^i}\, \xi(t). \qquad \text{......(12)}$$

Part II. Construction of the Complementary Function and of a Particular Integral

5·10. The Complementary Function. The notation and terminology summarised in the present section should be carefully noted as it will hereafter be adopted as the standard.

The number of the dependent variables is m and the n roots of $\Delta(\lambda) = 0$ are $\lambda_1, \lambda_2, \ldots, \lambda_n$. These roots are not necessarily all distinct, but different suffixes are used to specify the full set of roots. Generally λ_r denotes a typical simple root, and λ_s represents a typical member of a set of s equal roots.

The complete complementary function of the differential equations $f(D)\,x - \xi = 0$ is constructed as the sum of arbitrary multiples of the n independent constituents corresponding to the n roots $\lambda_1, \lambda_2, \ldots, \lambda_n$. Subject to the conventions just explained regarding the notation for roots, each root will contribute one constituent.

(*a*) *Simple Roots.* The constituent appropriate to the typical simple root λ_r is denoted by

$$x = \{k_{1r}, k_{2r}, \ldots, k_{mr}\}\, e^{\lambda_r t} \equiv k_r e^{\lambda_r t},$$

and the column of constants k_r can be chosen proportional to any non-vanishing column of the adjoint $F(\lambda_r)$ of $f(\lambda_r)$. This matrix will be a product of the type $k_r \kappa_r$.

(*b*) *Sets of Equal Roots.* If λ_s is one of a set of s equal roots, the constituent appropriate to λ_s is written

$$x = \{k_{1s}(t), k_{2s}(t), \ldots, k_{ms}(t)\}\, e^{\lambda_s t} \equiv k_s(t)\, e^{\lambda_s t},$$

and the s columns of the type $k_s(t)$ appropriate to the whole set of equal roots may be chosen proportional to any s linearly independent columns of the set of matrices (see § 5·7)

$$U_0(t, \lambda_s) = F(\lambda_s),$$

$$U_1(t, \lambda_s) = \overset{(1)}{F}(\lambda_s) + tF(\lambda_s),$$

$$U_2(t, \lambda_s) = \overset{(2)}{F}(\lambda_s) + 2t\overset{(1)}{F}(\lambda_s) + t^2F(\lambda_s),$$

$$\cdots\cdots\cdots\cdots\cdots\cdots\cdots\cdots\cdots\cdots\cdots\cdots\cdots\cdots\cdots\cdots\cdots\cdots$$

$$U_{s-1}(t, \lambda_s) = \overset{(s-1)}{F}(\lambda_s) + (s-1)\,t\overset{(s-2)}{F}(\lambda_s) + \frac{(s-1)(s-2)}{\lfloor 2}t^2\overset{(s-3)}{F}(\lambda_s) + \ldots + t^{s-1}F(\lambda_s).$$

The elements $k_{is}(t)$ are thus polynomials in t of degree $s-1$ at most. In the particular case where $s = 1$ (and the root concerned is thus simple) they are constants, and the brackets indicating their dependence on t may be omitted.

The columns $k_s(t)$ define, as it were, the modes of the constituents in their relation to purely exponential laws. The column appropriate to the typical root λ_s (whether simple or multiple) will be spoken of as the *modal column* corresponding to the root λ_s, and any element of a modal column will be referred to as a *modal coefficient*.

Since $k_s(t)\,e^{\lambda_s t}$ by definition satisfies the differential equations $f(D)\,x = 0$, it follows that the modal columns have the property*

$$f(\lambda_s + D)\,k_s(t) = 0.$$

The $(\overline{m}, n)$ matrix formed from the modal columns is of great importance in the further theory. It will be spoken of as the *modal matrix* and it will be denoted by

$$k(t) = \begin{bmatrix} k_{11}(t) & k_{12}(t) & \ldots & k_{1n}(t) \\ k_{21}(t) & k_{22}(t) & \ldots & k_{2n}(t) \\ \cdots & \cdots & \cdots & \cdots \\ k_{m1}(t) & k_{m2}(t) & \ldots & k_{mn}(t) \end{bmatrix}. \qquad \ldots\ldots(1)$$

In view of the remarks in the footnote to § 5·7 it is seen that in the set of s modal columns appropriate to any set of s roots equal to λ_s it is always legitimate to replace t by $(t-\tau)$, where τ is arbitrary. Hence in particular a modal matrix $k(t)$ may always be replaced by $k(t-\tau)$.

* Equation (5·6·1) is thus the particular case of this equation for $s = 1$.

If $M(t)$ represents the $(\overline{n}, n)$ diagonal matrix

$$M(t) = \begin{bmatrix} e^{\lambda_1 t} & 0 & \dots & 0 \\ 0 & e^{\lambda_2 t} & \dots & 0 \\ \dots & \dots & \dots & \dots \\ 0 & 0 & \dots & e^{\lambda_n t} \end{bmatrix}, \qquad \dots\dots(2)$$

and c is a column of n arbitrary constants $c_1, c_2, \dots, c_n$, the complete complementary function is expressible as

$$x = k(t)\, M(t)\, c.$$

With regard to the matrices $U_i(t, \lambda_s)$, it will be noted that $F(\lambda_s)$ appears as the coefficient of the highest power of t in every matrix of the set. Hence if $U_0(t, \lambda_s)$ contains any one column—say the first—which is not null, then the corresponding (first) columns in each of the succeeding matrices $U_1, U_2, \dots, U_{s-1}$ must contain $t, t^2, \dots, t^{s-1}$ for highest powers. It is evident therefore that these s columns must be linearly distinct, and they will accordingly yield the required s modal columns appropriate to the whole set of s equal roots.

If, on the other hand, $F(\lambda_s)$ is null (i.e. if $f(\lambda_s)$ is multiply degenerate), then the last term will be absent from $U_i(t, \lambda_s)$, and the coefficient of t^{i-1} (now the highest power of t) will be $\overset{(1)}{F}(\lambda_s)$. By the same argument as before, any one non-vanishing column of $\overset{(1)}{F}(\lambda_s)$ will then give rise to $s-1$ modal columns; and the additional modal column required must then be sought from the first other non-zero column of the family U_2, U_3, etc. which is linearly distinct from the $s-1$ columns already used. The extension to more complicated cases will be obvious.

Examples

(i) *Second-Order System.* Suppose

$$f(D)\, x \equiv \begin{bmatrix} D^2 - 5D + 3, & 4D^2 - 5D \\ 2D^2 - D, & -D^2 - D + 3 \end{bmatrix} \begin{bmatrix} x_1 \\ x_2 \end{bmatrix} = 0.$$

Here $\Delta(\lambda) = -9(\lambda+1)(\lambda-1)^3$, and

$$F(\lambda) = \begin{bmatrix} -\lambda^2 - \lambda + 3, & -4\lambda^2 + 5\lambda \\ -2\lambda^2 + \lambda, & \lambda^2 - 5\lambda + 3 \end{bmatrix}.$$

The roots of $\Delta(\lambda) = 0$ are denoted by $\lambda_1 = -1$ and $\lambda_2 = \lambda_3 = \lambda_4 = 1$. The first column of $F(\lambda_1)$ is $\{3, -3\}$, so that we may choose $k_1 = \{1, -1\}$.

Again, $F(\lambda_2)$ contains the non-zero (first) column $\{1, -1\}$, and the corresponding (first) columns of $\overset{(1)}{F}(\lambda_2)$ and $\overset{(2)}{F}(\lambda_2)$ are $\{-3, -3\}$ and $\{-2, -4\}$, respectively. Hence we may choose

$$k_2(t) = \{1, -1\},$$
$$k_3(t) = \{-3, -3\} + t\{1, -1\},$$
$$k_4(t) = \{-2, -4\} + 2t\{-3, -3\} + t^2\{1, -1\}.$$

The complementary function is accordingly

$$\begin{bmatrix} x_1 \\ x_2 \end{bmatrix} = \begin{bmatrix} 1, & 1, & -3+t, & -2-6t+t^2 \\ -1, & -1, & -3-t, & -4-6t-t^2 \end{bmatrix} \begin{bmatrix} e^{-t} & 0 & 0 & 0 \\ 0 & e^t & 0 & 0 \\ 0 & 0 & e^t & 0 \\ 0 & 0 & 0 & e^t \end{bmatrix} \begin{bmatrix} c_1 \\ c_2 \\ c_3 \\ c_4 \end{bmatrix}.$$

If this result be denoted by $x(t) = k(t)\,M(t)\,c$, the reader can readily verify that other possible forms of the complementary function are $x(t) = k(t-\tau)\,M(t)\,c'$ and $x(t) = k(t-\tau)\,M(t-\tau)\,c''$, where c' and c'' are new sets of constants, and τ is arbitrary.

(ii) *Form of Solution when* $\Delta(\lambda)$ *Vanishes Identically.* If $\Delta(\lambda)$ reduces to a constant, other than zero, so that the degree in λ of $\Delta(\lambda)$ is zero, no arbitrary constants can enter into the solution of $f(D)\,x = 0$. The general solution then is $x = 0$. On the other hand, if $\Delta(\lambda)$ vanishes identically, the degree in λ of $\Delta(\lambda)$ is indeterminate, so that the presence of arbitrary constants in the solution is not precluded. This case, which arises when the m equations of the system $f(D)\,x = 0$ are not distinct, is abnormal. It corresponds to the case in which one or more ciphers are present in the principal diagonal in Smith's canonical form for λ-matrices (see § 3·15).

A simple example is the system

$$h(D)\,x = \begin{bmatrix} D, & 0, & 0 \\ 0, & D(D-1), & 0 \\ 0, & 0, & 0 \end{bmatrix} \begin{bmatrix} x_1 \\ x_2 \\ x_3 \end{bmatrix} = 0,$$

for which $\Delta(\lambda) \equiv 0$. The general solution is clearly

$$x_1 = c_1,$$
$$x_2 = c_2 + c_3 e^t,$$
$$x_3 = \text{arbitrary, say } \psi(t).$$

Consider now an equivalent system $f(D)y = 0$, say

$$f(D)y \equiv Ah(D)By = 0,$$

in which $|A|$ and $|B|$ are assumed to be non-zero constants. For this new system $\Delta(\lambda) \equiv 0$ as before, and (as can be directly verified) all the first minors of $|f(\lambda)|$ contain $\lambda^2(\lambda-1)$ as a common factor. The general solution of $f(D)y = 0$ will accordingly contain one arbitrary function, arising from the fact that $|f(\lambda)| \equiv 0$. In addition there will be three constituents, not included under the arbitrary function, and involving three arbitrary constants in all, originating from the common factor of the first minors of $|f(\lambda)|$.

More generally, if the λ-matrix $f(\lambda)$ is of rank r, the general solution contains $n-r$ arbitrary functions. Further, if the common factor of the minors of order r is of degree p in λ, then there are p additional constituents of normal type corresponding to this common factor. As an illustration, consider the system for which

$$f(\lambda) \equiv \begin{bmatrix} \lambda, & \lambda, & \lambda^2+1 \\ \lambda^2-1, & \lambda^2+1, & \lambda \\ \lambda^3, & \lambda^3+2\lambda, & 2\lambda^2+1 \end{bmatrix}.$$

Here $f(\lambda)$ is of rank 2, and the first minors in $|f(\lambda)|$ have no common factor. The general solution contains one arbitrary function and no constituents of normal type, and may in fact be written

$$y_1 = (D^4+D^2+1)\psi(t),$$
$$y_2 = -(D^4-D^2-1)\psi(t),$$
$$y_3 = -2D\psi(t).$$

The differential operators are here respectively the cofactors of the elements in the first row of $|f(D)|$.

As a second illustration, take

$$f(\lambda) \equiv \begin{bmatrix} -1, & \lambda^2(\lambda^2-1), & \lambda^3(\lambda^2-1) \\ -1, & \lambda(\lambda^2-1)(\lambda+1), & \lambda^2(\lambda^2-1)(\lambda+1) \\ \lambda, & \lambda^3(\lambda^2-1), & \lambda^4(\lambda^2-1) \end{bmatrix}.$$

This matrix is also of rank 2, but the first minors contain $\lambda(\lambda^2-1)$ as a common factor. The general solution is

$$y_1 = 0,$$
$$y_2 = c_1+2c_3e^{-t}-D\psi(t),$$
$$y_3 = c_1+c_2e^t+c_3e^{-t}+\psi(t).$$

5·11. Construction of a Particular Integral. Suppose the differential equations to be $f(D)x - \xi(t) = 0$, and denote a particular integral as $P(t)$. Then evidently we can choose

$$P(t) = f^{-1}(D)\,\xi(t) = \frac{F(D)}{\Delta(D)}\,\xi(t). \qquad \text{......(1)}$$

The interpretation of this symbolic solution will now be considered.

(*a*) *Functions* $\xi(t)$ *Exponential.* A simple case frequently arising is where $\xi(t)$ can be represented over the range of t under consideration by a sum of the type

$$\xi(t) = \Sigma e^{\theta t}\rho.$$

The indices θ, and the columns of m constants ρ, may be real or complex. The part of the particular integral arising from the typical term of the series then is

$$P(t) = \frac{F(D)}{\Delta(D)}\,e^{\theta t}\rho = e^{\theta t}\frac{F(\theta)}{\Delta(\theta)}\rho. \qquad \text{......(2)}$$

This solution fails in the exceptional case where θ is a root of $\Delta(\lambda) = 0$ (e.g. resonant forced oscillations of an undamped dynamical system). Suppose $\theta = \lambda_r$, where for simplicity λ_r is assumed to be an unrepeated root of $\Delta(\lambda) = 0$. The equations of which a particular integral is required are now

$$f(D)x = e^{\lambda_r t}\rho. \qquad \text{......(3)}$$

Assume $P(t) = W_1(t, \lambda_r)\,b$, where W_1 is the matrix defined in § 5·7, and b is a column of constants left free for choice. On substitution of this value for $P(t)$ in (3) and application of (5·7·1), we obtain the condition

$$b = \frac{\rho}{\Delta(\lambda_r) + t\overset{(1)}{\Delta}(\lambda_r)}.$$

Now $\Delta(\lambda_r) = 0$, but since λ_r is here assumed not repeated $\overset{(1)}{\Delta}(\lambda_r) \neq 0$. Hence $b = \rho/\overset{(1)}{\Delta}(\lambda_r)$. The particular integral sought is accordingly

$$P(t) = \frac{e^{\lambda_r t}}{\overset{(1)}{\Delta}(\lambda_r)}\,[\overset{(1)}{F}(\lambda_r) + tF(\lambda_r)]\,\rho.$$

Since by (2), $\dfrac{e^{\theta t}F(\theta)}{\Delta(\theta)}\rho$ satisfies the differential equations

$$f(D)x - e^{\theta t}\rho = 0,$$

it follows by differentiation with respect to θ that the particular integral of the equations

$$f(D)x - t^q e^{\theta t}\rho = 0$$

is

$$P(t) = \frac{\partial^q}{\partial\theta^q}\left[\frac{e^{\theta t}F(\theta)}{\Delta(\theta)}\right]\rho = e^{\theta t}\left(\frac{\partial}{\partial\theta} + t\right)^q \frac{F(\theta)}{\Delta(\theta)}\rho.$$

(*b*) *Functions* $\xi(t)$ *General*. When the functions $\xi(t)$ are general, the usual procedure is to make use of certain expansions in partial fractions.

For example, if the roots λ_r are all distinct, we can use the simple identity

$$\frac{1}{\Delta(D)} = \sum_{r=1}^{n} \frac{1}{\Delta^{(1)}(\lambda_r)(D-\lambda_r)},$$

and express (1) as

$$P(t) = F(D) \sum_{r=1}^{n} \frac{1}{\Delta^{(1)}(\lambda_r)(D-\lambda_r)} \xi(t).$$

The particular integral given by this method is thus

$$P(t) = F(D) \sum_{r=1}^{n} \frac{e^{\lambda_r t}}{\Delta^{(1)}(\lambda_r)} \int_{t_0}^{t} e^{-\lambda_r t} \xi(t)\, dt. \qquad \text{......(4)}$$

The lower limit of integration gives rise to terms which can be included in the complementary function, and it is often omitted for convenience.

When repeated roots of $\Delta(\lambda) = 0$ occur, the part of the expression for $1/\Delta(D)$ in partial fractions arising from a root λ_s of multiplicity s is of the form $\sum_{q=1}^{s} \frac{\alpha_q}{(D-\lambda_s)^q}$, where the quantities α are constants which are readily found by the usual methods. The corresponding terms of the particular integral are then

$$\sum_{q=1}^{s} \alpha_q F(D)\, e^{\lambda_s t} Q^q e^{-\lambda_s t} \xi(t),$$

where Q^q denotes q repeated integrations with respect to t.

Another possible treatment is to express the complete operator $F(D)/\Delta(D)$ in partial fractions. In the special case where the degree in D of $F(D)$ does not exceed that of $\Delta(D)$, the operational form of the expansion corresponding to (5·9·4) can conveniently be used for this purpose. This yields

$$P(t) = \frac{F(0)}{\Delta(0)} \xi(t) + \sum_{r=1}^{n} \frac{F(\lambda_r)}{\lambda_r \Delta^{(1)}(\lambda_r)} \cdot \frac{D}{D-\lambda_r} \xi(t).$$

But

$$\frac{D}{D-\lambda_r} \xi(t) = e^{\lambda_r t}(1+\lambda_r Q)\, e^{-\lambda_r t} \xi(t)$$

$$= \xi(t) + \lambda_r e^{\lambda_r t} \int_{t_0}^{t} e^{-\lambda_r t} \xi(t)\, dt, \qquad \text{......(5)}$$

so that

$$P(t) = \frac{F(0)}{\Delta(0)} \xi(t) + \sum_{r=1}^{n} \frac{F(\lambda_r)}{\lambda_r \Delta^{(1)}(\lambda_r)} \xi(t) + \sum_{r=1}^{n} \frac{e^{\lambda_r t}}{\Delta^{(1)}(\lambda_r)} F(\lambda_r) \int_{t_0}^{t} e^{-\lambda_r t} \xi(t)\, dt. \qquad \text{......(6)}$$

If (5·9·4) is applied to evaluate $F(\infty)/\Delta(\infty)$, equation (6) can be written

$$P(t) = \frac{F(\infty)}{\Delta(\infty)}\xi(t) + \sum_{r=1}^{n} \frac{e^{\lambda_r t}}{\overset{(1)}{\Delta}(\lambda_r)} F(\lambda_r) \int_{t_0}^{t} e^{-\lambda_r t}\xi(t)\,dt. \quad ......(7)$$

As a particular application of (7) suppose $\xi(t) = e^{\theta t}\rho$. Then

$$P(t) = e^{\theta t}\frac{F(\infty)}{\Delta(\infty)}\rho + \sum_{r=1}^{n} \frac{e^{\theta t_0}(e^{\theta(t-t_0)} - e^{\lambda_r(t-t_0)})}{(\theta-\lambda_r)\overset{(1)}{\Delta}(\lambda_r)} F(\lambda_r)\rho. \quad ......(8)$$

The terms involving $e^{\lambda_r t}$ may be included under the complementary function and therefore omitted from (8). The particular integral can then, with a little reduction, be identified with (2). However, with regard to (8) and the more general formula (7), it may be noted that if (as normally) the degree in D of $F(D)$ is *less* than that of $\Delta(D)$, then $F(\infty)/\Delta(\infty) = 0$. In this case the particular integral as given has the convenient property that $P(t_0) = 0$.

CHAPTER VI

LINEAR ORDINARY DIFFERENTIAL EQUATIONS WITH CONSTANT COEFFICIENTS (*continued*)

PART I. BOUNDARY PROBLEMS

6·1. Preliminary Remarks. In Chapter v it has been shown that the general solution of a system of linear differential equations $f(D)x-\xi=0$ in m dependent variables contains n arbitrary constants, where n is the degree in λ of the determinant $\Delta(\lambda)\equiv|f(\lambda)|$. We shall now consider the question as to how the values of these constants are to be determined in order that the solution may satisfy any assigned supplementary conditions.*

A "supplementary" or "boundary" condition usually consists of a linear relation connecting the values of the dependent variables x, and possibly their derivatives up to a certain order, at one or more given points t_0, t_1, etc. of the range of variation of t. The problem is said to be a *one-point boundary problem*, a *two-point boundary problem*, and so on, according to the number of different points t_0, t_1, etc. concerned. Attention will here be restricted to one-point and two-point boundary problems.

EXAMPLES

(i) *Simple One-Point and Two-Point Boundary Problems.* If the differential equations are

$$\begin{bmatrix} 17D^2-17D+6, & 7D^2-9D+4 \\ 7D^2+D+10, & 3D^2-D+4 \end{bmatrix}\begin{bmatrix} x_1 \\ x_2 \end{bmatrix}=0, \qquad \text{......(1)}$$

the general solution, which involves four arbitrary constants, is

$$\{x_1(t), x_2(t)\}=c_1e^t\{1,-3\}+c_2e^{-t}\{1,-2\}+c_3e^{2t}\{7,-20\}+c_4e^{4t}\{8,-21\}.$$

If the values of x_1, x_2 and of their first derivatives $\dot{x}_1$, $\dot{x}_2$ at say $t=0$ are assigned, we have a simple one-point boundary problem; whereas if the values of x_1, x_2 at $t=0$ and $t=1$ are given, we have a two-point boundary problem. In either case the four constants of integration c_1, c_2, c_3, c_4 are uniquely determinable.

* The differential equations, considered in conjunction with the supplementary conditions, are sometimes referred to as a "linear differential system". The differential equations themselves are then spoken of as the "system of linear differential equations".

(ii) *General One-Point Boundary Problems.* The supplementary conditions are not always of the simple form given in example (i). As an illustration of a rather more general class of one-point boundary problem we may suppose the differential equations to be (1) and the supplementary conditions to be

$$\left.\begin{aligned} 3\dot{x}_1(t_0)-2x_1(t_0) \qquad\qquad - x_2(t_0) &= 7,\\ 2x_1(t_0)+2\dot{x}_2(t_0)+ x_2(t_0) &= 5,\\ \dot{x}_1(t_0)-2x_1(t_0)-3\dot{x}_2(t_0)-2x_2(t_0) &= -1,\\ x_1(t_0)+2\dot{x}_2(t_0)+ x_2(t_0) &= 6. \end{aligned}\right\} \qquad \ldots\ldots(2)$$

It is of course possible here to solve these algebraic equations and to derive explicitly the initial values of $x_1, x_2, \dot{x}_1, \dot{x}_2$. The problem is then reduced to the simple type already considered in example (i). However, this process, which amounts to a conversion of a given set of boundary conditions to an equivalent simpler set, may not always be convenient. It is sometimes preferable to deal directly with the boundary conditions in their original form. To express (2) concisely, we introduce the symbol D_0 to mean the differential operator

$$D_0 \equiv \frac{d}{dt_0} \equiv \left(\frac{d}{dt}\right)_{t=t_0}.$$

The conditions can then be written as

$$\begin{bmatrix} 3D_0-2, & -2 \\ 2, & 2D_0+1 \\ D_0-2, & -3D_0-2 \\ 1, & 2D_0+1 \end{bmatrix} \begin{bmatrix} x_1(t_0) \\ x_2(t_0) \end{bmatrix} = \begin{bmatrix} 7 \\ 5 \\ -1 \\ 6 \end{bmatrix}. \qquad \ldots\ldots(3)$$

With a more general one-point boundary problem the elements of the $(\overline{4}, 2)$ matrix on the left of (3) would be replaced by given polynomials in D_0, and a given column would be assigned on the right. Methods of solution with the boundary conditions in the unreduced form are given in § 6·4.

6·2. Characteristic Numbers.

In § 6·1 it was assumed that the number of the supplementary conditions exactly equals the number n of the free constants in the general solution. It was also tacitly assumed that in such cases the values of those constants would be uniquely determinable. However, if the differential equations themselves contain one or more variable parameters, free for choice, it may be possible, for particular values of these parameters, to obtain solutions in which all the free constants are not uniquely fixed, even though just n supplementary conditions are assigned. Alternatively, it may be possible—for special values of the parameters—to satisfy more than n supplementary conditions. The study of these particular values, which are known as *characteristic numbers*, is of fundamental importance in many physical applications, such as the determination of the natural frequencies of vibration of dynamical systems.

Example

Consider the transverse oscillations of a string stretched between the points $y = 0$ and $y = l$. Let μ denote the mass per unit length of the stretched string and T the tension: then the transverse displacement z at any time t satisfies the differential equation

$$\frac{\partial^2 z}{\partial y^2} = a^2 \frac{\partial^2 z}{\partial t^2},$$

where $a^2 = \mu/T$. Now assume a free vibration of the string to be $z = Z \sin \omega t$, where ω is a parameter left free for choice and Z is a function of y only which satisfies the equation

$$\frac{d^2 Z}{dy^2} + (a\omega)^2 Z = 0 \qquad \text{......(1)}$$

and the two supplementary conditions $Z = 0$ at $y = 0$ and $y = l$. The general solution of (1), which contains two arbitrary constants, is

$$Z = c_1 e^{ia\omega y} + c_2 e^{-ia\omega y},$$

and the supplementary conditions require that

$$\left.\begin{aligned} c_1 + c_2 &= 0, \\ c_1 e^{ia\omega l} + c_2 e^{-ia\omega l} &= 0. \end{aligned}\right\} \qquad \text{......(2)}$$

In general, the only solution of equations (2) is $c_1 = c_2 = 0$. However, in the present case ω is free for choice, and whenever it is a root of $\sin(a\omega l) = 0$, equations (2) are satisfied provided $c_1 = -c_2$. The characteristic numbers ω are accordingly π/al, $2\pi/al$, etc., and the solution corresponding to the rth of these numbers is $Z = c \sin(r\pi y/l)$, where c is a free constant.

6·3. Notation for One-Point Boundary Problems. Attention will here be restricted to the type of boundary problem in which no free parameters occur in the differential equations and the number of the assigned conditions exactly equals the number of the free constants in the general solution.

The simplest, and normal, type of problem is where the differential equations are of order N and the determinantal equation $\Delta(\lambda) = 0$ has Nm roots, so that $n = Nm$. In a one-point boundary problem it is then usual to have assigned the initial values of the m variables x and of their derivatives up to, and including, the order $N-1$. The problem will be said to be of *standard type* when the differential equations and the boundary conditions have the foregoing special features.

In the general one-point boundary problem the assigned conditions consist of n independent linear combinations of the values at $t = t_0$ of the dependent variables x, and (or) of the derivatives of these variables up to a given order s. Such a general set of n conditions may be represented by (see example (ii) of § 6·1)

$$\phi(D_0)\,x(t_0) = \begin{bmatrix} \phi_{11}(D_0), \phi_{12}(D_0), \ldots, \phi_{1m}(D_0) \\ \phi_{21}(D_0), \phi_{22}(D_0), \ldots, \phi_{2m}(D_0) \\ \cdots\cdots\cdots\cdots\cdots\cdots \\ \phi_{n1}(D_0), \phi_{n2}(D_0), \ldots, \phi_{nm}(D_0) \end{bmatrix} \begin{bmatrix} x_1(t_0) \\ x_2(t_0) \\ \cdots\cdots \\ x_m(t_0) \end{bmatrix} = \begin{bmatrix} \Phi_1 \\ \Phi_2 \\ \cdots \\ \Phi_n \end{bmatrix}. \qquad \ldots\ldots(1)$$

The $(\bar{n}, m)$ matrix $\phi(D_0)$ has for elements polynomials* of D_0 of degree s at most, while Φ_1, Φ_2, etc. are given constants.

The set of conditions (1) may be expressed alternatively as

$$(B_0 D_0^s + B_1 D_0^{s-1} + \ldots + B_{s-1} D_0 + B_s)\,x(t_0) = \Phi,$$

in which B_0, B_1, etc. are $(\bar{n}, m)$ matrices of given constants. Such a set of boundary conditions may conveniently be referred to as being of *order* s. If $s > N-1$, it is clearly always possible to reduce the order of the supplementary conditions to $N-1$ by successive substitutions from the differential equations.

When the problem is of standard type $\phi(D_0)$ is an $(\overline{Nm}, m)$ matrix, which can be partitioned into square matrices of order m and conveniently represented by

$$\phi(D_0) = \{I_m, I_m D_0, \ldots, I_m D_0^{N-1}\}. \qquad \ldots\ldots(2)$$

In § 6·4 a direct solution of the differential equations will be obtained which is applicable to the general one-point boundary problem. When the problem is of standard type a special form of the solution can be used which is often more convenient than the direct solution (see §§ 6·5 and 6·6).

* In connection with boundary problems of this general type it is important to observe that the choice of the polynomials $\phi_{ij}(D_0)$ and of the constants Φ_i is not altogether unrestricted. The boundary conditions must be compatible with the differential equations.

EXAMPLES

(a) Problems of Standard Type

(i) *Single Differential Equation.* If the given differential equation is

$$D^n x + p_1 D^{n-1} x + \dots + p_n x = \xi(t),$$

then $m = 1$ and $N = n$. Hence (1) and (2) give

$$\phi(D_0)\,x(t_0) = \{1, D_0, \dots, D_0^{n-1}\}\,x(t_0) = \{\Phi_1, \Phi_2, \dots, \Phi_n\}.$$

(ii) *Linear Dynamical Equations of Lagrangian Type* (see example (ii) of § 5·5). If there are m generalised coordinates q, then $N = 2$ and $n = 2m$. Hence, assuming given initial values Φ for the sets of quantities q and $\dot{q}$, we have

$$\phi(D_0)\,q(t_0) = \{I_m, I_m D_0\}\,q(t_0) = \Phi.$$

(b) Problems of General Type

(iii) *A System of Fourth Order.* Consider the homogeneous equations

$$f(D)\,x \equiv \begin{bmatrix} D+2, & D^3+2D^2-1 \\ D^2, & D^4-1 \end{bmatrix} \begin{bmatrix} x_1 \\ x_2 \end{bmatrix} = 0. \qquad \text{......(3)}$$

Here $m = 2$ and $N = 4$: but $\Delta(\lambda) = \lambda^2 - \lambda - 2$, so that $n = 2(\neq mN)$. Only two boundary conditions can be assigned in this case, and the problem is not of standard type. The simplest set of supplementary conditions here would be $x_1(t_0) = \Phi_1$ and $x_2(t_0) = \Phi_2$, corresponding to $\phi(D_0) = I_2$. For a standard problem with $m = 2$ and $N = 4$, we should require $n = 2 \times 4 = 8$, and the values of x and of its first seven derivatives at $t = t_0$ would be assigned.

(iv) *Change of Order of Boundary Conditions.* The formulae of § 5·9 can be applied to change the order of a set of one-point boundary conditions $\phi(D_0)\,x(t_0) = \Phi$. In (5·9·8) write ϕ for G, t_0 for t, D_0 for D: then

$$\Phi = \phi(D_0)\,x(t_0) = \sum_{r=1}^{n} \frac{C_r[f(D_0) - f(\lambda_r)]}{D_0 - \lambda_r}\,x(t_0) + \frac{\phi(D_0)\,F(D_0)}{\Delta(D_0)}\,\xi(t_0)$$

$$- \sum_{r=1}^{n} \frac{C_r}{D_0 - \lambda_r}\,\xi(t_0), \qquad \text{......(4)}$$

where $C_r = \phi(\lambda_r)\,F(\lambda_r)/\overset{(1)}{\Delta}(\lambda_r)$. Since $f(D_0) - f(\lambda_r)$ is exactly divisible by $D_0 - \lambda_r$ and the quotient is of degree $N-1$ in D_0, the given boundary conditions are expressed by (4) as an equivalent set of order $N-1$.

For example, let the differential equations be (3). Here $\lambda_1 = 2$ and $\lambda_2 = -1$, and it is readily verified that (4) yields the relation

$$\phi(D_0)\,x(t_0) = \tfrac{1}{3}\phi(2)\{15, -4\}[-D_0-1, -D_0^3-D_0^2]\,x(t_0)$$
$$-\tfrac{1}{3}\phi(-1)\{0, -1\}[-D_0+2, -D_0^3+2D_0^2]\,x(t_0). \quad \text{......(5)}$$

With equations (3) only two supplementary conditions can be assigned. If the order of these exceeds 3, they are reducible by (5) to an equivalent pair of order 3. Suppose, however, that the given conditions are simply $x_1(t_0) = \Phi_1$ and $x_2(t_0) = \Phi_2$, corresponding to $\phi(D_0) = I_2$. Equation (5) would then lead to an *increase* in order of the boundary conditions. Thus, with $\phi = I$ in (5), we obtain at $t = t_0$

$$\{x_1, x_2\} = \tfrac{1}{3}\{15, -4\}(-D_0x_1 - x_1 - D_0^3x_2 - D_0^2x_2)$$
$$-\tfrac{1}{3}\{0, -1\}(-D_0x_1 + 2x_1 - D_0^3x_2 + 2D_0^2x_2),$$

or

$$\left.\begin{aligned} 5D_0x_1 + 6x_1 + 5D_0^3x_2 + 5D_0^2x_2 &= 0, \\ D_0x_1 + 2x_1 + D_0^3x_2 + 2D_0^2x_2 - x_2 &= 0. \end{aligned}\right\} \quad \text{......(6)}$$

It is an instructive exercise to show that these equations are a necessary consequence of the differential equations (3). As already explained in § 5·9, equation (5·9·8) is not an identity unless the quotient Q is absent in (5·9·2). In the present example such a quotient would arise because the degree in D of $F(D)$ exceeds that of $\Delta(D)$: and it would actually give rise on the right-hand side of (5) to an additional term

$$\begin{bmatrix} D_0^2+D_0+3, & -D_0-3 \\ -1, & 0 \end{bmatrix}\begin{bmatrix} D_0+2, & D_0^3+2D_0^2-1 \\ D_0^2, & D_0^4-1 \end{bmatrix} x(t_0). \quad \text{......(7)}$$

The relation (5), as thus modified, would have been an identity for all values of x. On the other hand, the matrix product (7) contains $f(D_0)\,x(t_0)$ as its last two factors, and this is null by (3): for this reason the product is omitted from (5). The conditions that the fully expanded product (7) shall be null are precisely the equations (6).

6·4. Direct Solution of the General One-Point Boundary Problem.

Before obtaining the direct solution in its most general form, we shall consider a simple case. Suppose the differential equations to be given by

$$A\ddot{q} + B\dot{q} + Cq = 0,$$

representing the small free motions of a dynamical system with m generalised coordinates q. Let the $n\,(=2m)$ roots λ_r of

$$\Delta(\lambda) \equiv |A\lambda^2 + B\lambda + C| = 0$$

be all distinct, and let the values $q(t_0)$, $\dot{q}(t_0)$ be assigned. Now the most general solution of the foregoing equations is expressible as

$$q(t) = kM(t)\,c, \qquad \text{......(1)}$$

where k and M are as defined in § 5·10, and c is a column of n arbitrary constants. Moreover, since in the present case the roots λ_r are assumed to be all distinct, the $(\overline{m}, n)$ modal matrix k contains only constant elements.

On differentiation with respect to t, equation (1) gives

$$\dot{q}(t) = k\Lambda M(t)\,c,$$

where Λ is the diagonal matrix of the roots λ_r. Hence when $t = t_0$ we have $q(t_0) = kM(t_0)\,c$ and $\dot{q}(t_0) = k\Lambda M(t_0)\,c$. These two relations can be combined into the single matrix equation

$$\{q(t_0), \dot{q}(t_0)\} = lM(t_0)\,c, \qquad \text{......(2)}$$

where l is the square matrix of order n defined by

$$l \equiv \begin{bmatrix} k \\ \hline k\Lambda \end{bmatrix} \equiv \begin{bmatrix} k_{11} & \cdots & k_{1n} \\ \cdots & \cdots & \cdots \\ k_{m1} & \cdots & k_{mn} \\ \lambda_1 k_{11} & \cdots & \lambda_n k_{1n} \\ \cdots & \cdots & \cdots \\ \lambda_1 k_{m1} & \cdots & \lambda_n k_{mn} \end{bmatrix}.$$

Solution of (2) for c yields

$$c = M(-t_0)\,l^{-1}\{q(t_0), \dot{q}(t_0)\},$$

and substitution of this value for c in (1) gives the direct solution

$$q(t) = kM(t-t_0)\,l^{-1}\{q(t_0), \dot{q}(t_0)\}.$$

The corresponding solution for a non-homogeneous system of general order, and for a general one-point boundary problem, can be obtained by a similar method. Assume the equations to be

$$f(D)\,x(t) - \xi(t) = 0, \qquad \text{......(3)}$$

and let the boundary conditions be

$$\phi(D_0)\,x(t_0) = \Phi. \qquad \text{......(4)}$$

The general solution may be written

$$x(t) = k(t)\,M(t)\,c + P(t). \qquad \text{......(5)}$$

The first term on the right is the complementary function obtained in § 5·10, while $P(t)$ denotes any particular integral as found by the methods of § 5·11. On premultiplication of both sides of (5) by $\phi(D)$ it is readily proved that

$$\phi(D)\,x(t) = l(t)\,M(t)\,c + \phi(D)\,P(t), \qquad \text{......(6)}$$

where $l(t)$ is the square matrix of order n

$$l(t) \equiv \begin{bmatrix} l_{11}(t),\ l_{12}(t),\ \ldots,\ l_{1n}(t) \\ l_{21}(t),\ l_{22}(t),\ \ldots,\ l_{2n}(t) \\ \cdots\cdots\cdots\cdots\cdots\cdots \\ l_{n1}(t),\ l_{n2}(t),\ \ldots,\ l_{nn}(t) \end{bmatrix}, \qquad \text{......(7)}$$

the sth column of which is given by

$$\{l_{is}(t)\} = \phi(\lambda_s + D)\,\{k_{is}(t)\}.$$

In (6), which is true for all values of t, substitute $t = t_0$ and premultiply throughout by $M(-t_0)\,l^{-1}(t_0)$. Then using (4), we obtain

$$c = M(-t_0)\,l^{-1}(t_0)\,\Phi - M(-t_0)\,l^{-1}(t_0)\,\phi(D_0)\,P(t_0). \qquad \text{......(8)}$$

Substitution of this value for c in (5) yields the *direct matrix solution**

$$x(t) = k(t)\,M(t-t_0)\,l^{-1}(t_0)\,\{\Phi - \phi(D_0)\,P(t_0)\} + P(t). \qquad \text{......(9)}$$

A more symmetrical equivalent is

$$x(t) - P(t) = k(t)\,M(t-t_0)\,l^{-1}(t_0)\,\phi(D_0)\,\{x(t_0) - P(t_0)\}. \qquad \text{...(10)}$$

When the boundary problem is of standard type,† so that $\phi(D_0)$ has the special form (6·3·2), the matrix l is expressible in the partitioned form

$$l(t) = \begin{bmatrix} k_1(t), & k_2(t), & \ldots, & k_n(t) \\ (\lambda_1 + D)\,k_1(t), & (\lambda_2 + D)\,k_2(t), & \ldots, & (\lambda_n + D)\,k_n(t) \\ \cdots & \cdots & \cdots & \cdots \\ (\lambda_1 + D)^{N-1}k_1(t), & (\lambda_2 + D)^{N-1}k_2(t), & \ldots, & (\lambda_n + D)^{N-1}k_n(t) \end{bmatrix}, \qquad \text{......(11)}$$

where each element is a column matrix. If further the roots of $\Delta(\lambda) = 0$ are all distinct, the elements $l_{is}(t)$ are all constants, and the brackets

* Other forms of this solution can be obtained by replacing $k(t)$ by $k(t-\tau)$, where τ is arbitrary, and defining the matrix $l(t)$ to correspond (see remarks following equation (5·10·1)).

† For an alternative method of solution for problems of standard type see § 6·5.

indicating the dependence of these coefficients on t can be omitted. The formula (11) then reduces to

$$l = \begin{bmatrix} k_1, & k_2, & \ldots, & k_n \\ \lambda_1 k_1, & \lambda_2 k_2, & \ldots, & \lambda_n k_n \\ \cdots & \cdots & \cdots & \cdots \\ \lambda_1^{N-1} k_1, & \lambda_2^{N-1} k_2, & \ldots, & \lambda_n^{N-1} k_n \end{bmatrix}. \qquad \ldots\ldots(12)$$

Examples

(i) *A System of Fourth Order.* Consider the homogeneous equations (6·3·3)

$$f(D)\,x \equiv \begin{bmatrix} D+2, & D^3+2D^2-1 \\ D^2, & D^4-1 \end{bmatrix} \begin{bmatrix} x_1 \\ x_2 \end{bmatrix} = 0.$$

Here, with $\lambda_1 = 2$ and $\lambda_2 = -1$,

$$F(\lambda_1) = \begin{bmatrix} 15 & -15 \\ -4 & 4 \end{bmatrix} = \begin{bmatrix} 15 \\ -4 \end{bmatrix} [1, -1];$$

$$F(\lambda_2) = \begin{bmatrix} 0 & 0 \\ -1 & 1 \end{bmatrix} = \begin{bmatrix} 0 \\ -1 \end{bmatrix} [1, -1],$$

giving

$$k = \begin{bmatrix} 15 & 0 \\ -4 & -1 \end{bmatrix}.$$

If the initial conditions are $x_1 = \Phi_1$ and $x_2 = \Phi_2$ at $t_0 = 0$, corresponding to $\phi(D_0) = I_2$, then by (7) $l = k$ and

$$l^{-1} = \begin{bmatrix} 15 & 0 \\ -4 & -1 \end{bmatrix}^{-1} = -\tfrac{1}{15}\begin{bmatrix} -1 & 0 \\ 4 & 15 \end{bmatrix}.$$

Hence the solution is

$$x = -\tfrac{1}{15}\begin{bmatrix} 15 & 0 \\ -4 & -1 \end{bmatrix} \begin{bmatrix} e^{2t} & 0 \\ 0 & e^{-t} \end{bmatrix} \begin{bmatrix} -1 & 0 \\ 4 & 15 \end{bmatrix} \begin{bmatrix} \Phi_1 \\ \Phi_2 \end{bmatrix}.$$

(ii) *Linear Dynamical Equations of Lagrangian Type.* If the values at $t = t_0$ of the m generalised coordinates q and of the m generalised velocities $\dot{q}$ are assigned, then

$$\phi(D_0) = \{I_m, I_m D_0\}.$$

In this case

$$l(t) = \begin{bmatrix} k_1(t), & k_2(t), & \ldots, & k_n(t) \\ (\lambda_1+D)\,k_1(t), & (\lambda_2+D)\,k_2(t), & \ldots, & (\lambda_n+D)\,k_n(t) \end{bmatrix}.$$

(iii) *A Modified Form of the Direct Solution.* If the value of c given by (8) is introduced in (6) instead of in (5), the result is

$$\phi(D)\,x(t) = l(t)\,M(t-t_0)\,l^{-1}(t_0)\,\{\Phi - \phi(D_0)\,P(t_0)\} + \phi(D)\,P(t),$$

which, on premultiplication by $l^{-1}(t)$ and rearrangement, gives

$$l^{-1}(t)\,\phi(D)\,\{x(t) - P(t)\} = M(t-t_0)\,l^{-1}(t_0)\,\phi(D_0)\,\{x(t_0) - P(t_0)\}. \qquad \text{......(13)}$$

Now choose as new dependent variables the set of n functions $\alpha(t)$ given by

$$\alpha(t) = l^{-1}(t)\,\phi(D)\,x(t),$$

and put

$$\beta(t) = l^{-1}(t)\,\phi(D)\,P(t).$$

Then the solution (13) is expressible in the simple form

$$\alpha(t) - \beta(t) = M(t-t_0)\,\{\alpha(t_0) - \beta(t_0)\}.$$

In the case of the Lagrangian equations considered in example (ii) $x = q$ and $\phi(D) = \{I_m, I_m D\}$: hence

$$\alpha(t) = l^{-1}(t)\,\{q(t), \dot{q}(t)\},$$

$$\beta(t) = l^{-1}(t)\,\{P(t), \dot{P}(t)\}.$$

In dynamical applications considered in Chapter XI the quantities $\alpha(t)$ are referred to as the *reducing variables*, since by the choice of these variables the equations of motion are reduced to a simple diagonal form.

6·5. Special Solution for Standard One-Point Boundary Problems. In numerical applications of the direct matrix solution the principal difficulty lies in the calculation of the inverse matrix $l^{-1}(t_0)$. When the differential equations and the boundary conditions are of standard type, the direct inversion of this matrix can be avoided by an alternative method of solution, which will be referred to as the *special form of solution*. The intial conditions in this case are expressed by

$$\phi(D_0)\,x(t_0) = \{I_m, I_m D_0, \ldots, I_m D_0^{N-1}\}\,x(t_0) = \Phi, \qquad \text{......(1)}$$

and if the roots of $\Delta(\lambda) = 0$ are all distinct (as will here be assumed) the special solution is

$$x(t) - P(t) = \sum_{r=1}^{n} \frac{e^{\lambda_r(t-t_0)}}{\overset{(1)}{\Delta}(\lambda_r)}\,F(\lambda_r)\,\frac{[f(D_0) - f(\lambda_r)]}{D_0 - \lambda_r}\,\{x(t_0) - P(t_0)\}. \qquad \text{......(2)}$$

Since $[f(D_0) - f(\lambda_r)]/(D_0 - \lambda_r)$ is a polynomial of degree $N-1$ in D_0, the initial values required for an application of (2) will all be known.

To prove the formula, assume a column of m quantities $y(t)$ to be defined by the relation

$$y(t)-P(t)=\sum_{r=1}^{n}\frac{e^{\lambda_r(t-t_0)}}{\overset{(1)}{\Delta}(\lambda_r)}F(\lambda_r)\frac{[f(D_0)-f(\lambda_r)]}{D_0-\lambda_r}\{x(t_0)-P(t_0)\}, \qquad \text{......(3)}$$

in which $P(t)=\dfrac{F(D)}{\Delta(D)}\xi(t)$. The right-hand side of (3) is identical with that of (2). Now, since every column of the matrix $e^{\lambda_r t}F(\lambda_r)$ is annihilated by the operator $f(D)$, it follows that

$$f(D)\,y(t)=\frac{f(D)\,F(D)}{\Delta(D)}\xi(t)=\xi(t).$$

Hence the functions $y(t)$ satisfy the given differential equations. Next, premultiply both sides of (3) by the operator $\phi(D)$ and after differentiation put $t=t_0$. The result can be expressed as

$$\phi(D_0)\,y(t_0)=\sum_{r=1}^{n}\frac{C_r[f(D_0)-f(\lambda_r)]}{D_0-\lambda_r}x(t_0)$$
$$+\frac{\phi(D_0)\,F(D_0)}{\Delta(D_0)}\xi(t_0)-\sum_{r=1}^{n}\frac{C_r}{D_0-\lambda_r}\xi(t_0), \qquad \text{......(4)}$$

where $C_r=\phi(\lambda_r)\,F(\lambda_r)/\overset{(1)}{\Delta}(\lambda_r)$. But by (6·3·4) the right-hand side of (4) is equal to $\phi(D_0)\,x(t_0)$, or to Φ. Hence the functions $y(t)$ satisfy the required boundary conditions. This establishes (2).

With a problem of standard type the degree of $F(\lambda)$ in λ is less* than that of $\Delta(\lambda)$. Hence a convenient form for the particular integral such that $P(t_0)=0$ is (see (5·11·7))

$$P(t)=\sum_{r=1}^{n}\frac{e^{\lambda_r t}}{\overset{(1)}{\Delta}(\lambda_r)}F(\lambda_r)\int_{t_0}^{t}e^{-\lambda_r t}\xi(t)\,dt. \qquad \text{......(5)}$$

If, further, we write $F(\lambda_r)=k_r\kappa_r$ and

$$f(D)=A_0D^N+A_1D^{N-1}+\ldots+A_{N-1}D+A_N,$$

the special solution may be expressed as

$$x(t)-P(t)=\sum_{r=1}^{n}\frac{k_r\kappa_r}{\overset{(1)}{\Delta}(\lambda_r)}[A_0\lambda_r^{N-1}+A_1\lambda_r^{N-2}+\ldots+A_{N-1},\ldots,A_0\lambda_r+A_1,A_0]$$
$$\times\{x(t_0),D_0x(t_0),\ldots,D_0^{N-1}x(t_0)\}\,e^{\lambda_r(t-t_0)}. \qquad \text{......(6)}$$

* Since $f(\lambda)F(\lambda)=\Delta(\lambda)I$ and $f(\lambda)$ is of degree N, while $\Delta(\lambda)$ is of degree $n=mN$, it follows that $F(\lambda)$ has degree $N(m-1)$.

The special solution, when applied to homogeneous equations, is essentially the same as the "method of isolation" given by Routh.* In § 6·9 it will be shown that (2) is also the matrix equivalent of the well-known solution obtained by Heaviside and others by operational methods.

Examples

(i) *Single Homogeneous Equation of Order n.* Let the given differential equation be

$$f(D)x = (A_0D^n + A_1D^{n-1} + \dots + A_n)x = 0.$$

In this case $\Delta(\lambda) = f(\lambda)$ and the "adjoint" $F(\lambda) = 1$. Hence (6) gives

$$x(t) = \sum_{r=1}^{n} \frac{e^{\lambda_r(t-t_0)}}{\overset{(1)}{f}(\lambda_r)} [A_0\lambda_r^{n-1} + A_1\lambda_r^{n-2} + \dots + A_{n-1}, \dots, A_0\lambda_r + A_1, A_0] \times \{x(t_0), D_0x(t_0), \dots, D_0^{n-1}x(t_0)\}.$$

(ii) *Linear Dynamical Equations of Lagrangian Type.* If there are m generalised coordinates q and if $f(D) = AD^2 + BD + C$, the formula (6) gives

$$q(t) - P(t) = \sum_{r=1}^{2m} \frac{k_r \kappa_r}{\overset{(1)}{\Delta}(\lambda_r)} [A\lambda_r + B, A]\{q(t_0), \dot{q}(t_0)\} e^{\lambda_r(t-t_0)}.$$

(iii) *Various Identities.* When the roots of $\Delta(\lambda) = 0$ are all distinct and $P(t_0) = 0$ the direct solution (6·4·10) equivalent to the special solution is

$$x(t) - P(t) = kM(t-t_0)\, l^{-1}\{x(t_0), D_0x(t_0), \dots, D_0^{N-1}x(t_0)\}, \quad \text{......(7)}$$

in which l is given by (6·4·12). Comparison of (6) and (7) yields the identity

$$kM(t-t_0)\, l^{-1} = \sum_{r=1}^{n} \frac{F(\lambda_r)}{\overset{(1)}{\Delta}(\lambda_r)} [A_0\lambda_r^{N-1} + A_1\lambda_r^{N-2} + \dots + A_{N-1}, \dots, A_0\lambda_r + A_1, A_0] e^{\lambda_r(t-t_0)}. \quad \text{......(8)}$$

Expanding the exponentials on both sides of (8) and identifying the terms involving the different powers of $t - t_0$, we can deduce that

$$k\begin{bmatrix} \psi(\lambda_1) & 0 & \dots & 0 \\ 0 & \psi(\lambda_2) & \dots & 0 \\ \dots & \dots & \dots & \dots \\ 0 & 0 & \dots & \psi(\lambda_n) \end{bmatrix} l^{-1}$$

$$\equiv \sum_{r=1}^{n} \frac{F(\lambda_r)}{\overset{(1)}{\Delta}(\lambda_r)} [A_0\lambda_r^{N-1} + \dots + A_{N-1}, \dots, A_0\lambda_r + A_1, A_0]\psi(\lambda_r), \quad \text{......(9)}$$

in which $\psi(\lambda)$ is any polynomial of λ.

* See Chap. VIII of Ref. 13.

6·6.* Confluent Form of the Special Solution. We shall next indicate briefly the modifications to the special form of solution due to the presence of a set of s roots equal to λ_s. For simplicity the discussion will be restricted to homogeneous systems.

As for (5·9·5), write $\Delta(\lambda) = (\lambda - \lambda_s)^s \Delta_s(\lambda)$, and denote by V_i the family of matrices

$$V_i(t-t_0, \lambda_s) = \frac{1}{\lfloor s-i} \left[\frac{\partial^{s-i}}{\partial \lambda^{s-i}} \frac{e^{\lambda(t-t_0)} F(\lambda)}{\Delta_s(\lambda)} \right]_{\lambda=\lambda_s} \qquad \text{......(1)}$$

for $i = 1, 2, \ldots, s$. Then it can be shown that the contribution to the special solution arising from the set of s equal roots is given by the column

$$y(t) = \sum_{i=1}^{s} V_i(t-t_0, \lambda_s)\vartheta_i(D_0, \lambda_s)\, x(t_0), \qquad \text{......(2)}$$

where ϑ_i is as defined by (5·9·11).

To prove this result, we note firstly that, as for the family of matrices $W(t, \lambda_s)$ defined in § 5·7, every column of every matrix of the family V_i satisfies the differential equations $f(D)\,x(t) = 0$. Hence $y(t)$ satisfies the same differential equations. Again, premultiplication of (2) by $\phi(D)$ yields

$$\phi(D)\, y(t) = \sum_{i=1}^{s} \frac{\phi(D)}{\lfloor s-i} \left[\frac{\partial^{s-i}}{\partial \lambda^{s-i}} \frac{e^{\lambda(t-t_0)} F(\lambda)}{\Delta_s(\lambda)} \right]_{\lambda=\lambda_s} \vartheta_i(D_0, \lambda_s)\, x(t_0)$$

$$= \sum_{i=1}^{s} \frac{1}{\lfloor s-i} \left[\frac{\partial^{s-i}}{\partial \lambda^{s-i}} \frac{e^{\lambda(t-t_0)} \phi(\lambda) F(\lambda)}{\Delta_s(\lambda)} \right]_{\lambda=\lambda_s} \vartheta_i(D_0, \lambda_s)\, x(t_0),$$

so that, when $t = t_0$,

$$\phi(D_0)\, y(t_0) = \sum_{i=1}^{s} \frac{1}{\lfloor s-i} \left[\frac{\partial^{s-i}}{\partial \lambda^{s-i}} \frac{\phi(\lambda) F(\lambda)}{\Delta_s(\lambda)} \right]_{\lambda=\lambda_s} \vartheta_i(D_0, \lambda_s)\, x(t_0).$$

On application of (5·9·8) and (5·9·12) with $\xi = 0$, $G = \phi$, $t = t_0$, and $D = D_0$, it is now easy to verify that the functions $y(t)$ given by (2) correctly contribute to the satisfaction of the boundary conditions.

EXAMPLE

Lagrangian Equations of Motion of a Linear Conservative System. The dynamical equations in this case are of the special type

$$f(D)\, q \equiv (AD^2 + E)\, q = 0, \qquad \text{......(3)}$$

in which A and E are both symmetrical matrices.

It will be shown in § 9·9 that for such a system, if $\Delta(\lambda) = 0$ has s roots equal to λ_s, then the corresponding derived adjoint matrices up to and including $\overset{(s-2)}{F}(\lambda_s)$ are all null. Accordingly (see (1)), $V_i(t-t_0, \lambda_s) = 0$ if $i > 1$, while

$$V_1(t-t_0, \lambda_s) \equiv \frac{1}{\underline{|s-1}} \left[\frac{\partial^{s-1}}{\partial\lambda^{s-1}} \frac{e^{\lambda(t-t_0)} F(\lambda)}{\Delta_s(\lambda)}\right]_{\lambda=\lambda_s}$$

$$= \frac{e^{\lambda_s(t-t_0)}}{\underline{|s-1}\,\Delta_s(\lambda_s)} \overset{(s-1)}{F}(\lambda_s).$$

Moreover, putting $i = 1$ in (5·9·11), and using (3), we find

$$\vartheta_1(D_0, \lambda_s)\, q(t_0) \equiv \frac{f(D)_0 - f(\lambda_s)}{D_0 - \lambda_s} q(t_0) = A\{\lambda_s q(t_0) + \dot{q}(t_0)\}.$$

The special solution is thus

$$q(t) = \Sigma \frac{e^{\lambda_s(t-t_0)}}{\underline{|s-1}\,\Delta_s(\lambda_s)} \overset{(s-1)}{F}(\lambda_s)\, A\{\lambda_s q(t_0) + \dot{q}(t_0)\}, \qquad \text{......(4)}$$

where Σ denotes summation for all *distinct* values of the roots of $\Delta(\lambda) = 0$.

It is to be noted that with the present system of equations the matrices $W_i(t, \lambda_s)$ appropriate to the roots λ_s (see § 5·7) are all null with the exception of $W_{s-1}(t, \lambda_s) \equiv e^{\lambda_s t} \overset{(s-1)}{F}(\lambda_s)$, so that the corresponding s constituent solutions are all obtained from W_{s-1}. The modal columns are thus independent of t, even when multiple roots occur. The derived adjoint matrix $\overset{(s-1)}{F}(\lambda_s)$ will here be of rank s, and will be expressible as a matrix product of the type

$$\overset{(s-1)}{F}(\lambda_s) = \boldsymbol{k}(\lambda_s)\, \boldsymbol{\varkappa}(\lambda_s),$$

where $\boldsymbol{k}(\lambda_s)$ is an $(\overline{m}, s)$ matrix with s linearly independent columns, which can be taken to be the modal columns appropriate to the roots λ_s, and $\boldsymbol{\varkappa}(\lambda_s)$ is an $(\bar{s}, m)$ matrix with s linearly independent rows.

As a numerical illustration suppose

$$f(\lambda) = \begin{bmatrix} 32\lambda^2+48, & -16\lambda^2, & 4\lambda^2 \\ -16\lambda^2, & 32\lambda^2+48, & 4\lambda^2 \\ 4\lambda^2, & 4\lambda^2, & 11\lambda^2+12 \end{bmatrix},$$

and for simplicity take $t_0 = 0$. Then

$$\Delta(\lambda) = 3 \times 48^2(\lambda^2+1)^2(\lambda^2+4)$$

and
$$F(\lambda) = 48(\lambda^2+1)\begin{bmatrix} 7\lambda^2+12, & 4\lambda^2, & -4\lambda^2 \\ 4\lambda^2, & 7\lambda^2+12, & -4\lambda^2 \\ -4\lambda^2, & -4\lambda^2, & 16\lambda^2+48 \end{bmatrix}.$$

Denoting the roots of $\Delta(\lambda) = 0$ as $\lambda_1 = \lambda_2 = +i$, $\lambda_3 = +2i$, $\lambda_4 = \lambda_5 = -i$, $\lambda_6 = -2i$, we have $\Delta_2(\lambda_2) = -36 \times 48^2$ and $\Delta_1(\lambda_3) = 108 \times 48^2 i$, while

$$\overset{(1)}{F}(\lambda_2) = 2 \times 48i \begin{bmatrix} 5 & -4 & 4 \\ -4 & 5 & 4 \\ 4 & 4 & 32 \end{bmatrix}$$

and

$$F(\lambda_3) = -3 \times 48 \begin{bmatrix} -16 & -16 & 16 \\ -16 & -16 & 16 \\ 16 & 16 & -16 \end{bmatrix}.$$

The required solution, on cancellation of some numerical factors, is thus

$$q(t) = -\frac{ie^{it}}{18 \times 48} \begin{bmatrix} 5 & -4 & 4 \\ -4 & 5 & 4 \\ 4 & 4 & 32 \end{bmatrix} \begin{bmatrix} 32 & -16 & 4 \\ -16 & 32 & 4 \\ 4 & 4 & 11 \end{bmatrix} \{iq(0) + \dot{q}(0)\}$$

$$+ \frac{ie^{2it}}{108} \begin{bmatrix} -1 & -1 & 1 \\ -1 & -1 & 1 \\ 1 & 1 & -1 \end{bmatrix} \begin{bmatrix} 32 & -16 & 4 \\ -16 & 32 & 4 \\ 4 & 4 & 11 \end{bmatrix} \{2iq(0) + \dot{q}(0)\}$$

+ the corresponding conjugates.

This reduces to

$$q(t) = \tfrac{1}{9} \begin{bmatrix} 5 & -4 & 1 \\ -4 & 5 & 1 \\ 4 & 4 & 8 \end{bmatrix} \{q(0)\cos t + \dot{q}(0)\sin t\}$$

$$+ \tfrac{1}{18} \begin{bmatrix} 4 & 4 & -1 \\ 4 & 4 & -1 \\ -4 & -4 & 1 \end{bmatrix} \{2q(0)\cos 2t + \dot{q}(0)\sin 2t\}.$$

Note that $\overset{(1)}{F}(\lambda_2)$ is expressible as the matrix product

$$96i \begin{bmatrix} 5 & -4 \\ -4 & 5 \\ 4 & 4 \end{bmatrix} \begin{bmatrix} 1 & 0 & 4 \\ 0 & 1 & 4 \end{bmatrix}.$$

6·7.* Notation and Direct Solution for Two-Point Boundary Problems.

(*a*) *Notation.* In the case of a two-point boundary problem the supplementary conditions consist of n independent linear relations connecting the values at two points $t = t_0$ and $t = t_1$ of the m variables

x and of their derivatives up to order s. Such a set of relations may be denoted by

$$\phi(D_0)\,x(t_0)+\psi(D_1)\,x(t_1) = \Phi, \qquad \ldots\ldots(1)$$

where $D_0 \equiv \left(\frac{d}{dt}\right)_{t=t_0}$ and $D_1 \equiv \left(\frac{d}{dt}\right)_{t=t_1}$, and ϕ, ψ are $(\bar{n}, m)$ matrices of polynomials in their respective operators.

Suppose, for example, the differential equations to be of second order, and let there be $n = 2m$ roots λ. Then if the values of the m variables x at both $t = t_0$ and $t = t_1$ are assigned, we have $\phi(D_0) = \{I_m, 0\}$ and $\psi(D_1) = \{0, I_m\}$.

(*b*) *Direct Matrix Solution.* The treatment follows the same lines as for the one-point boundary problem (see § 6·4). Thus commencing with the general solution (6·4·5) we construct equation (6·4·6) as before, and the additional relation

$$\psi(D)\,x(t) = \mathbf{l}(t)\,M(t)\,c+\psi(D)\,P(t), \qquad \ldots\ldots(2)$$

in which $\mathbf{l}(t)$ is a square matrix of order n, similar to $l(t)$ but having for its sth column

$$\{\mathbf{l}_{is}\} = \psi(\lambda_s+D)\,\{k_{is}(t)\}.$$

Substituting $t = t_0$ in (6·4·6) and $t = t_1$ in (2) and adding, we obtain

$$\begin{aligned}\Phi &\equiv \phi(D_0)\,x(t_0)+\psi(D_1)\,x(t_1)\\ &= [l(t_0)\,M(t_0)+\mathbf{l}(t_1)\,M(t_1)]\,c+\phi(D_0)\,P(t_0)+\psi(D_1)\,P(t_1).\end{aligned}$$

This yields

$$c = L^{-1}\Phi - L^{-1}\{\phi(D_0)\,P(t_0)+\psi(D_1)\,P(t_1)\}, \qquad \ldots\ldots(3)$$

where L denotes the square matrix of order n

$$L(t_0, t_1) \equiv l(t_0)\,M(t_0)+\mathbf{l}(t_1)\,M(t_1).$$

Substitution of the value for c given by (3) in (6·4·5) yields the solution

$$x(t) = k(t)\,M(t)\,L^{-1}\{\Phi-\phi(D_0)\,P(t_0)-\psi(D_1)\,P(t_1)\}+P(t). \qquad \ldots\ldots(4)$$

PART II. SYSTEMS OF FIRST ORDER

6·8. Preliminary Remarks. The case where the differential equations are of first order is of particular importance not only on account of its comparative simplicity but also because special methods can be used which are not directly applicable to systems of higher order. Further, as shown in § 5·5, a system of general order can always be converted into a system of first order by a suitable choice of new variables.

The general first-order system in n dependent variables $y_1, y_2, \ldots, y_n$ will be written

$$f(D)\,y \equiv (vD - u)\,y = \eta(t), \qquad \ldots\ldots(1)$$

where v and u are square matrices of order n having constant elements. It will be assumed throughout that the matrix v is not singular, and that consequently the determinantal equation $\Delta(\lambda) \equiv | v\lambda - u | = 0$ has n roots. If v were singular, say of rank r, then $n-r$ of the equations (1) could be replaced by purely algebraic relations. These could be used to eliminate $n-r$ of the variables y, and the system would then be reduced to one involving only r equations and having for the coefficient of D a non-singular matrix (compare § 5·5).

In the special case where v is the unit matrix, equation (1) simplifies to

$$f(D)\,y \equiv (ID - u)\,y = \eta(t). \qquad \ldots\ldots(2)$$

Such a system will be described as being of *simple first-order form*. A system of the type (1) is always reducible to the simple form by premultiplication of the equation by v^{-1}.

In equation (1) there are n dependent variables y, whereas in presenting the theory of differential equations of general order we assumed m variables x. It is useful to preserve this distinction in the notation because in actual applications it often happens that the first-order system for solution in the variables y is derived (by methods such as those described in § 5·5) from a system of higher order but containing a fewer number m of variables x. The number of the new variables y will generally equal the degree n in λ of the determinantal equation of the system of higher order.

Attention will be restricted mainly to methods applicable to simple first-order systems. However, at the outset we shall apply the special method of solution given in § 6·5 to the general first-order system, and trace its connection with the method of Heaviside.

6·9. Special Solution of the General First-Order System, and its Connection with Heaviside's Method. In the present application of (6·5·2) there are n dependent variables y and n roots λ_r (assumed all distinct); moreover, $f(D) = vD - u$. Hence, if the form (6·5·5) is adopted for the particular integral, the special solution is

$$y(t) = \sum_{r=1}^{n} \frac{e^{\lambda_r(t-t_0)}}{\overset{(1)}{\Delta}(\lambda_r)} F(\lambda_r) \left\{ vy(t_0) + e^{\lambda_r t_0} \int_{t_0}^{t} e^{-\lambda_r t} \eta(t)\, dt \right\}. \quad \text{......(1)}$$

We shall now briefly relate this solution to that obtained by Heaviside and others.* The first rule in the application of Heaviside's method to the solution of the system $(vD-u)y = \eta$ is to treat D temporarily as a *constant*, and to solve for y_1, y_2, etc., the purely algebraic system of equations

$$(vD-u)y = vDy(t_0) + \eta(t).$$

When expressed by matrices the result of this preliminary operation is clearly

$$y = \frac{DF(D)}{\Delta(D)} vy(t_0) + \frac{F(D)}{\Delta(D)} \eta(t). \quad \text{......(2)}$$

The next step in the method is to reinstate the operational significance of D on the right-hand side of each of the equations contained in (2), and to expand the operators concerned in powers of D^{-1} or Q, which is now interpreted to mean integration with respect to t between the limits t_0 and t. To evaluate the results use is made of the identity (compare (5·9·4))

$$\frac{h(D)}{\Delta(D)} = \frac{h(0)}{\Delta(0)} + \sum_{r=1}^{n} \frac{h(\lambda_r)}{\lambda_r \overset{(1)}{\Delta}(\lambda_r)} \cdot \frac{D}{D-\lambda_r}, \quad \text{......(3)}$$

in which $h(D)$ is a typical polynomial in D of degree not exceeding that of $\Delta(D)$.

So far as concerns its final outcome, this treatment is equivalent to a direct application to (2) of the two matrix expansions

$$\frac{F(D)}{\Delta(D)} = \frac{F(0)}{\Delta(0)} + \sum_{r=1}^{n} \frac{F(\lambda_r)}{\lambda_r \overset{(1)}{\Delta}(\lambda_r)} \cdot \frac{D}{D-\lambda_r},$$

and

$$\frac{DF(D)}{\Delta(D)} = \sum_{r=1}^{n} \frac{F(\lambda_r)}{\overset{(1)}{\Delta}(\lambda_r)} \cdot \frac{D}{D-\lambda_r},$$

* See, for example, Refs. 16 and 17.

which are obvious deductions from (3).* When used in conjunction with (2), these identities yield

$$y(t) = \sum_{r=1}^{n} \frac{F(\lambda_r)}{\overset{(1)}{\Delta}(\lambda_r)} \cdot \frac{D}{D-\lambda_r} vy(t_0) + \left[\frac{F(0)}{\Delta(0)} + \sum_{r=1}^{n} \frac{F(\lambda_r)}{\lambda_r \overset{(1)}{\Delta}(\lambda_r)} \cdot \frac{D}{D-\lambda_r}\right] \eta(t).$$

Noting that $D/(D-\lambda_r)$, when applied as an operator to unity, yields $e^{\lambda_r(t-t_0)}$ (see (5·11·5) with $\xi = 1$), and using (5·11·6), we derive the solution

$$y(t) = \sum_{r=1}^{n} \frac{e^{\lambda_r(t-t_0)}}{\overset{(1)}{\Delta}(\lambda_r)} F(\lambda_r)\, vy(t_0) + P(t). \qquad \text{......(4)}$$

This agrees with (1), since in (4) the form of particular integral adopted is such that $P(t_0) = 0$.

Example

Special Solution of a System of Four Equations. Suppose the differential equations to be

$$\begin{bmatrix} 3 & -2 & 0 & -1 \\ 0 & 2 & 2 & 1 \\ 1 & -2 & -3 & -2 \\ 0 & 1 & 2 & 1 \end{bmatrix} Dy - \begin{bmatrix} 33 & 20 & 18 & 7 \\ -24 & -16 & -9 & -9 \\ 43 & 28 & 11 & 20 \\ -21 & -14 & -4 & -11 \end{bmatrix} y = \begin{bmatrix} -2 \\ 0 \\ 3 \\ -1 \end{bmatrix} e^{3t}, \qquad \text{......(5)}$$

with $y = \{1, 0, 0, 0\}$ at $t_0 = 0$. Here

$$f(\lambda) = \begin{bmatrix} 3\lambda-33, & -2\lambda-20, & -18, & -\lambda-7 \\ 24, & 2\lambda+16, & 2\lambda+9, & \lambda+9 \\ \lambda-43, & -2\lambda-28, & -3\lambda-11, & -2\lambda-20 \\ 21, & \lambda+14, & 2\lambda+4, & \lambda+11 \end{bmatrix},$$

and on reduction

$$\Delta(\lambda) = \lambda^4 - 5\lambda^2 + 4 = (\lambda^2-4)(\lambda^2-1).$$

The roots are accordingly $\lambda = \pm 2$ and ± 1, and these give

$$\overset{(1)}{\Delta}(2) = 12; \quad \overset{(1)}{\Delta}(-2) = -12; \quad \overset{(1)}{\Delta}(1) = -6; \quad \overset{(1)}{\Delta}(-1) = 6.$$

The corresponding adjoint matrices are found to be

$$F(2) = 12\{1, 0, -1, -1\}[1, 2, 0, -1],$$
$$F(-2) = 12\{1, -1, -1, -1\}[0, 1, 1, 1],$$
$$F(1) = 6\{1, 1, -2, -2\}[-3, -2, 10, 18],$$
$$F(-1) = 6\{1, 1, -2, -3\}[-3, -2, 9, 16].$$

* Note that with a first-order system the degree in D of $DF(D)$ cannot exceed that of $\Delta(D)$.

Since $vy(t_0) = \{3, 0, 1, 0\}$, the part of the special solution (1) independent of $\eta(t)$ can be written down at once as

$$3e^{2t}\{1, 0, -1, -1\}+e^{-2t}\{-1, 1, 1, 1\}+e^{t}\{-1, -1, 2, 2\}.$$

Next $$\int_0^t e^{-\lambda_r t}\eta(t)\,dt = \{-2, 0, 3, -1\}\frac{e^{(3-\lambda_r)t}-1}{3-\lambda_r}.$$

Hence, after a little reduction

$$\sum_{r=1}^{4}\frac{e^{\lambda_r t}}{\overset{(1)}{\Delta}(\lambda_r)}F(\lambda_r)\int_0^t e^{-\lambda_r t}\eta(t)\,dt$$
$$=(-e^{3t}+e^{2t})\{1, 0, -1, -1\}+(0{\cdot}4e^{3t}-0{\cdot}4e^{-2t})\{-1, 1, 1, 1\}$$
$$+(9e^{3t}-9e^{t})\{-1, -1, 2, 2\}+(4{\cdot}25e^{3t}-4{\cdot}25e^{-t})\{1, 1, -2, -3\}.$$

The required complete solution is accordingly

$$y(t)=(4e^{2t}-e^{3t})\{1, 0, -1, -1\}+(0{\cdot}6e^{-2t}+0{\cdot}4e^{3t})\{-1, 1, 1, 1\}$$
$$+(-8e^{t}+9e^{3t})\{-1, -1, 2, 2\}+(-4{\cdot}25e^{-t}+4{\cdot}25e^{3t})\{1, 1, -2, -3\}. \quad \text{......(6)}$$

6·10. Determinantal Equation, Adjoint Matrices, and Modal Columns for the Simple First-Order System.

(*a*) *Determinantal Equation.* The determinantal equation corresponding to (6·8·2) is

$$\Delta(\lambda)\equiv|\lambda I-u|\equiv\lambda^n+p_1\lambda^{n-1}+\ldots+p_{n-1}\lambda+p_n=0,$$

which is the characteristic equation of the matrix u. Moreover, the n roots $\lambda_1, \lambda_2, \ldots, \lambda_n$ are in this case the latent roots of u. Hence many of the results obtained in Chapter III bear intimately on the theory of the simple first-order system.

(*b*) *The Adjoint and Derived Matrices.* The adjoint matrix corresponding to any simple latent root (say λ_1) is given by (see (3·8·9))

$$F(\lambda_1)=(-1)^{n-1}f(\lambda_2)f(\lambda_3)\ldots f(\lambda_n).$$

Further, if $\lambda_1=\lambda_2=\ldots=\lambda_s$, then (see (3·8·10))

$$\frac{1}{\underline{|p}}\overset{(p)}{F}(\lambda_s)=(-1)^{n-p-1}[f(\lambda_s)]^{s-p-1}f(\lambda_{s+1})f(\lambda_{s+2})\ldots f(\lambda_n), \quad \text{......(1)}$$

where $p\leqslant s-1$.

For general values of λ, the adjoint matrix is given by the formulae (see (3·8·7) and (5·9·3))

$$F(\lambda)=u^{n-1}+(\lambda+p_1)u^{n-2}+\ldots+(\lambda^{n-1}+p_1\lambda^{n-2}+\ldots+p_{n-1})I$$
$$=\sum_{r=1}^{n}\frac{\Delta(\lambda)F(\lambda_r)}{(\lambda-\lambda_r)\overset{(1)}{\Delta}(\lambda_r)}.$$

(c) *Matrices $U_p(t, \lambda_s)$ of* § 5·10 (b). The formula (1) can be used to derive expressions for the matrices U_0, U_1, U_2, etc., which give the modal columns when repeated roots occur. If, as above, λ_s represents a root of multiplicity s (so that $\lambda_1 = \lambda_2 = \ldots = \lambda_s$), then

$$(-1)^{n-p-1}\frac{1}{\underline{|p}}U_p(t, \lambda_s)$$

$$= \left[I_n - tf(\lambda_s) + \frac{t^2}{\underline{|2}}f^2(\lambda_s) + \ldots + (-1)^p\frac{t^p}{\underline{|p}}f^p(\lambda_s)\right][f(\lambda_s)]^{s-p-1}\Pi,$$

where $$\Pi \equiv f(\lambda_{s+1})f(\lambda_{s+2})\ldots f(\lambda_n)$$

and $p \leqslant s-1$.

(d) *Modal Columns.* The s modal columns appropriate to a set of s roots equal to λ_s are given by any s linearly independent columns of the set of matrices $U_0, U_1, \ldots, U_{s-1}$. The typical modal column satisfies the relation

$$[I(D+\lambda_s) - u]\, k_s(t) = 0,$$

from which it follows that the modal matrix $k(t)$ has the property

$$\frac{dk(t)}{dt} + k(t)\,\Lambda = uk(t),$$

where Λ is the diagonal matrix of the latent roots. The last equation can be written

$$u = k(t)\,\Lambda k^{-1}(t) + \frac{dk(t)}{dt}k^{-1}(t). \qquad \ldots\ldots(2)$$

In the particular case where the latent roots of u are all distinct, the modal matrix is independent of t, and (2) then gives the usual collineatory transformation $u = k\Lambda k^{-1}$.

6·11. General, Direct, and Special Solutions of the Simple First-Order System.

(a) *General Solution.* When the expression on the right of (6·10·2) is substituted for u in (6·8·2) the resulting equation can be written briefly

$$\frac{dy}{dt} - k\Lambda k^{-1}y - \frac{dk}{dt}k^{-1}y = \eta.$$

On premultiplication by $M(-t)\,k^{-1}$ this gives

$$\frac{d}{dt}\{M(-t)\,k^{-1}y\} = M(-t)\,k^{-1}\eta,$$

and direct integration yields the general solution

$$y(t) = k(t)\,M(t)\left\{c + \int_{t_0}^{t} M(-t)\,k^{-1}(t)\,\eta(t)\,dt\right\}, \qquad \ldots\ldots(1)$$

where c denotes a column of n arbitrary constants. It should be noted that the particular integral in (1) vanishes at $t = t_0$. When the latent roots of u are all distinct an alternative form of the particular integral is

$$P(t) = \sum_{r=1}^{n} \frac{F(\lambda_r)}{\overset{(1)}{\Delta}(\lambda_r)} e^{\lambda_r t} \int_{t_0}^{t} e^{-\lambda_r t} \eta(t)\, dt. \qquad \text{......(2)}$$

(b) *Direct Solution.* The direct solution for general one-point and two-point boundary problems can be obtained from (6·4·10) and (6·7·4). When the values of y at $t = t_0$ are assigned (standard one-point boundary problem) a convenient form of the direct solution is (compare (1))

$$y(t) = k(t)\, M(t) \left\{ M(-t_0)\, k^{-1}(t_0)\, y(t_0) + \int_{t_0}^{t} M(-t)\, k^{-1}(t)\, \eta(t)\, dt \right\}. \qquad \text{......(3)}$$

If, further, the latent roots of u are all distinct, the modal matrix will be constant, and its reciprocal can readily be calculated by use of (3·8·12).

(c) *Special Solution (Standard One-Point Boundary Problem).* The special solution appropriate to the case of distinct latent roots is at once deducible from (6·9·1). The formula can be written

$$y(t) = \sum_{r=1}^{n} e^{\lambda_r (t-t_0)}\, Z_0(\lambda_r) \left\{ y(t_0) + e^{\lambda_r t_0} \int_{t_0}^{t} e^{-\lambda_r t} \eta(t)\, dt \right\}, \qquad \text{......(4)}$$

in which $Z_0(\lambda_r) \equiv F(\lambda_r)/\overset{(1)}{\Delta}(\lambda_r)$ (see also § 3·9).

Example

Direct Solution. If the equations (6·9·5) are premultiplied by

$$v^{-1} = \begin{bmatrix} 1, & 1, & -2, & -4 \\ 0, & 1, & 0, & -1 \\ -1, & -1, & 3, & 6 \\ 2, & 1, & -6, & -10 \end{bmatrix},$$

the result is
$$Dy - \begin{bmatrix} 7, & 4, & 3, & 2 \\ -3, & -2, & -5, & 2 \\ -6, & -4, & 0, & -4 \\ -6, & -4, & 1, & -5 \end{bmatrix} y = \begin{bmatrix} -4 \\ 1 \\ 5 \\ -12 \end{bmatrix} e^{3t}.$$

The direct solution (3) of this simple first-order system will now be obtained for the initial conditions $y = \{1, 0, 0, 0\}$ at $t_0 = 0$.

The latent roots of u in this case are $\lambda = \pm 2, \pm 1$, and

$$\overset{(1)}{\Delta}(2) = 12; \qquad F(2) = 12\{1, 0, -1, -1\}[3, 1, 2, 0];$$
$$\overset{(1)}{\Delta}(1) = -6; \qquad F(1) = 6\{1, 1, -2, -2\}[1, 0, 2, -1];$$
$$\overset{(1)}{\Delta}(-1) = 6; \qquad F(-1) = 6\{1, 1, -2, -3\}[0, 0, 1, -1];$$
$$\overset{(1)}{\Delta}(-2) = -12; \quad F(-2) = 12\{1, -1, -1, -1\}[1, 1, 1, 0].$$

Hence
$$k = \begin{bmatrix} 1, & 1, & 1, & 1 \\ 0, & 1, & 1, & -1 \\ -1, & -2, & -2, & -1 \\ -1, & -2, & -3, & -1 \end{bmatrix},$$

and by (3·8·12)
$$k^{-1} = \begin{bmatrix} 3, & 1, & 2, & 0 \\ -1, & 0, & -2, & 1 \\ 0, & 0, & 1, & -1 \\ -1, & -1, & -1, & 0 \end{bmatrix}.$$

The term independent of $\eta(t)$ in (3) is thus

$$\begin{bmatrix} 1, & 1, & 1, & 1 \\ 0, & 1, & 1, & -1 \\ -1, & -2, & -2, & -1 \\ -1, & -2, & -3, & -1 \end{bmatrix} \begin{bmatrix} e^{2t}, & 0, & 0, & 0 \\ 0, & e^{t}, & 0, & 0 \\ 0, & 0, & e^{-t}, & 0 \\ 0, & 0, & 0, & e^{-2t} \end{bmatrix} \begin{bmatrix} 3, & 1, & 2, & 0 \\ -1, & 0, & -2, & 1 \\ 0, & 0, & 1, & -1 \\ -1, & -1, & -1, & 0 \end{bmatrix} \begin{bmatrix} 1 \\ 0 \\ 0 \\ 0 \end{bmatrix}$$

$$= \begin{bmatrix} 1, & 1, & 1, & 1 \\ 0, & 1, & 1, & -1 \\ -1, & -2, & -2, & -1 \\ -1, & -2, & -3, & -1 \end{bmatrix} \begin{bmatrix} 3e^{2t} \\ -e^{t} \\ 0 \\ -e^{-2t} \end{bmatrix}. \qquad \text{......(5)}$$

To evaluate the particular integral, we note firstly that

$$k^{-1}\eta(t) = \{-1, -18, 17, -2\}e^{3t}.$$

On premultiplication by $M(-t)$ and integration

$$\int_0^t M(-t)\, k^{-1}\eta(t)\, dt = \int_0^t \{-e^{t}, -18e^{2t}, 17e^{4t}, -2e^{5t}\}\, dt$$
$$= \{-e^{t}+1, -9e^{2t}+9, 4{\cdot}25e^{4t}-4{\cdot}25, -0{\cdot}4e^{5t}+0{\cdot}4\}$$

Hence
$$M(t)\int_0^t M(-t)\, k^{-1}\eta(t)\, dt = \begin{bmatrix} -e^{3t}+e^{2t} \\ -9e^{3t}+9e^{t} \\ 4{\cdot}25e^{3t}-4{\cdot}25e^{-t} \\ -0{\cdot}4e^{3t}+0{\cdot}4e^{-2t} \end{bmatrix}.$$

Addition of this column to the column postmultiplying k in (5) gives the complete solution

$$y(t) = \begin{bmatrix} 1, & 1, & 1, & 1 \\ 0, & 1, & 1, & -1 \\ -1, & -2, & -2, & -1 \\ -1, & -2, & -3, & -1 \end{bmatrix} \begin{bmatrix} 4e^{2t} - e^{3t} \\ 8e^{t} - 9e^{3t} \\ -4{\cdot}25e^{-t} + 4{\cdot}25e^{3t} \\ -0{\cdot}6e^{-2t} - 0{\cdot}4e^{3t} \end{bmatrix}.$$

This agrees with (6·9·6).

6·12. Power Series Solution of Simple First-Order Systems. The great attractions of this method are its simplicity and the fact that it altogether avoids the generally troublesome problem of solution of the determinantal equation. On the other hand, in numerical applications, the process often has the disadvantage of slow convergence.

In § 2·7 it was shown that the matrix series

$$e^{ut} \equiv I + ut + \frac{u^2t^2}{\lfloor 2} + \dots$$

is absolutely and uniformly convergent, and that $De^{ut} = ue^{ut}$. It immediately follows that the complementary function of the differential equations (6·8·2) is given by

$$y = e^{ut}c, \qquad \dots\dots(1)$$

where c is a column of n arbitrary constants.

To obtain a particular integral premultiply (6·8·2) by e^{-ut} and write the result as $D(e^{-ut}y) = e^{-ut}\eta(t)$. A particular integral which vanishes at $t = t_0$ is accordingly

$$P(t) = e^{ut}\int_{t_0}^{t} e^{-ut}\eta(t)\,dt.$$

This leads to the general solution

$$y(t) = e^{ut}\left\{c + \int_{t_0}^{t} e^{-ut}\eta(t)\,dt\right\}. \qquad \dots\dots(2)$$

In a one-point boundary problem* where the values $y(t_0)$ are assigned, we have

$$y(t) = e^{u(t-t_0)}y(t_0) + e^{ut}\int_{t_0}^{t} e^{-ut}\eta(t)\,dt. \qquad \dots\dots(3)$$

In particular if $\eta(t) = e^{\theta t}\rho$, the last equation yields

$$y(t) = e^{u(t-t_0)}y(t_0) + e^{\theta t_0}(I_n e^{\theta(t-t_0)} - e^{u(t-t_0)})\,(\theta I_n - u)^{-1}\rho.$$

The following identity, obtained by comparison of (3) and (6·11·3), may be noted:

$$\exp u(t-t_0) = k(t)\,M(t-t_0)\,k^{-1}(t_0). \qquad \dots\dots(4)$$

* The case of a two-point boundary problem is dealt with in § 6·13.

Example

Computation of a Series Solution. As a very simple illustration of the method of computation, we assume the system to be

$$\begin{bmatrix}\dot{y}_1\\ \dot{y}_2\end{bmatrix}=\begin{bmatrix}0, & 1\\ -100, & 0\end{bmatrix}\begin{bmatrix}y_1\\ y_2\end{bmatrix},$$

and take for initial conditions $y_1 = 0$ and $y_2 = 10$ at $t_0 = 0$. The exact solution in this case is evidently $y_1 = \sin 10t$ with $y_2 = 10\cos 10t$: this will provide a comparison with the series solution.

Applying (3), with $\eta(t) = 0$, we have

$$y(t) = e^{ut}\{0, 10\},$$

where
$$u = \begin{bmatrix}0, & 1\\ -100, & 0\end{bmatrix}.$$

If $t = 0{\cdot}01$,

$$e^{0\cdot01u} = \begin{bmatrix}1, & 0\\ 0, & 1\end{bmatrix}+\begin{bmatrix}0, & 0{\cdot}01\\ -1, & 0\end{bmatrix}+\tfrac{1}{2}\begin{bmatrix}-0{\cdot}01, & 0\\ 0, & -0{\cdot}01\end{bmatrix}+\tfrac{1}{6}\begin{bmatrix}0, & -0{\cdot}0001\\ 0{\cdot}01, & 0\end{bmatrix}$$

$$+\tfrac{1}{24}\begin{bmatrix}0{\cdot}0001, & 0\\ 0, & 0{\cdot}0001\end{bmatrix}+\tfrac{1}{120}\begin{bmatrix}0, & 0{\cdot}000001\\ -0{\cdot}0001, & 0\end{bmatrix}+\dots$$

$$=\begin{bmatrix}0{\cdot}9950042, & 0{\cdot}0099833\\ -0{\cdot}9983341, & 0{\cdot}9950042\end{bmatrix}.$$

Hence
$$y(0{\cdot}01) = e^{0\cdot01u}\{0, 10\} = \{0{\cdot}099833, 9{\cdot}950042\},$$
$$y(0{\cdot}02) = e^{0\cdot01u}y(0{\cdot}01) = \{0{\cdot}198669, 9{\cdot}800667\},$$

and so on. The results of this iterative process may conveniently be summarised in tabular form as follows:

t	0	0·01	0·02	0·03	0·04	0·05
$y_1(t)$	0	0·099833	0·198669	0·295519	0·389417	0·479424
$y_2(t)$	10	9·950042	9·800667	9·553367	9·210614	8·775831

t	0·06	0·07	0·08	0·09	0·10
$y_1(t)$	0·564641	0·644216	0·717354	0·783325	0·841469
$y_2(t)$	8·253363	7·648430	6·967077	6·216112	5·403037

The results for $t = 0{\cdot}1$ agree to five figures with the values for sin 1·0 and 10 cos 1·0 as given by standard Tables.

As an alternative, a smaller interval might have been used at the outset. For instance, if $t = 0{\cdot}001$,

$$e^{0\cdot001u} = \begin{bmatrix}1, 0\\0, 1\end{bmatrix} + \begin{bmatrix}0, & 0\cdot001\\-0\cdot1, 0 & \end{bmatrix} + \tfrac{1}{2}\begin{bmatrix}-0\cdot0001, & 0\\0, & -0\cdot0001\end{bmatrix}$$
$$+\tfrac{1}{6}\begin{bmatrix}0, & -0\cdot0000001\\0\cdot00001, & 0\end{bmatrix} + \dots$$
$$= \begin{bmatrix}0\cdot9999500, 0\cdot0010000\\-0\cdot0999983, 0\cdot9999500\end{bmatrix}.$$

Forming the tenth power of this matrix, we obtain

$$e^{0\cdot01u} = \begin{bmatrix}0\cdot9950041, & 0\cdot0099835\\-0\cdot9983339, & 0\cdot9950041\end{bmatrix},$$

which agrees well with the value obtained above.

The choice of step is a matter for judgment, and will depend largely upon the magnitudes of the elements of the matrix u.

6·13.* Power Series Solution of the Simple First-Order System for a Two-Point Boundary Problem. The solution of the simple first-order system in power series, given in § 6·12, can be extended to two-point boundary problems as follows. For simplicity assume the equations to be homogeneous ($\eta = 0$) and let the assigned values be $\{y_1(t_0), y_2(t_0), \dots, y_s(t_0)\}$ and $\{y_{s+1}(t_1), y_{s+2}(t_1), \dots, y_n(t_1)\}$. Arrange the n dependent variables into the corresponding subsets

$$Y_s \equiv \{y_1, \dots, y_s\},$$
$$Y_{n-s} \equiv \{y_{s+1}, \dots, y_n\},$$

and express the matrix $e^{u(t-t_0)}$ in the partitioned form

$$\begin{bmatrix}\alpha(t, t_0), \beta(t, t_0)\\\gamma(t, t_0), \delta(t, t_0)\end{bmatrix}$$

where $\alpha, \beta, \gamma, \delta$ are, respectively, submatrices of types $(\bar{s}, s)$, $(\bar{s}, n-s)$, $(\overline{n-s}, s)$, $(\overline{n-s}, n-s)$. Then equation (6·12·3), with $\eta = 0$, gives

$$\begin{bmatrix}Y_s(t)\\Y_{n-s}(t)\end{bmatrix} = \begin{bmatrix}\alpha(t, t_0), \beta(t, t_0)\\\gamma(t, t_0), \delta(t, t_0)\end{bmatrix}\begin{bmatrix}Y_s(t_0)\\Y_{n-s}(t_0)\end{bmatrix}, \qquad \dots\dots(1)$$

so that, in particular

$$Y_{n-s}(t_1) = \gamma(t_1, t_0)\,Y_s(t_0) + \delta(t_1, t_0)\,Y_{n-s}(t_0).$$

This yields $\quad Y_{n-s}(t_0) = \delta^{-1}(t_1, t_0)\,\{Y_{n-s}(t_1) - \gamma(t_1, t_0)\,Y_s(t_0)\}, \qquad \dots\dots(2)$

which expresses the unknown set of values $Y_{n-s}(t_0)$ in terms of the known values $Y_s(t_0)$ and $Y_{n-s}(t_1)$. The solution required is then given by (1) in conjunction with (2).

CHAPTER VII

NUMERICAL SOLUTIONS OF LINEAR ORDINARY DIFFERENTIAL EQUATIONS WITH VARIABLE COEFFICIENTS

7·1. Range of the Chapter. The general theory of linear ordinary differential equations with variable coefficients covers an immense field which is quite beyond the scope of this book. In the present chapter our principal purpose will be to indicate how matrices can be applied usefully in the approximate solution of such differential equations.

Amongst the various special methods discussed, those described in § 7·9 and exemplified in § 7·10 are particularly powerful: one important field of application is to problems in mechanics which involve the determination of natural frequencies. The method of mean coefficients, described in § 7·11, is somewhat laborious, but it leads to good approximations even when the true solution is highly oscillatory. Examples of the use of this method are given in §§ 7·12–7·15.

7·2. Existence Theorems and Singularities. Linear differential equations with variable coefficients are rarely soluble by exact or elementary methods, and it is usually necessary to resort to numerical approximations. In a one-point boundary problem, for example, where the values of the dependent variable (or variables) and of the derivatives up to a certain order are specified at some datum point, say $t = t_0$, the normal procedure is to try a development of the solution in the form of a series of ascending powers of $t - t_0$. It is obviously of great assistance to know beforehand whether such a form of solution is justified, and, if so, the range of convergence. This information is supplied by "existence theorems", which specifically concern the conditions to be satisfied in order that solutions of differential equations may exist, and the ranges of validity of such solutions.*

In connection with any given linear differential equation, it is usual to describe as the *singularities* or the *singular points* those points for which the conditions necessary for the establishment of the relevant existence theorem are violated. All other points are generally referred to as *ordinary points*.

* For a detailed discussion of these questions the reader should consult a standard work on differential equations, e.g. Ref. 5. For an elementary exposition Ref. 18 may be recommended.

Examples

(i) *Single Homogeneous Linear Differential Equation of Order n.* Let the given equation be

$$\phi_0(t)\, D^n x + \phi_1(t)\, D^{n-1} x + \ldots + \phi_{n-1}(t)\, Dx + \phi_n(t)\, x = 0,$$

where $D \equiv d/dt$.

Suppose firstly that we are concerned with real quantities only (equation in real domain). Then the existence theorem for this equation states that if the coefficients $\phi_i(t)$ are continuous functions of t in the interval $a \leqslant t \leqslant b$, and $\phi_0(t)$ does not vanish in that interval, a unique solution expressible in power series exists which is continuous in (a, b) and which yields assigned values for $x(t_0)$, $\overset{(1)}{x}(t_0), \ldots, \overset{(n-1)}{x}(t_0)$ at any given point t_0 of (a, b). If the coefficients are all finite, one-valued and continuous throughout (a, b), the only singular points which can occur in (a, b) are the zeros of $\phi_0(t)$: all other points are ordinary.

When the variables are complex, let t_0 be a point of the Argand diagram (t-plane), other than a zero of $\phi_0(t)$, in the neighbourhood of which all the coefficients $\phi_i(t)$ are analytic* functions of t. Then a unique series solution in powers of $t - t_0$ exists which satisfies the given conditions at $t = t_0$. This series converges absolutely and uniformly at least within the circle having t_0 for centre and passing through that singularity of the set of coefficients $\phi_1/\phi_0, \phi_2/\phi_0, \ldots, \phi_n/\phi_0$ which lies nearest to t_0.

(ii) *The Standard System of First-Order Equations.* If the system of equations $Dy = u(t)\, y + \eta(t)$ is in the real domain, and the elements $u_{ij}(t)$ and $\eta_i(t)$ are continuous in the interval (a, b), then a continuous solution $y(t)$ exists which is unique in (a, b) and which yields given values $y(t_0)$ at a given point t_0 of the interval. Moreover, if the elements concerned are continuous for all positive and negative values of t (e.g. when u_{ij} and η_i are polynomials in t), then the solution will be continuous for all real values of t.

Problems frequently arise in which the elements u_{ij} and η_i satisfy the condition of continuity in (a, b) and are at the same time functions of a parameter, say K, real or complex. If K is complex, we shall assume it to be restricted to such a region R of the Argand diagram that u_{ij} and η_i are analytic functions of K at each point of R. In this

* That is, the coefficients are single-valued, continuous, and admit a unique derivative—namely, a derivative independent of the direction of approach.

case a unique solution $y(t, K)$ exists which is continuous with respect to t and analytic in K. For example, if the coefficients are integral functions (or polynomials) of K, the solution will itself be integral in K and uniformly convergent for all values of K when $a \leqslant t \leqslant b$. This solution will accordingly be of the type

$$y(t, K) = U_0(t) + U_1(t)\,K + U_2(t)\,K^2 + \ldots,$$

and if the assigned conditions at $t = t_0$ do not themselves involve K, the first term $U_0(t)$ of the series must alone satisfy the appropriate initial conditions, while the remaining terms $U_1(t)$, $U_2(t)$, etc. vanish at $t = t_0$. This often provides a powerful method of solution.

7·3. Fundamental Solutions of a Single Linear Homogeneous Equation. Let the given equation be

$$\phi_0(t)\,D^n x + \phi_1(t)\,D^{n-1}x + \ldots + \phi_{n-1}(t)\,Dx + \phi_n(t)\,x = 0,$$

and consider firstly the solution relative to an ordinary point, namely, a point t_0 other than a zero of $\phi_0(t)$. If the values $x(t_0)$, $\overset{(1)}{x}(t_0)$, ..., $\overset{(n-1)}{x}(t_0)$ are assigned, the required solution may be taken as

$$x(t) = X_0(t)\,x(t_0) + X_1(t)\,\overset{(1)}{x}(t_0) + \ldots + X_{n-1}(t)\,\overset{(n-1)}{x}(t_0),$$

in which $X_0(t)$, $X_1(t)$, ..., $X_{n-1}(t)$ are n special solutions, known as the *fundamental solutions*. These solutions satisfy the n distinct sets of initial conditions

$$X_0(t_0) = 1, \quad \overset{(1)}{X}_0(t_0) = 0, \quad \overset{(2)}{X}_0(t_0) = 0, \quad \ldots, \quad \overset{(n-1)}{X}_0(t_0) = 0;$$

$$X_1(t_0) = 0, \quad \overset{(1)}{X}_1(t_0) = 1, \quad \overset{(2)}{X}_1(t_0) = 0, \quad \ldots, \quad \overset{(n-1)}{X}_1(t_0) = 0;$$

and so on. The fundamental solution X_r will be a series of the form

$$X_r(t) = \frac{(t-t_0)^r}{\underline{|r}} + a_{rn}\frac{(t-t_0)^n}{\underline{|n}} + a_{r,n+1}\frac{(t-t_0)^{n+1}}{\underline{|n+1}} + \ldots.$$

The ordinary procedure is to substitute this series in the given equation, to arrange in ascending powers of $t-t_0$, and to equate to zero the coefficients of the successive powers. The constants a_{rn}, $a_{r,n+1}$, etc. are then found from recurrence relations.

In dealing with a singularity, it is usual as a preliminary step to transform the independent variable in such a way that the singularity is brought to the origin $t = 0$; for example, in the discussion of a singularity at infinity, the appropriate substitution is $t = 1/\tau$. This procedure clearly involves no loss of generality. Proceeding then by the well-known method of Frobenius we assume a series solution

$$x(t,\rho) = \sum_{\nu=0}^{\infty} c_\nu t^{\rho+\nu},$$

in which the exponent ρ is not necessarily a positive integer. The trial solution is substituted in the differential equation, and in the first place the coefficient, say $P(\rho)$, of the term of lowest degree in t is equated to zero. The nature of the singularity depends on the nature of the roots of the *indicial equation*

$$P(\rho) = 0.$$

If the indicial equation has n roots, the singularity is said to be *regular*. On the other hand, if $P(\rho)$ is of degree less than n, or is independent of ρ, the singularity is classed as *irregular*.

If, as will be assumed, the singularity is regular, it will be possible to write the differential equation in the form

$$f(D)\,x \equiv t^n D^n x + t^{n-1} P_1(t)\, D^{n-1} x + \ldots + t P_{n-1}(t)\, Dx + P_n(t)\, x = 0,$$

where the functions $P_i(t)$ are all finite, one-valued and continuous in an interval embracing $t = 0$ if t is real, or else are analytic near $t = 0$ if t is complex.

In the case of a regular singularity, provided that no two of the n exponents differ by zero or an integer, there will be n fundamental solutions of the assumed type, and the coefficients of each series will be determinable directly by recurrence relations. If there are multiple roots, or roots differing by integers, there will still be n fundamental solutions, but some of them will involve logarithmic elements.*

7·4. Systems of Simultaneous Linear Differential Equations. Adopting a notation similar to that used in § 5·1, we may write a system of m linear differential equations of order N connecting the m dependent variables $x_1, x_2, \ldots, x_m$ with t as

$$(\phi_0 D^N + \phi_1 D^{N-1} + \ldots + \phi_{N-1} D + \phi_N)\, x = \xi(t).$$

The coefficients ϕ_i are square matrices of order m, having for elements assigned functions of t. A system of the foregoing type is clearly reducible (in an indefinitely large number of ways) by methods similar to those illustrated in § 5·5 to a system of, say, n differential equations of the first order. In accordance with the notation of § 6·8, the most

* For a description of the treatment in these cases see § 16·3 of Ref. 5.

general system of n linear equations of the first order may be expressed as

$$(vD-u)\,y = \eta(t). \qquad \text{......(1)}$$

Here v and u are square matrices of order n whose elements are given functions of t. When $v = I_n$ the system is referred to as simple. For ranges of t in which v is not singular (1) can be reduced to a simple system by premultiplication throughout by v^{-1}.

EXAMPLE

Reduction of a Single Linear Equation to a First-Order System. To illustrate the methods of reduction, consider a single linear equation of order n

$$\phi_0(t)\,D^n x + \phi_1(t)\,D^{n-1}x + \ldots + \phi_n(t)\,x = \xi(t).$$

Writing (as for example (i) of § 5·5)

$$y = \{x, Dx, \ldots, D^{n-1}x\},$$

we derive the system of n first-order equations

$$\begin{bmatrix} 1 & 0 & \ldots & 0 & 0 \\ 0 & 1 & \ldots & 0 & 0 \\ \cdots & \cdots & \cdots & \cdots & \cdots \\ 0 & 0 & \ldots & 1 & 0 \\ 0 & 0 & \ldots & 0 & \phi_0 \end{bmatrix} Dy = \begin{bmatrix} 0 & 1 & 0 & \ldots & 0 \\ 0 & 0 & 1 & \ldots & 0 \\ \cdots & \cdots & \cdots & \cdots & \cdots \\ 0 & 0 & 0 & \ldots & 1 \\ -\phi_n & -\phi_{n-1} & -\phi_{n-2} & \cdots & -\phi_1 \end{bmatrix} y + \begin{bmatrix} 0 \\ 0 \\ \cdots \\ 0 \\ \xi(t) \end{bmatrix}.$$

For all ordinary points (i.e. for all points other than the zeros of $\phi_0(t)$) this equation can be replaced by the simple system

$$Dy = \begin{bmatrix} 0 & 1 & 0 & \ldots & 0 \\ 0 & 0 & 1 & \ldots & 0 \\ \cdots & \cdots & \cdots & \cdots & \cdots \\ 0 & 0 & 0 & \ldots & 1 \\ -\phi_n/\phi_0 & -\phi_{n-1}/\phi_0 & -\phi_{n-2}/\phi_0 & \cdots & -\phi_1/\phi_0 \end{bmatrix} y + \begin{bmatrix} 0 \\ 0 \\ \cdots \\ 0 \\ \xi/\phi_0 \end{bmatrix}.$$

The following is an alternative scheme of transformation which has been used by Baker* in relation to the Peano-Baker method of integration (see § 7·5). Retaining his notation, we suppose the given (homogeneous) equation to be

$$D^n x = \frac{P_{n-1}}{\phi_n} D^{n-1}x + \frac{P_{n-2}}{\phi_{n-1}\phi_n} D^{n-2}x + \ldots + \frac{P_0}{\phi_1\phi_2\cdots\phi_n}\,x,$$

and put $y_1 = x;\ y_2 = \phi_1 Dx;\ \ldots;\ y_n = \phi_1\phi_2\cdots\phi_{n-1}D^{n-1}x.$

* See p. 343 of Ref. 6.

This transformation of the variables yields the first-order system

$$Dy = \begin{bmatrix} 0 & 1/\phi_1 & 0 & 0 & \dots & 0 & 0 \\ 0 & H_1 & 1/\phi_2 & 0 & \dots & 0 & 0 \\ 0 & 0 & H_2 & 1/\phi_3 & \dots & 0 & 0 \\ \dots & \dots & \dots & \dots & \dots & \dots & \dots \\ 0 & 0 & 0 & 0 & \dots & H_{n-2} & 1/\phi_{n-1} \\ P_0/\phi_n & P_1/\phi_n & P_2/\phi_n & P_3/\phi_n & \dots & P_{n-2}/\phi_n & P_{n-1}/\phi_n + H_{n-1} \end{bmatrix} y, \qquad \dots\dots(2)$$

in which

$$H_s \equiv \sum_{r=1}^{s} \overset{(1)}{\phi_r} / \phi_r.$$

For instance, if $\phi_1 = \phi_2 = \dots = \phi_n = \phi$, then $H_s = s\overset{(1)}{\phi}/\phi$, and (2) simplifies to

$$Dy = \frac{1}{\phi}\begin{bmatrix} 0 & 1 & 0 & 0 & \dots & 0 & 0 \\ 0 & \overset{(1)}{\phi} & 1 & 0 & \dots & 0 & 0 \\ 0 & 0 & 2\overset{(1)}{\phi} & 1 & \dots & 0 & 0 \\ \dots & \dots & \dots & \dots & \dots & \dots & \dots \\ 0 & 0 & 0 & 0 & \dots & (n-2)\overset{(1)}{\phi} & 1 \\ P_0 & P_1 & P_2 & P_3 & \dots & P_{n-2} & P_{n-1} + (n-1)\overset{(1)}{\phi} \end{bmatrix} y.$$

An important special case is where $\phi = t$ and the functions P_{n-1}, P_{n-2}, etc. are polynomials in t. The given differential equation then is

$$t^n D^n x = P_{n-1} t^{n-1} D^{n-1} x + P_{n-2} t^{n-2} D^{n-2} x + \dots + P_0 x,$$

and the equivalent first-order system is

$$tDy = \begin{bmatrix} 0 & 1 & 0 & 0 & \dots & 0 & 0 \\ 0 & 1 & 1 & 0 & \dots & 0 & 0 \\ 0 & 0 & 2 & 1 & \dots & 0 & 0 \\ \dots & \dots & \dots & \dots & \dots & \dots & \dots \\ 0 & 0 & 0 & 0 & \dots & n-2 & 1 \\ P_0 & P_1 & P_2 & P_3 & \dots & P_{n-2} & P_{n-1} + n - 1 \end{bmatrix} y.$$

7·5. The Peano-Baker Method of Integration. This method of integration of the simple first-order system of differential equations was introduced by Peano in 1888, but the more recent developments are due to Baker.* The principle of the method is extremely simple.

* Historical notes are given in Ref. 19: see also Ref. 20 and § 16·5 of Ref. 5.

Considering firstly homogeneous systems, we suppose the given equations to be

$$Dy = u(t)\,y,$$

and assume a set of values for y to be assigned at $t = t_0$. Now it has been proved in § 2·11 that the matrizant $\Omega_{t_0}^{t}(u)$ has the fundamental property

$$D\Omega_{t_0}^{t}(u) = u\Omega_{t_0}^{t}(u).$$

Further, when $t = t_0$, $\Omega_{t_0}^{t}(u)$ reduces to I_n. Hence

$$y(t) \equiv \Omega_{t_0}^{t}(u)\,y(t_0) \qquad \text{......(1)}$$

satisfies the given system of equations and yields the required values $y(t_0)$ at $t = t_0$. This is the Peano-Baker form of solution. A feature of the solution is the wide range of its validity when t is complex; it extends to all paths of integration which do not encounter any of the barriers referred to in § 2·10. On the other hand its usefulness for practical computation is apt to be limited by slow convergence.

The matrizant method of solution can readily be applied to the non-homogeneous system

$$Dy = u(t)\,y + \eta(t).$$

For, when premultiplied throughout by $\Omega^{-1}(u)$, this equation can be expressed as

$$D(\Omega^{-1}y) = \Omega^{-1}\eta.$$

Hence a particular integral which vanishes at $t = t_0$ is

$$P(y) = \Omega(u)\,Q_{t_0}^{t}\{\Omega^{-1}(u)\,\eta(t)\}.$$

The solution required is accordingly

$$y(t) = \Omega(u)\,y(t_0) + \Omega(u)\,Q_{t_0}^{t}\{\Omega^{-1}(u)\,\eta(t)\}.$$

By expression of the matrizant in the appropriate partitioned form and a treatment similar to that adopted in § 6·13, it is possible also to obtain a formal solution for the case of a two-point boundary problem. The detailed construction of these formulae may be left to the reader.

7·6. Various Properties of the Matrizant. A useful property* of the matrizant is immediately deducible from equation (7·5·1). Suppose t_1 to be any point of the interval (t_0, t). Then

$$y(t_1) = \Omega_{t_0}^{t_1}(u)\,y(t_0)$$

and similarly

$$y(t) = \Omega_{t_1}^{t}(u)\,y(t_1).$$

It follows that

$$\Omega_{t_0}^{t}(u) = \Omega_{t_1}^{t}(u)\,\Omega_{t_0}^{t_1}(u).$$

* Equation (1) was originally given in Ref. 21. Equation (2) is due to Baker (Refs. 6 and 20); and equation (3) is equivalent to one given by Darboux (Ref. 22).

More generally, if the complete interval (t_0, t) is divided into any number of smaller intervals (t_0, t_1), (t_1, t_2), ..., (t_{s-1}, t), then

$$\Omega_{t_0}^{t}(u) = \Omega_{t_{s-1}}^{t}(u)\,\Omega_{t_{s-2}}^{t_{s-1}}(u)\ldots\Omega_{t_1}^{t_2}(u)\,\Omega_{t_0}^{t_1}(u). \qquad \text{......(1)}$$

A further property is expressed by the identity

$$\Omega(u+v) = \Omega(u)\,\Omega(V), \qquad \text{......(2)}$$

in which $$V = \Omega^{-1}(u)\,v\Omega(u).$$

To prove this, consider the system of equations $Dy = (u+v)\,y$, and let the values $y(t_0)$ be assigned. Then if $y = \Omega_{t_0}^{t}(u)\,Y$, the system reduces to $DY = VY$. Since $y(t_0) = Y(t_0)$, the solution of this last system of equations can be written $Y(t) = \Omega(V)\,y(t_0)$, or $y(t) = \Omega(u)\,\Omega(V)\,y(t_0)$. On comparison of this with the solution $y(t) = \Omega(u+v)\,y(t_0)$, obtained from the equations in their original form, the required result follows.

Finally, it may be noted that

$$\log \Delta = \int_{t_0}^{t} (u_{11} + u_{22} + \ldots + u_{nn})\,dt, \qquad \text{......(3)}$$

where $\Delta \equiv |\,\Omega(u)\,|$. The method of proof is general, but for brevity assume u to be a square matrix of order 2. By differentiation of Δ with respect to t,

$$\overset{(1)}{\Delta} = \begin{vmatrix} \overset{(1)}{\Omega}_{11} & \Omega_{12} \\ \overset{(1)}{\Omega}_{21} & \Omega_{22} \end{vmatrix} + \begin{vmatrix} \Omega_{11} & \overset{(1)}{\Omega}_{12} \\ \Omega_{21} & \overset{(1)}{\Omega}_{22} \end{vmatrix}.$$

Since $\overset{(1)}{\Omega} = u\Omega$, the preceding equation can be written

$$\overset{(1)}{\Delta} = \begin{vmatrix} u_{11}\Omega_{11} + u_{12}\Omega_{21}, & \Omega_{12} \\ u_{21}\Omega_{11} + u_{22}\Omega_{21}, & \Omega_{22} \end{vmatrix} + \begin{vmatrix} \Omega_{11}, & u_{11}\Omega_{12} + u_{12}\Omega_{22} \\ \Omega_{21}, & u_{21}\Omega_{12} + u_{22}\Omega_{22} \end{vmatrix}$$
$$= (u_{11} + u_{22})\,\Delta.$$

This yields on integration $\log \Delta = \int_{t_0}^{t} (u_{11} + u_{22})\,dt$.

7·7. A Continuation Formula. Suppose the solution of a system of linear differential equations to be required over the interval (t_0, t). If all points of the interval are ordinary, it is theoretically possible to obtain the solution by use of the fundamental series solutions relative to the initial point t_0. On the other hand, for purposes of practical computation, it may sometimes be preferable to employ the series relative to t_0 only up to some intermediate point of the range, say t_1, and to continue the solution thereafter by the use of series appropriate to the point t_1 or to some other point of the interval (t_1, t). More generally, if $n-1$ successive intermediate points $t_1, t_2, \ldots, t_{n-1}$ are taken, so that the whole interval (t_0, t) is divided into n steps, it is possible to base the computation in a typical step on the series relative to a suitable point of that step. The solution is carried over from step to step by identification of the initial conditions for any step with the end conditions for the preceding step. This fitting together of a sequence of solutions will be referred to as the *method of continuation*. We shall now obtain a matrix formula which expresses the process in a concise form.

The treatment is applicable in principle to linear differential equations of any order, but for simplicity we shall consider specifically the simple system of first-order homogeneous equations,

$$Dy = u(t)\,y. \qquad \text{......(1)}$$

A solution of the equations, valid in each step, will be assumed known. If T_s denotes some chosen point of the sth step (t_{s-1}, t_s), then the solution appropriate to that step may be supposed expressed either as a series of powers of $t-T_s$, or in terms of the matrizant, or in any other convenient form. The solution to be used in the sth step will be denoted by

$$y(t) = H_s(t)\,y(t_{s-1}). \qquad \text{......(2)}$$

Here H_s is a square matrix of order n, which is assumed to reduce to the unit matrix I_n when $t = t_{s-1}$. In the particular case where the point T_s is chosen at the initial point t_{s-1} of the step, the jth column of H would represent the fundamental solution appropriate to the special set of conditions $y_j(t_{s-1}) = 1$ and $y_\nu(t_{s-1}) = 0$ for $\nu \neq j$. On the other hand, if T_s is situated elsewhere in the step, the columns of H_s are suitable linear combinations of the fundamental solutions relative to T_s.

On application of (2) to the n assumed steps, we have the sequence of relations

$$y(t_1) = H_1(t_1)\,y(t_0),$$
$$y(t_2) = H_2(t_2)\,y(t_1),$$
$$\cdots\cdots\cdots\cdots\cdots\cdots\cdots\cdots$$
$$y(t_{n-1}) = H_{n-1}(t_{n-1})\,y(t_{n-2}),$$
$$y(t) = H_n(t)\,y(t_{n-1}).$$

Hence $$y(t) = H_n(t)\,H_{n-1}(t_{n-1})\,H_{n-2}(t_{n-2})\dots H_1(t_1)\,y(t_0),$$

or, say,* $$y(t) = \prod_n^1 (H_s)\,y(t_0). \qquad \text{......(3)}$$

The process of continuation adopted is thus represented by a chain of matrix multiplications. The reader should particularly note that the individual matrices of the chain will in general differ from step to step.

In the construction of the formula (3) it has been supposed that no singularity is encountered in the interval of t under consideration, but this restriction is not essential. The point T_s of the typical step can be chosen to be a singularity, provided the set (or sets) of solutions appropriate to that singularity are used.

In applications, the method of computation will depend to some extent upon the problem. For example, if the values of $y(t_0)$ are numerically assigned, the least laborious procedure is to compute the matrix chain $\Pi(H)$ from right to left; for then the product $H_1(t_1)\,y(t_0)$ can be calculated as a single column, which on premultiplication by $H_2(t_2)$ again yields a single column, and so on. Hence in this case multiplications of square matrices can be avoided. On the other hand, if—as often happens—the solution is required for arbitrary values of $y(t_0)$, the direct multiplications of the square matrices must be effected. If many matrices are involved, it is wise to compute subproducts in the manner described in example (vi) of § 1·5. This greatly facilitates the correction of errors, and is also valuable if a possible increase of the number of steps in certain parts of the range of integration is in view.

Some simple examples of the continuation formula follow.

EXAMPLES

(i) *Linear Equations with Constant Coefficients.* Assume the equations (1) to have constant coefficients, and choose $T_s = t_{s-1}$ in every step. Then $H_s(t) = \exp u(t-t_{s-1})$ (which is consistent with the condition $H_s(t_{s-1}) = I_n$). Hence the solution (3) is

$$y(t) = \exp u(t-t_{n-1})\exp u(t_{n-1}-t_{n-2})\dots\exp u(t_1-t_0)\,y(t_0). \qquad \text{......(4)}$$

* The notation $\prod_n^1 (H_s)$ is used to imply a product taken in the order $H_n\,H_{n-1}\dots H_1$.

If $\tau_s \equiv t_s - t_{s-1}$ denotes the total difference of t in the sth step, the formula may be written

$$y(t) = \prod_n^1 (\exp u\tau_s)\, y(t_0)$$

or

$$y(t) = \prod_n^1 \left(I + u\tau_s + \frac{u^2\tau_s^2}{\lfloor 2} + \dots\right) y(t_0).$$

If the steps are all equal and very small, this leads to the theorem

$$\exp ut \equiv \lim_{n\to\infty} \left(I + \frac{ut}{n}\right)^n.$$

(ii) *Matrizant Continuation.* Assume the equations to have variable coefficients and as before choose $T_s = t_{s-1}$ in every step. Then, adopting the matrizant form of solution, we have $H_s(t) = \Omega_{t_{s-1}}^t(u)$ with $H_s(t_{s-1}) = I_n$. The formula corresponding to (3) is thus

$$y(t) = \Omega_{t_{n-1}}^t(u)\, \Omega_{t_{n-2}}^{t_{n-1}}(u) \dots \Omega_{t_0}^{t_1}(u)\, y(t_0).$$

If the steps are so small that only first-order terms in $\tau_s = t_s - t_{s-1}$ need be retained, then approximately

$$\Omega_{t_{s-1}}^{t_s}(u) = I + \int_0^{\tau_s} u(t_{s-1} + \tau)\, d\tau + \dots = I + u(t_{s-1})\,\tau_s.$$

The solution is thus exhibited as the limiting value, when the number of steps is indefinitely increased, of the matrix product

$$y(t) = \prod_n^1 [I + u(t_{s-1})\,\tau_s]\, y(t_0).$$

7·8. Solution of the Homogeneous First-Order System of Equations in Power Series. Let the given system of equations be

$$v(t)\, Dy = u(t)\, y, \qquad \dots\dots(1)$$

and assume that over the range (t_0, t) of integration concerned the elements of the matrices $v(t)$ and $u(t)$, if not polynomials in $t - t_0 \equiv \tau$, are at all events expansible in Taylor's series, so that

$$v(t) \equiv v(t_0) + \tau \overset{(1)}{v}(t_0) + \frac{\tau^2}{\lfloor 2} \overset{(2)}{v}(t_0) + \dots,$$

$$u(t) \equiv u(t_0) + \tau \overset{(1)}{u}(t_0) + \frac{\tau^2}{\lfloor 2} \overset{(2)}{u}(t_0) + \dots.$$

Unless the matrix $v(t_0)$ is singular, these expansions may be written alternatively as

$$\left.\begin{aligned} v^{-1}(t_0)\, v(t) &= I_n + V_1\tau + V_2\tau^2 + \dots, \\ v^{-1}(t_0)\, u(t) &= U_0 + U_1\tau + U_2\tau^2 + \dots. \end{aligned}\right\} \qquad \dots\dots(2)$$

in which V_i and U_i are matrices of assigned constants.

As a trial solution assume

$$y(t) = S(\tau)\,y(t_0) \equiv (I_n + A_1\tau + A_2\tau^2 + \dots)\,y(t_0). \quad \dots\dots(3)$$

Here A_1, A_2, etc. are square matrices of constants to be determined, and the values $y(t_0)$ are regarded as arbitrarily assigned. On premultiplication of (1) by $v^{-1}(t_0)$ and substitution from (2) and (3), we obtain

$$(I_n + V_1\tau + V_2\tau^2 + \dots)(A_1 + 2A_2\tau + 3A_3\tau^2 + \dots)\,y(t_0)$$
$$= (U_0 + U_1\tau + U_2\tau^2 + \dots)(I_n + A_1\tau + A_2\tau^2 + \dots)\,y(t_0).$$

This must be identically satisfied for all sets of values $y(t_0)$. Equating to zero the total coefficients of the separate powers of τ, we therefore derive a set of recurrence relations from which the matrices A_i can be calculated. Thus

$$A_1 = U_0,$$
$$2A_2 = (U_0 - V_1)\,A_1 + U_1,$$
$$3A_3 = (U_0 - 2V_1)\,A_2 + (U_1 - V_2)\,A_1 + U_2,$$

and so on.

If $v(t) = I_n$, so that (1) is a simple system, there is the simplification $V_1 = V_2 = V_3 = \dots = 0$. The recurrence relations then give

$$A_1 = U_0,$$
$$2A_2 = U_0A_1 + U_1 = U_0^2 + U_1,$$
$$3A_3 = U_0A_2 + U_1A_1 + U_2 = U_0^3 + U_0U_1 + U_1U_0 + U_2,$$

and so on.

It should be noted that the columns of the matrix $S(\tau)$ give the fundamental solutions relative to the point t_0. The jth column of that matrix will be the fundamental solution appropriate to the special set of initial conditions $y_j(t_0) = 1$ and $y_\nu(t_0) = 0$ for $\nu \neq j$.

Example

Solution of a Single Equation. Consider the general solution of the equation $D^2x + tx = 0$ in the vicinity of $t = 0$. If $x = y_1$, and $Dx = y_2$, the equivalent first-order system is

$$Dy = (U_0 + U_1 t)\,y,$$

where $$U_0 = \begin{bmatrix} 0 & 1 \\ 0 & 0 \end{bmatrix} \quad \text{and} \quad U_1 = \begin{bmatrix} 0 & 0 \\ -1 & 0 \end{bmatrix}.$$

Hence

$$A_1 = U_0 = \begin{bmatrix} 0 & 1 \\ 0 & 0 \end{bmatrix},$$

$$2A_2 = \begin{bmatrix} 0 & 1 \\ 0 & 0 \end{bmatrix}^2 + \begin{bmatrix} 0 & 0 \\ -1 & 0 \end{bmatrix} = \begin{bmatrix} 0 & 0 \\ -1 & 0 \end{bmatrix},$$

$$3A_3 = \tfrac{1}{2}\begin{bmatrix} 0 & 1 \\ 0 & 0 \end{bmatrix}\begin{bmatrix} 0 & 0 \\ -1 & 0 \end{bmatrix} + \tfrac{1}{2}\begin{bmatrix} 0 & 0 \\ -1 & 0 \end{bmatrix}\begin{bmatrix} 0 & 1 \\ 0 & 0 \end{bmatrix} = \tfrac{1}{2}\begin{bmatrix} -1 & 0 \\ 0 & -2 \end{bmatrix},$$

and so on. This yields

$$S(t) = \begin{bmatrix} 1 & 0 \\ 0 & 1 \end{bmatrix} + \begin{bmatrix} 0 & 1 \\ 0 & 0 \end{bmatrix} t + \begin{bmatrix} 0 & 0 \\ -1 & 0 \end{bmatrix} \frac{t^2}{|\underline{2}} + \begin{bmatrix} -1 & 0 \\ 0 & -2 \end{bmatrix} \frac{t^3}{|\underline{3}} + \begin{bmatrix} 0 & -2 \\ 0 & 0 \end{bmatrix} \frac{t^4}{|\underline{4}}$$

$$+ \begin{bmatrix} 0 & 0 \\ 4 & 0 \end{bmatrix} \frac{t^5}{|\underline{5}} + \begin{bmatrix} 4 & 0 \\ 0 & 10 \end{bmatrix} \frac{t^6}{|\underline{6}} + \begin{bmatrix} 0 & 10 \\ 0 & 0 \end{bmatrix} \frac{t^7}{|\underline{7}} + \begin{bmatrix} 0 & 0 \\ -28 & 0 \end{bmatrix} \frac{t^8}{|\underline{8}}$$

$$+ \begin{bmatrix} -28 & 0 \\ 0 & -80 \end{bmatrix} \frac{t^9}{|\underline{9}} + \begin{bmatrix} 0 & -80 \\ 0 & 0 \end{bmatrix} \frac{t^{10}}{|\underline{10}} + \begin{bmatrix} 0 & 0 \\ 280 & 0 \end{bmatrix} \frac{t^{11}}{|\underline{11}} + \dots.$$

The fundamental solution appropriate, for instance, to the initial conditions $y_1(0) = 1$ and $y_2(0) = 0$ is thus

$$y_1 = 1 - \frac{t^3}{|\underline{3}} + \frac{4t^6}{|\underline{6}} - \frac{28t^9}{|\underline{9}} + \dots = 1 - \frac{t^3}{6} + \frac{t^6}{180} - \frac{t^9}{12960} + \dots,$$

$$y_2 = -\frac{t^2}{|\underline{2}} + \frac{4t^5}{|\underline{5}} - \frac{28t^8}{|\underline{8}} + \frac{280t^{11}}{|\underline{11}} - \dots = -\frac{t^2}{2} + \frac{t^5}{30} - \frac{t^8}{1440} + \frac{t^{11}}{142560} - \dots.$$

7·9. Collocation and Galerkin's Method. These methods of approximation* are powerful and relatively simple, and they are particularly valuable for problems involving the determination of characteristic numbers (see § 6·2).

A single differential equation of order n without variable parameters will first be considered, namely

$$f(D)\, x - \xi(t) = 0, \qquad \dots\dots(1)$$

in which $$f(D) \equiv \phi_0(t)\, D^n + \phi_1(t)\, D^{n-1} + \dots + \phi_n(t).$$

It will be supposed that an approximate solution of (1) in the interval $t = 0$ to $t = T$ is required. The boundary conditions must be such as to render the solution unique, and they may be assumed to consist of n non-homogeneous linear equations connecting the values of x, Dx,

* For a fuller discussion of the methods, see Refs. 23 and 24. The original papers by Galerkin describing his method are not readily accessible, but brief accounts are contained in various Russian publications, e.g. Refs. 25 and 26.

D^2x, etc. appropriate to, say, p specified points of the interval $(0, T)$. These p-point boundary conditions are thus of the type

$$B_i = b_i, \qquad \text{......(2)}$$

for $i = 1, 2, \ldots, n$, where the expressions B_i are linear and homogeneous in the values of x, Dx, D^2x, etc. corresponding to the boundary stations, and the constants b_i are not all zero.

The sth approximation to the solution, say $\boldsymbol{x}$, is assumed to have the form

$$\boldsymbol{x} = X_0(t) + X_1(t)\,c_1 + X_2(t)\,c_2 + \ldots + X_s(t)\,c_s,$$

where X_0 is any function of t which satisfies the conditions (2), X_1, X_2, etc. are any convenient functions each of which satisfies the simpler conditions $B_i = 0$ for $i = 1, 2, \ldots, n$, and c_1, c_2, etc. are constants left free for choice. When the functions X are chosen in this way, $\boldsymbol{x}$ necessarily satisfies the conditions (2) for all values of the constants c. The methods of approximation to be described only differ in the way in which these constants are determined. The approximate representation of the solution can be written concisely as

$$\boldsymbol{x} = X_0 + Xc, \qquad \text{......(3)}$$

where X denotes the row of the s functions $X_1, X_2, \ldots, X_s$, and c denotes the columns of the s constants $c_1, c_2, \ldots, c_s$.

Let $\epsilon(t)$ denote the function obtained when the expression (3) is substituted for x in the left-hand side of (1), so that

$$\epsilon(t) \equiv f(D)\,X(t)\,c + f(D)\,X_0(t) - \xi(t).$$

The quantity $\epsilon(t)$ is the error in the differential equation due to the approximation, and for a good approximation this error should be small throughout the interval $(0, T)$.

In the first and more obvious method, which will for convenience be termed *collocation*, the constants c are determined to give zero error at s selected points $t_1, t_2, \ldots, t_s$ of the interval. If $Y_j(t) \equiv f(D)\,X_j(t)$ and

$$\Theta \equiv \begin{bmatrix} Y_1(t_1) & Y_2(t_1) & \ldots & Y_s(t_1) \\ Y_1(t_2) & Y_2(t_2) & \ldots & Y_s(t_2) \\ \ldots & \ldots & \ldots & \ldots \\ Y_1(t_s) & Y_2(t_s) & \ldots & Y_s(t_s) \end{bmatrix},$$

the s equations for the constants in this case are given by

$$\Theta c = \{\xi(t_i) - Y_0(t_i)\}. \qquad \text{......(4)}$$

In an alternative method, due to Galerkin, the constants c are chosen so that s distinct weighted means of the error, namely $\int_0^T \epsilon(t)\, X_j(t)\, dt$, for $j=1, 2, \ldots, s$, are zero. The equations for the constants may then be expressed as

$$(Q_0^T X' f(D)\, X)\, c = Q_0^T X'(\xi - f(D)\, X_0). \qquad \ldots\ldots(5)$$

It can be shown that as s is increased the values of the constants c determined by equations (4) and (5) tend to equality.* Hence, when (as normally) the sequence of representations x obtained by either method converges, then also the sequence given by the other method converges to the same limit. From the computational standpoint the method of collocation has the advantage of great simplicity, since it avoids the labour of evaluation of integrals: on the other hand, Galerkin's method is generally the more rapidly convergent.

If the differential equation contains a variable parameter and is homogeneous, and if in addition the boundary conditions are homogeneous, so that $b_i = 0$ in (2) for $i = 1, 2, \ldots, n$, there may be characteristic values of the parameter for which solutions of the equation exist other than $y = 0$. With such problems the function X_0 will be absent from (3), and $\xi(t) = 0$. The terms on the right of (4) and (5) accordingly vanish, and the approximations to the characteristic numbers are given in each case by the condition of compatibility of s homogeneous linear algebraic equations. It may be noted that, in exceptional cases, even when the true characteristic numbers are wholly real, the earlier approximations to these numbers may be complex. However, if the representations of the solution are convergent, the imaginary parts in the approximation will either vanish or tend to zero as s is increased.

To illustrate the application of the methods to problems involving characteristic numbers, consider the single homogeneous differential equation of order n,

$$f(D)\, x + K\psi(D)\, x = 0, \qquad \ldots\ldots(6)$$

where K denotes the variable parameter. The form of approximate solution assumed in this case is $x = Xc$, and each of the s functions in X is chosen to satisfy the n homogeneous boundary conditions $B = 0$. The Galerkin equations, for instance, then are

$$(Q_0^T X' f(D)\, X)\, c + K(Q_0^T X' \psi(D)\, X)\, c = 0,$$

or, say,

$$(v + Kw)\, c = 0,$$

* For a formal proof, see Ref. 23.

in which v, w are square matrices of order s. Accordingly, the set of permissible values $K_1, K_2, \ldots, K_s$ given by the assumed approximation are the latent roots of the matrix u, where

$$u = -w^{-1}v. \qquad \ldots\ldots(7)$$

The methods can also be applied for the approximate solution of systems of linear ordinary (or even partial) differential equations. In such applications an approximate representation of the type $x = X_0 + Xc$, involving s constants c, is assumed for each of the m dependent variables. The leading term X_0 in each case must be a function of the independent variable (or variables), chosen to satisfy the boundary conditions $B = b$ completely, and the remaining functions X must satisfy $B = 0$. In the collocation method the complete domain of integration (assumed finite) is divided into s regions $\alpha_1, \alpha_2, \ldots, \alpha_s$, and a convenient point in each region is adopted for collocation. The assumed expressions for the dependent variable are then substituted in the differential equations, and the errors in the differential equations are made zero at each of the collocation points. The final outcome is a system of ms simultaneous linear algebraic equations for the m sets of s unknown constants c. If Galerkin's method is employed, the error ϵ_i corresponding to the ith differential equation is multiplied in turn by the functions X appropriate to the ith dependent variable, and the integrals of these products, taken over the domain of integration, are equated to zero. This process again yields ms algebraic equations for the unknown constants.

Consider, for example, the system of m homogeneous second-order ordinary differential equations $f(D)x + Kx = 0$, where $f(D)$ is now assumed to be a square matrix of order m, the elements of which are quadratic functions of D with variable coefficients. The scalar parameter K will be determined so that the m variables x_i vanish at both $t = 0$ and $t = T$: this defines a set of boundary conditions appropriate to a second-order system. In this case the approximate form of solution adopted can be expressed by partitioned matrices as

$$x = \begin{bmatrix} X_1, & 0, & \ldots, & 0 \\ 0, & X_2, & \ldots, & 0 \\ \ldots & \ldots & \ldots & \ldots \\ 0, & 0, & \ldots, & X_m \end{bmatrix} \begin{bmatrix} c_1 \\ c_2 \\ \ldots \\ c_m \end{bmatrix} \equiv Xc,$$

where X_i now denotes a *row* of s linearly independent functions each of which vanishes at $t = 0$ and $t = T$, and c_i is a *column* of s arbitrary constants. Now the column of the errors in this case is

$$\{\epsilon_i\} = f(D)\,Xc + KXc,$$

and Galerkin's equations may be expressed as

$$(Q_0^T X' f(D)\,X)\,c + K(Q_0^T X' X)\,c = 0,$$

or say $(V + KW)\,c = 0$, in which V, W are square matrices of order ms. Hence the numbers K are the latent roots of $-W^{-1}V$.

It should be noted that the approximations given by collocation will (normally) be unaffected by multiplication of the differential equations by arbitrary factors, but that this is not true for the Galerkin method. However, in mechanical problems, there is an optimum way of applying Galerkin's method which renders it equivalent to the use of Lagrange's equations (see Ref. 24).

7·10. Examples of Numerical Solution by Collocation and Galerkin's Method. The examples will be restricted to simple differential equations with known exact solutions, so that in each case tests of the accuracy of the approximate solutions will be possible.

(i) *First-Order Equation with Constant Coefficients.* Suppose the differential equation to be $(D-1)\,x = 0$ with the boundary condition $x = 1$ at $t = 0$. The exact solution is $x = \exp t$.

For the approximate solution in the interval (0, 1) assume

$$x = 1 + c_1 t + c_2 t^2 + \ldots + c_s t^s,$$

so that $X_j = t^j$ and $Y_j \equiv f(D)\,X_j = jt^{j-1} - t^j$. Hence, if the method of collocation is used, the equations for the constants are (see 7·9·4)

$$\Theta c = \{-Y_0(t_i)\} = \{1\},$$

where $\Theta_{ij} \equiv Y_j(t_i) = jt_i^{j-1} - t_i^j$. If $s = 4$, and if the points chosen for collocation are $t_1 = 0$, $t_2 = \frac{1}{3}$, $t_3 = \frac{2}{3}$, $t_4 = 1$, then

$$\begin{bmatrix} 1 & 0 & 0 & 0 \\ \frac{2}{3} & \frac{5}{9} & \frac{8}{27} & \frac{11}{81} \\ \frac{1}{3} & \frac{8}{9} & \frac{28}{27} & \frac{80}{81} \\ 0 & 1 & 2 & 3 \end{bmatrix} \begin{bmatrix} c_1 \\ c_2 \\ c_3 \\ c_4 \end{bmatrix} = \begin{bmatrix} 1 \\ 1 \\ 1 \\ 1 \end{bmatrix}.$$

These yield $c_1 = 1$, $c_2 = 0{\cdot}5078$, $c_3 = 0{\cdot}1406$, $c_4 = 0{\cdot}0703$.

The corresponding equations by Galerkin's method are (see 7·9·5)

$$Q_0^1\{t, t^2, t^3, t^4\}\,[1-t,\, 2t-t^2,\, 3t^2-t^3,\, 4t^3-t^4]\,c = Q_0^1\{t, t^2, t^3, t^4\},$$

and these reduce to

$$\begin{bmatrix} \frac{1}{6} & \frac{5}{12} & \frac{11}{20} & \frac{19}{30} \\ \frac{1}{12} & \frac{3}{10} & \frac{13}{30} & \frac{11}{21} \\ \frac{1}{20} & \frac{7}{30} & \frac{5}{14} & \frac{25}{56} \\ \frac{1}{30} & \frac{4}{21} & \frac{17}{56} & \frac{7}{18} \end{bmatrix} \begin{bmatrix} c_1 \\ c_2 \\ c_3 \\ c_4 \end{bmatrix} = \begin{bmatrix} \frac{1}{2} \\ \frac{1}{3} \\ \frac{1}{4} \\ \frac{1}{5} \end{bmatrix},$$

giving $c_1 = 0{\cdot}9975$, $c_2 = 0{\cdot}5149$, $c_3 = 0{\cdot}1325$, $c_4 = 0{\cdot}0732$.

The two different approximate solutions, and the approximation given by the terms in Taylor's series for $\exp t$ up to t^4, are compared below. The results given by Taylor's series are the least accurate.

Fourth Approximations to True Solution

Collocation:	$1+1{\cdot}0000t+0{\cdot}5078t^2+0{\cdot}1406t^3+0{\cdot}0703t^4$
Galerkin's method:	$1+0{\cdot}9975t+0{\cdot}5149t^2+0{\cdot}1325t^3+0{\cdot}0732t^4$
Taylor's Series:	$1+1{\cdot}0000t+0{\cdot}5000t^2+0{\cdot}1667t^3+0{\cdot}0417t^4$

Comparison of Numerical Values

t	0	0·1	0·2	0·3	0·4	0·5	0·6	0·7	0·8	0·9	1·0
$\exp t$	1·0	1·105	1·221	1·350	1·492	1·649	1·822	2·014	2·226	2·460	2·718
Collocation	1·0	1·105	1·221	1·350	1·492	1·649	1·822	2·014	2·226	2·460	2·719
Galerkin	1·0	1·105	1·221	1·350	1·492	1·649	1·822	2·014	2·225	2·459	2·718
Taylor	1·0	1·105	1·221	1·350	1·492	1·648	1·821	2·012	2·222	2·454	2·708

(ii) *Perfect Flow Across an Annulus.* A simple problem of perfect fluid flow in two dimensions will next be considered. The space between two concentric circular cylinders is assumed filled with fluid: the inner boundary $r = 1$ is at rest, while the outer boundary $r = 2$ is (instantaneously) moving with velocity U in the direction of the coordinate axis OX. If ψ denotes the stream-function, the mathematical problem is defined by the equation $\dfrac{\partial^2\psi}{\partial x^2}+\dfrac{\partial^2\psi}{\partial y^2} = 0$, in conjunction with the boundary conditions $\psi = 0$ at $r = 1$ and $\psi = -Uy$ at $r = 2$.

For the approximate treatment it will be convenient to write $\rho = \log r$ and $\alpha = \log 2$, and to express the differential equation as $\dfrac{\partial^2\psi}{\partial \rho^2}+\dfrac{\partial^2\psi}{\partial \theta^2} = 0$. Hence if we write $-\psi/U = R\sin\theta$, where R is a function of ρ to be found, we require

$$\frac{d^2R}{d\rho^2} - R = 0,$$

and the conditions to be satisfied are $R = 0$ at $\rho = 0$ and $R = 2$ at $\rho = \alpha$. The exact solution is $R = \frac{8}{3}\sinh\rho$, and this will be compared with approximate results given by collocation and by the Galerkin method.

For the approximate solution $\boldsymbol{R}$ we shall assume

$$\boldsymbol{R} = \frac{2}{\alpha}\rho + \rho(\rho-\alpha)(c_1 + c_2\rho + c_3\rho^2).$$

Then, if the collocation method is applied, and the three positions for collocation are chosen to be $\rho = 0$, $\rho = \alpha/2$ and $\rho = \alpha$, the values of the constants are readily found to be

$$c_1 = 0{\cdot}314268, \quad c_2 = 0{\cdot}453391, \quad c_3 = 0{\cdot}039681.$$

The corresponding values given by Galerkin's method work out as

$$c_1 = 0{\cdot}316111, \quad c_2 = 0{\cdot}448040, \quad c_3 = 0{\cdot}039563.$$

The two approximations are compared below with the exact solution $R = \frac{8}{3}\sinh\rho$.

ρ	0	0·1	0·2	0·3	0·4	0·5	0·6	0·69315
Exact Solution	0	0·26712	0·53691	0·81205	1·09533	1·38960	1·69774	2
Collocation	0	0·26718	0·53698	0·81208	1·09529	1·38949	1·69766	2
Galerkin	0	0·26711	0·53691	0·81206	1·09533	1·38958	1·69775	2

(iii) *Symmetrical Vibrations of an Annular Membrane.* The methods will next be applied to find rough values for the frequencies of vibration in the fundamental mode and the first and second overtones in the case of a uniform membrane bounded by the concentric circles $r = 1$ and $r = 2$.

Let T denote the tension per unit length, m the mass per unit area, and z the normal displacement at radius r and time t. Then the equation of vibration is

$$\frac{\partial^2 z}{\partial t^2} = \frac{T}{m}\left(\frac{\partial^2 z}{\partial r^2} + \frac{1}{r}\frac{\partial z}{\partial r}\right).$$

Assume a normal mode of vibration to be $z = R\sin\omega t$, and write $K = \omega^2 m/T$; then (see equation (7·9·6))

$$f(D)R + K\psi(D)R \equiv \frac{d^2R}{dr^2} + \frac{1}{r}\frac{dR}{dr} + KR = 0, \qquad \text{......(1)}$$

and K is to be determined by the condition of compatibility of the solution of (1) with the boundary conditions $R = 0$ at $r = 1$ and $r = 2$. The exact general solution in terms of Bessel functions is

$$R = AJ_0(r\sqrt{K}) + BY_0(r\sqrt{K}),$$

and in order that the boundary conditions may be satisfied K must be a root of the equation

$$J_0(\sqrt{K})\,Y_0(2\sqrt{K}) - J_0(2\sqrt{K})\,Y_0(\sqrt{K}) = 0.$$

The three lowest roots are found from tables of Bessel functions to be $K_1 = 9{\cdot}753$, $K_2 = 39{\cdot}35$, $K_3 = 88{\cdot}72$, corresponding to the fundamental and the first and second overtones.

For the approximations, the form of solution assumed is

$$\boldsymbol{R} = (r-1)(r-2)(c_1 + c_2 r + \ldots + c_s r^{s-1}),$$

and it should be noted that whereas this is a polynomial expression in r, the exact solution involves $Y_0(r)$ and therefore the logarithmic element $\log r$ which itself is not expansible by a Taylor's series in r. In the case of the collocation method the positions for collocation are chosen to be equally spaced in the interval $(1, 2)$ and to include the extremes, so that for instance $r_1 = 1$, $r_2 = 1{\cdot}5$, $r_3 = 2$ when $s = 3$. When Galerkin's method is used the equations can be treated conveniently as explained for (7·9·6). Some of the results obtained by the two methods compare as follows:

Characteristic Numbers for Vibrating Membrane

Method		K_1	K_2	K_3
Exact Solution		9·753	39·35	88·72
Collocation	$s=3$	9·51	—	—
	$s=4$	9·721	36·1	—
Galerkin	$s=2$	9·87	41·6	—
	$s=3$	9·752	41·79	101·5

It was pointed out in § 7·9 that the approximations given by the Galerkin method will be affected to some extent if the differential equation is multiplied throughout by an arbitrary function. To illustrate this, a few additional results obtained by Galerkin's method may be cited. A different representation is taken, with $s = 2$, namely

$$\boldsymbol{R} = c_1(r-1)(r-2) + c_2(r^2-1)(r^2-4),$$

and the differential equation (1) is assumed to be multiplied throughout by a function $f(r)$. The results for different functions $f(r)$ are as follows:

$f(r)$	1	r^2	r^4	r^6	$r^4 - 2{\cdot}25r^2$	Correct values
K_1	9·86	9·73	9·49	9·24	13·6	9·753
K_2	42·2	43·6	46·2	51·8	42·2	39·35

7·11. The Method of Mean Coefficients. The method to be described is applicable in principle to systems of equations of any order, but for simplicity it will here be assumed that the equations have been reduced to the simple first-order form

$$Dy = u(t)\, y.$$

An approximate solution will be obtained valid for the range (t_0, t) and for an arbitrary set of initial values $y(t_0)$.

If $t_1, t_2, \ldots, t_{n-1}$ denote any $n-1$ successive intermediate points of the range (t_0, t), then the exact solution is (see example (ii) of § 7·7)

$$y(t) = \Omega_{t_{n-1}}^{t}(u)\, \Omega_{t_{n-2}}^{t_{n-1}}(u) \ldots \Omega_{t_1}^{t_2}(u)\, \Omega_{t_0}^{t_1}(u)\, y(t_0).$$

As an approximation we shall now substitute for each variable element of u in the typical interval $\tau_p = t_p - t_{p-1}$ an average value taken over that interval, and determine the matrizant $\Omega_{t_{p-1}}^{t_p}$ on the assumption that the elements of u have these constant values. Thus let $U(p)$ denote the matrix of these average values in the pth interval; then since $U(p)$ is a matrix of constants (see example, § 2·11)

$$\Omega_{t_{p-1}}^{t_p} U(p) = \exp U(p)\, \tau_p.$$

Hence the complete approximate solution is

$$y(t) = [\exp (t - t_{n-1})\, U(n)]\, [\exp \tau_{n-1}\, U(n-1)] \ldots [\exp \tau_1\, U(1)]\, y(t_0).$$

For conciseness we shall write this as

$$y(t) = E_n(t - t_{n-1})\, E_{n-1}(\tau_{n-1})\, E_{n-2}(\tau_{n-2}) \ldots E_1(\tau_1)\, y(t_0). \quad \ldots\ldots(1)$$

The choice of the average values for the elements of u in any interval can to some extent be left to the judgment of the computer. If the intervals are small, sufficiently accurate values may often be found by inspection. In general, however, a good average value for u_{ij} in the interval (t_{p-1}, t_p) is provided by the arithmetic mean

$$U_{ij} = \frac{1}{t_p - t_{p-1}} \int_{t_{p-1}}^{t_p} u_{ij}(t)\, dt.$$

The integration may be carried out analytically or by Simpson's rule, according to convenience.

The individual matrices of the product (1) may themselves be evaluated approximately by expansion in power series (see example, § 6·12), or they may be computed exactly by the use of Sylvester's theorem or by (6·12·4). For instance, if the latent roots of the typical matrix U are all distinct, then

$$\exp \tau U = kM(\tau)\, k^{-1} = \sum_{r=1}^{n} \frac{e^{\lambda_r \tau}}{\overset{(1)}{\Delta}(\lambda_r)} F(\lambda_r).$$

As a simple example suppose the system to consist of only two equations, so that

$$U \equiv \begin{bmatrix} U_{11} & U_{12} \\ U_{21} & U_{22} \end{bmatrix}.$$

If the latent roots of U are complex, say $\lambda_1 = \mu + i\omega$ and $\lambda_2 = \mu - i\omega$, then (see § 6·10 (*b*))

$$E = \exp \tau U = \frac{U - \lambda_2 I}{\lambda_1 - \lambda_2} e^{\lambda_1 \tau} + \frac{U - \lambda_1 I}{\lambda_2 - \lambda_1} e^{\lambda_2 \tau},$$

which reduces to

$$E = e^{\mu\tau}\left(I \cos \omega\tau + \frac{U - \mu I}{\omega} \sin \omega\tau\right). \qquad \text{......(2)}$$

The case of real roots can be dealt with by writing $i\omega$ for ω in the preceding formulae.

The numerical solution of differential equations by the method of mean coefficients is illustrated in some detail in §§ 7·12–7·15.* The equations chosen are also soluble in terms of known functions, so that the accuracy of the approximate solution can in each case be verified. Some equations with highly oscillatory solutions are included, and these provide a particularly severe test of the method.

7·12. Solution by Mean Coefficients: Example No. 1. The equation is assumed to be

$$D^2x + (16\pi^2 e^{-2t} - \tfrac{1}{4})\, x = 0, \qquad \text{......(1)}$$

and a rough solution is required for the interval $t = 0$ to $t = 4$, the assigned conditions being $x = 1{\cdot}0$ and $Dx = 0{\cdot}5$ at $t = 0$.

In this first example the successive stages in the work will be explained in some detail.

Write $x = y_1$ and $Dx = y_2$; then (1) is equivalent to the simple first-order system

$$Dy = \begin{bmatrix} 0 & 1 \\ u_{21} & 0 \end{bmatrix} \begin{bmatrix} y_1 \\ y_2 \end{bmatrix} \equiv uy,$$

in which $u_{21} = \frac{1}{4} - 16\pi^2 e^{-2t}$. Here u_{21} is the only variable element of u, and its mean values in the successive steps will be obtained from the formula

$$U_{21}(p) = \frac{1}{t_p - t_{p-1}} \int_{t_{p-1}}^{t_p} u_{21}\, dt.$$

* The examples are taken, in revised form, from Ref. 21.

For the step (t_{p-1}, t_p), the equation to be used is

$$Dy = \begin{bmatrix} 0 & 1 \\ U_{21}(p) & 0 \end{bmatrix} y = U(p)\,y,$$

the solution of which is (see (7·11·1))

$$y(t) = E_p(t-t_{p-1})\,y(t_{p-1}). \qquad \text{......(2)}$$

The exponential can be evaluated easily since the latent roots $\mu \pm i\omega$ of $U(p)$ are $\pm\sqrt{U_{21}(p)}$, and are either both purely imaginary or both real according as $U_{21}(p) < 0$ or > 0. Hence in the formula (7·11·2) we write $\mu = 0$, and $i\omega = \sqrt{U_{21}(p)}$, so that ω may be either real or imaginary. The solution (2) thus reduces to

$$y(t) = \begin{bmatrix} \cos\omega(t-t_{p-1}), & \dfrac{1}{\omega}\sin\omega(t-t_{p-1}) \\ -\omega\sin\omega(t-t_{p-1}), & \cos\omega(t-t_{p-1}) \end{bmatrix} y(t_{p-1}).$$

The solution for the first step $(0, t_1)$ is $y(t) = E_1(t)\,y(0)$, and the terminal conditions, which are the initial conditions for the second step, are therefore $y(t_1) = E_1(t_1)\,y(0)$. Hence in the second step the solution is $y(t) = E_2(t-t_1)\,y(t_1) = E_2(t-t_1)\,E_1(t_1)\,y(0)$. Proceeding in this way, we find for the complete solution in the nth step

$$y(t) = E_n(t-t_{n-1})\,E_{n-1}(t_{n-1}-t_{n-2})\ldots E_2(t_2-t_1)\,E_1(t_1)\,y(0).$$

The computations are summarised in Tables 7·12·1 and 7·12·2. The choice of the steps is a matter for judgment: obviously, the more rapid the rate of variation of the coefficients, the closer should be the spacing. Steps of 0·1 have been adopted from $t = 0$ to $t = 1$, steps of 0·2 from $t = 1$ to $t = 2$, and steps of 0·5 from $t = 2$ to $t = 4$. Table 7·12·2 gives the matrices E and the initial and terminal values of y appropriate to the successive steps. For the first step, the initial conditions are $y = \{1{\cdot}0, 0{\cdot}5\}$. The initial conditions for the second step (i.e. the values of y_1 and y_2 at $t = 0{\cdot}1$) are found by computing the product

$$\begin{bmatrix} 0{\cdot}36673 & 0{\cdot}07783 \\ -11{\cdot}1202 & 0{\cdot}36673 \end{bmatrix} \begin{bmatrix} 1{\cdot}0 \\ 0{\cdot}5 \end{bmatrix} = \begin{bmatrix} 0{\cdot}4056 \\ -10{\cdot}9368 \end{bmatrix},$$

and so on.

The computations actually given yield the values of y_1 and y_2 (i.e. the original x and Dx) at the end-points of the steps. If intermediate values for the pth step are required, these can be calculated from (2), $y(t_{p-1})$ being now known.

The exact solution of (1) corresponding to the assigned conditions at $t = 0$ is

$$x = e^{t/2}\cos(4\pi e^{-t}).$$

The graph of this function is shown in Fig. 7·12·1 and the isolated points marked are the approximate values of $y_1 (= x)$ as computed in Table 7·12·2 (i.e. the upper figures in the third column). The accuracy, even in the present rough application, is seen to be very satisfactory.

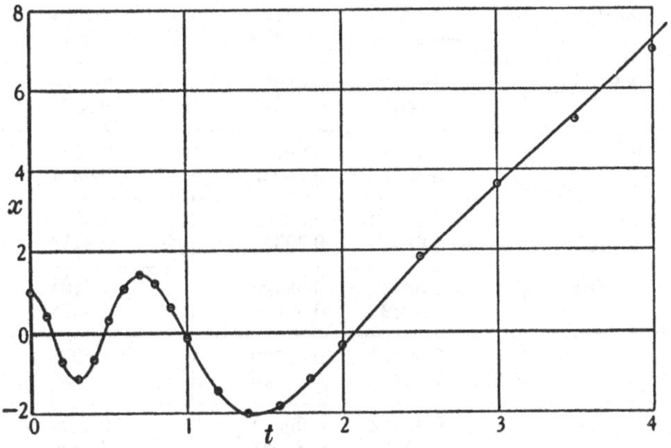

Fig. 7·12·1

Table 7·12·1. Example No. 1

Computation of Elements of Matrices E_p

Step	$U_{21}(p)$	ω	$\cos\omega\tau$	$\sin\omega\tau$	$\frac{1}{\omega}\sin\omega\tau$	$-\omega\sin\omega\tau$
0 –0·1	−142·88	11·953	0·36673	0·93033	0·07783	−11·1202
0·1–0·2	−116·93	10·813	0·47015	0·88259	0·08162	−9·5438
0·2–0·3	−95·69	9·782	0·55851	0·82950	0·08480	−8·1142
0·3–0·4	−78·30	8·849	0·63339	0·77383	0·08745	−6·8473
0·4–0·5	−64·06	8·004	0·69644	0·71762	0·08966	−5·7437
0·5–0·6	−52·41	7·239	0·74922	0·66231	0·09149	−4·7945
0·6–0·7	−42·86	6·547	0·79326	0·60889	0·09301	−3·9862
0·7–0·8	−35·044	5·920	0·82983	0·55800	0·09426	−3·3032
0·8–0·9	−28·646	5·352	0·86016	0·51003	0·09529	−2·7298
0·9–1·0	−23·409	4·838	0·88523	0·46517	0·09614	−2·2506
1·0–1·2	−17·368	4·167	0·67229	0·74029	0·17763	−3·0851
1·2–1·4	−11·553	3·399	0·77770	0·62864	0·18495	−2·1368
1·4–1·6	−7·665	2·7685	0·85058	0·52584	0·18994	−1·4558
1·6–1·8	−5·055	2·2484	0·90059	0·43467	0·19332	−0·97727
1·8–2·0	−3·306	1·8183	0·93460	0·35570	0·19562	−0·64677
2·0–2·5	−1·5783	1·2563	0·80911	0·58765	0·46776	−0·73828
2·5–3·0	−0·4225	0·6500	0·94764	0·31933	0·49125	−0·20758
3·0–3·5	+0·0025	−0·0505i	1·00032	−0·02525i	0·50000	+0·00127
3·5–4·0	+0·1589	−0·3986i	1·01993	−0·20063i	0·50334	+0·07997

Table 7·12·2. Example No. 1

Computation of the Matrix Chain $E_p E_{p-1} E_{p-2} \dots E_1 y(0)$

Step	Matrix $E_p(\tau_p)$		Column $y(t_{p-1})$
0·0–0·1	0·36673 −11·1202	0·07783 0·36673	1·0 0·5
0·1–0·2	0·47015 −9·5438	0·08162 0·47015	0·4056 −10·9368
0·2–0·3	0·55851 −8·1142	0·08480 0·55851	−0·7019 −9·0134
0·3–0·4	0·63339 −6·8473	0·08745 0·63339	−1·1564 0·6617
0·4–0·5	0·69644 −5·7437	0·08966 0·69644	−0·6746 8·3373
0·5–0·6	0·74922 −4·7945	0·09149 0·74922	0·2777 9·6811
0·6–0·7	0·79326 −3·9862	0·09301 0·79326	1·0938 5·9218
0·7–0·8	0·82983 −3·3032	0·09426 0·82983	1·4184 0·3374
0·8–0·9	0·86016 −2·7298	0·09529 0·86016	1·2088 −4·4053
0·9–1·0	0·88523 −2·2506	0·09614 0·88523	0·6200 −7·0890
1·0–1·2	0·67229 −3·0851	0·17763 0·67229	−0·1327 −7·6708
1·2–1·4	0·77770 −2·1368	0·18495 0·77770	−1·4518 −4·7476
1·4–1·6	0·85058 −1·4558	0·18994 0·85058	−2·0071 −0·5900
1·6–1·8	0·90059 −0·97727	0·19332 0·90059	−1·8193 2·4201
1·8–2·0	0·93460 −0·64677	0·19562 0·93460	−1·1706 3·9575
2·0–2·5	0·80911 −0·73828	0·46776 0·80911	−0·3199 4·4558
2·5–3·0	0·94764 −0·20758	0·49125 0·94764	1·8254 3·8414
3·0–3·5	1·00032 0·00127	0·50000 1·00032	3·6169 3·2613
3·5–4·0	1·01993 0·07997	0·50334 1·01993	5·2487 3·2670
—	— —	— —	6·9977 3·7518

7·13. Example No. 2. Assume the equation to be

$$D^2x+tx = 0, \qquad \qquad ......(1)$$

and let the solution be required for the interval $t = 0$ to $t = 8{\cdot}4$, the conditions at $t = 0$ being arbitrary.

The given equation is equivalent to the first-order system

$$Dy = \begin{bmatrix} 0 & 1 \\ -t & 0 \end{bmatrix} y \equiv uy,$$

where $y = \{x, Dx\}$. In this case the series solution already computed in the example, § 7·8, will be used from $t = 0$ to $t = 1{\cdot}8$. From $t = 1{\cdot}8$ to $t = 8{\cdot}4$ the method of mean coefficients will be applied.

Table 7·13·1, which gives details of the computation of the elements of the matrices E_p, is similar to Table 7·12·1, but in Table 7·13·2 the arrangement differs somewhat from that of Table 7·12·2. Since the initial conditions are here not numerically assigned, the continued product of the square matrices E_p is computed, and the result is multiplied by the (arbitrary) column matrix $y(0)$. The individual square matrices E_p are given in the second column of Table 7·13·2, while the continued products of these matrices, taken up to the beginnings of the successive steps, are entered in the third column. Thus, the first matrix in the third column is the unit matrix; the second is the product of the first matrix in the second column and this unit matrix; the third matrix in the third column is the product of the second matrices in the second and third columns, and so on. The last matrix in the third column is the continued product of all the matrices E_p from $t = 1{\cdot}8$ to $t = 8{\cdot}4$.

The values of y_1 and y_2 at the beginning of the steps, corresponding to known values at $t = 1{\cdot}8$, can now be written down. For instance, if $y = \{\xi, \eta\}$ at $t = 1{\cdot}8$, the value of y at $t = 6{\cdot}8$ is

$$y = \begin{bmatrix} -0{\cdot}53939, & -0{\cdot}39768 \\ 1{\cdot}24864, & -0{\cdot}93330 \end{bmatrix} \begin{bmatrix} \xi \\ \eta \end{bmatrix}.$$

As a particular case suppose that $y = \{1, 0\}$ at $t = 0$. The series solution computed in § 7·8 then gives $y(1{\cdot}8) = \{0{\cdot}2023, -1{\cdot}0623\}$. These values in turn provide the terminal values for y_1 as given in the last column of Table 7·13·2.

The exact solution of (1), with the initial conditions $\{x, Dx\} = \{1, 0\}$ at $t = 0$, is expressible in terms of Bessel functions as

$$1{\cdot}230x = \sqrt{\tfrac{1}{3}t}\, J_{\frac{1}{3}}(\tfrac{2}{3}t^{\frac{3}{2}}) - \sqrt{t}\, Y_{\frac{1}{3}}(\tfrac{2}{3}t^{\frac{3}{2}}).$$

The true and approximate solutions are compared in Fig. 7·13·1.

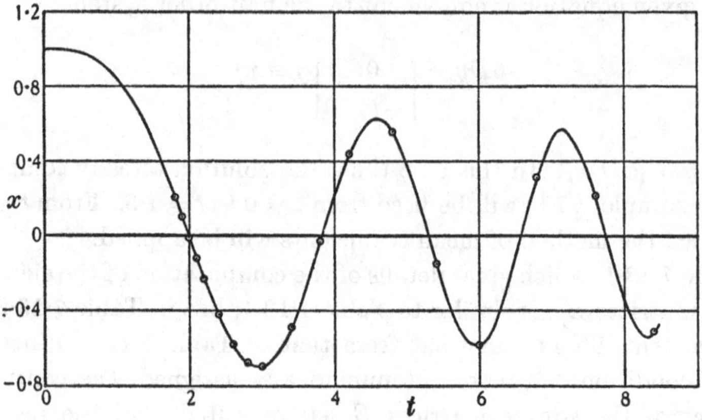

Fig. 7·13·1

Table 7·13·1. Example No. 2

Computation of Elements of Matrices E_p

Step	$U_{21}(p)$	ω	$\cos \omega\tau$	$\sin \omega\tau$	$\frac{1}{\omega} \sin \omega\tau$	$-\omega \sin \omega\tau$
1·8–1·9	−1·85	1·36015	0·99077	0·13559	0·09969	−0·18442
1·9–2·0	−1·95	1·39642	0·99027	0·13919	0·09968	−0·19437
2·0–2·1	−2·05	1·43178	0·98976	0·14269	0·09966	−0·20430
2·1–2·2	−2·15	1·46629	0·98927	0·14610	0·09964	−0·21422
2·2–2·4	−2·3	1·51657	0·95436	0·29868	0·19694	−0·45297
2·4–2·6	−2·5	1·58114	0·95042	0·31099	0·19669	−0·49172
2·6–2·8	−2·7	1·64317	0·94649	0·32275	0·19642	−0·53033
2·8–3·0	−2·9	1·70294	0·94255	0·33404	0·19615	−0·56885
3·0–3·4	−3·2	1·78885	0·75473	0·65603	0·36673	−1·17354
3·4–3·8	−3·6	1·89737	0·72556	0·68816	0·36269	−1·30569
3·8–4·2	−4·0	2·0	0·69671	0·71736	0·35868	−1·43472
4·2–4·8	−4·5	2·12132	0·29361	0·95592	0·45062	−2·02781
4·8–5·4	−5·1	2·25832	0·21413	0·97680	0·43253	−2·20593
5·4–6·0	−5·7	2·38747	0·13787	0·99045	0·41485	−2·36467
6·0–6·8	−6·4	2·52982	−0·43772	0·89911	0·35540	−2·27459
6·8–7·6	−7·2	2·68328	−0·54453	0·83874	0·31258	−2·25058
7·6–8·4	−8·0	2·82843	−0·63804	0·77000	0·27224	−2·17789

Table 7·13·2. Example No. 2

Computation of the Matrix Chain $E_p E_{p-1} \ldots E_1$

Step	Matrix E_p		Product $E_p E_{p-1} \ldots E_1$		Column $y(t_{p-1})$
1·8–1·9	0·99077	0·09969	1·0	0	0·2023
	−0·18442	0·99077	0	1·0	−1·0623
1·9–2·0	0·99027	0·09968	0·99077	0·09969	0·0945
	−0·19437	0·99027	−0·18442	0·99077	—
2·0–2·1	0·98976	0·09966	0·96275	0·19748	−0·0150
	−0·20430	0·98976	−0·37520	0·96175	—
2·1–2·2	0·98927	0·09964	0·91550	0·29131	−0·1243
	−0·21422	0·98927	−0·56805	0·91156	—
2·2–2·4	0·95436	0·19694	0·84908	0·37901	−0·2309
	−0·45297	0·95436	−0·75807	0·83937	—
2·4–2·6	0·95042	0·19669	0·66103	0·52702	−0·4261
	−0·49172	0·95042	−1·10808	0·62938	—
2·6–2·8	0·94649	0·19642	0·41031	0·62468	−0·5806
	−0·53033	0·94649	−1·37818	0·33903	—
2·8–3·0	0·94255	0·19615	0·11765	0·65785	−0·6750
	−0·56885	0·94255	−1·52203	−0·01040	—
3·0–3·4	0·75473	0·36673	−0·18766	0·61802	−0·6945
	−1·17354	0·75473	−1·50151	−0·38402	—
3·4–3·8	0·72556	0·36269	−0·69228	0·32561	−0·4859
	−1·30569	0·72556	−0·91301	−1·01510	—
3·8–4·2	0·69671	0·35868	−0·83343	−0·13192	−0·0285
	−1·43472	0·69671	0·24146	−1·16166	—
4·2–4·8	0·29361	0·45062	−0·49405	−0·50857	0·4403
	−2·02781	0·29361	1·36397	−0·62007	—
4·8–5·4	0·21413	0·43253	0·46958	−0·42874	0·5504
	−2·20593	0·21413	1·40231	0·84922	—
5·4–6·0	0·13787	0·41485	0·70709	0·27551	−0·1496
	−2·36467	0·13787	−0·73558	1·12761	—
6·0–6·8	−0·43772	0·35540	−0·20767	0·50577	−0·5793
	−2·27459	−0·43772	−1·77345	−0·49603	—
6·8–7·6	−0·54453	0·31258	−0·53939	−0·39768	0·3133
	−2·25058	−0·54453	1·24864	−0·93330	—
7·6–8·4	−0·63804	0·27224	0·68401	−0·07518	0·2182
	−2·17789	−0·63804	0·53402	1·40322	—
8·4	—	—	−0·29104	0·42998	−0·5156
	—	—	−1·83042	−0·73158	—

7·14. Example No. 3. As a more complicated example we shall next choose the equation

$$D^2x+\left(16\cos\frac{\pi t}{2}\right)Dx+\left(64\pi^2+64\cos^2\frac{\pi t}{2}-4\pi\sin\frac{\pi t}{2}\right)x=0,$$

and obtain the solution for the interval $t=0$ to $t=4$.

The equivalent first-order system is

$$Dy=\begin{bmatrix}0, & 1\\ -64\pi^2-64\cos^2\frac{\pi t}{2}+4\pi\sin\frac{\pi t}{2}, & -16\cos\frac{\pi t}{2}\end{bmatrix}y\equiv uy,$$

where
$$y=\{x, Dx\}.$$

In the present case the method of mean coefficients will be used throughout. The matrix u here has two variable elements u_{21} and u_{22}, and the latent roots are found to be complex, say $\lambda=\mu\pm i\omega$. The formula (7·11·2) gives

$$y(t_p)=e^{\mu\tau}\begin{bmatrix}-\frac{\mu}{\omega}\sin\omega\tau+\cos\omega\tau, & \frac{1}{\omega}\sin\omega\tau\\ \frac{U_{21}}{\omega}\sin\omega\tau, & \frac{(U_{22}-\mu)}{\omega}\sin\omega\tau+\cos\omega\tau\end{bmatrix}y(t_{p-1}),$$

or

$$y(t_p)=e^{\mu\tau}\begin{bmatrix}-\frac{\mu}{\omega}\sin\omega\tau+\cos\omega\tau, & \frac{1}{\omega}\sin\omega\tau\\ -\frac{(\mu^2+\omega^2)}{\omega}\sin\omega\tau, & \frac{\mu}{\omega}\sin\omega\tau+\cos\omega\tau\end{bmatrix}y(t_{p-1}),$$

since $U_{21}=-(\mu^2+\omega^2)$ and $U_{22}=2\mu$.

The computations, which are otherwise similar to those for Example No. 2, are summarised in Tables 7·14·1 and 7·14·2. The fourth column of Table 7·14·2, headed $\Pi e^{\mu\tau}$, represents the contributions to the matrix chain $E_p E_{p-1} \ldots E_1$ arising from the scalar multiplier $e^{\mu\tau}$ which appears in the solution. It is obviously convenient to effect these particular multiplications independently.

Table 7·14·2 may be regarded as completing the formal numerical solution. When any initial conditions $y(0)$ are assigned, the values of y at the ends of the steps are at once deducible as explained in connection with Example No. 2. For instance, at $t=2{\cdot}8$,

$$y(2{\cdot}8)=127{\cdot}0\begin{bmatrix}0{\cdot}5890, & 0{\cdot}03761\\ -20{\cdot}54, & 0{\cdot}3859\end{bmatrix}y(0)=\begin{bmatrix}74{\cdot}79, & 4{\cdot}776\\ -2608, & 49{\cdot}00\end{bmatrix}y(0).$$

The calculations have been completed in detail for the particular case $y(0) = \{1, -8\}$. It can be verified that the exact solution appropriate to these initial conditions is

$$x \equiv y_1 = \exp\left(-\frac{16}{\pi}\sin\frac{\pi t}{2}\right) \times \cos 8\pi t,$$

and the graph of this function is given in Fig. 7·14·1. It oscillates throughout the time range and the amplitude varies considerably.

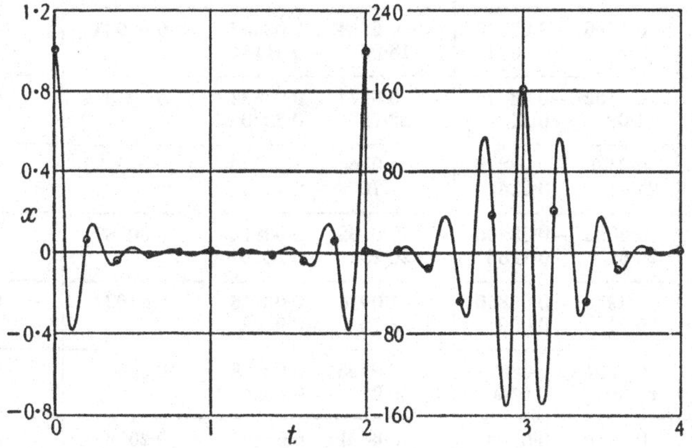

Fig. 7·14·1

Table 7·14·1. Example No. 3

Computation of Elements of Matrices E_p

Step	U_{21}	U_{22}	μ	ω	$e^{\mu\tau}$	$-\frac{\mu}{\omega}\sin\omega\tau + \cos\omega\tau$	$\frac{1}{\omega}\sin\omega\tau$	$\frac{U_{21}}{\omega}\sin\omega\tau$	$\frac{\mu}{\omega}\sin\omega\tau + \cos\omega\tau$
0 –0·2	−691·6	−15·74	−7·87	25·09	0·2072	0·0030	−0·03799	26·27	0·6010
0·2–0·4	−676·5	−14·20	−7·10	25·02	0·2417	0·0160	−0·03827	25·89	0·5594
0·4–0·6	−654·8	−11·27	−5·63	24·96	0·3240	0·0594	−0·03850	25·21	0·4930
0·6–0·8	−634·0	−7·23	−3·62	24·92	0·4853	0·1286	−0·03866	24·51	0·4084
0·8–1·0	−621·4	−2·49	−1·25	24·90	0·7796	0·2152	−0·03875	24·08	0·3120
1·0–1·2	−621·4	2·49	1·25	24·90	1·2827	0·3120	−0·03875	24·08	0·2152
1·2–1·4	−634·0	7·23	3·62	24·92	2·0606	0·4084	−0·03866	24·51	0·1286
1·4–1·6	−654·8	11·27	5·63	24·96	3·0864	0·4930	−0·03850	25·21	0·0594
1·6–1·8	−676·5	14·20	7·10	25·02	4·1371	0·5594	−0·03827	25·89	0·0160
1·8–2·0	−691·6	15·74	7·87	25·09	4·8259	0·6010	−0·03799	26·27	0·0030
2·0–2·2	−695·5	15·74	7·87	25·17	4·8259	0·6129	−0·03769	26·21	0·0197
2·2–2·4	−687·8	14·20	7·10	25·25	4·1371	0·5959	−0·03738	25·71	0·0651
2·4–2·6	−672·5	11·27	5·63	25·31	3·0864	0·5525	−0·03710	24·95	0·1347
2·6–2·8	−656·3	7·23	3·62	25·36	2·0606	0·4857	−0·03691	24·22	0·2185
2·8–3·0	−646·1	2·49	1·25	25·39	1·2827	0·4028	−0·03680	23·78	0·3108
3·0–3·2	−646·1	−2·49	−1·25	25·39	0·7796	0·3108	−0·03680	23·78	0·4028
3·2–3·4	−656·3	−7·23	−3·62	25·36	0·4853	0·2185	−0·03691	24·22	0·4857
3·4–3·6	−672·5	−11·27	−5·63	25·31	0·3240	0·1347	−0·03710	24·95	0·5525
3·6–3·8	−687·8	−14·20	−7·10	25·25	0·2417	0·0651	−0·03738	25·71	0·5959
3·8–4·0	−695·5	−15·74	−7·87	25·17	0·2072	0·0197	−0·03769	26·21	0·6129

Table 7·14·2. Example No. 3

Computation of the Matrix Chain $E_p E_{p-1} \dots E_1$

Step	Individual Matrices		Matrix Product		$\Pi e^{\mu\tau}$	$y(t_{p-1})$
0 –0·2	0·0030 26·27	−0·03799 0·6010	1·0 0	0 1·0	1·0	1·0 —
0·2–0·4	0·0160 25·89	−0·03827 0·5594	0·0030 26·27	−0·03799 0·6010	0·2072	0·06359 —
0·4–0·6	0·0594 25·21	−0·03850 0·4930	−1·0053 14·77	−0·02361 −0·6474	0·05008	−0·04089 —
0·6–0·8	0·1286 24·51	−0·03866 0·4084	−0·6284 −18·06	0·02352 −0·9144	0·01623	−0·01325 —
0·8–1·0	0·2152 24·08	−0·03875 0·3120	0·6174 −22·78	0·03838 0·2030	0·007874	0·00244 —
1·0–1·2	0·3120 24·08	−0·03875 0·2152	1·0156 7·76	0·00039 0·9875	0·006139	0·00622 —
1·2–1·4	0·4084 24·51	−0·03866 0·1286	0·0162 26·13	−0·03814 0·2219	0·007874	0·00253 —
1·4–1·6	0·4930 25·21	−0·03850 0·0594	−1·0036 3·76	−0·02416 −0·9063	0·01623	−0·01315 —
1·6–1·8	0·5594 25·89	−0·03827 0·0160	−0·6395 −25·08	0·02298 −0·6629	0·05008	−0·04123 —
1·8–2·0	0·6010 26·27	−0·03799 0·0030	0·6021 −16·96	0·03822 0·5843	0·2072	0·06139 —
2·0–2·2	0·6129 26·21	−0·03769 0·0197	1·0062 15·77	0·00077 1·0058	1·0	1·000 —
2·2–2·4	0·5959 25·71	−0·03738 0·0651	0·0223 26·68	−0·03744 0·0400	4·826	1·553 —
2·4–2·6	0·5525 24·95	−0·03710 0·1347	−0·9840 2·31	−0·02381 −0·9600	19·96	−15·84 —
2·6–2·8	0·4857 24·22	−0·03691 0·2185	−0·6294 −24·24	0·02246 −0·7234	61·62	−49·86 —
2·8–3·0	0·4028 23·78	−0·03680 0·3108	0·5890 −20·54	0·03761 0·3859	127·0	36·58 —
3·0–3·2	0·3108 23·78	−0·03680 0·4028	0·9931 7·62	0·00095 1·0143	162·9	160·5 —
3·2–3·4	0·2185 24·22	−0·03691 0·4857	0·0282 26·69	−0·03703 0·4312	127·0	41·19 —
3·4–3·6	0·1347 24·95	−0·03710 0·5525	−0·9790 13·65	−0·02401 −0·6874	61·62	−48·49 —
3·6–3·8	0·0651 25·71	−0·03738 0·5959	−0·6383 −16·88	0·02227 −0·9788	19·96	−16·30 —
3·8–4·0	0·0197 26·21	−0·03769 0·6129	0·5894 −26·47	0·03804 −0·0107	4·826	1·376 —
—	— —	— —	1·0093 −0·78	0·00115 0·9905	1·0	1·000 —

Between $t = 0$ and $t = 2$, the ordinate is small: on the other hand, it grows to large values of the order ± 150 between $t = 2$ and $t = 4$. Consequently, in order to obtain a reasonable graphical representation, the scale has been magnified 200 times for the first half of the range. The approximate results computed by postmultiplication of the matrix chain by $\{1, -8\}$ are marked in the diagram, and they accord well with the curve.

7·15. Example No. 4. We shall next solve by the method of mean coefficients the problem treated by other methods in example (iii) of § 7·10.

The first-order system equivalent to (7·10·1) is

$$\frac{dy}{dr} = \begin{bmatrix} 0 & 1 \\ -K & -1/r \end{bmatrix} y \equiv uy,$$

where $y_1 = R$ and $y_2 = dR/dr$. To obtain a preliminary estimate of the characteristic numbers K a constant mean value for $1/r$, namely $\frac{2}{3}$, will be assigned throughout the interval $r = 1$ to $r = 2$. The solution of the equation is then (see (7·11·2))

$$y(2) = e^{\mu}\begin{bmatrix} -\dfrac{\mu}{\omega}\sin\omega + \cos\omega, & \dfrac{1}{\omega}\sin\omega \\ -\dfrac{K}{\omega}\sin\omega, & -\dfrac{1}{\omega}(\frac{2}{3}+\mu)\sin\omega + \cos\omega \end{bmatrix} y(1), \quad \text{......(1)}$$

where $\mu \pm i\omega$ are the latent roots of u. The boundary conditions are $y_1(2) = y_1(1) = 0$, and the first of the two equations implicit in (1) therefore yields

$$0 = \frac{e^{\mu}}{\omega}\sin\omega \,.\, y_2(1),$$

whence $\sin\omega = 0$ or $\omega = n\pi$. But the latent roots of u, with $1/r = \frac{2}{3}$, are $-\frac{1}{3} \pm i\sqrt{K - \frac{1}{9}}$. Hence $\omega = n\pi = \sqrt{K - \frac{1}{9}}$, or $K = n^2\pi^2 + \frac{1}{9}$, and rough approximations for K are therefore π^2, $4\pi^2$, $9\pi^2$, etc.

Proceeding next to a more refined approximation, we assume the membrane to be divided into five sections by circles of radii 1·2, 1·4, 1·6 and 1·8 and use an average value for the radius of each section. If also trial values for K are adopted, namely 9·8 and 9·9, 39·4 and 39·5, 88·8 and 88·9, then for each value of K the elements of the successive matrices corresponding to the five sections are all numerically determinate. The value of y_2 at $r = 1$ is arbitrary: hence we may adopt as initial conditions $y(1) = \{0, 1\}$ and deduce the end conditions $y(2)$.

Table 7·15·1. Example No. 4

Computation of $y_1(2)$ for various values of K

Step	$K = 9{\cdot}8$		$K = 9{\cdot}9$	
	Matrix	$\{y_1, y_2\}$	Matrix	$\{y_1, y_2\}$
1·0–1·2	0·8211 0·1713 −1·6784 0·6670	0 1	0·8193 0·1712 −1·6947 0·6652	0 1
1·2–1·4	0·8195 0·1737 −1·7021 0·6875	0·1713 0·6670	0·8177 0·1736 −1·7186 0·6858	0·1712 0·6652
1·4–1·6	0·8184 0·1753 −1·7183 0·7027	0·2562 0·1670	0·8166 0·1752 −1·7349 0·7009	0·2555 0·1620
1·6–1·8	0·8174 0·1767 −1·7321 0·7149	0·2389 −0·3229	0·8156 0·1767 −1·7489 0·7131	0·2370 −0·3297
1·8–2·0	0·8166 0·1778 −1·7425 0·7241	0·1382 −0·6446	0·8148 0·1776 −1·7584 0·7225	0·1350 −0·6496
2·0	— — — —	−0·0018 —	— — — —	−0·0054 —

Step	$K = 39{\cdot}4$		$K = 39{\cdot}5$	
	Matrix	$\{y_1, y_2\}$	Matrix	$\{y_1, y_2\}$
1·0–1·2	0·3847 0·1386 −5·4626 0·2238	0 1	0·3472 0·1386 −5·4728 0·2225	0 1
1·2–1·4	0·3429 0·1405 −5·5359 0·2362	0·1386 0·2238	0·3415 0·1404 −5·5463 0·2348	0·1386 0·2225
1·4–1·6	0·3387 0·1419 −5·5915 0·2451	0·0790 −0·7144	0·3374 0·1418 −5·6020 0·2438	0·0786 −0·7165
1·6–1·8	0·3354 0·1430 −5·6327 0·2525	−0·0746 −0·6168	0·3340 0·1429 −5·6433 0·2511	−0·0751 −0·6150
1·8–2·0	0·3328 0·1438 −5·6666 0·2580	−0·1132 0·2645	0·3314 0·1437 −5·6773 0·2566	−0·1130 0·2694
2·0	— — — —	0·0004 —	— — — —	0·0013 —

Step	$K = 88{\cdot}8$		$K = 88{\cdot}9$	
	Matrix	$\{y_1, y_2\}$	Matrix	$\{y_1, y_2\}$
1·0–1·2	−0·2387 0·0924 −8·2050 −0·3219	0 1	−0·2396 0·0923 −8·2061 −0·3228	0 1
1·2–1·4	−0·2491 0·0937 −8·3207 −0·3203	0·0924 −0·3219	−0·2501 0·0936 −8·3218 −0·3213	0·0923 −0·3228
1·4–1·6	−0·2568 0·0945 −8·3960 −0·3191	−0·0532 −0·6657	−0·2578 0·0945 −8·3971 −0·3202	−0·0533 −0·6644
1·6–1·8	−0·2629 0·0953 −8·4634 −0·3182	−0·0492 0·6591	−0·2638 0·0952 −8·4645 −0·3191	−0·0490 0·6603
1·8–2·0	−0·2675 0·0959 −8·5144 −0·3175	0·0757 0·2067	−0·2686 0·0958 −8·5155 −0·3183	0·0758 0·2041
2·0	— — — —	−0·0004 —	— — — —	−0·0008 —

Linear interpolation between the values of K can then be used to give the value of K for which $y_1(2) = 0$.

The results are summarised in Table 7·15·1. The actual computations of the elements of the matrices E_p are not tabulated; they are similar to those of Example No. 3 for each value of K adopted, except that in this case the scalar factor $e^{0 \cdot 2\mu}$ is absorbed into the matrices for convenience. Since the initial conditions $y(1)$ are prescribed, the columns y are computed in succession until $y_1(2)$ is deduced. This differs slightly from zero for each approximate value of K chosen. Interpolation to make $y_1(2) = 0$ gives the values $K = 9 \cdot 750$, 39·36, 88·70, which agree very closely with the values 9·753, 39·35, 88·72 found from the tables of Bessel functions.

The mode of displacement, which is given by $y_1(1)$, $y_1(1 \cdot 2)$, $y_1(1 \cdot 4)$, $y_1(1 \cdot 6)$, $y_1(1 \cdot 8)$ and $y_1(2)$, can be deduced by interpolation, or by recalculation using the more exact values of K.

CHAPTER VIII

KINEMATICS AND DYNAMICS OF SYSTEMS

PART I. FRAMES OF REFERENCE AND KINEMATICS

8·1. Frames of Reference. Before taking up the subject of dynamics, we shall first deal with changes of reference axes, and with the analysis of motion quite apart from the question of the causation of the motion.

The positions which the points of any geometrical system, or the particles of any material system, occupy at any instant, relative to a

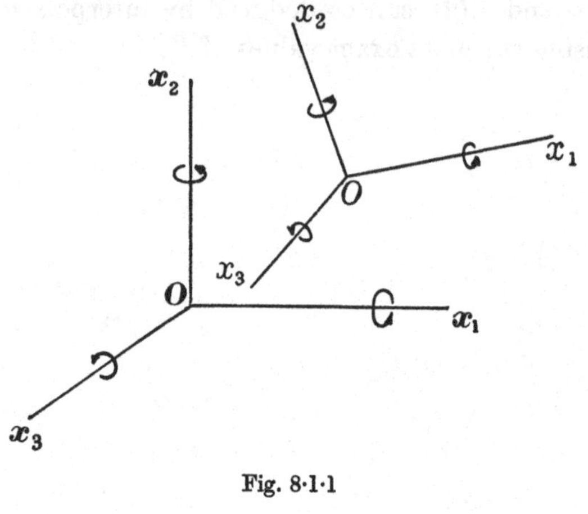

Fig. 8·1·1

datum body (e.g. the earth) can be specified by the Cartesian coordinates of these points or particles referred to a system of rectangular axes $\boldsymbol{Ox_1}$, $\boldsymbol{Ox_2}$, $\boldsymbol{Ox_3}$—or frame of reference $\boldsymbol{O(x_1, x_2, x_3)}$—fixed in that body. In usual practical applications the specification of position with reference to such a fixed frame of reference is treated as absolute; and it is convenient to regard the time-rates of change of the coordinates as defining "absolute velocity" and "absolute acceleration".

It is frequently helpful to adopt in conjunction with the fixed axes one or more auxiliary frames of reference, such as $O(x_1, x_2, x_3)$ in Fig. 8·1·1, the absolute positions of which are changing in some specified manner with time. For the sake of definiteness we shall suppose all sets of axes used to be right-handed, as represented in the diagram.

The conventions regarding the signs of rotations are as follows. An axis of rotation is regarded as possessing a definite sense, and the convention is adopted that the positive sense of rotation about the axis would appear clockwise to an observer looking along the axis in the positive sense. For example, the positive sense of Ox_1 is from O to x_1, and a positive rotation about Ox_1 would move x_2 towards x_3.

The Cartesian coordinates of a general point P, referred to the fixed axes, will hereafter be denoted as $\boldsymbol{x}_1$, $\boldsymbol{x}_2$, $\boldsymbol{x}_3$; while those of the same point, relative to the auxiliary axes, will be x_1, x_2, x_3. At the outset we shall consider the simple case of two-dimensional motion.

8·2. Change of Reference Axes in Two Dimensions. Here the position of the auxiliary frame of reference at any instant t can be specified completely by the Cartesian coordinates $(\boldsymbol{a}_1, \boldsymbol{a}_2)$ of O referred to the fixed axes, and by the inclination ϕ of Ox_1 to $\boldsymbol{Ox}_1$.

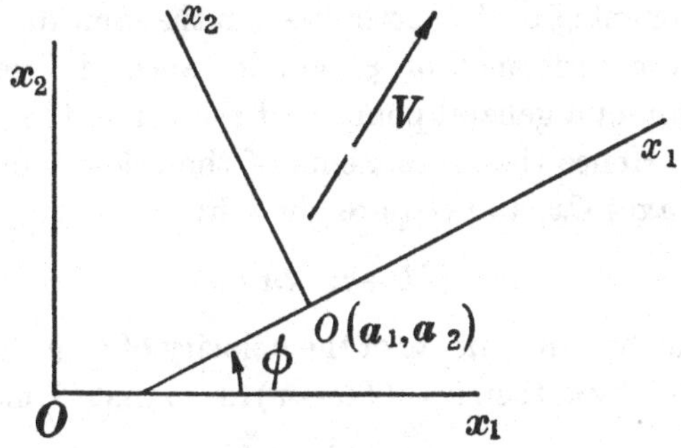

Fig. 8·2·1

If $\boldsymbol{\xi} \equiv \{\boldsymbol{\xi}_1, \boldsymbol{\xi}_2\}$ and $\xi \equiv \{\xi_1, \xi_2\}$ denote, respectively, the columns of the components of any vector $\boldsymbol{V}$ in the plane, measured parallel to the fixed and the auxiliary axes, then

$$\xi = \boldsymbol{l}\boldsymbol{\xi}, \qquad \text{......(1)}$$

in which

$$\boldsymbol{l} \equiv \begin{bmatrix} \cos\phi & \sin\phi \\ -\sin\phi & \cos\phi \end{bmatrix}.$$

The matrix $\boldsymbol{l}$ is orthogonal (see § 1·17) since

$$\boldsymbol{l}^{-1} = \begin{bmatrix} \cos\phi & -\sin\phi \\ \sin\phi & \cos\phi \end{bmatrix} = \boldsymbol{l}'.$$

Also

$$\boldsymbol{l}\frac{d\boldsymbol{l}^{-1}}{dt} = -\frac{d\boldsymbol{l}}{dt}\boldsymbol{l}^{-1} = \dot{\phi}\begin{bmatrix} \cos\phi & \sin\phi \\ -\sin\phi & \cos\phi \end{bmatrix}\begin{bmatrix} -\sin\phi & -\cos\phi \\ \cos\phi & -\sin\phi \end{bmatrix} = \begin{bmatrix} 0 & -p \\ p & 0 \end{bmatrix},$$

......(2)

where $p \equiv \dot{\phi}$ denotes the angular velocity of the moving frame. For brevity we shall write

$$\varpi \equiv \begin{bmatrix} 0 & -p \\ p & 0 \end{bmatrix}.$$

On differentiation with respect to t equation (1) yields

$$\frac{d\xi}{dt} = \boldsymbol{l}\frac{d\boldsymbol{\xi}}{dt} + \frac{d\boldsymbol{l}}{dt}\boldsymbol{\xi},$$

whence by (2) and (1)

$$\boldsymbol{l}\frac{d\boldsymbol{\xi}}{dt} = \frac{d\xi}{dt} + \varpi\xi = (ID+\varpi)\,\xi.$$

Now by (1) $\boldsymbol{l}\dfrac{d\boldsymbol{\xi}}{dt}$ represents the components of the time-rate of change of the vector $\boldsymbol{V}$, measured in the directions of the moving axes at time t. In other words the components of the time-rate of change of any vector measured in the directions of the moving axes are obtained from the components ξ of the vector itself in the same directions by the operation $ID+\varpi$ performed on ξ. For instance, if $\boldsymbol{V}$ represents the absolute position of a general point P of the plane, then $x = \boldsymbol{l}(\boldsymbol{x}-\boldsymbol{a})$, or $\boldsymbol{l}\boldsymbol{x} = \boldsymbol{l}\boldsymbol{a}+x$. Hence the components of the velocity of P measured parallel to the axes Ox_1 and Ox_2 are given by

$$v = (ID+\varpi)\,(\boldsymbol{l}\boldsymbol{a}+x). \qquad \text{......(3)}$$

In particular, if the components of the velocity of O in the same directions are denoted by u, then $u = (ID+\varpi)\,\boldsymbol{l}\boldsymbol{a}$, so that (3) may be written

$$v = u+(ID+\varpi)\,x.$$

Similarly, if the components of the acceleration of P measured parallel to the moving axes are α, then

$$\alpha = (ID+\varpi)\,v.$$

These formulae can, of course, also be established by quite elementary methods, but the treatment adopted has certain attractions when generalized to three dimensions.

EXAMPLES

(i) *Matrices Representing Finite Rotations.* If ϕ is increased to $\phi+\omega$, then $\boldsymbol{l}$ becomes

$$\boldsymbol{l}_1 \equiv \begin{bmatrix} \cos(\phi+\omega) & \sin(\phi+\omega) \\ -\sin(\phi+\omega) & \cos(\phi+\omega) \end{bmatrix} = \begin{bmatrix} \cos\omega & \sin\omega \\ -\sin\omega & \cos\omega \end{bmatrix}\boldsymbol{l}.$$

Hence premultiplication of $\boldsymbol{l}$ by the matrix

$$\rho(\omega) \equiv \begin{bmatrix} \cos\omega & \sin\omega \\ -\sin\omega & \cos\omega \end{bmatrix}$$

is equivalent to a rotation of the frame $O(x_1, x_2)$ through an angle ω about the normal axis through O. More generally, any succession of such rotations $\omega_1, \omega_2, \ldots, \omega_n$ will be represented by

$$\boldsymbol{l}_1 = \rho(\omega_n)\,\rho(\omega_{n-1})\ldots\rho(\omega_2)\,\rho(\omega_1)\,\boldsymbol{l} = \rho(\omega_n+\omega_{n-1}+\ldots+\omega_2+\omega_1)\,\boldsymbol{l}.$$

If $\omega_n = \omega_{n-1} = \ldots = \omega_1 = 2s\pi/n$, where s is any integer, then after the complete cycle of rotations the frame is returned to its original position and $[\rho(2s\pi/n)]^n = I_2$. Hence the matrices $\rho(2s\pi/n)$ are all nth roots of I_2.

(ii) *Determination of the Axes of a Central Conic.* The equation of a conic referred to axes $\boldsymbol{O}(\boldsymbol{x}_1, \boldsymbol{x}_2)$ through the centre may be written $\boldsymbol{x}'a\boldsymbol{x} = 1$, where a is a symmetrical square matrix. When referred to any other frame of reference $O(x_1, x_2)$ through $\boldsymbol{O}$ the equation becomes $x'bx = 1$, where b denotes the symmetrical matrix $\boldsymbol{l}a\boldsymbol{l}^{-1}$ or $\boldsymbol{l}a\boldsymbol{l}'$ (see also § 1·15). If $\boldsymbol{l}$ is chosen so that b is a diagonal matrix, say

$$b = \begin{bmatrix} 1/\alpha_1^2 & 0 \\ 0 & 1/\alpha_2^2 \end{bmatrix},$$

then α_1 and α_2 will be the semi-axes. In this case

$$a = \boldsymbol{l}^{-1}\begin{bmatrix} 1/\alpha_1^2 & 0 \\ 0 & 1/\alpha_2^2 \end{bmatrix}\boldsymbol{l},$$

so that $1/\alpha_1^2 \equiv \lambda_1$ and $1/\alpha_2^2 \equiv \lambda_2$ are the latent roots of a (see § 3·6). Moreover $\boldsymbol{l}_{11}$ and $\boldsymbol{l}_{12}$ will be proportional to the elements in any column (or row) of the adjoint of $\lambda_1 I - a$. The directions of the axes are then given by $\cot\phi = \boldsymbol{l}_{11}/\boldsymbol{l}_{12}$.

For instance, if the given conic is

$$3x_1^2 + 2x_1x_2 + 3x_2^2 = 8,$$

then
$$a = \begin{bmatrix} \frac{3}{8} & \frac{1}{8} \\ \frac{1}{8} & \frac{3}{8} \end{bmatrix},$$

and the latent roots of this matrix are found to be $\lambda_1 = \frac{1}{2}$ and $\lambda_2 = \frac{1}{4}$. Hence the semi-axes are $\alpha_1 = \sqrt{2}$ and $\alpha_2 = 2$. Again, the adjoint of $\lambda I - a$ is

$$\tfrac{1}{8}\begin{bmatrix} 8\lambda-3 & 1 \\ 1 & 8\lambda-3 \end{bmatrix},$$

and the rows are proportional to $[1, 1]$ for $\lambda_1 = \frac{1}{2}$ and to $[-1, 1]$ for $\lambda_2 = \frac{1}{4}$. Hence, if the axis $\boldsymbol{O}\boldsymbol{x}_1$ is taken to correspond to $\lambda_1 = \frac{1}{2}$, we have $\cot\phi = 1$. The inclination of the minor axis is thus 45° to $\boldsymbol{O}\boldsymbol{x}_1$.

8·3. Angular Coordinates of a Three-Dimensional Moving Frame of Reference. To specify the (absolute) orientation of a frame of reference in three dimensions three independent parameters are required. These may be chosen in various ways. The method which is usual in investigations concerning the motion of aeroplanes* is a modification of the system of "angular coordinates" originally introduced by Euler.

As we are at present concerned only with the question of orientation, we may temporarily suppose $\mathbf{O}$ and O in Fig. 8·1·1 to be coincident.

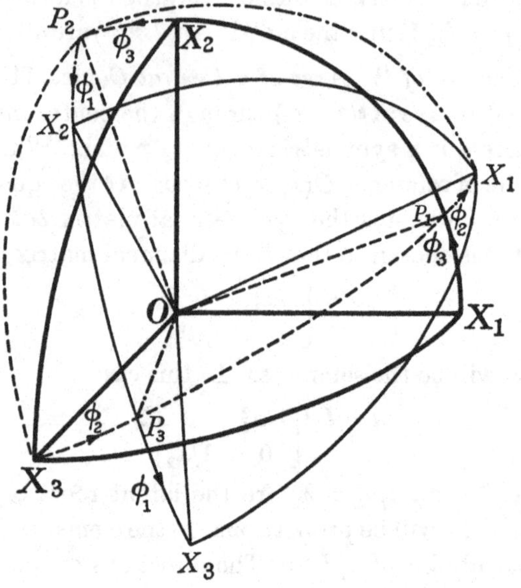

Fig. 8·3·1

With $\mathbf{O}$ as centre, draw a sphere of unit radius to intersect the fixed and the moving sets of axes in $\mathbf{X}_1, \mathbf{X}_2, \mathbf{X}_3$ and X_1, X_2, X_3 (see Fig. 8·3·1). In the original system of Eulerian coordinates the frame $\mathbf{O}(\mathbf{X}_1, \mathbf{X}_2, \mathbf{X}_3)$ is viewed as displaced to the position $\mathbf{O}(X_1, X_2, X_3)$ by three successive rotations which are not actually represented in Fig. 8·3·1. Firstly, a rotation about $\mathbf{OX}_3$ brings $\mathbf{X}_2$ to some intermediate position X_2'; a rotation about $\mathbf{O}X_2'$ next brings $\mathbf{X}_3$ to X_3; lastly a rotation about $\mathbf{O}X_3$ brings the frame to its final position. In the modified system (specified in Fig. 8·3·1) the successive rotations are ϕ_3, ϕ_2, ϕ_1, respectively, about the successive *carried* positions of the axes $\mathbf{OX}_3$, $\mathbf{OX}_2$, $\mathbf{OX}_1$. For brevity

* See, for instance, p. 251 of Ref. 27.

a succession of rotations such as that represented in Fig. 8·3·1 will be called a sequence of rotations ϕ_3, ϕ_2, ϕ_1 about the carried axes taken in the order OX_3, OX_2, OX_1.

8·4. The Orthogonal Matrix of Transformation. The components $\boldsymbol{\xi}_1$, $\boldsymbol{\xi}_2$, $\boldsymbol{\xi}_3$ and ξ_1, ξ_2, ξ_3 of any vector measured respectively parallel to the fixed axes and to the moving axes in Fig. 8·1·1 will be connected by the linear substitution

$$\xi = \boldsymbol{l}\boldsymbol{\xi}, \qquad \ldots\ldots(1)$$

in which

$$\boldsymbol{l} \equiv \begin{bmatrix} l_{11} & l_{12} & l_{13} \\ l_{21} & l_{22} & l_{23} \\ l_{31} & l_{32} & l_{33} \end{bmatrix},$$

and l_{s1}, l_{s2}, l_{s3} denote the direction cosines of the typical moving axis OX_s referred to the fixed axes. More generally, if $\boldsymbol{M}$ is a $(\overline{3}, n)$ matrix representing the components of any n vectors parallel to the fixed axes, and if M is the corresponding matrix of vector components parallel to the moving axes, then $M = \boldsymbol{l}\boldsymbol{M}$. In particular, suppose M to be the unit matrix I_3 corresponding to the three unit vectors parallel to the moving axes. Then it is easy to see from Fig. 8·1·1 that $\boldsymbol{M} = \boldsymbol{l}'$, so that $I_3 = \boldsymbol{l}\boldsymbol{l}'$. The matrix of transformation is accordingly orthogonal. Particular properties to be noted are that $|\boldsymbol{l}| = 1$, and that, since $\boldsymbol{l}'$ is also the adjoint of $\boldsymbol{l}$, every element of $\boldsymbol{l}$ equals its own cofactor.

8·5. Matrices Representing Finite Rotations of a Frame of Reference. Suppose the frame $O(x_1, x_2, x_3)$ in Fig. 8·1·1 to be rotated about the axis Ox_1 through an angle ω to a new position $O(x_1, x_2^*, x_3^*)$, and let $\boldsymbol{l}$ become

$$\boldsymbol{l}_1 = \begin{bmatrix} l_{11} & l_{12} & l_{13} \\ l_{21}^* & l_{22}^* & l_{23}^* \\ l_{31}^* & l_{32}^* & l_{33}^* \end{bmatrix}.$$

Then since Ox_2^* is perpendicular to Ox_1, and inclined at angles ω and $\frac{1}{2}\pi - \omega$ to Ox_2 and Ox_3, respectively, we have by the properties of direction cosines

$$\boldsymbol{l}\{l_{21}^*, l_{22}^*, l_{23}^*\} = \{0, \cos\omega, \sin\omega\},$$

and similarly

$$\boldsymbol{l}\{l_{31}^*, l_{32}^*, l_{33}^*\} = \{0, -\sin\omega, \cos\omega\}.$$

It follows that

$$\boldsymbol{l}\boldsymbol{l}_1' = \boldsymbol{l}\begin{bmatrix} l_{11} & l_{21}^* & l_{31}^* \\ l_{12} & l_{22}^* & l_{32}^* \\ l_{13} & l_{23}^* & l_{33}^* \end{bmatrix} = \begin{bmatrix} 1 & 0 & 0 \\ 0 & \cos\omega & -\sin\omega \\ 0 & \sin\omega & \cos\omega \end{bmatrix},$$

whence by transposition and use of the orthogonal property,

$$\boldsymbol{l}_1 = \begin{bmatrix} 1 & 0 & 0 \\ 0 & \cos\omega & \sin\omega \\ 0 & -\sin\omega & \cos\omega \end{bmatrix} \boldsymbol{l}.$$

Writing this relation as $\boldsymbol{l}_1 = \rho_1(\omega)\,\boldsymbol{l}$, ...(1)

we see that the premultiplication of $\boldsymbol{l}$ by $\rho_1(\omega)$ corresponds to the rotation of the frame $O(x_1, x_2, x_3)$ through an angle ω about the axis Ox_1. Similarly, the premultipliers

$$\rho_2(\omega) \equiv \begin{bmatrix} \cos\omega & 0 & -\sin\omega \\ 0 & 1 & 0 \\ \sin\omega & 0 & \cos\omega \end{bmatrix}$$

and

$$\rho_3(\omega) \equiv \begin{bmatrix} \cos\omega & \sin\omega & 0 \\ -\sin\omega & \cos\omega & 0 \\ 0 & 0 & 1 \end{bmatrix}$$

correspond to rotations ω of the frame about Ox_2 and Ox_3, respectively. The matrices ρ_1, ρ_2, ρ_3 have the characteristic orthogonal property $\rho^{-1} = \rho'$; also it can be verified that

$$\rho_1 \frac{d\rho_1'}{d\omega} = \begin{bmatrix} 0 & 0 & 0 \\ 0 & 0 & -1 \\ 0 & 1 & 0 \end{bmatrix}; \qquad \rho_2 \frac{d\rho_2'}{d\omega} = \begin{bmatrix} 0 & 0 & 1 \\ 0 & 0 & 0 \\ -1 & 0 & 0 \end{bmatrix};$$

$$\rho_3 \frac{d\rho_3'}{d\omega} = \begin{bmatrix} 0 & -1 & 0 \\ 1 & 0 & 0 \\ 0 & 0 & 0 \end{bmatrix}. \qquad \text{......(2)}$$

If the frame is given a sequence of three rotations ω_3, ω_2, ω_1, the first being about Ox_3 and the others being about the successive carried positions of Ox_2 and Ox_1, then

$$\boldsymbol{l}_1 = \rho_1(\omega_1)\,\rho_2(\omega_2)\,\rho_3(\omega_3)\,\boldsymbol{l}, \qquad \text{......(3)}$$

where $\boldsymbol{l}_1$ now is the matrix of transformation for the final position. The order of these rotations is, of course, not commutative in general. However, in the special case where the angles of rotation are infinitesimally small, say $\delta\omega_3$, $\delta\omega_2$, $\delta\omega_1$, then the last equation gives

$$\boldsymbol{l} + \delta\boldsymbol{l} = \begin{bmatrix} 1 & \delta\omega_3 & 0 \\ -\delta\omega_3 & 1 & 0 \\ 0 & 0 & 1 \end{bmatrix} \begin{bmatrix} 1 & 0 & -\delta\omega_2 \\ 0 & 1 & 0 \\ \delta\omega_2 & 0 & 1 \end{bmatrix} \begin{bmatrix} 1 & 0 & 0 \\ 0 & 1 & \delta\omega_1 \\ 0 & -\delta\omega_1 & 1 \end{bmatrix} \boldsymbol{l},$$

or
$$\delta \boldsymbol{l} = \begin{bmatrix} 0 & \delta\omega_3 & -\delta\omega_2 \\ -\delta\omega_3 & 0 & \delta\omega_1 \\ \delta\omega_2 & -\delta\omega_1 & 0 \end{bmatrix} \boldsymbol{l}.$$

The order in which such infinitesimal rotations are performed is thus immaterial. If the total time of the operation is δt, then in the limit (see also (8·2·2))

$$\boldsymbol{l}\frac{d\boldsymbol{l}^{-1}}{dt} = -\frac{d\boldsymbol{l}}{dt}\boldsymbol{l}^{-1} = \begin{bmatrix} 0 & -p_3 & p_2 \\ p_3 & 0 & -p_1 \\ -p_2 & p_1 & 0 \end{bmatrix} \equiv \varpi, \text{ say,} \qquad \text{......(4)}$$

in which p_1, p_2, p_3 are the instantaneous angular velocities of the moving frame about the axes Ox_1, Ox_2, Ox_3, respectively.

Examples

(i) *A Property of the Matrix* ϖ. It can at once be verified that, if y_1, y_2, y_3 are arbitrary, then

$$\begin{bmatrix} 0 & -p_3 & p_2 \\ p_3 & 0 & -p_1 \\ -p_2 & p_1 & 0 \end{bmatrix}\begin{bmatrix} y_1 \\ y_2 \\ y_3 \end{bmatrix} + \begin{bmatrix} 0 & -y_3 & y_2 \\ y_3 & 0 & -y_1 \\ -y_2 & y_1 & 0 \end{bmatrix}\begin{bmatrix} p_1 \\ p_2 \\ p_3 \end{bmatrix} = 0. \qquad \text{......(5)}$$

This identity will be useful later.

(ii) *Postmultiplications Representing Rotations.* Equation (1) can be written alternatively as $\boldsymbol{l}_1 = \boldsymbol{l}\sigma_1(\omega)$, where

$$\sigma_1(\omega) \equiv \boldsymbol{l}'\rho_1(\omega)\,\boldsymbol{l}. \qquad \text{......(6)}$$

Direct multiplication of the product $\boldsymbol{l}'\rho_1(\omega)\,\boldsymbol{l}$ yields the somewhat cumbrous expression

$$\sigma_1(\omega) = \cos\omega\, I_3 + (1-\cos\omega)\{l_{11}, l_{12}, l_{13}\}[l_{11}, l_{12}, l_{13}] + \sin\omega \begin{bmatrix} 0 & l_{13} & -l_{12} \\ -l_{13} & 0 & l_{11} \\ l_{12} & -l_{11} & 0 \end{bmatrix}.$$

A more elegant form, whose verification may be left to the reader, is

$$\sigma_1(\omega) = (B')^{-1}B, \qquad \text{......(7)}$$

where
$$B \equiv \begin{bmatrix} 1 & l_{13}\tan\frac{1}{2}\omega & -l_{12}\tan\frac{1}{2}\omega \\ -l_{13}\tan\frac{1}{2}\omega & 1 & l_{11}\tan\frac{1}{2}\omega \\ l_{12}\tan\frac{1}{2}\omega & -l_{11}\tan\frac{1}{2}\omega & 1 \end{bmatrix}.$$

(iii) *Rotation of Frame $O(x_1, x_2, x_3)$ about a General Axis through O.* Suppose the given general axis to be OX_1, and let OX_2, OX_3 be any two further axes which together with OX_1 form an orthogonal frame whose matrix is $[\boldsymbol{L}_{ij}]$. Now imagine the frame $O(x_1, x_2, x_3)$ to be rigidly attached to the frame $O(X_1, X_2, X_3)$, and take points on the axes Ox_1, Ox_2, Ox_3 at unit distance from O. Then denoting the *columns* of the coordinates of these points relative to the frame $O(X_1, X_2, X_3)$ as ξ, η, ζ, we have by an obvious application of (8·4·1)

$$[\xi, \eta, \zeta] = \boldsymbol{L}\boldsymbol{l}'.$$

But the sets of coordinates ξ, η, ζ are unchanged when the frame $O(X_1, X_2, X_3)$ is given a rotation ω about OX_1. Hence also

$$[\xi, \eta, \zeta] = \rho_1(\omega)\,\boldsymbol{L}\boldsymbol{l}_1',$$

where $\boldsymbol{l}_1$ is the matrix appropriate to the new position of the frame $O(x_1, x_2, x_3)$. It follows that $\boldsymbol{L}\boldsymbol{l}' = \rho_1(\omega)\,\boldsymbol{L}\boldsymbol{l}_1'$, which gives

$$\boldsymbol{l}_1 = \boldsymbol{l}\boldsymbol{L}'\rho_1(\omega)\,\boldsymbol{L}. \qquad \text{......(8)}$$

Hence on application of (6) and (7)

$$\boldsymbol{l}_1 = \boldsymbol{l}(B')^{-1}B, \qquad \text{......(9)}$$

where
$$B \equiv \begin{bmatrix} 1 & \boldsymbol{L}_{13}\tan\tfrac{1}{2}\omega & -\boldsymbol{L}_{12}\tan\tfrac{1}{2}\omega \\ -\boldsymbol{L}_{13}\tan\tfrac{1}{2}\omega & 1 & \boldsymbol{L}_{11}\tan\tfrac{1}{2}\omega \\ \boldsymbol{L}_{12}\tan\tfrac{1}{2}\omega & -\boldsymbol{L}_{11}\tan\tfrac{1}{2}\omega & 1 \end{bmatrix}. \qquad \text{......(10)}$$

Note that if the frame $O(X_1, X_2, X_3)$ coincides with $O(x_1, x_2, x_3)$, so that $\boldsymbol{L} = \boldsymbol{l}$, then the formula (8) immediately reduces to (1).

(iv) *Rodrigues' Formula.* This formula states that if a rigid body is turned through an angle ω about an axis through the origin $\boldsymbol{O}$ whose direction cosines are $(\boldsymbol{L}_{11}, \boldsymbol{L}_{12}, \boldsymbol{L}_{13})$, and if $(\boldsymbol{x}_1, \boldsymbol{x}_2, \boldsymbol{x}_3)$ and $(\boldsymbol{x}_1^*, \boldsymbol{x}_2^*, \boldsymbol{x}_3^*)$ are the coordinates, referred to fixed axes, of any point P of the body before and after the rotation, then

$$B'\boldsymbol{x} = B\boldsymbol{x}^*,$$

where B is as defined by (10). This formula can be deduced as follows. Suppose (x_1, x_2, x_3) to be the coordinates of P referred to any orthogonal frame of reference fixed in the body and passing through $\boldsymbol{O}$. Then if $\boldsymbol{l}$ is the matrix of this frame, we have by (8·4·1) and (9)

$$x = \boldsymbol{l}\boldsymbol{x} = \boldsymbol{l}_1\boldsymbol{x}^* = \boldsymbol{l}(B')^{-1}B\boldsymbol{x}^*.$$

Rodrigues' formula immediately follows.

8·6. Matrix of Transformation and Instantaneous Angular Velocities Expressed in Angular Coordinates. We shall next apply (8·5·3) to the particular sequence of rotations represented by Fig. 8·3·1. If $\boldsymbol{l}$ now denotes the matrix for the final position, then clearly

$$\boldsymbol{l} = \rho_1(\phi_1)\,\rho_2(\phi_2)\,\rho_3(\phi_3) \qquad \text{......(1)}$$

and

$$\boldsymbol{l}^{-1} = \rho_3'(\phi_3)\,\rho_2'(\phi_2)\,\rho_1'(\phi_1). \qquad \text{......(2)}$$

The product (1) yields

$$\boldsymbol{l} \equiv \begin{bmatrix} c_2c_3, & c_2s_3, & -s_2 \\ -c_1s_3+s_1s_2c_3, & c_1c_3+s_1s_2s_3, & s_1c_2 \\ s_1s_3+c_1s_2c_3, & -s_1c_3+c_1s_2s_3, & c_1c_2 \end{bmatrix}, \qquad \text{......(3)}$$

where c_i, s_i are abbreviations for $\cos\phi_i$, $\sin\phi_i$. This formula gives the direction cosines of the moving axes explicitly in terms of the angular coordinates of the frame.

Corresponding formulae for the instantaneous angular velocities p_1, p_2, p_3 (see § 8·5) can be obtained as follows. Using (8·5·4) in conjunction with (1) and (2), we have

$$\varpi = \boldsymbol{l}\frac{d\boldsymbol{l}^{-1}}{dt} = \rho_1\rho_2\rho_3\left(\dot{\phi}_3\frac{d\rho_3'}{d\phi_3}\rho_2'\rho_1' + \dot{\phi}_2\rho_3'\frac{d\rho_2'}{d\phi_2}\rho_1' + \dot{\phi}_1\rho_3'\rho_2'\frac{d\rho_1'}{d\phi_1}\right),$$

which on application of (8·5·2) gives

$$\varpi = \dot{\phi}_3\rho_1\rho_2\begin{bmatrix} 0 & -1 & 0 \\ 1 & 0 & 0 \\ 0 & 0 & 0 \end{bmatrix}\rho_2'\rho_1' + \dot{\phi}_2\rho_1\begin{bmatrix} 0 & 0 & 1 \\ 0 & 0 & 0 \\ -1 & 0 & 0 \end{bmatrix}\rho_1' + \dot{\phi}_1\begin{bmatrix} 0 & 0 & 0 \\ 0 & 0 & -1 \\ 0 & 1 & 0 \end{bmatrix}.$$

After some reduction this yields

$$\varpi = \begin{bmatrix} 0, & -c_2c_1\dot{\phi}_3+s_1\dot{\phi}_2, & c_2s_1\dot{\phi}_3+c_1\dot{\phi}_2 \\ c_2c_1\dot{\phi}_3-s_1\dot{\phi}_2, & 0, & s_2\dot{\phi}_3-\dot{\phi}_1 \\ -c_2s_1\dot{\phi}_3-c_1\dot{\phi}_2, & -s_2\dot{\phi}_3+\dot{\phi}_1, & 0 \end{bmatrix}. \qquad \text{......(4)}$$

Hence on comparison of (4) and (8·5·4)

$$\left.\begin{aligned} p_1 &= \dot{\phi}_1 - \dot{\phi}_3\sin\phi_2, \\ p_2 &= \dot{\phi}_2\cos\phi_1 + \dot{\phi}_3\sin\phi_1\cos\phi_2, \\ p_3 &= -\dot{\phi}_2\sin\phi_1 + \dot{\phi}_3\cos\phi_1\cos\phi_2. \end{aligned}\right\} \qquad \text{......(5)}$$

These relations may be written conveniently

$$p = R\dot{\phi}, \qquad \text{......(6)}$$

where
$$R \equiv \begin{bmatrix} 1 & 0 & -\sin\phi_2 \\ 0 & \cos\phi_1 & \sin\phi_1\cos\phi_2 \\ 0 & -\sin\phi_1 & \cos\phi_1\cos\phi_2 \end{bmatrix}.$$
The reciprocal of R is
$$R^{-1} = \sec\phi_2 \begin{bmatrix} \cos\phi_2 & \sin\phi_1\,\sin\phi_2 & \cos\phi_1\sin\phi_2 \\ 0 & \cos\phi_1\cos\phi_2 & -\sin\phi_1\cos\phi_2 \\ 0 & \sin\phi_1 & \cos\phi_1 \end{bmatrix}.$$

Similar methods may be used with any other system of angular coordinates.

8·7. Components of Velocity and Acceleration.

As in § 8·4 let $\boldsymbol{\xi}$ and ξ represent, respectively, the components of any vector measured parallel to the fixed axes $\boldsymbol{Ox_1}$, $\boldsymbol{Ox_2}$, $\boldsymbol{Ox_3}$ and to the moving axes Ox_1, Ox_2, Ox_3. Then by differentiation of (8·4·1) with respect to t and application of (8·5·4) it is seen exactly as in § 8·2 that the components of the time-rate of change of the vector, measured in the directions of the moving axes at time t, are given by the operation $ID+\varpi$ performed on ξ.

In particular, let the column x denote the coordinates of a general point or particle P at any instant relative to the moving axes; let u, v denote respectively the components of the absolute velocity of O and of P; and let α denote the components of the acceleration of P, all measured in the directions of the moving axes. Then as in § 8·2
$$v = u + (ID+\varpi)\,x \qquad \text{......(1)}$$
and
$$\alpha = (ID+\varpi)\,v = (ID+\varpi)\,u + (ID+\varpi)^2\,x. \qquad \text{......(2)}$$

If P is a point of a rigid body and the axes $O(x_1, x_2, x_3)$ are fixed in the body (i.e. are *body axes*), then the coordinates x do not change as the body moves, and equations (1) and (2) reduce to
$$v = u + \varpi x, \qquad \text{......(3)}$$
$$\alpha = (ID+\varpi)\,u + (D\varpi + \varpi^2)\,x. \qquad \text{......(4)}$$
Equation (3) can be written alternatively as (see (8·5·5))
$$v = u - \chi p, \qquad \text{......(5)}$$
where
$$\chi \equiv \begin{bmatrix} 0 & -x_3 & x_2 \\ x_3 & 0 & -x_1 \\ -x_2 & x_1 & 0 \end{bmatrix}.$$

Expressions for the matrices $\boldsymbol{l}$ and ϖ in terms of the angular coordinates ϕ of the moving axes are given in § 8·6.

EXAMPLES

(i) *Kinetic Energy of a Rigid Body*. The kinetic energy T is given by

$$2T = \Sigma_0 \boldsymbol{m}(v_1^2+v_2^2+v_3^2) = \Sigma_0 \boldsymbol{m} v'v,$$

where $\boldsymbol{m}$ is the mass of a typical particle and Σ_0 denotes summation for all the particles. If the moving axes are fixed in the body, then by (5)

$$\begin{aligned} 2T &= \Sigma_0 \boldsymbol{m}(u'-p'\chi')(u-\chi p) \\ &= \Sigma_0 \boldsymbol{m}(u'u-2u'\chi p-p'\chi^2 p), \end{aligned}$$

since $\chi' = -\chi$ and $u'\chi p = p'\chi' u$. If, further, O is the centre of mass, so that $\Sigma_0 \boldsymbol{m}\chi = 0$, then

$$2T = \boldsymbol{M}u'u+p'Jp,$$

where $\boldsymbol{M}$ is the total mass and J denotes the symmetrical square matrix

$$J \equiv -\Sigma_0 \boldsymbol{m}\chi^2 \equiv \begin{bmatrix} \Sigma_0 \boldsymbol{m}(x_2^2+x_3^2) & -\Sigma_0 \boldsymbol{m}x_1x_2 & -\Sigma_0 \boldsymbol{m}x_1x_3 \\ -\Sigma_0 \boldsymbol{m}x_2x_1 & \Sigma_0 \boldsymbol{m}(x_3^2+x_1^2) & -\Sigma_0 \boldsymbol{m}x_2x_3 \\ -\Sigma_0 \boldsymbol{m}x_3x_1 & -\Sigma_0 \boldsymbol{m}x_3x_2 & \Sigma_0 \boldsymbol{m}(x_1^2+x_2^2) \end{bmatrix}. \quad \text{......(6)}$$

(ii) *Angular Momenta.* If h_1, h_2, h_3 are the angular momenta about body axes $O(x_1, x_2, x_3)$ through the centre of mass, then it is readily shown that

$$h = Jp,$$

where J is given by (6). The components of the time-rate of the angular momentum are accordingly

$$\vartheta = (ID+\varpi)\,Jp. \quad \text{......(7)}$$

(iii) *The Principal Axes and Moments of Inertia.* Suppose the moments and products of inertia of a rigid body appropriate to the axes $O\boldsymbol{x}_1$, $O\boldsymbol{x}_2$, $O\boldsymbol{x}_3$ (see Fig. 8·1·1) to be known. It is required to find the corresponding constants for another set of axes Ox_1, Ox_2, Ox_3, derived by the matrix of transformation $\boldsymbol{l}$. In particular it is required to find the principal axes, for which all the products of inertia vanish. Write

$$\boldsymbol{G} \equiv \begin{bmatrix} \Sigma_0 \boldsymbol{m}\boldsymbol{x}_1^2 & \Sigma_0 \boldsymbol{m}\boldsymbol{x}_1\boldsymbol{x}_2 & \Sigma_0 \boldsymbol{m}\boldsymbol{x}_1\boldsymbol{x}_3 \\ \Sigma_0 \boldsymbol{m}\boldsymbol{x}_2\boldsymbol{x}_1 & \Sigma_0 \boldsymbol{m}\boldsymbol{x}_2^2 & \Sigma_0 \boldsymbol{m}\boldsymbol{x}_2\boldsymbol{x}_3 \\ \Sigma_0 \boldsymbol{m}\boldsymbol{x}_3\boldsymbol{x}_1 & \Sigma_0 \boldsymbol{m}\boldsymbol{x}_3\boldsymbol{x}_2 & \Sigma_0 \boldsymbol{m}\boldsymbol{x}_3^2 \end{bmatrix},$$

and let G be the corresponding matrix with the coordinates x substituted for $\boldsymbol{x}$. Then

$$G = \Sigma_0 \boldsymbol{m}xx' = \Sigma_0 \boldsymbol{m}\boldsymbol{l}\boldsymbol{x}\boldsymbol{x}'\boldsymbol{l}^{-1} = \boldsymbol{l}\boldsymbol{G}\boldsymbol{l}^{-1}.$$

If the new axes are principal axes, G reduces to the diagonal matrix

$$\begin{bmatrix} \Sigma_0 m x_1^2 & 0 & 0 \\ 0 & \Sigma_0 m x_2^2 & 0 \\ 0 & 0 & \Sigma_0 m x_3^2 \end{bmatrix} \equiv \begin{bmatrix} \frac{1}{2}(A_2+A_3-A_1) & 0 & 0 \\ 0 & \frac{1}{2}(A_3+A_1-A_2) & 0 \\ 0 & 0 & \frac{1}{2}(A_1+A_2-A_3) \end{bmatrix},$$

where A_1, A_2, A_3 are the moments of inertia about the principal axes. Hence the diagonal elements in the last matrix are the latent roots λ_1, λ_2, λ_3 of G, so that $A_1 = \lambda_2+\lambda_3$, $A_2 = \lambda_3+\lambda_1$ and $A_3 = \lambda_1+\lambda_2$. Moreover, the first row of elements in l (which yield the direction cosines of the principal axis corresponding to A_1) will be proportional to any row of the adjoint of $\lambda_1 I - G$. The other two rows of l are given similarly.

A similar method can be used to find the axes of a central quadric (see also example (ii) of § 8·2). As a numerical illustration suppose the given quadric to be $x'ux = 1$, where

$$u = \begin{bmatrix} 1 & 0·5 & 0·5 \\ 0·5 & 1 & 0·5 \\ 0·5 & 0·5 & 1 \end{bmatrix}.$$

The latent roots of u are $\lambda_1 = 0·5$, $\lambda_2 = 0·5$ and $\lambda_3 = 2$, so that the equation represents an ellipsoid the semi-axes of which are $\sqrt{2}$, $\sqrt{2}$ and $1/\sqrt{2}$. The direction cosines of the short principal axis are $(1/\sqrt{3}, 1/\sqrt{3}, 1/\sqrt{3})$, which are taken proportional to the first row of the adjoint of $\lambda_3 I - u$. For the equal roots λ_1 and λ_2, the adjoint $\lambda I - u$ is null, and the directions of the corresponding equal axes are obviously indeterminate.

(iv) *Homogeneous Strain.* Let x represent the coordinates of a point and write

$$X = ax. \qquad \text{......(8)}$$

where a is a square matrix of given constants. Then the field of points X is said to be derived from the field x by a *homogeneous strain*. An equivalent statement is that X is a *linear vector function* of x.

Take three points at unit distance from the origin, and let the three lines joining the points to the origin be at right angles. Then the three columns formed by the coordinates of the points can be combined into a single orthogonal square matrix $S(x)$. A problem which sometimes arises is to find $S(\boldsymbol{x})$ so that the square matrix $S(\boldsymbol{X})$ of the transformed points shall have the general orthogonal property

$$S'(\boldsymbol{X})\,S(\boldsymbol{X}) = d, \qquad \text{......(9)}$$

where d is a diagonal matrix. In other words, it is required to find a set of three mutually perpendicular directions which remain mutually perpendicular after the strain.

By equation (8) it follows that $S(\boldsymbol{X}) = aS(\boldsymbol{x})$. Hence the condition (9) becomes

$$S'(\boldsymbol{x})\,a'aS(\boldsymbol{x}) = S^{-1}(\boldsymbol{x})\,bS(\boldsymbol{x}) = d,$$

where b is the symmetrical matrix $a'a$. Accordingly $S(\boldsymbol{x})$ can be identified with the modal matrix k of b, and d is the diagonal matrix of the latent roots of b.

In the special case where a is symmetrical, $b = a^2$, and then b and a have the same modal matrix. Hence $\boldsymbol{X} = \lambda\boldsymbol{x}$, if $\boldsymbol{x}$ is identified with a column of k, and λ is the corresponding latent root. Thus the lines joining the points $\boldsymbol{x}$ and $\boldsymbol{X}$ to the origin have the same directions, so that the set of three mutually perpendicular lines have the same directions before and after the strain. In this case the strain is said to be *pure*. In general (8) represents a pure strain combined with a rotation.

8·8. Kinematic Constraint of a Rigid Body. The position of a rigid body at any instant is defined uniquely by the position of any convenient rectangular frame of reference $O(x_1, x_2, x_3)$ fixed in that body. Now the position of the body axes can be specified by six parameters—for instance, the Cartesian coordinates $\boldsymbol{a}$ of O, referred to fixed axes, and three angular coordinates ϕ. Hence, unless definite relations are assigned between the six parameters, the rigid body has six *degrees of freedom*. These degrees of freedom may be taken to correspond to the positional coordinates just mentioned, or to any other equivalent set. If, on the other hand, relations are assigned between the positional coordinates, the body will be subject to *geometric* or *kinematic constraint*, and it will then have less than six degrees of freedom.

A very simple example of constraint would be fixture of one point of the body. Again, a point might be restricted to lie on a fixed surface, or to lie on a surface or curved guide which is itself forced to move in a prescribed manner. These constraints can all be represented by functional relations connecting the positional coordinates and possibly also the time variable t.

More general types of constraint can occur in which sliding or rolling contact is imposed between the body and fixed or movable guides.

The restricting conditions may then consist of relations involving the time-rates of change of the positional coordinates. When these relations can be integrated without further knowledge of the motion, so that they may be replaced by equivalent conditions explicitly connecting the positional coordinates and possibly the time variable, the constraint is said to be *holonomous*.* When the kinematic conditions are not explicitly integrable the constraint is *non-holonomous*.

Example

Non-Holonomous Constraint. Suppose a sphere of radius R and centre O to roll (without sliding) on the fixed plane $\mathbf{x}_3 = R$. Choose auxiliary rectangular axes $O(x_1, x_2, x_3)$ carried by the sphere and initially coincident with the fixed frame of reference $\mathbf{O}(\mathbf{x}_1, \mathbf{x}_2, \mathbf{x}_3)$. At time t let O be at $(\mathbf{a}_1, \mathbf{a}_2, 0)$ and let the angular coordinates of the sphere be ϕ_1, ϕ_2, ϕ_3 (see Fig. 8·3·1). Then if $\mathbf{x}$ and x denote the columns of the coordinates of a general point P of the sphere at time t, referred to the fixed axes and the body axes respectively, we have by (8·4·1) $x = \mathbf{l}(\mathbf{x}-\mathbf{a})$, where $\mathbf{l}$ is given by (8·6·3). Moreover, since P moves with the sphere we have by differentiation $0 = \mathbf{l}(\dot{\mathbf{x}}-\dot{\mathbf{a}})+\dot{\mathbf{l}}(\mathbf{x}-\mathbf{a})$. Now if the point considered is in contact with the plane at time t, then $\mathbf{x} = \{\mathbf{a}_1, \mathbf{a}_2, R\}$. Moreover, since there is no slip at the point of contact, the total velocity components of P must vanish, so that $\dot{\mathbf{x}} = 0$. Hence the equation of constraint is

$$\mathbf{l}\{\dot{\mathbf{a}}_1, \dot{\mathbf{a}}_2, 0\} = \dot{\mathbf{l}}\{0, 0, R\}.$$

If (8·6·3) is used to express $\mathbf{l}$ and $\dot{\mathbf{l}}$ in terms of the angular coordinates, the last equation yields the two independent conditions

$$\dot{\mathbf{a}}_1 \cos\phi_3 + \dot{\mathbf{a}}_2 \sin\phi_3 = -R\dot{\phi}_2,$$
$$\dot{\mathbf{a}}_1 \sin\phi_3 - \dot{\mathbf{a}}_2 \cos\phi_3 = -R\dot{\phi}_1 \cos\phi_2.$$

These are non-integrable, so that the constraint imposed is non-holonomous. The sphere has three degrees of freedom so long as it remains in contact with the plane and no slipping occurs. If there were nothing to prevent the sphere from leaving the plane on one side, the constraint would be described as "one-sided".

8·9. Systems of Rigid Bodies and Generalised Coordinates.

The conception of positional coordinates can be generalised to apply to any system of rigid bodies or particles. Such a system may include

* Also *holonomic.* The term was introduced by Hertz to denote a constraint expressed by integral—as distinct from differential—relations.

certain bodies which, though possibly kinematically constrained, nevertheless have no direct geometric linkage with other members of the system: more usually the various bodies will be connected together in some way. It is evident that any possible configuration of the whole system can be specified completely by means of a suitable set of parameters or *generalised coordinates*, whose number will depend on the number of the bodies and of the kinematic constraints.

The equations which represent the constraints mathematically will in general consist of a set of relations between the generalised coordinates, and the time-rates of change of these coordinates. We shall, however, be mainly concerned with cases in which these equations are directly integrable, so that they yield an equivalent set of relations explicitly connecting only the generalised coordinates and possibly also the time variable t. Restricting attention to such holonomous systems, we see that the configuration at any instant can be completely specified by a set of s generalised coordinates, in conjunction with, say, r independent relations connecting these coordinates and possibly also involving t. The total number of effectively independent generalised coordinates will thus be $m = s-r$, and will equal the total number of degrees of freedom of the system.

In the case of the single rigid body considered in § 8·7, a typical point P of the body was identified by means of its Cartesian coordinates x, referred to a set of axes carried by the body. The Cartesian coordinates $\boldsymbol{x}$ of P, referred to the fixed frame of reference $\boldsymbol{O}(\boldsymbol{x}_1, \boldsymbol{x}_2, \boldsymbol{x}_3)$, were then expressible in terms of x and the positional coordinates $\boldsymbol{a}$ and ϕ of the body. Similarly, with a system of rigid bodies, it will be possible to identify a particular point or particle P of the system by a set of parameters (corresponding to x) and to express the Cartesian coordinates $\boldsymbol{x}$ of P, referred to fixed axes, in terms of those parameters and the generalised coordinates of the system. Suppose the set of parameters relevant to P to be denoted for convenience by the single symbol α, and as usual let $q_1, q_2, \ldots, q_m$ be the m effectively independent generalised coordinates of the system. Then the functional dependence of $\boldsymbol{x}$ on α, q (and possibly also t) will be expressible as

$$\boldsymbol{x}_1 = f_1(q_1, q_2, \ldots, q_m, t, \alpha),$$
$$\boldsymbol{x}_2 = f_2(q_1, q_2, \ldots, q_m, t, \alpha),$$
$$\boldsymbol{x}_3 = f_3(q_1, q_2, \ldots, q_m, t, \alpha),$$

or more briefly as
$$\boldsymbol{x} = f(q_1, q_2, \ldots, q_m, t, \alpha), \qquad \ldots\ldots(1)$$

where x, f denote column matrices. These equations will be referred to as the *geometrical equations* of the system.

Since the parameters α are assumed to be invariable with time, equation (1) yields on total differentiation with respect to t

$$\dot{x} = \frac{\partial f}{\partial t} + \sum_{i=1}^{m} \dot{q}_i \frac{\partial f}{\partial q_i}. \qquad \text{......(2)}$$

The quantities $\dot{q}_i$ are termed the *generalised components of velocity* or simply the *generalised velocities*. Similarly, the quantities $\ddot{q}_i$ are spoken of as the *generalised accelerations*.

Part II. Statics and Dynamics of Systems

8·10. Virtual Work and the Conditions of Equilibrium.

The conditions of equilibrium of a system are most conveniently obtained by an application of the principle of virtual work.

If a system of forces which act on a single particle is in equilibrium, then no work will be done in any displacement of the particle, provided that the forces remain constant in magnitude and direction during the displacement. Conversely, if the work done, as calculated on the foregoing assumption, is zero in any possible displacement of the particle, then the forces acting on the particle must be in equilibrium. It is true that in most actual cases the forces do vary as their point of application is moved, but for the purpose exclusively of a test for equilibrium they must be supposed to remain constant. Any actual changes in the forces due to motion of the point of application affect stability, but not equilibrium.

Next consider a system of particles, connected or unconnected. It would be possible to test the equilibrium of each particle separately by a calculation of the work done in an arbitrary displacement. Since in such a calculation the forces, if any, between the particles themselves would have to be included, the method would offer no practical advantage. However, suppose all the displacements considered to be infinitesimal and such that they do not violate any of the rigid connections or frictionless constraints of the system: then it is easily shown* that the forces corresponding to such connections or constraints do not enter into the expression for the virtual work of the system as a whole.

* See, for instance, p. 167 of Ref. 28.

The time variable does not enter into the discussion of static equilibrium. Any given configuration of the system is assumed to be completely specified by the m generalised coordinates q, and our object is to decide whether that particular configuration satisfies the conditions of equilibrium. Suppose X_1, X_2, X_3 to represent the components of a force applied at the point whose coordinates are x_1, x_2, x_3 and let X, x denote the corresponding columns. Then the total work done in the displacement is given by

$$\delta W = \Sigma_0 X' \delta x,$$

where Σ_0 denotes summation extending to all the particles and applied forces. If the virtual displacement given corresponds to increments δq_i of the generalised coordinates, then by (8·9·1)

$$\delta x = \sum_{i=1}^{m} \delta q_i \frac{\partial f}{\partial q_i},$$

and the total work can accordingly be expressed as

$$\delta W = P' \delta q = (\delta q') P,$$

where δq denotes the column of the increments δq_i, and P represents the column of the quantities

$$P_i = \Sigma_0 X' \frac{\partial f}{\partial q_i}.$$

For equilibrium δW must be zero for all possible virtual displacements, i.e. for all ratios of the increments δq_i. The condition for equilibrium is thus

$$P = 0.$$

It is usual to call P_i the *generalised component of force* (or more briefly the *generalised force*) corresponding to q_i, but it should be noted that P_i has not necessarily the physical dimensions of a force. For instance, if q_i denotes an angle, P_i has the dimensions of a moment.

8·11. Conservative and Non-Conservative Fields of Force.

Certain of the forces applied to the system may have the distinctive property that the work done by them when the system is given a displacement depends solely on the initial and the final configurations. Such forces are said to be *conservative*.

Suppose the generalised coordinates q to have the values Q for some datum configuration. Then the work done by the conservative forces when the system is displaced from q to Q will be a definite function V of the coordinates q which is termed the *potential energy*. It follows that the work done by such forces in an arbitrary infinitesimal displacement δq is given by $\delta W = -(\delta q')\left\{\frac{\partial V}{\partial q_i}\right\}$, where $\left\{\frac{\partial V}{\partial q_i}\right\}$ denotes the column $\left\{\frac{\partial V}{\partial q_1}, \frac{\partial V}{\partial q_2}, \ldots, \frac{\partial V}{\partial q_m}\right\}$. Hence if the part of the typical generalised force arising from the non-conservative forces be $\boldsymbol{P}_i$, then the total generalised force will be given by

$$P_i = \boldsymbol{P}_i - \frac{\partial V}{\partial q_i}. \qquad \text{......(1)}$$

If the field of force is wholly conservative, then the conditions of equilibrium are $\left\{\frac{\partial V}{\partial q_i}\right\} = 0$. These require the potential energy to be stationary. If further the potential energy is a minimum for the equilibrium position, this position is stable.†

An important special case is that in which V is a homogeneous quadratic function of the coordinates q. The potential energy will be expressible in this way when the values of the coordinates appropriate to the position of equilibrium under zero external load are chosen to be zero and the displacements from this position are always small. Then

$$V = \tfrac{1}{2}q'Eq,$$

where E is a symmetrical square matrix. Hence (compare (2·8·3))

$$\left\{\frac{\partial V}{\partial q_i}\right\} = Eq, \qquad \text{......(2)}$$

so that the conditions of equilibrium are briefly

$$Eq = \boldsymbol{P}.$$

In systems where aerodynamical forces play a part, it usually happens that, while these forces are not conservative, yet they are linearly expressible in terms of the displacements by means of *aerodynamical derivatives*.‡ Suppose the contribution of these forces to $\boldsymbol{P}$ to be written $-Wq+w$, and let $\boldsymbol{P}^*$ be the part of $\boldsymbol{P}$ due to the remaining forces, if any. Then the typical condition of equilibrium is

$$Cq = w + \boldsymbol{P}^*,$$

where $C \equiv W + E$. The square matrix C (which is not in general symmetrical) is termed the *stiffness matrix*, and its elements are the *stiffness coefficients*, or simply the *stiffnesses*. Each stiffness is the sum of two terms, namely an *aerodynamical stiffness* which originates from the air forces, and an *elastic stiffness* which actually represents the

† See, for instance, § 86 of Ref. 29. ‡ See also § 9·5.

influence of all the conservative forces. Provided C is not singular† the last equation may be premultiplied by $\Phi \equiv C^{-1}$ to give

$$q = \Phi(w+P^*).$$

The matrix Φ is called the *flexibility matrix*, and its elements are the *flexibility coefficients*, or the *flexibilities*.‡ It will be noted that w represents the generalised wind forces appropriate to the configuration for which the coordinates q are zero.

Examples

(i) *Reciprocal Theorem for a Conservative System.* Suppose any two *external* load systems $\mathfrak{P}(1)$ and $\mathfrak{P}(2)$ to be applied, and let the corresponding displacements be $q(1)$ and $q(2)$. Then the conditions for equilibrium after either displacement are $\mathfrak{P} - \left\{\dfrac{\partial V}{\partial q_i}\right\} = 0$. Hence by (2), since E is symmetrical,

$$q'(1)\,\mathfrak{P}(2) = q'(1)\,E\,q(2) = q'(2)\,E\,q(1) = q'(2)\,\mathfrak{P}(1).$$

These reciprocal relations are valid for any number of degrees of freedom, and the theorem is in fact true for a continuous elastic body.

(ii) *Principal Directions of Loading at a Point of an Elastic Body or Structure.* A principal direction at a point P is such that a load applied to P in that direction displaces P in the same direction. Let unit force applied in the direction $\boldsymbol{Ox}_1$ give component displacements $\boldsymbol{\delta}_{11}$, $\boldsymbol{\delta}_{21}$, $\boldsymbol{\delta}_{31}$: and similarly, let the displacements due to unit forces parallel to $\boldsymbol{Ox}_2$, $\boldsymbol{Ox}_3$ be respectively $\boldsymbol{\delta}_{12}$, $\boldsymbol{\delta}_{22}$, $\boldsymbol{\delta}_{32}$ and $\boldsymbol{\delta}_{13}$, $\boldsymbol{\delta}_{23}$, $\boldsymbol{\delta}_{33}$. By Hooke's law a force with components $\boldsymbol{X}_1$, $\boldsymbol{X}_2$, $\boldsymbol{X}_3$ will produce a displacement whose components $\boldsymbol{d}_1$, $\boldsymbol{d}_2$, $\boldsymbol{d}_3$ are given by $\{\boldsymbol{d}_i\} = [\boldsymbol{\delta}_{ij}]\{\boldsymbol{X}_i\}$, or say $\boldsymbol{d} = \boldsymbol{\delta X}$. Now for a load in a principal direction we require $\boldsymbol{d} = \lambda \boldsymbol{X}$, where λ is a scalar multiplier: hence

$$(\lambda I - \boldsymbol{\delta})\boldsymbol{X} = 0.$$

It can be shown that the principle of the conservation of energy will be violated unless $\boldsymbol{\delta}$ is symmetrical. For example, if the point P be made to describe a closed rectangular path with its sides parallel to $\boldsymbol{Ox}_1$ and $\boldsymbol{Ox}_2$, then the whole work done in the cycle of displacement will not be zero unless $\boldsymbol{\delta}_{12} = \boldsymbol{\delta}_{21}$. It then follows from the orthogonal property of the modal matrix of a symmetrical matrix (see example

† When C is singular there is no position of equilibrium unless $w+P^* = 0$, in which case the system is neutral.

‡ The flexibilities are also referred to as *influence numbers* in the theory of beams.

(v) of § 3·8) that there are three mutually perpendicular principal directions at any point of the structure. The proof given shows that this conclusion follows merely from Hooke's law and the principle of the conservation of energy.

8·12. Dynamical Systems. By a dynamical system is meant one or more rigid bodies, which, ideally, may be mere particles. The bodies may be quite independent, or they may be subject to constraints of the types already discussed in Part I of this chapter. They may also influence one another by direct communication of momentum by impact, or by communication of momentum at a distance by gravitational or electro-magnetic forces. In the last category may also be included the effects of ideal massless elastic links, and—with certain restrictions—the reactions of a perfect fluid in which the bodies are immersed.

A continuous elastic body is not a dynamical system according to the foregoing definition. Nevertheless it is possible to consider such a body as the limit of a dynamical system having an indefinitely large number of particles with ideal massless elastic links between the particles. Moreover, it is often convenient to suppose a continuous elastic body to be replaced by a body having only a finite number of definite modes of deformation. Such a body is said to be *semi-rigid*: it is a true dynamical system possessing a finite number of degrees of freedom. For example, a semi-rigid cantilever beam capable of bending according to a given law of flexural curvature can be imagined as a system of infinitesimal rigid rods so geared together that their displacements accord with the given law; such a semi-rigid beam would have one degree of freedom. Naturally a semi-rigid body will not behave in the same way as its elastic counterpart in all circumstances, but its behaviour may sometimes be exactly the same, provided that the modes of distortion are properly chosen. The chief value of the conception of the semi-rigid body is that it permits approximate calculations of reasonable accuracy in cases where exact treatment is not mathematically feasible. For instance, the differential equations governing the oscillations of an aeroplane wing moving through the air are extremely complicated, but the stability of the wing can be investigated approximately on the supposition that the elastic wing is a semi-rigid body possessing only two degrees of freedom corresponding to flexure and torsion.*

* See p. 7 of Ref. 30.

8·13. Equations of Motion of an Aeroplane. Before considering dynamical systems comprising several bodies, we shall express in matrix form the equations of motion of a single rigid body. The method of construction of the equations is quite general, but for the sake of definiteness we shall suppose the body to be a (rigid) aeroplane in flight. The notation differs in some respects from that which is standard in aerodynamics,* but the necessary scheme of conversion will be stated at the end of the section.

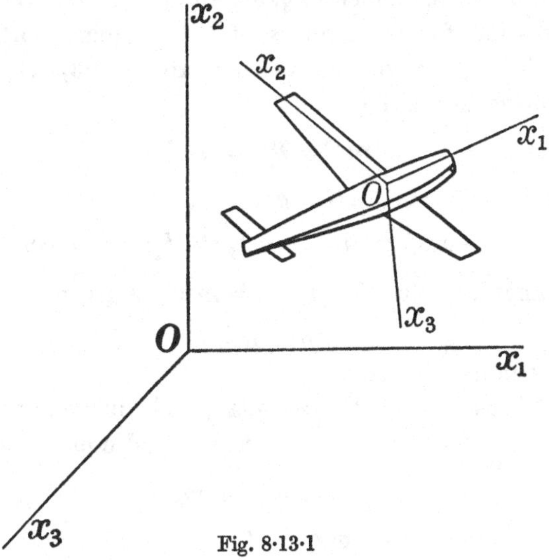

Fig. 8·13·1

In Fig. 8·13·1 the frame $\boldsymbol{O}(\boldsymbol{x}_1, \boldsymbol{x}_2, \boldsymbol{x}_3)$ is fixed relative to the earth with $\boldsymbol{Ox}_3$ vertically downwards, while Ox_1, Ox_2, Ox_3 are body axes through the centre of mass O of the aeroplane. The reader should imagine that he is viewing the aeroplane in the diagram from below. The axis Ox_1 is longitudinal and directed forwards, Ox_2 is drawn laterally to starboard, while Ox_3 is in the plane of symmetry and downwards. The six parameters used to describe the position of the aeroplane at any instant are to be regarded as the three rectangular coordinates of O referred to the fixed axes, and the three angular coordinates ϕ_1, ϕ_2, ϕ_3 defined in § 8·3.

Let $\boldsymbol{m}$ denote the total mass of the aeroplane, and let $\boldsymbol{m}G$, $\boldsymbol{m}X$ denote, respectively, the columns of the components of the weight and of the total remaining applied forces, measured in the directions

* Chap. v of Ref. 27.

of the body axes. Also let L be the column of the moments about the same axes, and u, p the columns of the components of the linear and the angular velocities.

Applying (8·7·4) and (8·7·7), and remembering that O is the centre of mass, we obtain the six equations of motion in the form

$$(ID+\varpi)\,u = X+G, \qquad \text{......(1)}$$

$$(ID+\varpi)\,Jp = L, \qquad \text{......(2)}$$

where ϖ and J are as defined in §§ 8·5 and 8·7, respectively.

The expression for G in terms of the angular coordinates* ϕ is given immediately by equations (8·4·1) and (8·6·3). Thus, if g is the acceleration due to gravity,

$$mG = l\{0, 0, mg\}, \qquad \text{......(3)}$$

or

$$G = gn,$$

where $n_1 = -\sin\phi_2;\quad n_2 = \cos\phi_2 \sin\phi_1;\quad n_3 = \cos\phi_2\cos\phi_1.$

Lastly, the angular velocities p are themselves given by

$$p = R\dot{\phi}, \qquad \text{......(4)}$$

where R is as defined in § 8·6.

As a special case assume the aeroplane to be in steady motion. Then u, p, X and L are columns of constants, and equations (1) and (2) reduce to

$$\varpi u = X+gn, \qquad \text{......(5)}$$

$$\varpi Jp = L. \qquad \text{......(6)}$$

From (5) it follows that n is a column of constants, so that ϕ_2 and ϕ_1 are constant. Equation (4) accordingly simplifies to

$$p = \Omega n, \qquad \text{......(7)}$$

where $\dot{\phi}_3 \equiv \Omega$ also is a constant. Hence in steady motion

$$p'p = \Omega^2 n'n = \Omega^2.$$

The simplest type of steady motion, of course, is that of steady rectilinear symmetrical flight. Then with $\Omega = 0$ and $\phi_1 = 0$, the conditions (5) and (6) reduce to

$$X = g\{\sin\phi_2, 0, \cos\phi_2\},$$
$$L = 0.$$

* In relation to the present problem the angular displacements ϕ_3, ϕ_2, ϕ_1 represent rotations in *yaw*, *pitch* and *roll*, effected successively about the carried positions of the body axes.

Addendum. The scheme of conversion to standard aeronautical symbols is as below:

Matrices Used		*Equivalent in Standard Symbols*
x	... ...	$\{x, y, z\}$
u	... ...	$\{u, v, w\}$
X	... ...	$\{X, Y, Z\}$
L	... ...	$\{L, M, N\}$
p	... ...	$\{p, q, r\}$
ϕ	... ...	$\{\phi, \theta, \psi\}$
ϖ	... ...	$\begin{bmatrix} 0 & -r & q \\ r & 0 & -p \\ -q & p & 0 \end{bmatrix}$
J	...	$\begin{bmatrix} A & -F & -E \\ -F & B & -D \\ -E & -D & C \end{bmatrix}$

8·14. Lagrange's Equations of Motion of a Holonomous System. Suppose x to represent the Cartesian coordinates, referred to fixed axes, of a typical particle of mass m. The system will be assumed holonomous, so that these coordinates will be expressible in terms of the generalised coordinates q_i, and possibly also the time variable t, by the geometrical equations (8·9·1). If X denotes the column of the components of force acting on m, then the equations of motion for the particle are expressed by

$$m\ddot{x} = X. \qquad \text{......(1)}$$

This equation states that the applied force is in equilibrium with the reversed effective force.*

Premultiply equation (1) by the row matrix

$$\frac{\partial f'}{\partial q_i} \equiv \left[\frac{\partial f_1}{\partial q_i}, \frac{\partial f_2}{\partial q_i}, \frac{\partial f_3}{\partial q_i}\right]$$

and sum for all particles of the system. Then

$$\Sigma_0 m \frac{\partial f'}{\partial q_i} \ddot{x} = \Sigma_0 \frac{\partial f'}{\partial q_i} X = \Sigma_0 X' \frac{\partial f}{\partial q_i}, \qquad \text{......(2)}$$

where Σ_0 denotes summation for all the particles.

* In engineering it is common to speak of the reversed effective force as the *inertia force.*

Consider, firstly, the expression on the right of this equation. This is formally identical with the quantity P_i which was derived in § 8·10 by an application of the principle of virtual work and which was termed the generalised component of force, provided it is understood that in the virtual displacements of the dynamical system the time variable t (which may enter explicitly into the geometrical equations of the system) is not varied. Forces which do no work* in the type of displacement contemplated, and which therefore do not appear in P_i, include forces between the particles of a rigid body, pressures or tensions in inextensible connecting members, the reactions at fixed pivots or fixed smooth (or in some cases perfectly rough) guides, and lastly the reactions at any smooth guides which provide kinematic constraint and are moved in some prescribed manner. These last forces, in particular, will be absent from P_i, since the guides concerned are not moved in the virtual displacement.

The expression on the left of equation (2) represents the appropriate generalised component of the effective forces on the system. In Lagrange's method it is very conveniently expressed in terms of the kinetic energy function. The essential step is that the total kinetic energy of the system, namely

$$T = \Sigma_0 \tfrac{1}{2}\boldsymbol{m}\dot{\boldsymbol{x}}'\dot{\boldsymbol{x}}, \qquad \text{......(3)}$$

must first be expressed as a function of the generalised velocities $\dot{q}$, the generalised coordinates q, and the time variable t. This can be done immediately by use of the relations (8·9·2). The resulting expression for the kinetic energy will clearly be quadratic (though not necessarily homogeneous) in the quantities, so that, say,

$$T = \tfrac{1}{2}\dot{q}'A\dot{q} + \dot{q}'\boldsymbol{A} + \boldsymbol{A}_0.$$

The symmetrical square matrix A, the column matrix $\boldsymbol{A}$, and the scalar $\boldsymbol{A}_0$ depend in general on the coordinates q and on t, but if the geometrical equations do not contain t explicitly $\boldsymbol{A}$ and $\boldsymbol{A}_0$ will be absent. In this special case T will be a homogeneous quadratic function of the variables $\dot{q}$, and its coefficients will depend only on the coordinates q.

Now by (3), $\dfrac{\partial T}{\partial \dot{q}_i} = \Sigma_0 \boldsymbol{m}\dot{\boldsymbol{x}}' \dfrac{\partial \dot{\boldsymbol{x}}}{\partial \dot{q}_i}$. But equations (8·9·2) give on differentiation $\dfrac{\partial \dot{\boldsymbol{x}}}{\partial \dot{q}_i} = \dfrac{\partial f}{\partial q_i}$, where the generalised velocities are treated for

* Forces which do no work are sometimes referred to as *constraints*.

the present as distinct variables, independent of the generalised coordinates. Hence

$$\frac{\partial T}{\partial \dot{q}_i} = \Sigma_0 m\dot{\boldsymbol{x}}' \frac{\partial f}{\partial q_i}.$$

Again, since obviously the differential operators $\frac{d}{dt}\left(\equiv \frac{\partial}{\partial t} + \sum_{i=1}^{m} \dot{q}_i \frac{\partial}{\partial q_i}\right)$ and $\frac{\partial}{\partial q_i}$ are permutable, it follows that

$$\frac{d}{dt}\left(\frac{\partial T}{\partial \dot{q}_i}\right) = \Sigma_0 m\ddot{\boldsymbol{x}}' \frac{\partial f}{\partial q_i} + \Sigma_0 m\dot{\boldsymbol{x}}' \frac{\partial \dot{\boldsymbol{x}}}{\partial q_i} = \Sigma_0 m \frac{\partial f'}{\partial q_i} \ddot{\boldsymbol{x}} + \frac{\partial T}{\partial q_i}.$$

Hence finally
$$\Sigma_0 m \frac{\partial f'}{\partial q_i} \ddot{\boldsymbol{x}} = \frac{d}{dt}\left(\frac{\partial T}{\partial \dot{q}_i}\right) - \frac{\partial T}{\partial q_i}.$$

Lagrange's equations of motion are thus derived in the form

$$\frac{d}{dt}\left\{\frac{\partial T}{\partial \dot{q}_i}\right\} - \left\{\frac{\partial T}{\partial q_i}\right\} = P. \qquad \text{......(4)}$$

If, as in § 8·11, V denotes the potential energy of the conservative applied forces, and $\boldsymbol{P}$ is the contribution to P due to the non-conservative forces, then (4) may be written

$$\frac{d}{dt}\left\{\frac{\partial L}{\partial \dot{q}_i}\right\} - \left\{\frac{\partial L}{\partial q_i}\right\} = \boldsymbol{P}, \qquad \text{......(5)}$$

in which $L \equiv T - V$. The quantity L, namely the difference of the kinetic and potential energies, is referred to as the *Lagrangian function* or the *kinetic potential.*

Examples

(i) *Expanded Form of Lagrange's Equations.* For simplicity assume the constraints to be independent of time. Then if, for instance, there are two degrees of freedom,

$$2T = A_{11}\dot{q}_1^2 + A_{22}\dot{q}_2^2 + 2A_{12}\dot{q}_1\dot{q}_2,$$

and on differentiation $\frac{\partial T}{\partial \dot{q}_1} = A_{11}\dot{q}_1 + A_{12}\dot{q}_2$, so that

$$\frac{d}{dt}\left(\frac{\partial T}{\partial \dot{q}_1}\right) - \frac{\partial T}{\partial q_1} = A_{11}\ddot{q}_1 + A_{12}\ddot{q}_2 + \left(\dot{q}_1 \frac{\partial A_{11}}{\partial q_1} + \dot{q}_2 \frac{\partial A_{11}}{\partial q_2}\right)\dot{q}_1$$
$$+ \left(\dot{q}_1 \frac{\partial A_{12}}{\partial q_1} + \dot{q}_2 \frac{\partial A_{12}}{\partial q_2}\right)\dot{q}_2 - \frac{1}{2}\left(\frac{\partial A_{11}}{\partial q_1}\dot{q}_1^2 + \frac{\partial A_{22}}{\partial q_1}\dot{q}_2^2 + 2\frac{\partial A_{12}}{\partial q_1}\dot{q}_1\dot{q}_2\right).$$

On rearrangement this gives the first dynamical equation in the explicit form

$$A_{11}\ddot{q}_1 + A_{12}\ddot{q}_2 + \frac{1}{2}\frac{\partial A_{11}}{\partial q_1}\dot{q}_1^2 + \left(\frac{\partial A_{12}}{\partial q_2} - \frac{1}{2}\frac{\partial A_{22}}{\partial q_1}\right)\dot{q}_2^2 + \frac{\partial A_{11}}{\partial q_2}\dot{q}_1\dot{q}_2 = P_1.$$

Similarly, noting that $A_{12} = A_{21}$, we have

$$A_{21}\ddot{q}_1 + A_{22}\ddot{q}_2 + \left(\frac{\partial A_{21}}{\partial q_1} - \frac{1}{2}\frac{\partial A_{11}}{\partial q_2}\right)\dot{q}_1^2 + \frac{1}{2}\frac{\partial A_{22}}{\partial q_2}\dot{q}_2^2 + \frac{\partial A_{22}}{\partial q_1}\dot{q}_1\dot{q}_2 = P_2.$$

More generally, if there are m degrees of freedom, the rth dynamical equation will be

$$\sum_{i=1}^{m} A_{ir}\ddot{q}_i + \sum_{i=1}^{m}\sum_{j=1}^{m} C_{ij}^{(r)}\dot{q}_i\dot{q}_j = P_r,$$

in which†

$$2C_{ij}^{(r)} \equiv \frac{\partial A_{ir}}{\partial q_j} + \frac{\partial A_{jr}}{\partial q_i} - \frac{\partial A_{ij}}{\partial q_r}.$$

(ii) *Discriminants of the Kinetic Energy Function.* In the case of a holonomous dynamical system with constraints independent of time the kinetic energy function is a homogeneous quadratic function of the type $T = \frac{1}{2}\dot{q}'A\dot{q}$. Since the kinetic energy is necessarily positive, it follows that T is a positive quadratic form, and that its discriminants are all necessarily positive (see § 1·16). It should be noted that no discriminant of the kinetic energy can vanish, unless the system has the unusual property that all its particles can be at rest while some of the generalised velocities are not zero.

8·15.* Ignoration of Coordinates. It sometimes happens that certain of the generalised coordinates do not appear explicitly in the Lagrangian function of a conservative system though the corresponding generalised velocities are present. Such coordinates are said to be *ignorable*.‡

It will be convenient to suppose that the system has m non-ignorable coordinates $q_1, q_2, \ldots, q_m$ and in addition k ignorable coordinates $\eta_1, \eta_2, \ldots, \eta_k$. Then by definition the function $L \equiv T - V$ is explicitly dependent only on the sets of quantities q, $\dot{q}$ and $\dot{\eta}$. If the constraints are independent of time, the kinetic energy will be a homogeneous quadratic function which can be represented by means of partitioned matrices as (see example (ii) of § 1·15)

$$T = \tfrac{1}{2}[\dot{q}', \dot{\eta}']\begin{bmatrix}\alpha, & \beta \\ \gamma, & \delta\end{bmatrix}\begin{bmatrix}\dot{q} \\ \dot{\eta}\end{bmatrix} = \tfrac{1}{2}\dot{q}'\alpha\dot{q} + \dot{\eta}'\gamma\dot{q} + \tfrac{1}{2}\dot{\eta}'\delta\dot{\eta}. \quad \text{......(1)}$$

The symmetrical square matrices α and δ are of the orders m and k respectively, while β and γ are respectively of the types $(\overline{m, k})$ and

† The quantity $C_{ij}^{(r)}$ is sometimes denoted by a "Christoffel's symbol" $\begin{bmatrix} ij \\ r \end{bmatrix}$, but this symbol is here avoided as it might lead to confusion with matrices.

‡ Other terms in use are *cyclic, absent* or *kinosthenic.* See § 38 of Ref. 31, and § 84 of Ref. 29.

$(\bar{k}, m)$ with $\gamma' = \beta$. All these matrices are independent of the coordinates η.

The k equations of motion appropriate to the ignorable coordinates are here simply $\frac{d}{dt}\left\{\frac{\partial T}{\partial \dot{\eta}_i}\right\} = 0$. These can be integrated immediately to give

$$\left\{\frac{\partial T}{\partial \dot{\eta}_i}\right\} \equiv \gamma\dot{q} + \delta\dot{\eta} = \sigma, \qquad \text{......(2)}$$

where σ denotes a column of k constants of integration. The velocities $\dot{\eta}$ can now be expressed in terms of q and $\dot{q}$. Thus

$$\dot{\eta} = \delta^{-1}\sigma - \delta^{-1}\gamma\dot{q}, \qquad \text{......(3)}$$

or, alternatively, since δ is symmetrical and $\gamma' = \beta$,

$$\dot{\eta}' = \sigma'\delta^{-1} - \dot{q}'\beta\delta^{-1}. \qquad \text{......(4)}$$

Next consider the m equations of motion appropriate to the non-ignorable coordinates, namely

$$\frac{d}{dt}\left\{\frac{\partial T}{\partial \dot{q}_i}\right\} - \left\{\frac{\partial}{\partial q_i}(T-V)\right\} = 0. \qquad \text{......(5)}$$

These equations contain explicitly not only the non-ignorable variables but also the quantities $\dot{\eta}$ and $\ddot{\eta}$. However, by means of (3) and (4) it is possible to eliminate $\dot{\eta}$ and $\ddot{\eta}$, and so to derive a set of m differential equations involving only the non-ignorable coordinates. In this way the dynamical system is effectively reduced to one having only m degrees of freedom. We shall now show how the equations appropriate to the reduced system can be constructed directly. Let

$$R = T_1 - \dot{\eta}'\sigma - V, \qquad \text{......(6)}$$

where T_1 is the function T expressed in terms of the variables q and $\dot{q}$, and $\dot{\eta}$ is expressed in terms of the same variables. Then

$$\left\{\frac{\partial R}{\partial \dot{q}_i}\right\} = \left\{\frac{\partial T_1}{\partial \dot{q}_i}\right\} - \left[\frac{\partial \dot{\eta}'}{\partial \dot{q}_i}\right]\sigma, \qquad \text{......(7)}$$

where $\left[\frac{\partial \dot{\eta}'}{\partial \dot{q}_i}\right] \equiv \left\{\frac{\partial}{\partial \dot{q}_i}\right\}\dot{\eta}'$ is an $(\bar{m}, k)$ matrix. But

$$\left\{\frac{\partial T_1}{\partial \dot{q}_i}\right\} = \left\{\frac{\partial T}{\partial \dot{q}_i}\right\} + \left[\frac{\partial \dot{\eta}'}{\partial \dot{q}_i}\right]\left\{\frac{\partial T}{\partial \dot{\eta}_i}\right\}.$$

Hence by (2) and (7) $\left\{\frac{\partial R}{\partial \dot{q}_i}\right\} = \left\{\frac{\partial T}{\partial \dot{q}_i}\right\}.$

Similarly,

$$\left\{\frac{\partial R}{\partial q_i}\right\} = \left\{\frac{\partial T_1}{\partial q_i}\right\} - \left[\frac{\partial \dot{\eta}'}{\partial q_i}\right]\sigma - \left\{\frac{\partial V}{\partial q_i}\right\} = \left\{\frac{\partial}{\partial q_i}(T-V)\right\}.$$

The required reduced form of the equations (5) is accordingly

$$\frac{d}{dt}\left\{\frac{\partial R}{\partial \dot{q}_i}\right\} - \left\{\frac{\partial R}{\partial q_i}\right\} = 0. \qquad \text{......(8)}$$

The explicit expression for R in terms of q and $\dot{q}$ is readily found as follows. By direct substitution for $\delta\dot{\eta}$ from (2) in (1) and use of (6) and (4) we have

$$\begin{aligned} R+V &= \tfrac{1}{2}\dot{q}'\alpha\dot{q} + \tfrac{1}{2}\dot{\eta}'\gamma\dot{q} - \tfrac{1}{2}\dot{\eta}'\sigma \\ &= \tfrac{1}{2}\dot{q}'\alpha\dot{q} + \tfrac{1}{2}(\sigma'\delta^{-1} - \dot{q}'\beta\delta^{-1})(\gamma\dot{q} - \sigma). \end{aligned}$$

Hence $\quad R = \tfrac{1}{2}\dot{q}'(\alpha - \beta\delta^{-1}\gamma)\dot{q} + \dot{q}'\beta\delta^{-1}\sigma - \tfrac{1}{2}\sigma'\delta^{-1}\sigma - V.$

It should be noted that although equations (8) are similar in general form to Lagrange's equations, yet the function R is not constructed in the usual way as the difference of a kinetic energy function and a potential function. It may contain terms which are linear in the velocities $\dot{q}$.

8·16.* The Generalised Components of Momentum and Hamilton's Equations. The kinetic energy of a single particle is $T = \tfrac{1}{2}m\dot{\boldsymbol{x}}'\dot{\boldsymbol{x}}$, and the three components of momentum, say $\boldsymbol{p}_1, \boldsymbol{p}_2, \boldsymbol{p}_3$, are given by

$$\boldsymbol{p} = m\dot{\boldsymbol{x}} = \left\{\frac{\partial T}{\partial \dot{\boldsymbol{x}}_i}\right\}. \qquad \text{......(1)}$$

Further, the kinetic energy, when expressed as a function of the components of momentum, takes the form

$$T_1 = \frac{1}{2m}\boldsymbol{p}'\boldsymbol{p}, \qquad \text{......(2)}$$

while the equations of motion of the particle (see (8·14·1)) become $\dot{\boldsymbol{p}} = \boldsymbol{X}$ in conjunction with (1). These equations illustrate in a very simple way Hamilton's equations of motion, which make use of the momenta—or in effect the generalised velocities—as auxiliary dependent variables. The construction of Hamilton's equations for a holonomous dynamical system will now be considered.

By analogy with (1) the *m generalised momenta* for a dynamical system are defined by

$$\{p_i\} = \left\{\frac{\partial T}{\partial \dot{q}_i}\right\}.$$

If the kinetic energy function is

$$T = \tfrac{1}{2}\dot{q}'A\dot{q} + \dot{q}'\boldsymbol{A} + \boldsymbol{A}_0, \qquad \text{......(3)}$$

then

$$p = A\dot{q} + \boldsymbol{A}. \qquad \text{......(4)}$$

When, as is normally the case, the symmetrical matrix A is non-singular, the preceding equation gives by inversion

$$\dot{q} = A^{-1}(p - \boldsymbol{A}). \qquad \text{......(5)}$$

By the use of (5) and its transposed the kinetic energy (3) can be written explicitly as a quadratic function of the variables p. Thus, if the function when so expressed is distinguished as T_1, we have

$$T_1 = \tfrac{1}{2}(p' - \boldsymbol{A}')A^{-1}(p - \boldsymbol{A}) + (p' - \boldsymbol{A}')A^{-1}\boldsymbol{A} + \boldsymbol{A}_0, \qquad \text{......(6)}$$

which is the generalisation of (2).

In Hamilton's equations the generalised momenta p and the coordinates q are used as dependent variables instead of $\dot{q}$ and q. Moreover, the Lagrangian function L is replaced by the *Hamiltonian function* H defined as

$$H = \dot{q}'p - T + V. \qquad \text{......(7)}$$

As with the kinetic energy, the function H may be expressed either in terms of p or of $\dot{q}$. In the first case, if (5) and (6) are used in conjunction with (7),

$$H = \tfrac{1}{2}(p' - \boldsymbol{A}')A^{-1}(p - \boldsymbol{A}) - \boldsymbol{A}_0 + V. \qquad \text{......(8)}$$

Similarly, if (3) and (4) are used in conjunction with (7),

$$H = T - \dot{q}'\boldsymbol{A} - 2\boldsymbol{A}_0 + V.$$

From the last equation it is seen that if the constraints are independent of the time (so that $\boldsymbol{A} = 0$ and $\boldsymbol{A}_0 = 0$), then H reduces to the total energy $T + V$.

To obtain the dynamical equations in Hamilton's form, we note firstly that the Lagrangian equations (8·14·5) may be written

$$\dot{p} - \left\{\frac{\partial T}{\partial q_i}\right\} = \boldsymbol{P} - \left\{\frac{\partial V}{\partial q_i}\right\}.$$

But by (3)

$$\frac{\partial T}{\partial q_i} = \tfrac{1}{2}\dot{q}'\frac{\partial A}{\partial q_i}\dot{q} + \dot{q}'\frac{\partial \boldsymbol{A}}{\partial q_i} + \frac{\partial \boldsymbol{A}_0}{\partial q_i},$$

which on application of (4) and use of the relation

$$A^{-1}\frac{\partial A}{\partial q_i} = -\frac{\partial A^{-1}}{\partial q_i}A$$

may be expressed as*

$$\frac{\partial T}{\partial q_i} = -\tfrac{1}{2}(p'-\boldsymbol{A}')\frac{\partial A^{-1}}{\partial q_i}(p-\boldsymbol{A})+(p'-\boldsymbol{A}')A^{-1}\frac{\partial \boldsymbol{A}}{\partial q_i}+\frac{\partial \boldsymbol{A}_0}{\partial q_i}$$
$$= \frac{\partial V}{\partial q_i}-\frac{\partial H}{\partial q_i}, \text{ by (8).} \qquad \text{......(9)}$$

Hence the first set of equations is†

$$\left\{\frac{\partial H}{\partial q_i}\right\} = \boldsymbol{P}-\dot{p}. \qquad \text{......(10)}$$

The second set of equations, obtained by differentiation of (8) and use of (5), is

$$\left\{\frac{\partial H}{\partial p_i}\right\} = A^{-1}(p-\boldsymbol{A}) = \dot{q}. \qquad \text{......(11)}$$

The relations (10) and (11) jointly are the complete Hamiltonian equations.

Examples

(i) *Reciprocal Property of a Dynamical System.* If the kinetic energy is a homogeneous quadratic function of the generalised velocities, then by (4) $p = A\dot{q}$. Suppose $\dot{q}(1)$ and $\dot{q}(2)$ to represent two different sets of generalised velocities of the system in one of its configurations, and let $p(1)$ and $p(2)$ be the corresponding momenta. Then since A is symmetrical

$$\dot{q}'(1)\,p(2) = \dot{q}'(1)\,A\dot{q}(2) = \dot{q}'(2)\,A\dot{q}(1) = \dot{q}'(2)\,p(1).$$

This reciprocal property should be compared with another given in example (i) of § 8·11.

(ii) *Equation of Energy.* If the constraints are independent of the time and the system is conservative, (10) and (11) give

$$\frac{d}{dt}(T+V) = \frac{dH}{dt} = \dot{p}'\left\{\frac{\partial H}{\partial p_i}\right\}+\dot{q}'\left\{\frac{\partial H}{\partial q_i}\right\} = \dot{p}'\dot{q}-\dot{q}'\dot{p} = 0.$$

Hence for such a system the total energy is constant.

* If, as is usual in aerodynamical applications, the generalised forces contain parts $\boldsymbol{P}$ dependent on the generalised velocity components $\dot{q}$, then these parts must be expressed in terms of the variables p, q and t by use of (5) before they are introduced in the Hamiltonian equations (10).

† The reader should be careful to note that in the differentiation $\frac{\partial T}{\partial q_i}$ on the left of equation (9) T is expressed as a function of $\dot{q}$, q and t, whereas in the differentiation $\frac{\partial H}{\partial q_i}$ on the right of the equation, H is a function of p, q and t.

8·17.* Lagrange's Equations with a Moving Frame of Reference. In some applications it is convenient to refer the displacements of the system to a base which is itself in motion. This moving frame of reference (not necessarily fixed in any body of the system) will as usual be denoted by $O(x_1, x_2, x_3)$, and the Cartesian coordinates of a typical point P relative to the frame will be represented by x. Then in place of the geometrical equations (8·9·1) we shall have for the typical point

$$x = f(q_1, q_2, \dots, q_m, t, \alpha).$$

The coordinates of P referred to a fixed base $\boldsymbol{O}(\boldsymbol{x}_1, \boldsymbol{x}_2, \boldsymbol{x}_3)$ will as previously be denoted by $\boldsymbol{x}$, while υ and α represent respectively the components of the total velocity and total acceleration of P, measured in the directions of the moving axes. Finally, u denotes the velocity components of O, measured in the same directions. The actual expressions for υ and α in terms of u, x and of the angular velocities p of the moving base are given by equations (8·7·1) and (8·7·2).

Now consider the respects in which the analysis of § 8·14 requires modification to allow for the motion of the base of reference. The equations of motion of the typical particle P may in the present case be taken as $\boldsymbol{m}\alpha = X$, where X represents the column of the components of force in the directions of the moving axes. A treatment similar to that adopted with (8·14·1) leads to an equation which it will be convenient to write as

$$\Sigma_0 \boldsymbol{m} \frac{\partial f'}{\partial q_i} \alpha = \Sigma_0 X' \frac{\partial f}{\partial q_i}. \qquad \text{......(1)}$$

The quantity on the right is as before denoted by P_i, and is referred to as the generalised force corresponding to q_i. If the system is given an infinitesimal virtual displacement involving a variation δq_i of q_i only, the work done by the external forces will be $P_i \delta q_i$. In this hypothetical displacement the frame of reference $O(x_1, x_2, x_3)$ is regarded as fixed.

To evaluate the quantity on the left of equation (1), first substitute for α from (8·7·2). The result is somewhat cumbrous, but it can be reduced conveniently by the introduction of a *centrifugal potential function* defined by

$$V_0 = \tfrac{1}{2}\Sigma_0 \boldsymbol{m} x' \varpi^2 x + \Sigma_0 \boldsymbol{m} x' \varpi u + \Sigma_0 \boldsymbol{m} x' \dot{u}.$$

Partial differentiation of this function yields

$$\frac{\partial V_0}{\partial q_i} = \Sigma_0 \boldsymbol{m} \frac{\partial f'}{\partial q_i} N,$$

where

$$N = \varpi^2 x + \varpi u + \dot{u}.$$

If further
$$J_i = \Sigma_0 m \frac{\partial f'}{\partial q_i} \dot{\varpi} x$$
and
$$G_i = 2\,\Sigma_0 m \frac{\partial f'}{\partial q_i} \varpi \dot{x},$$
the left-hand side of (1) can be written
$$\frac{\partial V_0}{\partial q_i} + J_i + G_i + \Sigma_0 m \frac{\partial f'}{\partial q_i} \ddot{x}.$$

The last term in this formula is expressible in terms of the kinetic energy of the motion relative to the base—namely, $\frac{1}{2}\Sigma_0 m\dot{x}'\dot{x}$—by the method of § 8·14. The quantity J_i represents an inertia effect dependent on the displacements, but it cannot be derived from a potential since the matrix $\dot{\varpi}$ is not symmetrical. Finally, the quantity G_i depends on $\dot{x}$ and is therefore a linear function of the generalised velocities. If the coefficient of $\dot{q}_j$ in G_i is denoted by G_{ij}, then
$$G_{ij} = 2\,\Sigma_0 m \frac{\partial f'}{\partial q_i} \varpi \frac{\partial f}{\partial q_j},$$
or on substitution for ϖ from (8·5·4)
$$G_{ij} = -2\,\Sigma_0 m\left(p_1 \frac{\partial(f_2, f_3)}{\partial(q_i, q_j)} + p_2 \frac{\partial(f_3, f_1)}{\partial(q_i, q_j)} + p_3 \frac{\partial(f_1, f_2)}{\partial(q_i, q_j)}\right).$$

Hence $G_{ij} = -G_{ji}$, and in particular $G_{ii} = 0$. The quantities G_{ij} are spoken of as the *gyrostatic coefficients*, and it is to be noted that on account of the properties just stated they do not appear in the expression for the total energy.

When the centrifugal potential V_0 is a homogeneous quadratic function, say $V_0 = \frac{1}{2}q'\sigma q$, the coefficients σ_{ij} are referred to as the *centrifugal stiffnesses*. They play exactly the same part as the gravitational and elastic stiffnesses.

Example

Suppose O to be always at rest, and let $p_1 = p_2 = 0$, while $p_3 = \Omega$ (a constant). Then
$$V_0 = \Sigma_0 \tfrac{1}{2} m x' \begin{bmatrix} 0, & -\Omega, & 0 \\ \Omega, & 0, & 0 \\ 0, & 0, & 0 \end{bmatrix}^2 x = -\tfrac{1}{2}\Omega^2 \Sigma_0 m(x_1^2 + x_2^2).$$

The typical generalised component of centrifugal force is thus

$$-\frac{\partial V_0}{\partial q_i} = \Omega^2 \Sigma_0 m\left(x_1 \frac{\partial f_1}{\partial q_i} + x_2 \frac{\partial f_2}{\partial q_i}\right),$$

and the typical centrifugal stiffness is

$$\sigma_{ij} = \frac{\partial^2 V_0}{\partial q_i \partial q_j} = -\Omega^2 \Sigma_0 m\left(\frac{\partial f_1}{\partial q_i}\frac{\partial f_1}{\partial q_j} + \frac{\partial f_2}{\partial q_i}\frac{\partial f_2}{\partial q_j} + f_1 \frac{\partial^2 f_1}{\partial q_i \partial q_j} + f_2 \frac{\partial^2 f_2}{\partial q_i \partial q_j}\right).$$

Finally, the typical gyrostatic coefficient is

$$G_{ij} = -2\Omega\, \Sigma_0 m \frac{\partial(f_1, f_2)}{\partial(q_i, q_j)}$$

and $J_i = 0$.

CHAPTER IX

SYSTEMS WITH LINEAR DYNAMICAL EQUATIONS

9·1. Introductory Remarks. The present Chapter deals with motions governed by linear ordinary differential equations with constant coefficients. The language of the dynamics of material systems will be used throughout, but the treatment can, for instance, be applied equally well to electrical systems. The discussion and exemplification of approximate numerical methods of solution is reserved for Chapter X.

9·2. Disturbed Motions. Except with very special systems or types of motion the differential equations which arise in dynamics are non-linear and do not admit exact solution. It is, however, sometimes possible to obtain particular solutions, such as those corresponding to equilibrium or steady motion. Then, if the system is supposed to be slightly disturbed from this known condition, the resulting small motion of deviation will be given by a set of linear differential equations. In the special case where the undisturbed state of the system is one of equilibrium or steady motion, the equations of disturbed motion will have constant coefficients and will be soluble by the methods of Chapters V and VI. In more general cases the equations will have for coefficients given functions of time, and they will thus be of the types considered in Chapter VII.

The disturbances just referred to may be of two kinds. They may be merely temporary, and represented by a set of initial conditions of motion which differ slightly from those corresponding to the undisturbed motion. The motion of deviation is then said to be *free*. On the other hand the disturbances may consist of small persistent forces which vary in some assigned way with time. In this case the motion of deviation is said to be *forced*.

The usual method of construction of the equations of deviation is, briefly, as follows. Let the set of values of the generalised coordinates corresponding to the undisturbed state of the system be represented by Q: these values are to be regarded as assigned functions of t (or possibly given constants). In the disturbed motion let the *deviations* (i.e. the increments of the generalised coordinates) be q. Then to form the equations of disturbed motion substitute in Lagrange's equations the values $Q+q$ for the coordinates, $Q+\dot{q}$ for the velocities, and so on, neglect all terms involving products of the quantities q or of their derivatives, and introduce the persistent disturbing forces, if any. In this way a set of linear differential equations in the deviations q is derived. These equations are, of course, only valid for the representation of deviant motions which are "small" in the sense implicit in the approximations used.

9·3. Conservative System Disturbed from Equilibrium. The simplest problem of the class just considered is that in which a conservative holonomous system with constraints independent of time receives a small disturbance from equilibrium. If, as will be assumed, the generalised coordinates are measured from this position of equilibrium, the quantities Q will all be zero.

Now use the expanded form of Lagrange's equations (see example (i) of § 8·14), and reject all the terms involving products of the deviations q or their derivatives. If there are no permanent disturbing forces, the equations are simply

$$A\ddot{q} = -\left\{\frac{\partial V}{\partial q_i}\right\},$$

where the elements of A are all constants. Further, since $q = 0$ is a position of equilibrium, the terms which are linear in q will be absent from V, so that this function may (to the order of approximation considered) be taken as

$$V = E_0 + \tfrac{1}{2}q'Eq.$$

Hence the required equations of free disturbed motion are

$$A\ddot{q} + Eq = 0, \qquad \text{......(1)}$$

in which A and E are both symmetrical matrices of constants.

If the disturbed motion is forced, the equations are of the more general type

$$A\ddot{q} + Eq = \xi(t),$$

in which $\xi(t)$ represents the column of the disturbing forces.

The principal diagonal elements of A, which obviously play a part analogous to ordinary moments of inertia, may for convenience be spoken of as the *generalised moments of inertia*. For a similar reason it is natural to refer to the elements of type A_{ij} $(i \neq j)$ as *generalised products of inertia*. Finally, the elements E_{ii} are described as the *direct elastic stiffnesses*, while those of type E_{ij} $(i \neq j)$ are called *elastic cross stiffnesses*.

Example

Transformation of the Coordinates. In the discussion of the Lagrangian equations of small motions it is sometimes convenient to transform the m generalised coordinates q to a new set, say $\mathbf{q}$. Suppose the two sets to be connected by the linear substitution $q = u\mathbf{q}$, where u is a non-singular square matrix of order m with given constant elements. Then

$$T = \tfrac{1}{2}\dot{q}'A\dot{q} = \tfrac{1}{2}\dot{\mathbf{q}}'u'Au\dot{\mathbf{q}}$$

and

$$V = E_0 + \tfrac{1}{2}\mathbf{q}'u'Eu\mathbf{q}.$$

The equation corresponding to (1) is accordingly

$$(u'Au)\ddot{\mathbf{q}} + (u'Eu)\mathbf{q} = 0.$$

It should be noted that $u'Au$ and $u'Eu$ are both symmetrical.

9·4.* Disturbed Steady Motion of a Conservative System with Ignorable Coordinates. The state of steady motion considered is that in which all the non-ignorable coordinates, and the generalised velocities corresponding to all the ignorable coordinates, have constant values. It has been shown in § 8·15 that when a system has ignorable coordinates the dynamical equations are reducible to a simpler set which involves only the non-ignorable coordinates. The final equations are similar in general form to Lagrange's equations, but the Lagrangian function L is replaced by a function R which may contain terms linear in the generalised velocities. It is the presence of these linear terms in the "modified" Lagrangian function R which distinguishes this problem of disturbed steady motion from that of disturbed equilibrium.†

As in § 8·15 we shall suppose that there are m non-ignorable coordinates q, and k ignorable coordinates η. Then if the kinetic energy function is represented by

$$T = \tfrac{1}{2}\dot{q}'\alpha\dot{q} + \dot{\eta}'\gamma\dot{q} + \tfrac{1}{2}\dot{\eta}'\delta\dot{\eta},$$

the k dynamical equations appropriate to the ignorable coordinates give on integration (see (8·15·2))

$$\gamma\dot{q} + \delta\dot{\eta} = \sigma, \qquad \text{......(1)}$$

while the remaining m equations for the system in its reduced form are (see (8·15·8))

$$\frac{d}{dt}\left\{\frac{\partial R}{\partial \dot{q}_i}\right\} - \left\{\frac{\partial R}{\partial q_i}\right\} = 0, \qquad \text{......(2)}$$

in which $R = \tfrac{1}{2}\dot{q}'(\alpha - \beta\delta^{-1}\gamma)\dot{q} + \dot{q}'\beta\delta^{-1}\sigma - \tfrac{1}{2}\sigma'\delta^{-1}\sigma - V.$

† Compare p. 195 of Ref. 31.

In steady motion the velocities $\dot{\eta}$ are all constant. Moreover, the non-ignorable coordinates also have constant values which, without loss of generality, may be taken to be zero. The quantities q then represent the *deviations* of the non-ignorable coordinates in the disturbed motion. Again, corresponding to the constant values of $\dot{\eta}$ and of q in the steady motion there will be a set of constants given by (1). These same constant values for σ are assumed to hold good for the disturbed motions to be considered.†

To derive the equations of free disturbed motion, neglect all terms in the expanded form of R which are above the second degree in the variables $\dot{q}$ and q. To this order of approximation R is quadratic in the variables $\dot{q}$ and q. It will be convenient to write

$$R = \tfrac{1}{2}\dot{q}'\mathfrak{A}\dot{q} + \dot{q}'\mathfrak{B}q - \tfrac{1}{2}q'\mathfrak{C}q + \dot{q}'\mathfrak{E} + q'\mathfrak{F},$$

in which now $\mathfrak{A}$, $\mathfrak{B}$, $\mathfrak{C}$, etc. are all matrices of constants. The square matrices $\mathfrak{A}$ and $\mathfrak{C}$ are symmetrical, but $\mathfrak{B}$ is in general not symmetrical. Then

$$\left\{\frac{\partial R}{\partial \dot{q}_i}\right\} = \mathfrak{A}\dot{q} + \mathfrak{B}q + \mathfrak{E} \quad \text{and} \quad \left\{\frac{\partial R}{\partial q_i}\right\} = \mathfrak{B}'\dot{q} - \mathfrak{C}q + \mathfrak{F}.$$

Hence equation (2) becomes

$$\mathfrak{A}\ddot{q} + (\mathfrak{B} - \mathfrak{B}')\dot{q} + \mathfrak{C}q - \mathfrak{F} = 0.$$

Since this is satisfied by $q = 0$, we require $\mathfrak{F} = 0$. Hence finally

$$\mathfrak{A}\ddot{q} + (\mathfrak{B} - \mathfrak{B}')\dot{q} + \mathfrak{C}q = 0.$$

This equation is of a rather more general form than that corresponding to disturbed equilibrium, in that it contains terms dependent on the generalised velocities. The elements of the skew symmetric matrix $\mathfrak{B} - \mathfrak{B}'$ are spoken of as *gyrostatic coefficients* (compare also § 8·17).

9·5. Small Motions of Systems Subject to Aerodynamical Forces. The aerodynamical forces acting on a system will not, in general, be conservative, and will, strictly speaking, depend on the whole previous history of the motion. When such a system is disturbed from a known condition (e.g. equilibrium or steady motion) it is usual to assume as an approximation that the differences between the values of the air forces in the disturbed and undisturbed states of the system depend linearly on the sets of quantities q, $\dot{q}$, $\ddot{q}$, at most, where q denotes the deviations of the generalised coordinates. Hence, if $\mathfrak{P}_i^*$ and $\mathfrak{P}_i$ represent respectively the typical generalised aerodynamical

† This implies that the disturbances do not introduce generalised forces corresponding to any of the ignorable coordinates.

forces for the disturbed and undisturbed states, then according to the foregoing approximation

$$\mathfrak{P}_i^* = \mathfrak{P}_i + \sum_{j=1}^{m} \left(\frac{\partial \mathfrak{P}_i}{\partial q_j} q_j + \frac{\partial \mathfrak{P}_i}{\partial \dot{q}_j} \dot{q}_j + \frac{\partial \mathfrak{P}_i}{\partial \ddot{q}_j} \ddot{q}_j \right).$$

The constant coefficients $\dfrac{\partial \mathfrak{P}_i}{\partial q_j}, \dfrac{\partial \mathfrak{P}_i}{\partial \dot{q}_j}, \dfrac{\partial \mathfrak{P}_i}{\partial \ddot{q}_j}$ may be referred to as the *generalised aerodynamical derivatives.* If the column of the generalised aerodynamical forces is expressed as

$$\mathfrak{P}^* = \mathfrak{P} - Wq - B\dot{q} - \bar{A}\ddot{q},$$

then
$$W_{ij} = -\frac{\partial \mathfrak{P}_i}{\partial q_j}; \quad B_{ij} = -\frac{\partial \mathfrak{P}_i}{\partial \dot{q}_j}; \quad \bar{A}_{ij} = -\frac{\partial \mathfrak{P}_i}{\partial \ddot{q}_j}.$$

To obtain the equations of disturbed equilibrium of an aerodynamical system the method of § 9·3 is followed, but the generalised forces $\mathfrak{P}^*$ are included. The final equations may be written

$$(A + \bar{A})\ddot{q} + B\dot{q} + (E + W)q = \xi(t).$$

The elements of $\bar{A}$, with their signs reversed, are called the *acceleration derivatives.* They are generally small, and are either treated as negligible or are assumed to be of such a nature that they can be absorbed into the generalised moments and products of inertia without destruction of the symmetry of the inertia matrix A. On this understanding it is usual to write the last equation as

$$A\ddot{q} + B\dot{q} + Cq = \xi(t), \qquad \text{......(1)}$$

where $C = E + W$. The elements of B are described as the *damping coefficients*, while those of W are the *aerodynamical stiffnesses.* In general neither the *damping matrix* B nor the *total stiffness matrix* C will be symmetrical. The principal diagonal elements of C are spoken of as the *direct stiffnesses*, while the remaining elements are *cross stiffnesses*; similarly, the principal diagonal elements of B are *direct dampings*, while the remainder are *cross dampings.* The coefficients of type A_{ij}, B_{ij}, C_{ij} $(i \neq j)$ are collectively called the *couplings.*

9·6. Free Disturbed Steady Motion of an Aeroplane. The equations in this case can be deduced directly from the equations of general motion of an aeroplane given in § 8·13. The symbols have the same meanings as before, except that they now represent incremental values in the disturbed motion: the corresponding values in the given steady motion are denoted by the same symbols with a cipher suffix. As is usual in the treatment of aeroplane stability,* the body axes are assumed to be principal axes, so that the inertia matrix J will have the diagonal form

$$J = \begin{bmatrix} J_1 & 0 & 0 \\ 0 & J_2 & 0 \\ 0 & 0 & J_3 \end{bmatrix}.$$

The scheme of conversion to standard aeronautical symbols will be considered later.

Equations (8·13·1) and (8·13·2) give immediately for the disturbed motion

$$(ID+\varpi+\varpi_0)(u+u_0) = X+X_0+G+G_0,$$

$$(ID+\varpi+\varpi_0)J(p+p_0) = L+L_0.$$

Omitting the terms which arise only in the steady motion and neglecting squares and products of the deviations, we obtain for the deviant motion

$$(ID+\varpi_0)u+\varpi u_0 = X+G, \qquad \text{......(1)}$$

$$(ID+\varpi_0)Jp+\varpi Jp_0 = L. \qquad \text{......(2)}$$

Now (see example (i) of § 8·5) $\varpi u_0 = -a_0 p$, where

$$a_0 \equiv \begin{bmatrix} 0 & -u_{30} & u_{20} \\ u_{30} & 0 & -u_{10} \\ -u_{20} & u_{10} & 0 \end{bmatrix}.$$

Similarly, $$\varpi_0 Jp+\varpi Jp_0 = -b_0 p,$$

where $$b_0 \equiv \begin{bmatrix} 0 & (J_2-J_3)p_{30} & (J_2-J_3)p_{20} \\ (J_3-J_1)p_{30} & 0 & (J_3-J_1)p_{10} \\ (J_1-J_2)p_{20} & (J_1-J_2)p_{10} & 0 \end{bmatrix}.$$

Again, expressing X and L in terms of aerodynamical derivatives, and neglecting acceleration derivatives, we may write

$$X = \Phi_{11}u+\Phi_{12}p,$$

$$L = \Phi_{21}u+\Phi_{22}p,$$

in which Φ_{11}, Φ_{12}, Φ_{21}, Φ_{22} are square matrices the (i,j)th elements of which are respectively the derivatives

$$\frac{\partial X_i}{\partial u_j},\quad \frac{\partial X_i}{\partial p_j},\quad \frac{\partial L_i}{\partial u_j},\quad \frac{\partial L_i}{\partial p_j}.$$

Next, by differentiation of the expression (8·13·3) for the components of the weight in a general motion, it readily follows that in

* See, for instance, Chap. x of Ref. 27.

the present deviant motion

$$G = gc_0\phi,$$

where
$$c_0 \equiv \begin{bmatrix} 0 & -\cos\phi_{20} & 0 \\ \cos\phi_{20}\cos\phi_{10} & -\sin\phi_{20}\sin\phi_{10} & 0 \\ -\cos\phi_{20}\sin\phi_{10} & -\sin\phi_{20}\cos\phi_{10} & 0 \end{bmatrix}.$$

On making these substitutions in (1) and (2) we derive the equations

$$(\Phi_{11}-\varpi_0-ID)\,u+(\Phi_{12}+a_0)\,p+gc_0\phi = 0, \qquad \text{......(3)}$$

$$J^{-1}\Phi_{21}u+(J^{-1}\Phi_{22}+J^{-1}b_0-ID)\,p = 0. \qquad \text{......(4)}$$

It remains, finally, to connect the deviations p with the deviations ϕ and their rates of change. The necessary relation, which is deducible by differentiation of (8·13·4) and use of (8·13·7), may be written

$$p = (R_0D+\Omega c_0)\,\phi,$$

where c_0 is as defined above, and R_0 is the matrix in (8·6·6) with the values ϕ_0 appropriate to the steady motion substituted for ϕ. A rather more convenient form is

$$R_0^{-1}p+(\Omega e_0-ID)\,\phi = 0, \qquad \text{......(5)}$$

in which
$$e_0 \equiv -R_0^{-1}c_0 = \sec\phi_{20}\begin{bmatrix} 0 & 1 & 0 \\ -\cos^2\phi_{20} & 0 & 0 \\ 0 & \sin\phi_{20} & 0 \end{bmatrix}.$$

The required equations of deviant motion are (3), (4) and (5). They may be written concisely by partitioned matrices as

$$\frac{d}{dt}\{u,p,\phi\} = \Upsilon\{u,p,\phi\}, \qquad \text{......(6)}$$

with
$$\Upsilon \equiv \begin{bmatrix} \Phi_{11}-\varpi_0 & \Phi_{12}+a_0 & gc_0 \\ J^{-1}\Phi_{21} & J^{-1}(\Phi_{22}+b_0) & 0 \\ 0 & R_0^{-1} & \Omega e_0 \end{bmatrix}.$$

The scheme of conversion to standard aeronautical symbols is as given in the addendum to § 8·13, with the simplifications $D=E=F=0$. The derivatives $\frac{\partial X_1}{\partial u_1}$, $\frac{\partial X_1}{\partial u_2}$, etc. correspond in the standard notation to $\frac{\partial X}{\partial u}$, $\frac{\partial X}{\partial v}$, etc., and these are usually abbreviated to X_u, X_v, etc. The λ-determinant of the system of equations (6)—after extraction of a factor λ indicating neutrality of the aeroplane with respect to the angle ψ—can be expressed in standard symbols as follows:

$$\Delta(\lambda) = \begin{vmatrix} X_u-\lambda, & X_v+r_0, & X_w-q_0, & X_p, & X_q-w_0, & X_r+v_0, & -g\cos\theta_0, & 0 \\ Y_u-r_0, & Y_v-\lambda, & Y_w+p_0, & Y_p+w_0, & Y_q, & Y_r-u_0, & -g\sin\theta_0\sin\phi_0, & g\cos\theta_0\cos\phi_0 \\ Z_u+q_0, & Z_v-p_0, & Z_w-\lambda, & Z_p-v_0, & Z_q+u_0, & Z_r, & -g\sin\theta_0\cos\phi_0, & -g\cos\theta_0\sin\phi_0 \\ \dfrac{L_u}{A}, & \dfrac{L_v}{A}, & \dfrac{L_w}{A}, & \dfrac{L_p}{A}-\lambda, & \dfrac{L_q}{A}+\left(\dfrac{B-C}{A}\right)r_0, & \dfrac{L_r}{A}+\left(\dfrac{B-C}{A}\right)q_0, & 0, & 0 \\ \dfrac{M_u}{B}, & \dfrac{M_v}{B}, & \dfrac{M_w}{B}, & \dfrac{M_p}{B}+\left(\dfrac{C-A}{B}\right)r_0, & \dfrac{M_q}{B}-\lambda, & \dfrac{M_r}{B}+\left(\dfrac{C-A}{B}\right)p_0, & 0, & 0 \\ \dfrac{N_u}{C}, & \dfrac{N_v}{C}, & \dfrac{N_w}{C}, & \dfrac{N_p}{C}+\left(\dfrac{A-B}{C}\right)q_0, & \dfrac{N_q}{C}+\left(\dfrac{A-B}{C}\right)p_0, & \dfrac{N_r}{C}-\lambda, & 0, & 0 \\ 0, & 0, & 0, & 0, & \cos\phi_0, & -\sin\phi_0, & -\lambda, & -\Omega\cos\theta_0 \\ 0, & 0, & 0, & 1, & \tan\theta_0\sin\phi_0, & \tan\theta_0\cos\phi_0, & \Omega\sec\theta_0, & -\lambda \end{vmatrix}$$

9·7. Review of Notation and Terminology for General Linear Systems. The few special problems already considered show how linear differential equations with constant coefficients arise naturally in dynamics. We shall now briefly summarise the essential formulae connected with the solution of such systems of dynamical equations.*

The Lagrangian equations for a linear system having m generalised coordinates $q_1, q_2, \ldots, q_m$ will be of the type

$$f(D)\,q \equiv (AD^2 + BD + C)\,q = \xi(t), \qquad \ldots\ldots(1)$$

where A, B, C are respectively the *inertia*, *damping* and *stiffness* matrices, and $\xi(t)$ denotes a column of assigned "forcing" functions. In a free motion of the system $\xi(t) = 0$.

The determinantal equation

$$\Delta(\lambda) \equiv |\, f(\lambda)\,| \equiv p_0\lambda^n + p_1\lambda^{n-1} + \ldots + p_{n-1}\lambda + p_n = 0$$

is in general of degree $n = 2m$, and the n roots (not necessarily all distinct) are denoted by $\lambda_1, \lambda_2, \ldots, \lambda_n$. The constituent solution appropriate to an unrepeated root λ_r is denoted by

$$q = \{k_{ir}\}\,e^{\lambda_r t} \equiv k_r e^{\lambda_r t},$$

and the column of constants k_r is referred to as the rth modal column. This column may be chosen proportional to any non-vanishing column of the adjoint $F(\lambda_r)$ of the matrix $f(\lambda_r)$. On the other hand, if λ_s represents a member of a set of s equal roots, the constituent solution appropriate to λ_s is written

$$q = \{k_{is}(t)\}\,e^{\lambda_s t} \equiv k_s(t)\,e^{\lambda_s t}.$$

In this case the s modal columns relevant to the complete set of roots equal to λ_s may be chosen proportional to any s linearly independent columns of the family of matrices

$$U_0(t, \lambda_s) = F(\lambda_s),$$

$$U_1(t, \lambda_s) = \overset{(1)}{F}(\lambda_s) + tF(\lambda_s),$$

....

$$U_{s-1}(t, \lambda_s) = \overset{(s-1)}{F}(\lambda_s) + (s-1)\,t\overset{(s-2)}{F}(\lambda_s) + \frac{(s-1)(s-2)}{\underline{|2}}\,t^2\overset{(s-3)}{F}(\lambda_s) + \ldots + t^{s-1}F(\lambda_s).$$

The modal coefficients $k_{is}(t)$ will be polynomials in t of degree $s-1$ at most.

* For the detailed exposition see Chaps. v and vi.

When $\xi(t)=0$ the most general solution of equations (1), derived as a linear superposition of arbitrary multiples of the constituent solutions, is

$$q = c_1 k_1(t)\, e^{\lambda_1 t} + c_2 k_2(t)\, e^{\lambda_2 t} + \dots + c_n k_n(t)\, e^{\lambda_n t},$$

in which c_1, c_2, etc. are arbitrary constants, real or complex. This solution may be written more concisely as

$$q = k(t)\, M(t)\, c.$$

Here $k(t)$ denotes the $(\overline{m}, n)$ modal matrix

$$k(t) \equiv \begin{bmatrix} k_{11}(t), & k_{12}(t), & \dots, & k_{1n}(t) \\ k_{21}(t), & k_{22}(t), & \dots, & k_{2n}(t) \\ \dots & \dots & \dots & \dots \\ k_{m1}(t), & k_{m2}(t), & \dots, & k_{mn}(t) \end{bmatrix}$$

and

$$M(t) \equiv \begin{bmatrix} e^{\lambda_1 t}, & 0, & \dots, & 0 \\ 0, & e^{\lambda_2 t}, & \dots, & 0 \\ \dots & \dots & \dots & \dots \\ 0, & 0, & \dots, & e^{\lambda_n t} \end{bmatrix},$$

while c represents the column of the arbitrary constants.

It is sometimes convenient to substitute for the Lagrangian equations (1) an equivalent set of equations of the first order,* obtained by the use of auxiliary variables. If the generalised momenta $p \equiv A\dot{q}$ are adopted as the auxiliary variables (see § 8·16), then the complete set of equations in the Hamiltonian form is

$$\begin{bmatrix} \dot{q} \\ \dot{p} \end{bmatrix} = \begin{bmatrix} 0, & A^{-1} \\ -C, & -BA^{-1} \end{bmatrix} \begin{bmatrix} q \\ p \end{bmatrix} + \begin{bmatrix} 0 \\ \xi(t) \end{bmatrix}.$$

A convenient alternative scheme is to adopt the generalised velocities $\dot{q}$ as the auxiliary variables. The equivalent linear system then is (see example (ii) of § 5·5)

$$\dot{y} = uy + \eta(t),$$

where $y = \begin{bmatrix} q \\ \dot{q} \end{bmatrix}$; $\eta(t) = \begin{bmatrix} 0 \\ A^{-1}\xi(t) \end{bmatrix}$; $u = \begin{bmatrix} 0, & I \\ -A^{-1}C, & -A^{-1}B \end{bmatrix}$.

9·8. General Nature of the Constituent Motions. (*a*) *Real Roots.* The modal column obtained from each real (simple or multiple) root of the determinantal equation of the set of equations (9·7·1) will

* For special methods of solution of first-order systems of equations, see Part II of Chap. VI.

consist of real elements. The corresponding constituent motion will be a *subsidence* or a *divergence* according as the root concerned is negative or positive. A zero root, if unrepeated, yields a constant constituent representing neutral equilibrium. On the other hand, a repeated zero root in general gives rise to a constituent involving polynomials in t, and thus to a growing motion.

(*b*) *Complex Roots.* The modal column derived from a complex root is, in general, complex. It is customary in dynamics to associate together the conjugate complex constituents obtained from conjugate complex pairs of roots, and so to derive equivalent pairs of real constituent motions. Suppose $\lambda_s = \mu + i\omega$ to be a root of multiplicity s, where i denotes $\sqrt{-1}$: then the conjugate complex root $\bar{\lambda}_s = \mu - i\omega$ also has multiplicity s. Hence for every modal column obtained from the set of s roots equal to λ_s there will be a corresponding conjugate complex modal column derived from the set of s roots equal to $\bar{\lambda}_s$. If $k_s(t) = \alpha(t) + i\beta(t)$ and $\bar{k}_s(t) = \alpha(t) - i\beta(t)$ are, respectively, the modal columns appropriate to λ_s and $\bar{\lambda}_s$, and if $\rho e^{i\tau}$ is a complex arbitrary constant, then the most general real linear combination of the two constituents concerned will be expressible as

$$q = \rho e^{\mu t} \cos(\omega t + \tau)\, \alpha(t) - \rho e^{\mu t} \sin(\omega t + \tau)\, \beta(t).$$

In the special case where the roots λ_s and $\bar{\lambda}_s$ are unrepeated, the elements of α and β are constants, and then the last equation may be written

$$q = \rho e^{\mu t}\{\gamma_j \cos(\omega t + \epsilon_j + \tau)\},$$

where

$$\{\gamma_j e^{i\epsilon_j}\} = \alpha + i\beta.$$

The motion in this case may be regarded as effectively a single oscillatory constituent having an arbitrary amplitude ρ and an arbitrary epoch τ. The motion will ultimately die away, or grow to an indefinitely large amplitude, according as the real part μ of the root λ_s is negative or positive.*

(*c*) *Purely Imaginary Roots.*† If damping forces are present in the system (i.e. if $B \neq 0$), the modal column corresponding to a purely imaginary root will as a general rule contain complex elements. The

* The general character of the roots of $\Delta(\lambda) = 0$ can be ascertained by the use of the test functions (see § 4·22 (*b*)).

† The important case in which $B = 0$ and C is symmetrical is discussed separately in § 9·9.

real constituents appropriate to such a root and its conjugate imaginary will thus be of the types considered in (*b*) with the simplification $\mu = 0$. Simple imaginary roots give rise to pure sinusoidal oscillations, while repeated imaginary roots will in general yield oscillations whose amplitudes grow in proportion to a polynomial in t.

The dynamical coefficients sometimes depend on one or more variable parameters. If the values of these parameters are such that one or more of the constituent motions of the system are simply sinusoidal, the state of the system is said to be *critical*, and the parameters have *critical values*. The critical state requires that $\pm i\omega$ shall be roots of $\Delta(\lambda) = 0$, so that

$$T_{n-1} = 0, \qquad \text{......(1)}$$

where T_{n-1} is the penultimate test determinant (see § 4·22 (*b*)). When the parameters are such that (1) is satisfied, they are either critical according to the definition, or else $\Delta(\lambda) = 0$ has equal and opposite roots. The transition from a completely stable state of a system to oscillatory instability is always indicated by the vanishing of T_{n-1}. A simple illustration of a critical parameter is the critical speed for flutter of an aerodynamical system (see § 12·1).

9·9. Modal Columns for a Linear Conservative System. It is a well-known theorem* that if the dynamical equations are of the type

$$f(D)\,q \equiv (AD^2 + E)\,q = 0, \qquad \text{......(1)}$$

in which A and E are both symmetrical, the roots of $\Delta(\lambda) = 0$ are either real or purely imaginary (i.e. they cannot be complex): they are all purely imaginary if the potential energy $V \equiv \frac{1}{2}q'Eq$ is a positive function.

The usual proof is as follows. Write $\lambda^2 = -z$ and

$$\Delta_m(z) \equiv \begin{vmatrix} A_{11}z - E_{11}, & A_{12}z - E_{12}, & \dots, & A_{1m}z - E_{1m}, & 0 \\ A_{21}z - E_{21}, & A_{22}z - E_{22}, & \dots, & A_{2m}z - E_{2m}, & 0 \\ \dots & \dots & \dots & \dots & \dots \\ A_{m1}z - E_{m1}, & A_{m2}z - E_{m2}, & \dots, & A_{mm}z - E_{mm}, & 0 \\ 0, & 0, & \dots, & 0, & 1 \end{vmatrix}. \qquad \text{......(2)}$$

Also, let $\Delta_{m-1}(z)$ be the determinant obtained when the first row and first column of Δ_m are erased, $\Delta_{m-2}(z)$ be the determinant formed when

* For historical notes see p. 183 of Ref. 31.

the first two rows and columns of Δ_m are erased, and so on until finally $\Delta_0 = 1$: the degrees in z of these determinants are represented by the suffices. Since the discriminants of the kinetic energy function are all necessarily positive (see example (ii) of § 8·14), the determinants $\Delta_0, \Delta_1, \ldots, \Delta_m$ are alternately positive and negative when $z = -\infty$, but are all positive when $z = +\infty$. From example (i) of § 1·17 it is seen that when z passes through a root of any intermediate determinant of the series, say Δ_i $(0 \neq i \neq m)$, then Δ_{i-1} and Δ_{i+1} necessarily have opposite signs. Now Δ_0 has the constant value 1, while the one root of the equation $\Delta_1 = 0$ is obviously real. Hence for this root $\Delta_2 < 0$. But $\Delta_2 > 0$ when $z = \pm\infty$, so that the equation $\Delta_2 = 0$ has two real roots which are separated by the root of $\Delta_1 = 0$. Similarly, it follows that $\Delta_3 = 0$ has three real roots separated by the roots of $\Delta_2 = 0$, and more generally the roots of $\Delta_i = 0$ are all real, and separate those of $\Delta_{i+1} = 0$. Since $z = -\lambda^2$, it follows that the $2m$ roots of the original determinantal equation are either real or purely imaginary.

If the potential energy is a positive function, the discriminants of V are all positive. Accordingly the signs of the determinants $\Delta_0(z), \Delta_1(z), \ldots, \Delta_m(z)$ are alternately positive and negative when $z = 0$, so that all the roots of $\Delta_m = 0$ are positive. By the same argument as before it follows that the roots of $\Delta(\lambda) = 0$ in this case are all purely imaginary.

Since the roots of $\Delta_i(z) = 0$ separate those of $\Delta_{i+1}(z) = 0$, it is seen that if $\Delta_m(z) = 0$ has s equal roots z_s, then $\Delta_{m-1}(z)$ has $s-1$ equal roots z_s, $\Delta_{m-2}(z) = 0$ has $s-2$ equal roots z_s, and so on, until finally z_s is a simple root of $\Delta_{m-s+1}(z) = 0$. Moreover, these conclusions are true for any other arrangement of the principal diagonal elements in (2). Now refer to example (i) of § 1·17. If the two determinants on the left of equation (1·17·1) contain respectively the factors $(z-z_s)^s$ and $(z-z_s)^{s-2}$, and if the first two on the right each contain $(z-z_s)^{s-1}$, then obviously the non-diagonal minor (there called A_{12}) must also contain the factor $(z-z_s)^{s-1}$. This shows that if z_s is an s-fold root of $\Delta_m(z) = 0$, then $(z-z_s)^{s-1}$ is a factor of every first minor of $\Delta_m(z)$. It follows that if λ_s, and therefore also $-\lambda_s$, are s-fold roots of $\Delta(\lambda) = 0$, then $(\lambda^2 - \lambda_s^2)^{s-1}$ is a factor of every first minor of $\Delta(\lambda)$. The adjoint matrix $F(\lambda)$ accordingly contains the same factor, so that $F(\lambda_s)$ and all the derived adjoint matrices up to and including $\overset{(s-2)}{F}(\lambda_s)$ are null. In this case also the matrices $U_i(t, \lambda_s)$ are all null with the exception of $U_{s-1}(t, \lambda_s)$, which reduces to the constant matrix $\overset{(s-1)}{F}(\lambda_s)$. Hence the following important theorem: *When a conservative system is disturbed from equilibrium the modal columns are (in general) all independent of t even when multiple roots of* $\Delta(\lambda) = 0$ *occur.*

The exception to the foregoing theorem arises when E is singular and $\lambda_s = -\lambda_s = 0$, so that there are in all $2s$ zero roots of $\Delta(\lambda) = 0$. In this case λ^{2s-2} is a factor of every first minor of $\Delta(\lambda)$, so that the adjoint matrix is of the form $\lambda^{2s-2}G(\lambda^2)$. Here all the derived adjoint matrices involved, with the exception of $\overset{(2s-2)}{F(\lambda_s)}$, are null, and the $2s$ modal columns corresponding to the $2s$ zero roots are obtained from U_{2s-2} and U_{2s-1}, which reduce respectively to $\overset{(2s-2)}{F(\lambda_s)}$ and $(2s-1)\,t\overset{(2s-2)}{F(\lambda_s)}$. If k_i is any one of the s linearly independent columns of $\overset{(2s-2)}{F(\lambda_s)}$, the corresponding two constituent motions are $(a_i + b_i t)\,k_i$, where a_i and b_i are arbitrary constants. This exceptional case, in which E is singular, will hereafter be assumed to be excluded.

When λ_r is a simple root the appropriate modal column k_r is given as usual by any non-vanishing column of the adjoint matrix $F(\lambda_r)$. Since the elements of this matrix are polynomials in λ_r^2, and since λ_r^2 is real, the modal column corresponding to λ_r is real, and is the same as for the root $-\lambda_r$. If there are s roots equal to λ_s, the corresponding s modal columns are given by the s linearly independent columns which the matrix $\overset{(s-1)}{F(\lambda_s)}$ necessarily contains. It is easy to see that, as for simple roots, the modal columns appropriate to the set of roots λ_s can always be taken to be real and equal to those for the roots $-\lambda_s$. Hence in all cases, if the roots are $\pm\lambda_1, \pm\lambda_2, \ldots, \pm\lambda_m$, and if k_0 is the square matrix of order m formed from the m modal columns appropriate to $\lambda_1, \lambda_2, \ldots, \lambda_m$, then k_0 also gives the modal columns appropriate to $-\lambda_1, -\lambda_2, \ldots, -\lambda_m$.

Examples

(i) *Roots all Distinct.* Suppose

$$f(\lambda) \equiv \begin{bmatrix} 44\lambda^2+108 & -8\lambda^2 & 0 \\ -8\lambda^2 & 35\lambda^2+54 & 12\lambda^2 \\ 0 & 12\lambda^2 & 28\lambda^2+36 \end{bmatrix}.$$

Then $\Delta(\lambda) = 16 \times 27 \times 81(\lambda^2+1)(\lambda^2+2)(\lambda^2+3)$, so that the roots of $\Delta(\lambda) = 0$, say $\lambda_1 = +i$, $\lambda_2 = +i\sqrt{2}$, $\lambda_3 = +i\sqrt{3}$, and $\lambda_4 = -i$, $\lambda_5 = -i\sqrt{2}$, $\lambda_6 = -i\sqrt{3}$, are all distinct.

The adjoint matrices corresponding to $\lambda_1, \lambda_2, \lambda_3$ are

$$F(\lambda_1) = 8\{1, -8, -12\}[1, -8, -12],$$
$$F(\lambda_2) = -16\{4, -5, 6\}[4, -5, 6],$$
$$F(\lambda_3) = 72\{4, 4, -3\}[4, 4, -3],$$

so that we may choose

$$k_0 = \begin{bmatrix} 1 & 4 & 4 \\ -8 & -5 & 4 \\ -12 & 6 & -3 \end{bmatrix}.$$

(ii) *Repeated Roots.* An illustration is provided by the system already considered in the example to § 6·6, for which

$$f(\lambda) = \begin{bmatrix} 32\lambda^2+48 & -16\lambda^2 & 4\lambda^2 \\ -16\lambda^2 & 32\lambda^2+48 & 4\lambda^2 \\ 4\lambda^2 & 4\lambda^2 & 11\lambda^2+12 \end{bmatrix},$$

and $$\Delta(\lambda) = 3 \times 48^2(\lambda^2+1)^2(\lambda^2+4).$$

If the roots of $\Delta(\lambda) = 0$ are denoted by $\lambda_1 = \lambda_2 = +i$, $\lambda_3 = +2i$, $\lambda_4 = \lambda_5 = -i$, $\lambda_6 = -2i$, then

$$F(\lambda_2) = 0 \quad \text{with} \quad \overset{(1)}{F}(\lambda_2) = 2 \times 48i \begin{bmatrix} 5 & -4 \\ -4 & 5 \\ 4 & 4 \end{bmatrix} \begin{bmatrix} 1 & 0 & 4 \\ 0 & 1 & 4 \end{bmatrix},$$

while $$F(\lambda_3) = 3 \times 48 \times 16\{1, 1, -1\}[1, 1, -1].$$

Hence we may choose $$k_0 = \begin{bmatrix} 5 & -4 & 1 \\ -4 & 5 & 1 \\ 4 & 4 & -1 \end{bmatrix}.$$

(iii) *Matrix E Singular.* Suppose

$$f(\lambda) = \begin{bmatrix} 3\lambda^2+1, & 2, & 3 \\ 2, & 12\lambda^2+4, & 6 \\ 3, & 6, & 27\lambda^2+9 \end{bmatrix}.$$

Then $\Delta(\lambda) = 12 \times 81\lambda^4(\lambda^2+1)$ and

$$F(\lambda) = 3\lambda^2 \begin{bmatrix} 108\lambda^2+72, & -18, & -12 \\ -18, & 27\lambda^2+18, & -6 \\ -12, & -6, & 12\lambda^2+8 \end{bmatrix}.$$

For the quadruple zero root $\lambda_1 = \lambda_2 = \lambda_3 = \lambda_4 = 0$, the three matrices $\overset{(1)}{F}(\lambda_4)$, $\overset{(3)}{F}(\lambda_4)$ and $F(\lambda_4)$ are null, but

$$\overset{(2)}{F}(\lambda_4) = 6\begin{bmatrix} 72 & -18 & -12 \\ -18 & 18 & -6 \\ -12 & -6 & 8 \end{bmatrix} = 12\begin{bmatrix} 3 & 6 \\ -3 & 3 \\ 1 & -4 \end{bmatrix}\begin{bmatrix} 6 & -3 & 0 \\ 3 & 0 & -1 \end{bmatrix}.$$

Hence the constituent motions appropriate to the four zero roots may be taken as $(a+bt)\{3, -3, 1\}$ and $(c+dt)\{6, 3, -4\}$.

9·10. The Direct Solution for a Linear Conservative System and the Normal Coordinates. From the discussion of § 9·9 it follows that if the roots of $\Delta(\lambda) = 0$ are denoted by $\pm\lambda_1, \pm\lambda_2, \ldots, \pm\lambda_m$, the complete motion of the system is represented by

$$q(t) = k_0 M_0(t)\, a + k_0 M_0(-t)\, b, \qquad \text{......(1)}$$

where a and b are two columns each containing m arbitrary constants of integration, k_0 is the matrix of the m modal columns appropriate to the roots $\lambda_1, \lambda_2, \ldots, \lambda_m$, and $M_0(t)$ is the diagonal matrix

$$\begin{bmatrix} e^{\lambda_1 t} & 0 & \cdots & 0 \\ 0 & e^{\lambda_2 t} & \cdots & 0 \\ \cdots & \cdots & \cdots & \cdots \\ 0 & 0 & \cdots & e^{\lambda_m t} \end{bmatrix}.$$

If the values of the generalised coordinates and velocities at $t = 0$ are assigned, and if Λ_0 denotes the diagonal matrix of the roots $\lambda_1, \lambda_2, \ldots, \lambda_m$, the direct solution is easily verified to be

$$q(t) = \tfrac{1}{2}k_0[M_0(t) + M_0(-t)]\, k_0^{-1}q(0) + \tfrac{1}{2}k_0[M_0(t) - M_0(-t)]\, \Lambda_0^{-1}k_0^{-1}\dot{q}(0).$$

The corresponding special form of the solution has already been obtained in equation (6·6·4).

If new generalised coordinates Q are adopted given by

$$q = k_0 Q,$$

the general solution (1) reduces to

$$Q = M_0(t)\, a + M_0(-t)\, b, \qquad \text{......(2)}$$

while the equations $A\ddot{q} + Eq = 0$ transform to (see example to § 9·3)

$$(k_0' A k_0)\, \ddot{Q} + (k_0' E k_0)\, Q = 0. \qquad \text{......(3)}$$

From (2) it is seen that when the generalised coordinates are chosen in this way a disturbance of the system restricted to any one of its coordinates Q_r produces motion in that particular coordinate only; in other words, the motions in the several degrees of freedom are uncoupled. The coordinates Q are usually termed the *normal* or *principal* coordinates.

Evidently, if d_0 denotes an arbitrary non-singular matrix, the system of differential equations

$$d_0(\ddot{Q}-\Lambda_0^2 Q)=0 \qquad \ldots\ldots(4)$$

will be satisfied by (2). Choosing

$$d_0 = k_0' A k_0, \qquad \ldots\ldots(5)$$

we then have also, by (3) and (4),

$$-d_0\Lambda_0^2 = k_0' E k_0. \qquad \ldots\ldots(6)$$

Now $k_0' A k_0$ and $k_0' E k_0$ are both symmetrical; hence by (5) and (6), $d_{ij} = d_{ji}$ and $-d_{ij}\lambda_j^2 = -d_{ji}\lambda_i^2$, where d_{ij} denotes the typical element of d_0. Accordingly, if the roots of $\Delta(\lambda)=0$ are all distinct, $d_{ij}=d_{ji}=0$, and d_0 will therefore be a diagonal matrix. The normal coordinates, which can clearly be taken in arbitrary multiples, may in this case be chosen to be

$$Q = d_0 k_0^{-1} q,$$

or, on application of (5), $\quad Q = k_0' A q. \qquad \ldots\ldots(7)$

When $\Delta(\lambda)=0$ has multiple roots, the matrix d_0 need not be diagonal. For instance, if $\lambda_1=\lambda_2$, then $d_{12}=d_{21}$, but these elements do not necessarily vanish. The diagonal matrix d_0 is in this case replaced by

$$\begin{bmatrix} d_{11} & d_{12} & 0 & \ldots & 0 \\ d_{21} & d_{22} & 0 & \ldots & 0 \\ 0 & 0 & d_{33} & \ldots & 0 \\ \ldots & \ldots & \ldots & \ldots & \ldots \\ 0 & 0 & 0 & \ldots & d_{mm} \end{bmatrix}.$$

More generally, if the m roots $\lambda_1, \lambda_2, \ldots, \lambda_m$ consist of a roots equal to λ_a, b roots equal to λ_b, and so on up to p roots equal to λ_p, where $a+b+\ldots+p=m$, then

$$d_0 = \begin{bmatrix} d_a, & 0, & \ldots, & 0 \\ 0, & d_b, & \ldots, & 0 \\ \ldots & \ldots & \ldots & \ldots \\ 0, & 0, & \ldots, & d_p \end{bmatrix},$$

where the typical square submatrix d_i is symmetrical, non-singular, and of order i. It is easy to see that the normal coordinates can be defined in accordance with the formula $Q = k_0' A q$ even when multiple roots occur. For

$$k_0^{-1} q = M(t)\, a + M(-t)\, b,$$

and since d_0 is obviously permutable with $M(t)$, it follows that

$$d_0 k_0^{-1} q = M(t)\, d_0 a + M(-t)\, d_0 b.$$

Hence

$$Q = M(t)\, \alpha + M(-t)\, \beta,$$

where α and β are columns of arbitrary constants. The coordinates Q, defined in accordance with the formula, are thus normal.

Examples

(i) *System with Roots all Distinct.* Suppose the given equations to be

$$\begin{bmatrix} 44 & -8 & 0 \\ -8 & 35 & 12 \\ 0 & 12 & 28 \end{bmatrix} \ddot{q} + \begin{bmatrix} 108 & 0 & 0 \\ 0 & 54 & 0 \\ 0 & 0 & 36 \end{bmatrix} q = 0.$$

Then, if the modal columns are chosen as in example (i) of § 9·9,

$$k_0 = \begin{bmatrix} 1 & 4 & 4 \\ -8 & -5 & 4 \\ -12 & 6 & -3 \end{bmatrix}.$$

Hence the normal coordinates may be taken as

$$Q = k_0' A q = \begin{bmatrix} 1 & -8 & -12 \\ 4 & -5 & 6 \\ 4 & 4 & -3 \end{bmatrix} \begin{bmatrix} 44 & -8 & 0 \\ -8 & 35 & 12 \\ 0 & 12 & 28 \end{bmatrix} q$$

$$= \begin{bmatrix} 108 & -432 & -432 \\ 216 & -135 & 108 \\ 144 & 72 & -36 \end{bmatrix} q.$$

As a verification of (5) and (6)

$$k_0' A k_0 = \begin{bmatrix} 108 & -432 & -432 \\ 216 & -135 & 108 \\ 144 & 72 & -36 \end{bmatrix} \begin{bmatrix} 1 & 4 & 4 \\ -8 & -5 & 4 \\ -12 & 6 & -3 \end{bmatrix}$$

$$= \begin{bmatrix} 8748 & 0 & 0 \\ 0 & 2187 & 0 \\ 0 & 0 & 972 \end{bmatrix} = d_0.$$

Similarly it is easily shown that

$$k_0' E k_0 = \begin{bmatrix} 8748 & 0 & 0 \\ 0 & 4374 & 0 \\ 0 & 0 & 2916 \end{bmatrix} = -\Lambda_0^2 d_0.$$

(ii) *System with Multiple Roots.* If the equations are

$$\begin{bmatrix} 32 & -16 & 4 \\ -16 & 32 & 4 \\ 4 & 4 & 11 \end{bmatrix} \ddot{q} + \begin{bmatrix} 48 & 0 & 0 \\ 0 & 48 & 0 \\ 0 & 0 & 12 \end{bmatrix} q = 0,$$

the roots of $\Delta(\lambda) = 0$ are $\lambda_1 = \lambda_2 = +i$, $\lambda_3 = +2i$, $\lambda_4 = \lambda_5 = -i$, and $\lambda_6 = -2i$. As in example (ii) of § 9·9

$$k_0 = \begin{bmatrix} 5 & -4 & -1 \\ -4 & 5 & -1 \\ 4 & 4 & 1 \end{bmatrix},$$

and the normal coordinates may be taken as

$$Q \equiv k_0' A q = \begin{bmatrix} 5 & -4 & 4 \\ -4 & 5 & 4 \\ -1 & -1 & 1 \end{bmatrix} \begin{bmatrix} 32 & -16 & 4 \\ -16 & 32 & 4 \\ 4 & 4 & 11 \end{bmatrix} q$$

$$= \begin{bmatrix} 240 & -192 & 48 \\ -192 & 240 & 48 \\ -12 & -12 & 3 \end{bmatrix} q.$$

In the present case

$$k_0' A k_0 = \begin{bmatrix} 2160 & -1728 & 0 \\ -1728 & 2160 & 0 \\ 0 & 0 & 27 \end{bmatrix} = d_0,$$

while

$$k_0' E k_0 = \begin{bmatrix} 2160 & -1728 & 0 \\ -1728 & 2160 & 0 \\ 0 & 0 & 108 \end{bmatrix} = -\Lambda_0^2 d_0.$$

(iii) *The Matrix $E^{-1}A$.* Since $k_0 M_0(t)$ is a solution of $A\ddot{q} + Eq = 0$, it follows that

$$A k_0 \Lambda_0^2 + E k_0 = 0.$$

When E is non-singular this may be written

$$k_0^{-1} E^{-1} A k_0 = -\Lambda_0^{-2},$$

so that the canonical form of $E^{-1}A$ is always diagonal, even when repeated roots occur. Hence $E^{-1}A$ has the modal matrix k_0 and the latent roots $-\lambda_1^2, -\lambda_2^2, \ldots, -\lambda_m^2$. This result will be used in Chapter X in connection with the solution by iterative methods of the dynamical equations in the form $E^{-1}A\ddot{q}+q=0$. The matrix $E^{-1}A$ is there referred to as the *dynamical matrix*.

9·11. Orthogonal Properties of the Modal Columns and Rayleigh's Principle for Conservative Systems. If λ_r and λ_s are any two different roots of $\Delta(\lambda)=0$, then, since $k_r e^{\lambda_r t}$ and $k_s e^{\lambda_s t}$ are solutions of $f(D)\,q \equiv (AD^2+E)\,q=0$, it follows that

$$f(\lambda_r)\,k_r = 0, \qquad \ldots\ldots(1)$$

$$f(\lambda_s)\,k_s = 0. \qquad \ldots\ldots(2)$$

Hence also

$$k_s' f(\lambda_r)\,k_r = 0, \qquad \ldots\ldots(3)$$

$$k_r' f(\lambda_s)\,k_s = 0. \qquad \ldots\ldots(4)$$

But, since in the case of a conservative system $f(\lambda_r)$ and $f(\lambda_s)$ are symmetrical, (3) may be transposed to give

$$k_r' f(\lambda_r)\,k_s = 0. \qquad \ldots\ldots(5)$$

On subtraction of (4) from (5) we have the result

$$(\lambda_r^2-\lambda_s^2)\,k_r'Ak_s = 0.$$

Hence if $\lambda_r \neq \pm\lambda_s$,

$$k_r'Ak_s = 0, \qquad \ldots\ldots(6)$$

so that also

$$k_r'Ek_s = 0. \qquad \ldots\ldots(7)$$

Equations (6) and (7) express the orthogonal properties of the modal columns.*

On premultiplying equation (1) by k_r' we have

$$\lambda_r^2 k_r'Ak_r = -k_r'Ek_r. \qquad \ldots\ldots(8)$$

Equation (8) can be regarded as defining λ_r as a function of the elements k_{ir} of the modal column k_r. On differentiation of (8) with respect to each of the elements k_{ir} in succession we derive the set of equations

$$\lambda_r\left\{\frac{\partial\lambda_r}{\partial k_{ir}}\right\} k_r'Ak_r = -\lambda_r^2 Ak_r - Ek_r = -f(\lambda_r)\,k_r = 0.$$

Since $k_r'Ak_r$ is an essentially positive function, it follows that

$$\left\{\frac{\partial\lambda_r}{\partial k_{ir}}\right\} = 0. \qquad \ldots\ldots(9)$$

The value of λ_r as defined by equation (8) is accordingly stationary for small variations of the modal columns from their true values. This

* See § 92 of Ref. 29.

constitutes Rayleigh's principle,* and by its aid very accurate values of frequencies of oscillatory constituents can be obtained from rather rough approximations to the corresponding modal column. If in (8) we regard k_r as defining an approximate mode, the frequency n_r will be given by the formula

$$4\pi^2 n_r^2 = k_r' E k_r / k_r' A k_r. \qquad \text{......(10)}$$

Examples

(i) *Orthogonal Properties of the Modal Columns.* Referring to the system considered in example (ii) of § 9·9, we have

$$A = \begin{bmatrix} 32 & -16 & 4 \\ -16 & 32 & 4 \\ 4 & 4 & 11 \end{bmatrix}; \qquad E = \begin{bmatrix} 48 & 0 & 0 \\ 0 & 48 & 0 \\ 0 & 0 & 12 \end{bmatrix}.$$

The two modal columns appropriate to the double roots $\pm i$ are here $k_1 = \{5, -4, 4\}$ and $k_2 = \{-4, 5, 4\}$, while the column appropriate to the simple roots $\pm 2i$ is $k_3 = \{1, 1, -1\}$.

Applying equations (6) and (7) we have

$$k_3' A k_2 = [1, 1, -1] \begin{bmatrix} 32 & -16 & 4 \\ -16 & 32 & 4 \\ 4 & 4 & 11 \end{bmatrix} \begin{bmatrix} -4 \\ 5 \\ 4 \end{bmatrix} = [12, 12, -3] \begin{bmatrix} -4 \\ 5 \\ 4 \end{bmatrix} = 0,$$

$$k_3' E k_2 = 12[1, 1, -1] \begin{bmatrix} 4 & 0 & 0 \\ 0 & 4 & 0 \\ 0 & 0 & 1 \end{bmatrix} \begin{bmatrix} -4 \\ 5 \\ 4 \end{bmatrix} = 12[4, 4, -1] \begin{bmatrix} -4 \\ 5 \\ 4 \end{bmatrix} = 0.$$

Similarly,

$$k_3' A k_1 = [12, 12, -3]\{5, -4, 4\} = 0,$$

$$k_3' E k_1 = 12[4, 4, -1]\{5, -4, 4\} = 0.$$

On the other hand

$$k_2' A k_1 = [-4, 5, 4] \begin{bmatrix} 32 & -16 & 4 \\ -16 & 32 & 4 \\ 4 & 4 & 11 \end{bmatrix} \begin{bmatrix} 5 \\ -4 \\ 4 \end{bmatrix} = [-4, 5, 4] \begin{bmatrix} 240 \\ -192 \\ 48 \end{bmatrix} \neq 0.$$

(ii) *A Generalisation of Rayleigh's Principle for Dissipative Systems of Special Type.* In the case of dissipative systems the equations of free motion have the form

$$f(D)q \equiv (AD^2 + BD + C)q = 0.$$

* For a fuller discussion of Rayleigh's principle see Ref. 32.

If λ_r is a simple root of $\Delta(\lambda) = 0$, and if k_r is the corresponding modal column, then $f(\lambda_r) k_r = 0$. On premultiplication of this equation by k_r' the result is
$$\lambda_r^2 k_r' A k_r + \lambda_r k_r' B k_r + k_r' C k_r = 0.$$
As for equation (8) we may now differentiate with respect to the elements k_{ir} of k_r in succession. When A, B, C are symmetrical the equations so obtained can be written
$$\left\{\frac{\partial \lambda_r}{\partial k_{ir}}\right\} (2\lambda_r k_r' A k_r + k_r' B k_r) = -2f(\lambda_r) k_r = 0.$$
Hence in this special case $\left\{\dfrac{\partial \lambda_r}{\partial k_{ir}}\right\} = 0.$

(iii) *Relations between Modal Columns for Dissipative Systems.* If k_r, k_s denote two modal columns distinct from k_t, then by appropriate premultiplication of the equation $f(\lambda_t) k_t = 0$ we have
$$\lambda_t^2 k_r' A k_t + \lambda_t k_r' B k_t + k_r' C k_t = 0, \qquad \text{......(11)}$$
$$\lambda_t^2 k_s' A k_t + \lambda_t k_s' B k_t + k_s' C k_t = 0. \qquad \text{......(12)}$$
Elimination of λ_t from (11) and (12) yields a relation between the modal columns k_r, k_s, k_t. Two similar relations can be obtained by interchange.

If r and t are interchanged in (11), then
$$\lambda_r^2 k_t' A k_r + \lambda_r k_t' B k_r + k_t' C k_r = 0, \qquad \text{......(13)}$$
and in the special case where A, B, C are all symmetrical the coefficients in (11) and (13) will be identical. Thus λ_r and λ_t will be the roots of the quadratic equation
$$\lambda^2 k_r' A k_t + \lambda k_r' B k_t + k_r' C k_t = 0. \qquad \text{......(14)}$$
As a numerical illustration suppose
$$f(\lambda) = \begin{bmatrix} 2 & 0 \\ 0 & 1 \end{bmatrix} \lambda^2 + \begin{bmatrix} 1 & 2 \\ 2 & 2 \end{bmatrix} \lambda + \begin{bmatrix} 1 & 0 \\ 0 & 3 \end{bmatrix},$$
with
$$\Delta(\lambda) = (\lambda^2+1)(2\lambda+3)(\lambda+1)$$
and
$$F(\lambda) = \begin{bmatrix} \lambda^2+2\lambda+3, & -2\lambda \\ -2\lambda, & 2\lambda^2+\lambda+1 \end{bmatrix}.$$
If $\lambda_1 = -1$, $\lambda_2 = -\frac{3}{2}$, $\lambda_3 = +i$, $\lambda_4 = -i$, then
$$k_1 = \{1, 1\}, \quad k_2 = \{3, 4\}, \quad k_3 = \{1+i, -i\}, \quad k_4 = \{1-i, i\}.$$
Hence, using k_1 and k_2, we have
$$k_1' A k_2 = 10, \quad k_1' B k_2 = 25, \quad k_1' C k_2 = 15.$$
Equation (14) is accordingly $10\lambda^2 + 25\lambda + 15 = 0$, the roots of which are $\lambda = -1$ and $\lambda = -\frac{3}{2}$.

9·12. Forced Oscillations of Aerodynamical Systems. The general theory of the forced oscillations of aerodynamical systems is rather complicated, and only a brief discussion can be attempted here. In recent years the subject has assumed importance in relation to such questions as the mass-balance of control surfaces of aeroplanes and the prediction of the critical speeds for wing flutter.†

A dynamical system can be forced either by the application of known small periodic forces or by the imposition of known small periodic movements at one or more given points of the system. The method of construction of the equations of motion will be sufficiently illustrated if we assume that a simply sinusoidal force or movement is imposed at a single point Q and in the direction of the fixed reference axis $\boldsymbol{Ox}_1$. Let the Cartesian coordinates of Q, referred to the fixed axes, be $\boldsymbol{x}_1$, $\boldsymbol{x}_2$, $\boldsymbol{x}_3$ at any instant of the motion, and be $\boldsymbol{x}_1^*$, $\boldsymbol{x}_2^*$, $\boldsymbol{x}_3^*$ when the system is in its mean position. Then if the generalised coordinates are measured from the mean position, the first geometrical equation (see § 8·9) yields, to the first order of small quantities, a relation of the type

$$\boldsymbol{x}_1 - \boldsymbol{x}_1^* = \eta_1 q_1 + \eta_2 q_2 + \dots + \eta_m q_m = \eta' q,$$

where the coefficients η_i are constants dependent on the position of Q in the system, and η, q denote column matrices.

Firstly, assume that a known force $\boldsymbol{F}_1 \sin \omega t$ acts at Q in the direction $\boldsymbol{Ox}_1$, and that no constraints are applied at Q in the directions $\boldsymbol{Ox}_2$ and $\boldsymbol{Ox}_3$. Then the corresponding set of generalised forces is

$$P = \boldsymbol{F}_1 \eta \sin \omega t,$$

and the equations of motion are therefore

$$f(D)\, q \equiv (AD^2 + BD + C)\, q = \rho \sin \omega t, \qquad \dots\dots(1)$$

where $\rho = \boldsymbol{F}_1 \eta$.

Next suppose that Q is given a known motion $\boldsymbol{d}_1 \sin \omega t$ in the direction $\boldsymbol{Ox}_1$, and that as before no constraints are imposed in the directions $\boldsymbol{Ox}_2$ and $\boldsymbol{Ox}_3$. Then the generalised forces corresponding to the unknown reaction $\boldsymbol{X}_1$ at Q are given by $P = \boldsymbol{X}_1 \eta$, while on account of the imposed motion at Q the generalised coordinates will be connected by the condition

$$\eta' q = \boldsymbol{d}_1 \sin \omega t. \qquad \dots\dots(2)$$

In this case the differential equations of motion are represented by

$$(AD^2 + BD + C)\, q = \boldsymbol{X}_1 \eta. \qquad \dots\dots(3)$$

† See, for example, Refs. 33, 34, 35. The discussion given in this section is based on Ref. 36.

On elimination of the unknown reaction $\boldsymbol{X}_1$ we obtain $m-1$ differential equations connecting the m generalised coordinates q. These, taken in conjunction with (2), suffice to determine the motion. The elimination can be effected conveniently as follows. Write

$$u = \begin{bmatrix} 1, & -\eta_2/\eta_1, & -\eta_3/\eta_1, & \dots, & -\eta_m/\eta_1 \\ 0, & 1, & 0, & \dots, & 0 \\ 0, & 0, & 1, & \dots, & 0 \\ \dots & \dots & \dots & \dots & \dots \\ 0, & 0, & 0, & \dots, & 1 \end{bmatrix}.$$

Then (2) is readily seen to be contained in the relation

$$q = u\left\{\frac{\boldsymbol{d}_1}{\eta_1}\sin\omega t, q_2, q_3, \dots, q_m\right\}.$$

Substituting this expression for q in (3), and premultiplying the equation throughout by u', we obtain

$$u'(AD^2+BD+C)\,u\left\{\frac{\boldsymbol{d}_1}{\eta_1}\sin\omega t, q_2, q_3, \dots, q_m\right\} = \{\boldsymbol{X}_1\eta_1, 0, \dots, 0\}. \quad \dots\dots(4)$$

The last $m-1$ of the scalar differential equations in (4) are independent of $\boldsymbol{X}_1$, and they involve all the generalised coordinates with the exception of q_1. They can evidently be represented by a matrix equation similar in form to (1), except that in general there are present forcing terms in quadrature with the imposed sinusoidal displacement of Q.

From the preceding discussion it will be clear that both of the cases considered will be covered if we assume the equations of motion to be of the type

$$(AD^2+BD+C)\,q = \rho\sin\omega t + \boldsymbol{\rho}\cos\omega t, \quad \dots\dots(5)$$

where ρ and $\boldsymbol{\rho}$ are columns of assigned constants.

The complete motion which follows any given initial conditions may be regarded as the superposition of a free motion and of a forced motion. If the real part of any root of $\Delta(\lambda)=0$ is positive, the free motion will in time completely mask the forced motion. On the other hand, if the real parts of all the roots are negative, the forced motion will ultimately predominate. In the discussion which follows only the forced motion is considered. For simplicity the roots of $\Delta(\lambda)=0$ are assumed to be all distinct.

Let the forced motion in the coordinate q_s be

$$q_s = R_s\sin(\omega t+\epsilon_s),$$

and write $\sigma \equiv \{R_j e^{i\epsilon_j}\}$ with $\bar{\sigma} \equiv \{R_j e^{-i\epsilon_j}\}$, where i denotes $\sqrt{-1}$. Then, on substituting this trial solution in (5), we see that the constants σ must be such that

$$\left.\begin{aligned} f(i\omega)\,\sigma &= \rho + i\mathfrak{p}, \\ f(-i\omega)\,\bar{\sigma} &= \rho - i\mathfrak{p}. \end{aligned}\right\} \qquad \text{......(6)}$$

Provided $\pm i\omega$ are not roots of $\Delta(\lambda) = 0$, the $2m$ scalar equations contained in (6) give the values of σ and $\bar{\sigma}$ uniquely. On solution of these equations the forced amplitude corresponding to q_s is given by

$$R_s = \text{mod}\,(E_s/\Delta(i\omega)), \qquad \text{......(7)}$$

where E_s is the determinant obtained when the sth column of $\Delta(i\omega)$ is replaced by $\rho + i\mathfrak{p}$. To obtain the expression for the forced amplitude in an explicit form, suppose the expanded form of the determinant $\Delta(\lambda)$ to be

$$p_0\lambda^n + p_1\lambda^{n-1} + \ldots + p_{n-1}\lambda + p_n,$$

where $n = 2m$. Then

$$\Delta(i\omega)\,\Delta(-i\omega) = P_0\Omega^n - P_1\Omega^{n-1} + \ldots - P_{n-1}\Omega + P_n,$$

where $\Omega \equiv \omega^2$, and the coefficients are conveniently expressible as

$$P_0 = p_0^2,$$
$$P_1 = [p_0, p_2]\{p_2, p_0\} - p_1^2,$$
$$P_2 = [p_0, p_2, p_4]\{p_4, p_2, p_0\} - [p_1, p_3]\{p_3, p_1\},$$

and so on. In the same way, if $\bar{E}_s$ denotes the conjugate of E_s, we can write

$$E_s\bar{E}_s = e_0\Omega^{n-2} - e_1\Omega^{n-3} + \ldots - e_{n-3}\Omega + e_{n-2},$$

where the coefficients e are real. Hence

$$\begin{aligned} R_s^2 &\equiv E_s\bar{E}_s/\Delta(i\omega)\,\Delta(-i\omega) \\ &= \frac{e_0\Omega^{n-2} - e_1\Omega^{n-3} + \ldots - e_{n-3}\Omega + e_{n-2}}{P_0\Omega^n - P_1\Omega^{n-1} + \ldots - P_{n-1}\Omega + P_n}. \end{aligned} \qquad \text{......(8)}$$

An alternative expression can be obtained if the roots $\lambda_1, \lambda_2, \ldots, \lambda_n$ of $\Delta(\lambda) = 0$ are known. For then

$$R_s^2 = E_s\bar{E}_s/P_0 \prod_{r=1}^{n} (\omega^2 + \lambda_r^2),$$

and on resolution into partial fractions, this gives

$$R_s^2 = \frac{1}{P_0}\sum_{r=1}^{n} \frac{\alpha_r}{\omega^2 + \lambda_r^2}, \qquad \text{......(9)}$$

where
$$\alpha_r = \frac{e_0(\lambda_r^2)^{n-2} + e_1(\lambda_r^2)^{n-3} + \ldots + e_{n-3}(\lambda_r^2) + e_{n-2}}{\prod_{j \neq r} (\lambda_j^2 - \lambda_r^2)}.$$

Since the expressions on the right of equations (8) and (9) are identical, and since the numerator in (8) is of degree $n-2$ in ω^2 while the denominator is of degree n, it follows that $\Sigma\,\alpha_r = 0$.

If the forcing frequency is regarded as a variable parameter, there will usually be certain values of this frequency for which the forced amplitude of the motion in any given coordinate q_s is stationary. The equation which gives these frequencies can be obtained by differentiation of (8) or (9) with respect to ω and use of the condition $\dfrac{\partial R_s}{\partial \omega} = 0$. Hence

$$\sum_{r=1}^{n} \frac{\alpha_r \omega}{(\omega^2+\lambda_r^2)^2} = 0.$$

This yields, apart from the trivial solutions $\omega = 0$ and $\omega = \infty$,

$$\sum_{r=1}^{n} \alpha_r \prod_{j \neq r} (\omega^2+\lambda_j^2)^2 = 0. \qquad \text{......(10)}$$

Since $\Sigma\,\alpha_r = 0$, equation (10) is in general of degree $4m-3$ in ω^2: only real positive roots ω^2 of this equation yield real values of the frequency. The maxima and the minima of forced amplitude necessarily occur in alternation, and since $\omega = \infty$ corresponds to a minimum ($R_s = 0$), it readily follows that there can be at most $2m-1$ maxima for a system having m degrees of freedom.

If two conjugate roots of $\Delta(\lambda) = 0$, say $\mu_r \pm i\omega_r$, have their real parts μ_r small, it is obvious from (9) that a pronounced peak value of R_s will be obtained for some value of ω close to ω_r. If the roots are actually critical, so that $\mu_r = 0$, R_s becomes infinitely great for $\omega = \omega_r$. This is the familiar phenomenon of *resonance*. For this condition the assumed form of solution fails, and in fact the forced motion is no longer simply sinusoidal but is constructed of constituents of the type $t \cos \omega t$ (see § 5·11 (*a*)). The amplitude of oscillation then grows without limit.

Example

Forced Oscillations of a Model Aeroplane Wing. Consider the forced flexure-torsion oscillations of a cantilever wing. The laws of bending and twisting are assumed to be invariable for the range of airspeeds to be considered, and the wing is accordingly treated as semi-rigid (see § 8·12). Fig. 9·12·1 is a diagram of the system. The generalised coordinates q_1, q_2 are chosen to be respectively the downward linear displacements at the points L, T of the wing tip which lie on the

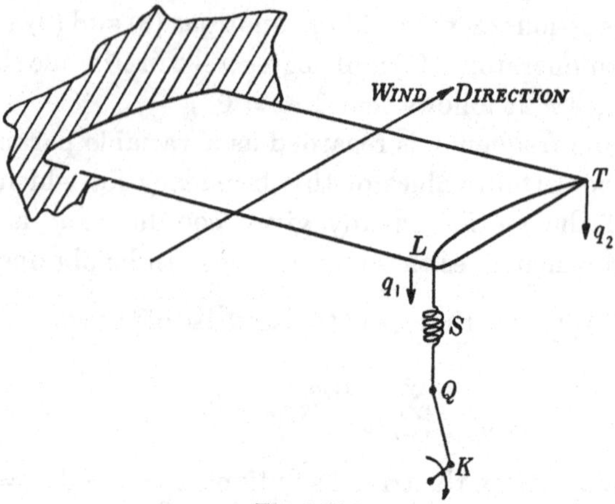

Fig. 9·12·1

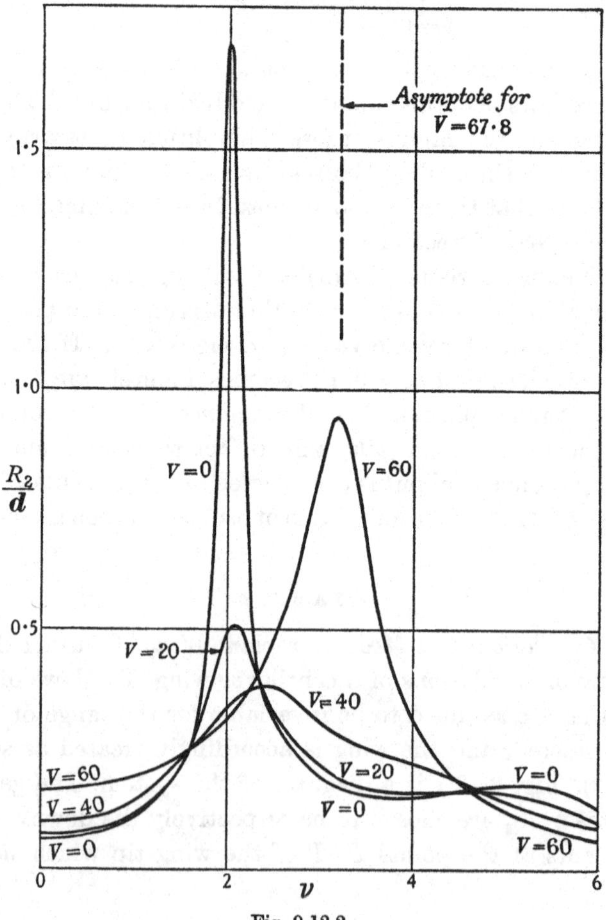

Fig. 9·12·2

leading and trailing edges. It is supposed that the wing is given forced sinusoidal oscillations by means of an oscillator K connected through a spring of stiffness σ_0 to the point L.

If the imposed motion at Q is $\mathbf{d}\sin\omega t$, the dynamical equations in the present simple case are readily shown to have the form

$$(AD^2+BD+C)\{q_1, q_2\} = \mathbf{d}\sigma_0 \sin\omega t\{1, 0\},$$

and the amplitudes R_1, R_2 of the motions in q_1, q_2 are given respectively by

$$\frac{1}{\sigma_0}\left(\frac{R_1}{\mathbf{d}}\right)^2 = \frac{(-A_{22}\omega^2+C_{22})^2+\omega^2 B_{22}^2}{(p_0\omega^4-p_2\omega^2+p_4)^2+\omega^2(p_1\omega^2-p_3)^2}$$

and

$$\frac{1}{\sigma_0}\left(\frac{R_2}{\mathbf{d}}\right)^2 = \frac{(-A_{21}\omega^2+C_{21})^2+\omega^2 B_{21}^2}{(p_0\omega^4-p_2\omega^2+p_4)^2+\omega^2(p_1\omega^2-p_3)^2}.$$

These formulae are applied in Ref. 36 to a light aeroplane wing of span 9 feet and chord 3 feet, for which the numerical data are as follows. The spring stiffness σ_0 is chosen to be 4 pounds per foot, and the terms independent of the airspeed V in the damping coefficients B_{ij} represent an allowance for elastic damping.

Dynamical Coefficients for Light Aeroplane Wing

Coefficient	Value $\times 9^2$	Coefficient	Value $\times 9^2$
A_{11}	5·443	A_{21}	2·907
B_{11}	$0{\cdot}4466V+29$	B_{21}	$-0{\cdot}3456V-9$
C_{11}	$-0{\cdot}633V^2+2502{\cdot}7$	C_{21}	$-0{\cdot}144V^2-614{\cdot}7$
A_{12}	2·907	A_{22}	4·743
B_{12}	$0{\cdot}8076V-9$	B_{22}	$0{\cdot}8064V+9$
C_{12}	$0{\cdot}633V^2-614{\cdot}7$	C_{22}	$0{\cdot}144V^2+1487{\cdot}7$

The results are given in Fig. 9·12·2, which shows the influence of forcing frequency $\nu\equiv\omega/2\pi$ and airspeed V on R_2. The critical speed for flutter of the wing works out at 67·8 feet per second, and the corresponding critical frequency is 3·2 cycles per second. It is seen that R_2 tends to become infinite for this frequency as the critical speed is approached.

CHAPTER X

ITERATIVE NUMERICAL SOLUTIONS OF LINEAR DYNAMICAL PROBLEMS

10·1. Introductory. In the present Chapter the iterative methods described in §§ 4·17 and 4·18 will be applied to obtain approximate solutions of problems relating to the small oscillations of dynamical systems. Iterative methods are particularly advantageous when the number of degrees of freedom is large, since the solution is obtained without expansion of the Lagrangian determinant. Moreover, they often lend themselves well to the approximate treatment of continuous systems such as tapered beams.

Conservative systems are considered in Part I, and the extension to dissipative systems is briefly dealt with in Part II.

PART I. SYSTEMS WITH DAMPING FORCES ABSENT

10·2. Remarks on the Underlying Theory. With a conservative system the Lagrangian equations of free motion have the form (see § 9·3)

$$A\ddot{q} + Eq = 0,$$

in which both the inertia matrix A and the stiffness matrix E are symmetrical. It will be assumed that both A and E are non-singular, as is usually the case. In § 9·9 it has been shown that the typical constituent motion may be taken as

$$q = (e^{\lambda_r t} + e^{-\lambda_r t})\, k_r,$$

and that λ_r^2 and all the elements of the modal column k_r are real. Moreover,

$$(A\lambda_r^2 + E)\, k_r = 0. \qquad \text{......(1)}$$

For the application of the iterative method it is convenient to write $-\lambda^2 = z = \varpi^{-1}$, and to express (1) in the alternative forms

$$(\varpi_r I - U)\, k_r = 0, \qquad \text{......(2)}$$

$$(z_r I - U^{-1})\, k_r = 0, \qquad \text{......(3)}$$

where $U \equiv E^{-1}A$. The matrix U will be spoken of as the *dynamical matrix*.* From (2) and (3) it follows that ϖ_r is a latent root of U, and that z_r is a latent root of U^{-1}.

* Note that this matrix is, in general, not symmetrical.

Equation (2) could be treated in the usual way by the direct solution of the characteristic equation of U. However, the iterative method is often preferable, more especially when the constituents are known to be all oscillatory and only the fundamental mode (which corresponds to the dominant root ϖ_r) is required. If the mode of highest frequency is required, the iterative method should be applied to equation (3). The details of the process of solution are amply explained in the examples* given later, but some preliminary general remarks may be useful.

(*a*) To construct the dynamical matrix U it is often advantageous to form the flexibility matrix $\Phi \equiv E^{-1}$ directly rather than to derive it by inversion of the stiffness matrix. The displacements q due to a set of generalised static loads W are given by

$$q = \Phi W.$$

Hence to find the element Φ_{ij} it is only necessary to obtain the value of q_i when $W_j = 1$ and all the remaining generalised loads are zero. This direct method for the construction of Φ is particularly valuable in problems on the vibrations of beams (see §§ 10·4, 10·6).

(*b*) When merely the fundamental frequency and mode are required, the solution is directly obtained by repeated premultiplications of an arbitrary column by U. If the frequencies and modes in the overtones are required, it is necessary to evaluate modified matrices by the methods of § 4·18, and to use these matrices for the iterations instead of U. The required row κ_r can be found very simply from the corresponding column k_r when the system is conservative. For equation (2) can be written $\Phi A k_r = \varpi_r k_r$, which yields on premultiplication by A and transposition

$$k_r' A \Phi A \equiv k_r' A U = \varpi_r k_r' A.$$

Hence κ_r may be chosen to be $k_r' A$. This conclusion accords with the orthogonal properties of the modes discussed in § 9·11.

(*c*) The work of iteration will be considerably shortened if by experience or intuition an initial column can be chosen which is not very different from the true mode.

(*d*) The case of equal roots, which is somewhat unusual in dynamics, requires special consideration. The form of a high power of a general matrix u which possesses a dominant latent root of multiplicity s has been given in § 4·15, and it is easy to see how such a root may be

* The examples in §§ 10·3, 10·4 and 10·6 are taken from Ref. 10, and that in § 10·11 from Ref. 11.

obtained from $u^m x_0$, where x_0 is arbitrary. However, in the case of the dynamical matrix U of a conservative system, there is a considerable simplification. It can readily be deduced from the results of § 9·9 that if ϖ_s is a latent root of U of multiplicity s, the derived adjoint matrices of $\varpi_s I - U$ up to $\overset{(s-2)}{F}(\varpi_s)$ all vanish. Hence the only term arising from the multiple root ϖ_s in Sylvester's expansion of U^m is proportional to $\varpi_s^m \overset{(s-1)}{F}(\varpi_s)$. Accordingly, the usual iterative method gives ϖ_s as though it were a simple root, and the s modes are proportional to the s linearly independent columns of $\overset{(s-1)}{F}(\varpi_s)$. If s different arbitrary columns are repeatedly premultiplied by U, s different possible modes will, in general, be obtained. It may also be noted that the canonical form of U is diagonal (see example (iii), § 9·10).

(*e*) When an accurate value of the fundamental frequency is required, but no great interest attaches to the mode, an application of Rayleigh's principle (see (9·11·9)) may be found advantageous. The principle states that small changes in the mode do not affect the frequency. Hence an approximate mode, given for instance by an uncompleted application of the iterative method, can be used to deduce a relatively accurate frequency. When the equations of motion are in the form (1), equation (9·11·10) may be applied. On the other hand, if the iterative method is being used, a convenient modification is obtained by premultiplication of (2) by k_r'. This yields

$$\varpi_r = k_r' U k_r / k_r' k_r. \quad \text{......(4)}$$

An illustration is given in § 10·4.

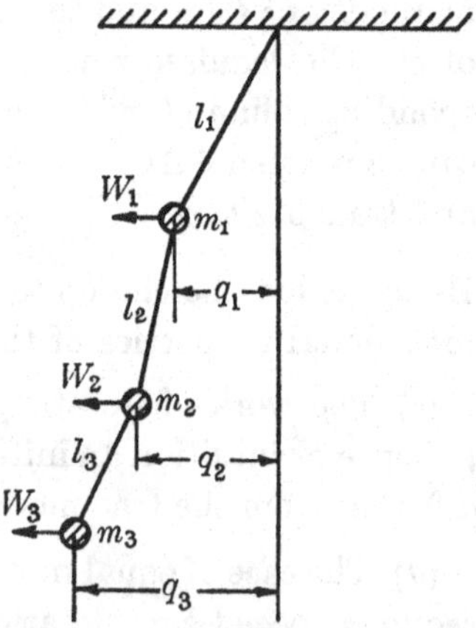

Fig. 10·3·1

10·3. Example No. 1: Oscillations of a Triple Pendulum.

To provide a simple example we shall first consider the oscillations of a triple pendulum under gravity in a vertical plane (see Fig. 10·3·1). The three coordinates q_1, q_2, q_3 will be taken to be the small horizontal displacements of the masses m_1, m_2, m_3, respectively, from the equilibrium position. Let unit force be applied horizontally to m_1; then the three masses will each be displaced a distance $a \equiv l_1/g(m_1+m_2+m_3)$, so that $\{a, a, a\} = \Phi\{1, 0, 0\} = \{\Phi_{11}, \Phi_{21}, \Phi_{31}\}$. When unit force is applied horizontally to m_2, m_1 will again be displaced a distance a, but m_2 and m_3 will each move a distance $a+b$, where $b \equiv l_2/g(m_2+m_3)$. In this manner the matrix Φ is readily found to be

$$\Phi = \begin{bmatrix} a, & a, & a \\ a, & a+b, & a+b \\ a, & a+b, & a+b+c \end{bmatrix},$$

where $c \equiv l_3/gm_3$.

If W denotes a set of horizontal forces applied to the masses, the displacements are given by $q = \Phi W$. Suppose now that W represents the inertia forces. Then

$$\begin{bmatrix} q_1 \\ q_2 \\ q_3 \end{bmatrix} = \begin{bmatrix} a, & a, & a \\ a, & a+b, & a+b \\ a, & a+b, & a+b+c \end{bmatrix} \begin{bmatrix} -m_1\ddot{q}_1 \\ -m_2\ddot{q}_2 \\ -m_3\ddot{q}_3 \end{bmatrix}$$

$$= -\begin{bmatrix} m_1 a, & m_2 a, & m_3 a \\ m_1 a, & m_2(a+b), & m_3(a+b) \\ m_1 a, & m_2(a+b), & m_3(a+b+c) \end{bmatrix} \begin{bmatrix} \ddot{q}_1 \\ \ddot{q}_2 \\ \ddot{q}_3 \end{bmatrix}.$$

The dynamical equations have here been constructed directly in the form $q = -U\ddot{q}$.

When the system is oscillating in a single mode, we may take q proportional to $k_r e^{\lambda_r t}$, and then

$$(I + \lambda_r^2 U) k_r = 0,$$

or

$$(\varpi_r I - U) k_r = 0. \qquad \text{......(1)}$$

As a numerical example suppose that $m_1 = m_2 = m_3 = m$ and $l_1 = l_2 = l_3 = l$. Then

$$U = \frac{l}{6g} \begin{bmatrix} 2 & 2 & 2 \\ 2 & 5 & 5 \\ 2 & 5 & 11 \end{bmatrix}.$$

If the scalar factor is absorbed into ϖ, so that now ϖ is taken to mean $-6g/\lambda^2 l$, equation (1) becomes

$$\varpi k = uk,$$

where u is the numerical part of U. This last equation is readily solved by the method of iteration. The dominant value of ϖ and the associated column k, which correspond to the fundamental oscillation, are quickly obtained when an arbitrary column, say $\{1, 1, 1\}$, is repeatedly premultiplied by u. Thus

$$u\{1, 1, 1\} = \{6, 12, 18\} = 18\{0{\cdot}3, 0{\cdot}6, 1\},$$
$$u\{0{\cdot}3, 0{\cdot}6, 1\} = \{4, 9, 15\} = 15\{0{\cdot}26, 0{\cdot}6, 1\},$$
$$u\{0{\cdot}26, 0{\cdot}6, 1\} = 14{\cdot}53\{0{\cdot}25688, 0{\cdot}58716, 1\},$$

and after nine steps in all,

$$u\{0{\cdot}254885, 0{\cdot}584225, 1\} = 14{\cdot}4309\{0{\cdot}254885, 0{\cdot}584225, 1\}.$$

The fundamental frequency is given by $\frac{1}{2\pi}\sqrt{\frac{6g}{l\varpi_1}}$, and a close approximation to its value is thus

$$\frac{1}{2\pi}\sqrt{\frac{6g}{14{\cdot}4309l}} = 0{\cdot}102624\sqrt{\frac{g}{l}}.$$

Moreover, the corresponding modal column k_1 is proportional to $\{0{\cdot}254885, 0{\cdot}584225, 1\}$. It will be observed that at each step in the iteration a homologous element in the column is for convenience reduced to unity.

If the overtones are required, the methods of § 4·18 may be applied. It is necessary in the first place to find the row κ_1, where $\varpi_1\kappa_1 = \kappa_1 u$. Since A is here a scalar matrix, we deduce at once that (see § 10·2 (*b*))

$$\kappa_1 = k_1' = [0{\cdot}254885, 0{\cdot}584225, 1].$$

Now in any oscillation from which the fundamental is absent $\kappa_1 q = 0$. Hence we may write

$$0{\cdot}254885q_1 + 0{\cdot}584225q_2 + q_3 = 0,$$

or

$$q_1 = -2{\cdot}29211q_2 - 3{\cdot}92334q_3. \qquad \text{......(2)}$$

Accordingly, when the fundamental is absent

$$\begin{bmatrix} q_1 \\ q_2 \\ q_3 \end{bmatrix} = \begin{bmatrix} 0 & -2{\cdot}29211 & -3{\cdot}92334 \\ 0 & 1 & 0 \\ 0 & 0 & 1 \end{bmatrix} \begin{bmatrix} q_1 \\ q_2 \\ q_3 \end{bmatrix}.$$

Substitution in the right-hand side of the equation $\varpi q = uq$ yields $\varpi q = vq$, where

$$v = \begin{bmatrix} 2 & 2 & 2 \\ 2 & 5 & 5 \\ 2 & 5 & 11 \end{bmatrix} \begin{bmatrix} 0 & -2{\cdot}29211 & -3{\cdot}92334 \\ 0 & 1 & 0 \\ 0 & 0 & 1 \end{bmatrix}$$
$$= \begin{bmatrix} 0, & -2{\cdot}58422, & -5{\cdot}84668 \\ 0, & 0{\cdot}41578, & -2{\cdot}84668 \\ 0, & 0{\cdot}41578, & 3{\cdot}15332 \end{bmatrix}.$$

This matrix is, of course, identical with that given by the formula (4·18·3) with $r = 1$, namely $v = u - \frac{1}{\kappa_{11}}\{u_{i1}\}\kappa_1$.

We now proceed by the iterative method, starting with an arbitrary column, say $\{1, 1, 1\}$. In the process of approximation it is evidently unnecessary to compute the leading element of any column, since this is always multiplied by a cipher in the succeeding step. In this way we find

$$v\{1,1,1\} = \{—, -2{\cdot}4309, 3{\cdot}5691\} = 3{\cdot}5691\{—, -0{\cdot}68110, 1\},$$

$$v\{—, -0{\cdot}68110, 1\} = \{—, -3{\cdot}1299, 2{\cdot}8701\}$$
$$= 2{\cdot}8701\{—, -1{\cdot}09049, 1\},$$

and after fifteen approximations the column repeats itself, the scalar factor being 2·6152. A computation of the leading element then gives

$$v\{-0{\cdot}95670, -1{\cdot}29429, 1\} = 2{\cdot}6152\{-0{\cdot}95670, -1{\cdot}29429, 1\}.$$

Hence in the first overtone the mode is $\{-0{\cdot}95670, -1{\cdot}29429, 1\}$, and the frequency is $\frac{1}{2\pi}\sqrt{\frac{6g}{2{\cdot}6152l}} = 0{\cdot}24107\sqrt{\frac{g}{l}}$.

To determine the second overtone we note that the condition for absence of the first overtone is $k_2'Aq = 0$, or

$$[-0{\cdot}95670, -1{\cdot}29429, 1]q = 0,$$

whence
$$q_1 = -1{\cdot}35287q_2 + 1{\cdot}04526q_3. \qquad \text{......(3)}$$

Elimination of q_1 between (2) and (3) yields

$$0 = 0{\cdot}93924q_2 + 4{\cdot}96860q_3,$$

or*
$$q_2 = -5{\cdot}2900q_3.$$

Hence for a motion in which the fundamental and the first overtone are both absent we may write

$$\begin{bmatrix} q_1 \\ q_2 \\ q_3 \end{bmatrix} = \begin{bmatrix} 1, & 0, & 0 \\ 0, & 0, & -5{\cdot}2900 \\ 0, & 0, & 1 \end{bmatrix} \begin{bmatrix} q_1 \\ q_2 \\ q_3 \end{bmatrix},$$

* This result could also have been obtained by repeated postmultiplication of an arbitrary row by v, which yields $[0, 1, 5{\cdot}2899]v = 2{\cdot}6152[0, 1, 5{\cdot}2899]$: see also footnote to § 4·18.

and making this substitution on the right-hand side of the equation $\varpi q = vq$, we obtain $\varpi q = wq$, where

$$w = \begin{bmatrix} 0, & -2{\cdot}58422, & -5{\cdot}84668 \\ 0, & 0{\cdot}41578, & -2{\cdot}84668 \\ 0, & 0{\cdot}41578, & 3{\cdot}15332 \end{bmatrix} \begin{bmatrix} 1, & 0, & 0 \\ 0, & 0, & -5{\cdot}2900 \\ 0, & 0, & 1 \end{bmatrix}$$

$$= \begin{bmatrix} 0, & 0, & 7{\cdot}8238 \\ 0, & 0, & -5{\cdot}0461 \\ 0, & 0, & 0{\cdot}9539 \end{bmatrix}.$$

The iterative process here yields, in one step, the result

$$w\{8{\cdot}2019, -5{\cdot}2900, 1\} = 0{\cdot}9539\{8{\cdot}2019, -5{\cdot}2900, 1\}.$$

The mode is thus proportional to $\{8{\cdot}2019, -5{\cdot}2900, 1\}$ and the frequency is found to be $0{\cdot}39916\sqrt{\frac{g}{l}}$.

The second overtone could alternatively have been deduced directly from the inverse of the matrix u. It is readily found that

$$u^{-1} = \tfrac{1}{6}\begin{bmatrix} 5 & -2 & 0 \\ -2 & 3 & -1 \\ 0 & -1 & 1 \end{bmatrix},$$

and the iterative process when applied to this matrix yields the result

$$u^{-1}\{8{\cdot}20180, -5{\cdot}28994, 1\} = 1{\cdot}048323\{8{\cdot}20180, -5{\cdot}28994, 1\}.$$

The mode agrees well with that found previously, while the frequency works out as

$$\frac{1}{2\pi}\sqrt{\frac{6 \times 1{\cdot}048323g}{l}} = 0{\cdot}399157\sqrt{\frac{g}{l}}.$$

10·4. Example No. 2: Torsional Oscillations of a Uniform Cantilever. An illustration will next be given of the use of the method in relation to continuous systems. As an approximation the given system is replaced by a finite system of rigid or semi-rigid* units, suitably interconnected. In order that this finite system may reproduce closely the behaviour of the continuous system, the number of such units must usually be large.

Suppose the given uniform cantilever to be divided into ten equal sections. These sections are assumed to be rigid, and to be interconnected by torsion springs, the stiffnesses of which are so chosen that

* For definition of "semi-rigid" systems see § 8·12.

the torsional stiffness for the complete finite system measured at the middle of each section agrees with that for the continuous beam at the corresponding position. The coordinates q are taken to be the angular displacements of the segments from the equilibrium positions. Let s be the span of the beam, C the torsional stiffness per unit length and P the polar moment of inertia for unit length.

The flexibility matrix is easily found to be

$$\Phi = \left(\frac{s}{C}\right)\begin{bmatrix} 0{\cdot}05 & 0{\cdot}05 & 0{\cdot}05 & 0{\cdot}05 & 0{\cdot}05 & 0{\cdot}05 & 0{\cdot}05 & 0{\cdot}05 & 0{\cdot}05 & 0{\cdot}05 \\ 0{\cdot}05 & 0{\cdot}15 & 0{\cdot}15 & 0{\cdot}15 & 0{\cdot}15 & 0{\cdot}15 & 0{\cdot}15 & 0{\cdot}15 & 0{\cdot}15 & 0{\cdot}15 \\ 0{\cdot}05 & 0{\cdot}15 & 0{\cdot}25 & 0{\cdot}25 & 0{\cdot}25 & 0{\cdot}25 & 0{\cdot}25 & 0{\cdot}25 & 0{\cdot}25 & 0{\cdot}25 \\ 0{\cdot}05 & 0{\cdot}15 & 0{\cdot}25 & 0{\cdot}35 & 0{\cdot}35 & 0{\cdot}35 & 0{\cdot}35 & 0{\cdot}35 & 0{\cdot}35 & 0{\cdot}35 \\ 0{\cdot}05 & 0{\cdot}15 & 0{\cdot}25 & 0{\cdot}35 & 0{\cdot}45 & 0{\cdot}45 & 0{\cdot}45 & 0{\cdot}45 & 0{\cdot}45 & 0{\cdot}45 \\ 0{\cdot}05 & 0{\cdot}15 & 0{\cdot}25 & 0{\cdot}35 & 0{\cdot}45 & 0{\cdot}55 & 0{\cdot}55 & 0{\cdot}55 & 0{\cdot}55 & 0{\cdot}55 \\ 0{\cdot}05 & 0{\cdot}15 & 0{\cdot}25 & 0{\cdot}35 & 0{\cdot}45 & 0{\cdot}55 & 0{\cdot}65 & 0{\cdot}65 & 0{\cdot}65 & 0{\cdot}65 \\ 0{\cdot}05 & 0{\cdot}15 & 0{\cdot}25 & 0{\cdot}35 & 0{\cdot}45 & 0{\cdot}55 & 0{\cdot}65 & 0{\cdot}75 & 0{\cdot}75 & 0{\cdot}75 \\ 0{\cdot}05 & 0{\cdot}15 & 0{\cdot}25 & 0{\cdot}35 & 0{\cdot}45 & 0{\cdot}55 & 0{\cdot}65 & 0{\cdot}75 & 0{\cdot}85 & 0{\cdot}85 \\ 0{\cdot}05 & 0{\cdot}15 & 0{\cdot}25 & 0{\cdot}35 & 0{\cdot}45 & 0{\cdot}55 & 0{\cdot}65 & 0{\cdot}75 & 0{\cdot}85 & 0{\cdot}95 \end{bmatrix}$$

and the inertia matrix is evidently $0{\cdot}1sP$ times the unit matrix I_{10}. The dynamical matrix U is thus simply the numerical matrix $u \equiv \left(\frac{C}{s}\right)\Phi$ multiplied by the scalar $0{\cdot}1s^2P/C$. Hence the iterative process, when applied to the matrix u, will yield the dominant value of

$$\varpi = -10C/s^2\lambda^2P.$$

To illustrate the convergence, the initial mode chosen in Table 10·4·1 differs widely from the true fundamental mode. For the purpose of comparison, the exact solution for the continuous beam is tabulated beside the solution obtained for the segmented beam.

Rayleigh's principle will next be applied to determine the frequency from an approximate mode. Assume the third column of Table 10·4·1 to be the approximate mode k_r: then the fourth column is Uk_r. Hence

$$k_r' Uk_r = (0{\cdot}0752 \times 0{\cdot}3139) + (0{\cdot}2241 \times 0{\cdot}9342) + \ldots + (1{\cdot}0 \times 4{\cdot}0100),$$

and
$$k_r' k_r = 0{\cdot}0752^2 + 0{\cdot}2241^2 + \ldots + 1{\cdot}0^2.$$

Accordingly

$$\varpi_r = k_r' Uk_r / k_r' k_r = 19{\cdot}8630/4{\cdot}8916 = 4{\cdot}0606.$$

This value for ϖ_r is identical with that found in the eighth column, and is very accurate for the segmented beam.

Table 10·4·1

Fundamental Torsional Oscillation of a Uniform Cantilever

Initial column	Iteration number								Exact solution for continuous beam
	1		2		3		4		
0·05	0·25	0·0752	0·3139	0·0783	0·3190	0·0786	0·3196	0·0787	0·0787
0·15	0·745	0·2241	0·9342	0·2330	0·9492	0·2340	0·9508	0·2342	0·2342
0·25	1·225	0·3684	1·5321	0·3821	1·5562	0·3837	1·5587	0·3839	0·3839
0·35	1·68	0·5053	2·0932	0·5220	2·1249	0·5239	2·1281	0·5241	0·5241
0·45	2·10	0·6316	2·6037	0·6493	2·6414	0·6512	2·6452	0·6514	0·6515
0·55	2·475	0·7444	3·0511	0·7609	3·0930	0·7626	3·0972	0·7627	0·7628
0·65	2·795	0·8406	3·4240	0·8539	3·4685	0·8551	3·4729	0·8553	0·8553
0·75	3·05	0·9173	3·7129	0·9259	3·7586	0·9267	3·7631	0·9267	0·9267
0·85	3·23	0·9714	3·9100	0·9751	3·9561	0·9753	3·9606	0·9754	0·9754
0·95	3·325	1·0	4·0100	1·0	4·0561	1·0	4·0606	1·0	1·0
ϖ	$\frac{3\cdot325}{0\cdot95} = 3\cdot50$		4·010		4·056		4·061		4·053

10·5. Example No. 3: Torsional Oscillations of a Multi-Cylinder Engine.

In this example a multi-cylinder engine is supposed to be coupled to an airscrew of very great inertia. The rate of revolution of the airscrew is assumed constant and the system can therefore be treated as though the airscrew were fixed. The problem is now very similar to that dealt with in § 10·4 except that the stiffnesses and inertias of the segments may be variable along the shaft.

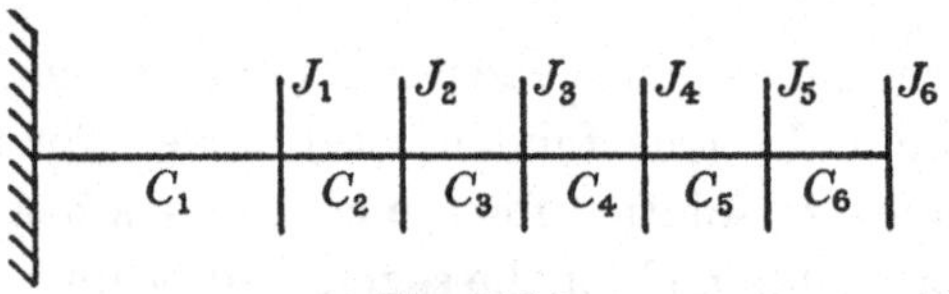

Fig. 10·5·1

Let the moments of inertia of the rotating masses, augmented by an allowance for the reciprocating masses,* be $J_1, J_2, \ldots, J_n$, and let the stiffnesses of the sections of the crankshaft be $C_1, C_2, \ldots, C_n$, as shown in Fig. 10·5·1 for the case $n = 6$. The n coordinates q are again chosen to be the angular displacements from the equilibrium positions. Suppose unit torque to be applied at J_1; then $q_1 = q_2 = \ldots = q_n = 1/C_1 = f_1$, say. If unit torque is applied at J_2 the displacements are $q_1 = f_1$ and

* This allowance corresponds to half of the reciprocating mass supposed situated at the centre of the crankpin.

$q_2 = q_3 = \ldots = q_n = f_1+f_2$, where $f_2 = 1/C_2$. Proceeding in this way we find that

$$\Phi = \begin{bmatrix} f_1, & f_1, & f_1, & \ldots, & f_1 \\ f_1, & f_1+f_2, & f_1+f_2, & \ldots, & f_1+f_2 \\ \cdots & \cdots & \cdots & \cdots & \cdots \\ f_1, & f_1+f_2, & f_1+f_2+f_3, & \ldots, & f_1+f_2+\ldots+f_n \end{bmatrix},$$

which gives

$$U = \begin{bmatrix} J_1f_1, & J_2f_1, & J_3f_1, & \ldots, & J_nf_1 \\ J_1f_1, & J_2(f_1+f_2), & J_3(f_1+f_2), & \ldots, & J_n(f_1+f_2) \\ \cdots & \cdots & \cdots & \cdots & \cdots \\ J_1f_1, & J_2(f_1+f_2), & J_3(f_1+f_2+f_3), & \ldots, & J_n(f_1+f_2+\ldots+f_n) \end{bmatrix}.$$

As a numerical example assume $n = 6$; $J_1 = J_2 = \ldots = J_6 = J$; and $C_2 = C_3 = \ldots = C_6 = C$, with $C_1 = 0{\cdot}5C$. Then

$$U = \frac{J}{C}\begin{bmatrix} 2 & 2 & 2 & 2 & 2 & 2 \\ 2 & 3 & 3 & 3 & 3 & 3 \\ 2 & 3 & 4 & 4 & 4 & 4 \\ 2 & 3 & 4 & 5 & 5 & 5 \\ 2 & 3 & 4 & 5 & 6 & 6 \\ 2 & 3 & 4 & 5 & 6 & 7 \end{bmatrix} = \frac{J}{C}u, \text{ say.}$$

Table 10·5·1

Fundamental Torsional Oscillation of a Six-Cylinder Engine

Initial column	Iteration number							
	1		2		3		4	
0	2	0·29	7·72	0·389	8·980	0·4014	9·1046	0·40285
0	3	0·43	11·29	0·568	13·081	0·5848	13·2555	0·58652
0	4	0·57	14·43	0·727	16·614	0·7427	16·8216	0·74431
0	5	0·71	17·00	0·856	19·420	0·8681	19·6450	0·86924
0	6	0·86	18·86	0·950	21·370	0·9553	21·6003	0·95575
1	7	1·0	19·86	1·0	22·370	1·0	22·6003	1·0

Iteration number					
5		6		7	
9·11734	0·40301	9·11870	0·40303	9·11884	0·40303
13·27316	0·58671	13·27504	0·58673	13·27523	0·58673
16·84246	0·74448	16·84467	0·74450	16·84489	0·74450
19·66745	0·86935	19·66982	0·86936	19·67005	0·86936
21·62320	0·95580	21·62562	0·95580	21·62585	0·95580
22·62320	1·0	22·62562	1·0	22·62585	1·0

Hence if $\varpi = -\frac{C}{\lambda^2 J}$ the equation to be solved is $\varpi q = uq$. The computations in the iterative process, with $\{0, 0, 0, 0, 0, 1\}$ chosen as the initial column, are summarised in Table 10·5·1.

It will be seen that seven iterations are sufficient to determine the mode to five significant figures, and the value of ϖ is then 22·626.

In a treatment of the same problem by Carter* an equation equivalent to the characteristic equation of u is given with certain terms kept general. With the data chosen for the present example his equation can be written

$$\varpi^6 - 27\varpi^5 + 105\varpi^4 - 140\varpi^3 + 81\varpi^2 - 21\varpi + 2 = 0,$$

and it can be verified that the dominant root is the value of ϖ just computed by the iterative process.

10·6. Example No. 4: Flexural Oscillations of a Tapered Beam. We shall next find the fundamental mode and frequency of flexural oscillation of a cantilever beam which closely resembles an untwisted airscrew blade. The beam chosen is such that the differential equation governing the flexural oscillations has a known exact solution, which will be compared later with the solution given by the iterative method. The symbols to be used are as follows:

A = sectional area at a current point,
B = flexural rigidity at a current point,
σ = constant density of the material,
s = span of the beam,
ηs = distance of a current point from the root,
q = amplitude of oscillation,
A_0, B_0 = values of A, B at root.

The beam is specified by the equations

$$A = A_0(1-\eta),$$

$$B = B_0 \frac{(1-\eta)^2(184+258\eta+222\eta^2+76\eta^3-75\eta^4)}{184(1+15\eta)}.$$

The displacement in the true fundamental mode is proportional to

$$6\eta^2 + 28\eta^3 - 15\eta^4,$$

and the corresponding frequency is given by

$$-\lambda^2 = 13{\cdot}6956\left(\frac{B_0}{s^4 \sigma A_0}\right).$$

* Ref. 37.

For the iterative treatment the beam is supposed to be divided into ten segments of equal length. Each segment is replaced by a particle of mass equal to that of the segment and situated near the appropriate centroid, the position being chosen to provide round numbers for the fraction of the span, namely 0·05, 0·15, 0·25, 0·35, 0·45, 0·55, 0·65, 0·75, 0·84, 0·93. The particles are assumed to be interconnected by springs in such a manner that their displacements under a system of static loads are identical with the displacements of the continuous beam at the corresponding positions under the same conditions of loading. Since the flexibility matrix is symmetrical, it is sufficient to consider the case where the load is external to the point at which the displacement is measured. Let unit load be applied at a point L distant $s\eta_L$ from the root; then the bending moment at a current point P distant $s\eta$ from the root (where $\eta < \eta_L$) is $s(\eta_L - \eta)$. The curvature at P is therefore $s(\eta_L - \eta)/B$. The bending of an element of length $s\,d\eta$ at P causes a rotation of the part of the beam external to P of amount $s^2(\eta_L - \eta)\,d\eta/B$, and the resulting displacement of a point D distant $s\eta_D$ from the root is $s^3(\eta_L - \eta)(\eta_D - \eta)\,d\eta/B$. The total displacement at D due to the unit load at L is thus

$$q = \int_0^{\eta_D} s^3(\eta_L - \eta)(\eta_D - \eta)\,d\eta/B,$$

where $\eta < \eta_D < \eta_L$. This integral is computed with the aid of Simpson's rule to give the flexibility coefficients.

The mass of any segment is given at once by $\int \sigma A s\,d\eta$ taken over the segment, and the inertia matrix consists simply of the diagonal matrix of the masses. The dynamical matrix is finally given by

$$U = \frac{s^4 \sigma A_0}{10^3 B_0} u,$$

where u is the numerical matrix

$$\begin{bmatrix}
0\cdot0048 & 0\cdot0179 & 0\cdot0278 & 0\cdot0338 & 0\cdot0374 & 0\cdot0378 & 0\cdot0350 & 0\cdot0290 & 0\cdot0195 & 0\cdot0072 \\
0\cdot0200 & 0\cdot1539 & 0\cdot2895 & 0\cdot3848 & 0\cdot4383 & 0\cdot4504 & 0\cdot4224 & 0\cdot3530 & 0\cdot2396 & 0\cdot0890 \\
0\cdot0351 & 0\cdot3281 & 0\cdot8018 & 1\cdot1895 & 1\cdot4256 & 1\cdot5089 & 1\cdot4399 & 1\cdot2190 & 0\cdot8340 & 0\cdot3122 \\
0\cdot0494 & 0\cdot5032 & 1\cdot3725 & 2\cdot3550 & 3\cdot0453 & 3\cdot3530 & 3\cdot2778 & 2\cdot8195 & 1\cdot9500 & 0\cdot7361 \\
0\cdot0646 & 0\cdot6774 & 1\cdot9440 & 3\cdot5990 & 5\cdot1304 & 5\cdot9868 & 6\cdot0473 & 5\cdot3138 & 3\cdot7247 & 1\cdot4204 \\
0\cdot0798 & 0\cdot8509 & 2\cdot5148 & 4\cdot8431 & 7\cdot3172 & 9\cdot1962 & 9\cdot7464 & 8\cdot8148 & 6\cdot2893 & 2\cdot4299 \\
0\cdot0950 & 1\cdot0260 & 3\cdot0855 & 6\cdot0872 & 9\cdot5029 & 12\cdot5311 & 14\cdot1421 & 13\cdot3540 & 9\cdot7689 & 3\cdot8417 \\
0\cdot1102 & 1\cdot2002 & 3\cdot6570 & 7\cdot3307 & 11\cdot6902 & 15\cdot8666 & 18\cdot6956 & 18\cdot7320 & 14\cdot2533 & 5\cdot7557 \\
0\cdot1235 & 1\cdot3574 & 4\cdot1700 & 8\cdot4500 & 13\cdot6570 & 18\cdot8680 & 22\cdot7941 & 23\cdot7555 & 19\cdot0581 & 8\cdot0196 \\
0\cdot1368 & 1\cdot5139 & 4\cdot6838 & 9\cdot5700 & 15\cdot6250 & 21\cdot8691 & 26\cdot8919 & 28\cdot7782 & 24\cdot0587 & 10\cdot8837
\end{bmatrix}$$

If $\varpi = -10^3 B_0/s^4 \sigma A_0 \lambda^2$, the equation for solution is $\varpi q = uq$, in which q here represents the column of the displacements of the ten particles.

The solution by iteration is shown in Table 10·6·1, where the arbitrary initial mode is chosen to be parabolic. Since the amplitudes near the tip are liable to rather larger errors than elsewhere for highly tapered beams, the amplitude at 0·75 of the span is chosen as the standard of reference, and is reduced to unity at each step. The fundamental mode and the corresponding value of ϖ are determined in five steps, and compare well with the exact solution for the continuous beam.

Table 10·6·1

Fundamental Flexural Oscillation of a Tapered Beam

η	Initial column	Iteration number			
		1		2	
0·05	$0{\cdot}00\dot{4}$	0·1358	0·00183	0·1288	0·00177
0·15	0·040	1·5916	0·02143	1·5502	0·02142
0·25	$0{\cdot}11\dot{1}$	5·5129	0·07420	5·2631	0·07271
0·35	$0{\cdot}21\dot{7}$	12·4446	0·16749	11·9308	0·16481
0·45	0·360	22·8059	0·30696	21·9603	0·30325
0·55	$0{\cdot}53\dot{7}$	36·7092	0·49410	35·5004	0·49037
0·65	$0{\cdot}75\dot{1}$	54·0074	0·72692	52·4388	0·72436
0·75	1·0000	74·2962	1·00000	72·3938	1·00000
0·84	1·2544	94·5355	1·27243	92·3589	1·27578
0·93	1·5376	115·9423	1·56055	113·5088	1·56794
ϖ		74·296		72·394	

Iteration number						Exact solution for continuous beam
3		4		5		
0·1284	0·00177	0·1284	0·00177	0·1284	0·00177	0·00177
1·5458	0·02138	1·5455	0·02138	1·5455	0·02138	0·02126
5·2493	0·07260	5·2483	0·07259	5·2482	0·07259	0·07221
11·9025	0·16460	11·9004	0·16459	11·9003	0·16459	0·16381
21·9144	0·30307	21·9110	0·30304	21·9109	0·30304	0·30182
35·4370	0·49007	35·4325	0·49005	35·4324	0·49004	0·48853
52·3618	0·72413	52·3565	0·72412	52·3564	0·72411	0·72279
72·3097	1·00000	72·3043	1·00000	72·3044	1·00000	1·00000
92·2751	1·27610	92·2702	1·27615	92·2705	1·27614	1·27964
113·4305	1·56868	113·4266	1·56875	113·4272	1·56875	1·57935
72·310		72·304		72·304		73·016

10·7. Example No. 5: Symmetrical Vibrations of an Annular Membrane. The transverse vibrations of an annular membrane have already been discussed in example (iii) of § 7·10 and in § 7·15. The same problem will now be solved by the iterative method.

If the annulus is bounded by the radii r_1, r_2, and if unit load is distributed evenly round a circle of radius r, then for $\rho \leqslant r$ the small displacement q of a circle of radius ρ is readily shown to be

$$q = \frac{\log (r_2/r) \log (\rho/r_1)}{2\pi T \log (r_2/r_1)},$$

where T is the tension per unit length of the membrane. In the problem previously considered $r_1 = 1$, $r_2 = 2$: hence

$$q = \frac{\log (2/r) \log \rho}{2\pi T \log 2} \quad (\rho \leqslant r). \qquad \text{......(1)}$$

Suppose the membrane to be divided into ten annuli of equal width, having mean radii 1·05, 1·15, 1·25, etc. Let each annulus be replaced by a massive circular ring at the mean radius of the annulus, and let these rings be interconnected by springs in such a manner that the displacements under a static load system equal the displacements of the continuous membrane under the same load system. The flexibility matrix is then determinable from the equation (1): the case $\rho \leqslant r$ need only be considered since the matrix is symmetrical. The inertia matrix is again the diagonal matrix of the masses and is easily found. The final equation is $\varpi p = uq$, in which $u \times 10^3$ is the matrix

$$\begin{bmatrix}
6·226 & 5·856 & 5·406 & 4·883 & 4·291 & 3·636 & 2·921 & 2·150 & 1·327 & 0·454 \\
5·347 & 16·776 & 15·487 & 13·987 & 12·292 & 10·415 & 8·367 & 6·160 & 3·802 & 1·301 \\
4·541 & 14·248 & 24·727 & 22·332 & 19·625 & 16·628 & 13·359 & 9·835 & 6·070 & 2·078 \\
3·798 & 11·915 & 20·678 & 30·034 & 26·394 & 22·363 & 17·967 & 13·227 & 8·164 & 2·794 \\
3·107 & 9·748 & 16·918 & 24·574 & 32·679 & 27·688 & 22·245 & 16·377 & 10·108 & 3·460 \\
2·463 & 7·727 & 13·410 & 19·477 & 25·902 & 32·658 & 26·237 & 19·316 & 11·922 & 4·081 \\
1·859 & 5·832 & 10·121 & 14·700 & 19·548 & 24·647 & 29·980 & 22·072 & 13·623 & 4·663 \\
1·290 & 4·048 & 7·025 & 10·204 & 13·569 & 17·108 & 20·810 & 24·665 & 15·223 & 5·211 \\
0·753 & 2·363 & 4·102 & 5·957 & 7·922 & 9·989 & 12·150 & 14·400 & 16·735 & 5·728 \\
0·245 & 0·768 & 1·332 & 1·935 & 2·573 & 3·244 & 3·946 & 4·677 & 5·435 & 6·219
\end{bmatrix}$$

and

$$\varpi = -\frac{10}{\lambda^2}\left(\frac{T}{m}\right)\frac{\log_{10} 2}{\log_e 10},$$

where m denotes the mass per unit area of the membrane.

In Table 10·7·1 the initial arbitrary column chosen is parabolic, and yields zero displacement at the edges of the membrane. The fundamental mode is proportional to the last column in the Table, and the corresponding value of ϖ is 0·13514. To provide a comparison with the results of §§ 7·10 and 7·15, make the substitution

$$K = \frac{10 \log_{10} 2}{\varpi \log_e 10}.$$

Then the value of K works out as 9·674, in comparison with the figure 9·753 obtained from tables of Bessel functions.

Table 10·7·1

Fundamental Transverse Vibration of an Annular Membrane

Initial column	Iteration number 1		Iteration number 2		Iteration number 3	
	$\times 10^{-3}$		$\times 10^{-3}$		$\times 10^{-3}$	
0·19	25·85	0·184	25·24	0·186	25·203	0·1862
0·52	71·69	0·511	70·02	0·515	69·875	0·5164
0·76	107·18	0·764	104·65	0·769	104·362	0·7712
0·92	130·38	0·929	127·01	0·934	126·523	0·9350
1·00	140·32	1·000	136·05	1·000	135·320	1·0000
1·00	136·95	0·976	131·85	0·969	130·887	0·9672
0·92	121·12	0·863	115·54	0·849	114·454	0·8458
0·76	94·55	0·674	89·24	0·656	88·213	0·6519
0·52	59·79	0·426	55·83	0·410	55·090	0·4071
0·19	20·25	0·144	18·76	0·138	18·494	0·1367

Iteration number 4		Iteration number 5		Iteration number 6	
$\times 10^{-3}$		$\times 10^{-3}$		$\times 10^{-3}$	
25·200	0·1864	25·202	0·1865	25·203	0·1865
69·864	0·5168	69·867	0·5170	69·869	0·5170
104·326	0·7718	104·326	0·7719	104·326	0·7720
126·437	0·9354	126·425	0·9355	126·423	0·9355
135·174	1·0000	135·147	1·0000	135·140	1·0000
130·685	0·9668	130·644	0·9667	130·633	0·9667
114·223	0·8450	114·173	0·8448	114·161	0·8448
87·996	0·6510	87·947	0·6508	87·935	0·6507
54·937	0·4064	54·903	0·4062	54·893	0·4062
18·438	0·1364	18·426	0·1363	18·423	0·1363

10·8. Example No. 6: A System with Two Equal Frequencies. Consider the system of light rigid rods shown in Fig. 10·8·1. The rods AB, CD, EF, each of unit length, swing about the axis AE under the constraint of springs of stiffnesses $\frac{1}{3}$, $\frac{4}{3}$, $\frac{1}{3}$, respectively, and are also jointed to the rigid rod BF. The lengths AC, CE, BD, and DF are all equal. Lastly, the rod DG, which is of unit length, swings about the axis BF under the constraint of a spring between CD and DG of stiffness $\frac{1}{9}$. Masses 4, 4, 1 are carried at B, F, G, respectively.

Take as generalised coordinates q_1, q_2, q_3 the linear displacements of B, F, G, respectively. Then the displacement of D is $\frac{1}{2}(q_1+q_2)$. In a general static displacement of the system, the elastic moments at A, C, E, D, will be $\frac{1}{3}q_1$, $\frac{2}{3}(q_1+q_2)$, $\frac{1}{3}q_2$, $\frac{1}{9}(q_3-q_1-q_2)$, and since the lever arms are all of unit length, the vertical forces at B, D, F, G are also $\frac{1}{3}q_1$, $\frac{2}{3}(q_1+q_2)$, $\frac{1}{3}q_2$, $\frac{1}{9}(q_3-q_1-q_2)$. To find the flexibility matrix, apply unit load at B, F, G in succession. When unit load is applied at B, we have by moments about AE,

$$1 = \tfrac{1}{3}q_1+\tfrac{2}{3}(q_1+q_2)+\tfrac{1}{3}q_2 = q_1+q_2,$$

while by moments about AB,

$$\tfrac{2}{3}q_2+\tfrac{2}{3}(q_1+q_2) = 0, \quad \text{or} \quad q_1+2q_2 = 0.$$

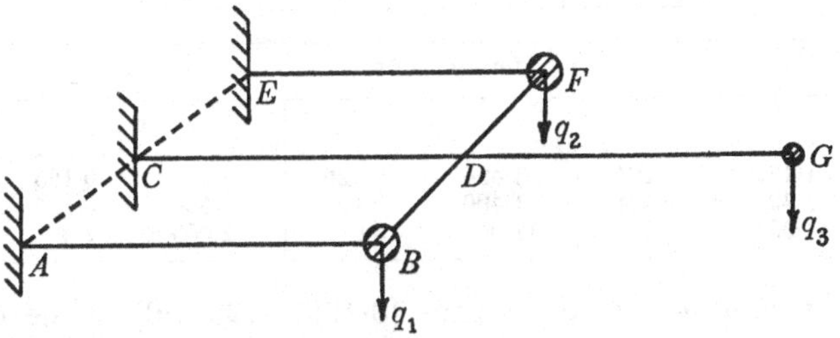

Fig. 10·8·1

Hence $q_1 = 2$, $q_2 = -1$, and since the moment at D is zero, $q_3 = q_1+q_2 = 1$. The displacements are thus $\{2, -1, 1\}$. Similarly, when unit load is applied at F, the displacements are $\{-1, 2, 1\}$. When unit load is applied at G, we have by moments about AE,

$$2 = \tfrac{1}{3}q_1+\tfrac{2}{3}(q_1+q_2)+\tfrac{1}{3}q_2 = q_1+q_2,$$

and, since the displacement is symmetrical, $q_1 = q_2 = 1$. Moreover, by moments about BF,

$$1 = \tfrac{1}{9}(q_3-q_1-q_2) \quad \text{or} \quad q_3 = 11.$$

Hence in this case the displacements are $\{1, 1, 11\}$. The flexibility matrix is thus

$$\Phi = \begin{bmatrix} 2 & -1 & 1 \\ -1 & 2 & 1 \\ 1 & 1 & 11 \end{bmatrix}.$$

The inertia matrix is evidently

$$A = \begin{bmatrix} 4 & 0 & 0 \\ 0 & 4 & 0 \\ 0 & 0 & 1 \end{bmatrix}.$$

Hence the dynamical matrix is

$$U = \Phi A = \begin{bmatrix} 8 & -4 & 1 \\ -4 & 8 & 1 \\ 4 & 4 & 11 \end{bmatrix}.$$

If the initial column for the iterative solution of $\varpi q = Uq$ is chosen to be $\{0, 0, 1\}$, the results are as tabulated below.

Initial column	Iteration number					
	1		2		3	
0	1	0·091	1·364	0·116	1·464	0·123
0	1	0·091	1·364	0·116	1·464	0·123
1	11	1·0	11·728	1·0	11·928	1·0

Iteration number					
4		5		6	
1·492	0·124	1·496	0·125	1·500	0·125
1·492	0·124	1·496	0·125	1·500	0·125
11·984	1·0	11·992	1·0	12·000	1·0

The corresponding mode is thus $\{0{\cdot}125, 0{\cdot}125, 1{\cdot}0\}$, while $\varpi = 12$. On the other hand, when $\{0, 1, 0\}$ is adopted as the initial column, the results are as follows:

Initial column	Iteration number							
	1		2		3		4	
0	−4	−1·0	−15·0	−1·0	−12·6	−1·0	−12·144	−1·0
1	8	2·0	21·0	1·4	16·2	1·286	15·288	1·259
0	4	1·0	15·0	1·0	12·6	1·0	12·144	1·0

Iteration number					
5		6		7	
−12·036	−1·0	−12·008	−1·0	−12·0	−1·0
15·072	1·252	15·016	1·250	15·0	1·250
12·036	1·0	12·008	1·0	12·0	1·0

Hence again $\varpi = 12$, but the mode is now $\{-1{\cdot}0, 1{\cdot}250, 1{\cdot}0\}$. The system therefore has two equal frequencies given by $\varpi = 12$, and the two corresponding modes may be arbitrary linear combinations of $\{0{\cdot}125, 0{\cdot}125, 1{\cdot}0\}$ and $\{-1{\cdot}0, 1{\cdot}250, 1{\cdot}0\}$. Subtraction of the second of these columns from the first yields $\{1{\cdot}125, -1{\cdot}125, 0\}$, so that the modes may conveniently be taken as $\{1, -1, 0\}$ and $\{1, 1, 8\}$. The first represents a torsional motion about the undeflected axis CDG, while the other is a pure flexural motion having the same frequency. From physical considerations it is now clear that the third mode is purely flexural and has a different frequency. This mode and frequency can readily be found from the conditions for absence of the two fundamentals, namely (see § 10·2 (*b*))

$$k_1'Aq \equiv [1, -1, 0]Aq = [4, -4, 0]q = 0,$$
$$k_2'Aq \equiv [1, 1, 8]Aq = [4, 4, 8]q = 0.$$

Hence $q_1 = q_2 = -q_3$, or the mode in the overtone is $k_3 = \{-1, -1, 1\}$. The relation $Uk_3 = \varpi_3 k_3$ now gives at once $\varpi_3 = 3$. It can in fact be verified that

$$U^m = \tfrac{1}{18}\begin{bmatrix} 1 & -1 & -1 \\ 1 & 1 & -1 \\ 8 & 0 & 1 \end{bmatrix}\begin{bmatrix} 12^m & 0 & 0 \\ 0 & 12^m & 0 \\ 0 & 0 & 3^m \end{bmatrix}\begin{bmatrix} 1 & 1 & 2 \\ -9 & 9 & 0 \\ -8 & -8 & 2 \end{bmatrix}$$
$$= \frac{12^m}{18}\begin{bmatrix} 1 & -1 \\ 1 & 1 \\ 8 & 0 \end{bmatrix}\begin{bmatrix} 1 & 1 & 2 \\ -9 & 9 & 0 \end{bmatrix} + \frac{3^m}{18}\begin{bmatrix} -1 \\ -1 \\ 1 \end{bmatrix}[-8 \quad -8 \quad 2],$$

and it will be observed that the canonical form of U is diagonal.

10·9. Example No. 7: The Static Twist of an Aeroplane Wing under Aerodynamical Load.

An illustration of the iterative solution of a statical problem may be useful. Suppose an aeroplane wing to be subjected to an aerodynamical twisting moment which varies with the angle of twist. It is required to determine the equilibrium configuration of the wing.

Suppose the wing to be divided into n segments. Let α_r be the initial mean incidence of the rth segment, measured from the position corresponding to zero aerodynamical moment, and let $\alpha_r + \theta_r$ be the incidence under load. The aerodynamical twisting moment is assumed to be $M_r = -W_r(\alpha_r + \theta_r)$, where $-W_r$ is a positive coefficient depending on chord, wind speed, etc. The flexibility matrix of the wing will be denoted by Φ, and $\Phi^{-1} = E$. Hence if the set of aerodynamical loads M is applied to the wing, and if W denotes the diagonal matrix of the coefficients W_r, the twists θ will be given by

$$\theta = \Phi M = -\Phi W(\alpha + \theta) = g + f\theta, \qquad \text{......(1)}$$

where $g = -\Phi W\alpha$ and $f = -\Phi W$.

This equation can be solved at once by means of the iterations $\theta(s+1) = g + f\theta(s)$ as described in § 4·13 (d), provided that the latent roots of f all have moduli less than unity. It will now be shown that if the equilibrium position is stable, this condition is satisfied.

Equation (1) can be written

$$(E+W)\theta + W\alpha = 0.$$

Choose new coordinates ϕ given by

$$(E+W)\phi = (E+W)\theta + W\alpha,$$

so that ϕ is measured from the equilibrium position. In any displacement ϕ the change in the potential energy is given by $\frac{1}{2}\phi'(E+W)\phi$; and if the energy is to be a minimum in the equilibrium position, which is the condition for stability (see § 8·11), then the quadratic form $\phi'(E+W)\phi$ must always be positive. In this case the arguments of § 9·9 can be applied to the determinant $|E\lambda + W|$. When $\lambda = 0$, since W is a diagonal matrix of negative quantities, the determinants Δ_i of § 9·9 are alternately positive and negative. When $\lambda = 1$, they are by hypothesis all positive. It follows that the roots of $|E\lambda + W| = 0$, which are the latent roots of $-E^{-1}W = -\Phi W = f$, are all real, positive, and less than unity. Hence if the equilibrium position is stable, the iterative process converges.

As a simple numerical example, imagine a wing to be divided into three sections. Let unit torque applied at sections 1, 2, 3, respectively, produce corresponding angular deflections (in degrees) of $\{1, 1, 1\}$, $\{1, 2, 2\}$, $\{1, 2, 3\}$. Further, let the chord and wind speed, etc., be such that $W_1 = -0{\cdot}1$, $W_2 = -0{\cdot}1$, $W_3 = -0{\cdot}05$ unit of torque per degree. Finally let the uniform initial incidence be one degree. Then

$$f = -\Phi W = \begin{bmatrix} 1 & 1 & 1 \\ 1 & 2 & 2 \\ 1 & 2 & 3 \end{bmatrix} \begin{bmatrix} 0{\cdot}1 & 0 & 0 \\ 0 & 0{\cdot}1 & 0 \\ 0 & 0 & 0{\cdot}05 \end{bmatrix} = \begin{bmatrix} 0{\cdot}1 & 0{\cdot}1 & 0{\cdot}05 \\ 0{\cdot}1 & 0{\cdot}2 & 0{\cdot}1 \\ 0{\cdot}1 & 0{\cdot}2 & 0{\cdot}15 \end{bmatrix},$$

$$g = -\Phi W\alpha = \begin{bmatrix} 0{\cdot}1 & 0{\cdot}1 & 0{\cdot}05 \\ 0{\cdot}1 & 0{\cdot}2 & 0{\cdot}1 \\ 0{\cdot}1 & 0{\cdot}2 & 0{\cdot}15 \end{bmatrix} \begin{bmatrix} 1 \\ 1 \\ 1 \end{bmatrix} = \begin{bmatrix} 0{\cdot}25 \\ 0{\cdot}4 \\ 0{\cdot}45 \end{bmatrix}.$$

Choosing $\theta(0) = g$, and calculating in succession the values of

$$\theta(s+1) = g + f\theta(s),$$

we obtain the following results:

$\theta(0)=g$	$f\theta(0)$	$\theta(1)=g+f\theta(0)$	$f\theta(1)$	$\theta(2)$	$f\theta(2)$	$\theta(3)$
0·25	0·087	0·337	0·120	0·370	0·132	0·382
0·4	0·150	0·550	0·206	0·606	0·227	0·627
0·45	0·172	0·622	0·237	0·687	0·261	0·711

$f\theta(3)$	$\theta(4)$	$f\theta(4)$	$\theta(5)$	$f\theta(5)$	$\theta(6)$	$f\theta(6)$	$\theta(7)$
0·136	0·386	0·138	0·388	0·139	0·389	0·139	0·389
0·235	0·635	0·238	0·638	0·239	0·639	0·239	0·639
0·270	0·720	0·274	0·724	0·275	0·725	0·275	0·725

Hence the solution is

$$\theta = \{0{\cdot}389, 0{\cdot}639, 0{\cdot}725\}.$$

It may be noted that the latent roots of f are 0·373, 0·050, and 0·027.

Part II. Systems with Damping Forces Present

10·10. Preliminary Remarks. The methods of Part I can be extended to the case where damping or "motional" forces are present. The first step is to replace the Lagrangian equations by an equivalent system of the first order. Usually the most convenient scheme of reduction is that given at the end of § 9·7. If the original equations are

$$(AD^2+BD+C)\,q = 0,$$

and if $y = \{q, \dot{q}\}$, the reduced equations of free motion will be

$$Dy = uy, \qquad \text{......(1)}$$

in which

$$u \equiv \begin{bmatrix} 0, & I \\ -A^{-1}C, & -A^{-1}B \end{bmatrix}.$$

If $y = e^{\lambda_s t} k_s$ denotes the typical constituent of (1), then

$$(\lambda_s I - u)\, k_s = 0,$$

and this equation can be treated, as before, by the iterative method. The modes of motion are obtained successively in descending order of the moduli of the latent roots of u (i.e. of the roots λ_s). If the modes are required in ascending order of the moduli, then the process should be applied to the equation $(\lambda_s^{-1} I - u^{-1})\, k_s = 0$.

Since the latent roots of u will frequently be complex, it may be necessary to apply the formulae given in § 4·20. If a pair of conjugate complex roots $\lambda_1, \lambda_2 = \mu \pm i\omega$ are dominant, then μ and ω can be determined from the relations

$$\mu^2+\omega^2 = \frac{E_{s+1}E_{s+3}-E_{s+2}^2}{E_sE_{s+2}-E_{s+1}^2},$$

$$2\mu = \frac{E_sE_{s+3}-E_{s+1}E_{s+2}}{E_sE_{s+2}-E_{s+1}^2},$$

where E_s, E_{s+1}, etc. are consecutive values of a homologous element of the column or row used in the iteration, and s is sufficiently large.

10·11. Example: The Oscillations of a Wing in an Airstream. The system to be considered is a model aeroplane wing placed at a small angle of incidence in an airstream. The three degrees of freedom assumed are wing flexure, wing twist, and motion of the aileron relative to the wing. When the wind speed is 12 feet per second the matrices of the dynamical coefficients are as follows:*

$$A = \begin{bmatrix} 17\cdot6 & 0\cdot128 & 2\cdot89 \\ 0\cdot128 & 0\cdot00824 & 0\cdot0413 \\ 2\cdot89 & 0\cdot0413 & 0\cdot725 \end{bmatrix} \times 10^{-3},$$

$$B = \begin{bmatrix} 7\cdot66 & 0\cdot245 & 2\cdot10 \\ 0\cdot0230 & 0\cdot0104 & 0\cdot0223 \\ 0\cdot600 & 0\cdot0756 & 0\cdot658 \end{bmatrix} \times 10^{-3},$$

$$C = \begin{bmatrix} 121 & 1\cdot89 & 15\cdot9 \\ 0 & 0\cdot0270 & 0\cdot0145 \\ 11\cdot9 & 0\cdot364 & 15\cdot5 \end{bmatrix} \times 10^{-3}.$$

It is found that

$$A^{-1} = \begin{bmatrix} 0\cdot1709 & 1\cdot063 & -0\cdot7417 \\ 1\cdot063 & 176\cdot5 & -14\cdot29 \\ -0\cdot7417 & -14\cdot29 & 5\cdot150 \end{bmatrix} \times 10^{3}.$$

Hence

$$u = \left[\begin{array}{ccc|ccc} 0 & 0 & 0 & 1 & 0 & 0 \\ 0 & 0 & 0 & 0 & 1 & 0 \\ 0 & 0 & 0 & 0 & 0 & 1 \\ \hline -11\cdot85 & -0\cdot08172 & 8\cdot764 & -0\cdot8885 & 0\cdot003147 & 0\cdot1054 \\ 41\cdot43 & -1\cdot573 & 202\cdot0 & -3\cdot628 & -1\cdot016 & 3\cdot235 \\ 28\cdot46 & -0\cdot08696 & -67\cdot82 & 2\cdot920 & -0\cdot05901 & -1\cdot512 \end{array}\right].$$

* Numerical data extracted from Table 24 of Ref. 30.

When the initial arbitrary column is chosen to be

$$y(0) = \{0, 0, 0, 0, 0, 1\},$$

the iterative process yields

$$y(1) = uy(0) = \{0, 0, 1, 0{\cdot}1054, 3{\cdot}235, -1{\cdot}512\},$$

$$y(2) = uy(1) = \{0{\cdot}1054, 3{\cdot}235, -1{\cdot}512, 8{\cdot}521, 193{\cdot}4, -65{\cdot}42\},$$

and so on. We now choose a homologous element from the successive columns, and derive (for instance) the Table given below. The method of tabulation is a modification of a scheme due to Aitken (see § 4·20).

No. of column s	Last element E_s	$F_s = E_s E_{s+2} - E_{s+1}^2$	$\frac{F_{s+1}}{F_s} = \mu^2 + \omega^2$	$G_s = E_s E_{s+3} - E_{s+1} E_{s+2}$	$\frac{G_s}{F_s} = 2\mu$
1	1	$-67{\cdot}71$	68·07	—	—
2	$-1{\cdot}512$	$-46{\cdot}09 \times 10^2$	71·23	—	—
3	$-6{\cdot}542 \times 10$	$-32{\cdot}83 \times 10^4$	71·52	—	—
4	$2{\cdot}176 \times 10^2$	$-23{\cdot}48 \times 10^6$	71·93	—	—
5	$4{\cdot}294 \times 10^3$	$-16{\cdot}89 \times 10^8$	72·00	—	—
6	$-2{\cdot}318 \times 10^4$	$-12{\cdot}16 \times 10^{10}$	72·02	—	—
7	$-2{\cdot}681 \times 10^5$	$-8{\cdot}758 \times 10^{12}$	72·04	—	—
8	$2{\cdot}144 \times 10^6$	$-6{\cdot}309 \times 10^{14}$	72·04	$1{\cdot}116 \times 10^{15}$	$-1{\cdot}769$
9	$1{\cdot}552 \times 10^7$	$-4{\cdot}545 \times 10^{16}$	—	—	—
10	$-1{\cdot}819 \times 10^8$	—	—	—	—
11	$-0{\cdot}7962 \times 10^9$	—	—	—	—

Accordingly, $\mu^2 + \omega^2 = 72{\cdot}04$ and $2\mu = -1{\cdot}769$, so that the dominant latent roots of the matrix u are λ_1 and its conjugate λ_2, where

$$\lambda_1 = \mu + i\omega = -0{\cdot}885 + i\,8{\cdot}443.$$

The modal column, say $k_1 = \xi + i\eta$, corresponding to the root λ_1, is readily found. Since k_1 is arbitrary to a complex scalar multiplier, we may choose for instance the rth column in the iteration to be half the sum of the conjugate complexes $\xi \pm i\eta$, i.e., ξ. Then the $(r+1)$th column will be $\mu\xi - \omega\eta$, and hence ξ and η are obtained. The first three elements of $\xi + i\eta$ define the amplitudes and epochs of the generalised coordinates q, and the last three those of the generalised velocities $\dot{q}$. Actually, since $\dot{q} = \lambda q$, the last three elements in any column are identical with the first three elements in the succeeding column; hence q may be determined from one column only. In the present instance, the 10th column is

$$10^6\{-2{\cdot}412, -39{\cdot}00, 15{\cdot}52, 26{\cdot}65, 528{\cdot}8, -181{\cdot}9\}.$$

On omission of the factor 10^6 this gives

$$\{\xi_1, \xi_2, \xi_3\} = \{-2{\cdot}412, -39{\cdot}00, 15{\cdot}52\},$$

while η is defined by

$$-0{\cdot}885\{-2{\cdot}412, -39{\cdot}00, 15{\cdot}52\} - 8{\cdot}443\{\eta_1, \eta_2, \eta_3\}$$
$$= \{26{\cdot}65, 528{\cdot}8, -181{\cdot}9\}.$$

Hence

$$k_1 = \{-2{\cdot}412 - i\,2{\cdot}904, -39{\cdot}00 - i\,58{\cdot}55, 15{\cdot}52 + i\,19{\cdot}92\},$$

or, since a scalar multiplier may be extracted,

$$k_1 = \{1{\cdot}0, 18{\cdot}53 + i\,1{\cdot}961, -6{\cdot}686 - i\,0{\cdot}2085\}.$$

This value can be verified by substitution in the equation

$$\lambda_1^2 A k_1 + \lambda_1 B k_1 + C k_1 = 0.$$

The remaining modes of motion can be found as usual by the use of the condition for absence of the dominant mode. If R_m and R_{m+1} are two successive rows obtained by repeated postmultiplication of an arbitrary row by u, then (see § 4·18) the conditions for absence of the dominant mode are $R_m y = 0$ and $R_{m+1} y = 0$. In the present example, if [0, 0, 0, 0, 0, 1] is used as an initial row, then the 9th and 10th rows found by the iterative process may be written

$$8{\cdot}273 \times 10^7[1, -0{\cdot}000781, -1{\cdot}910, -0{\cdot}004369, -0{\cdot}001575, 0{\cdot}1877],$$
$$44{\cdot}09 \times 10^7[1, -0{\cdot}002531, -2{\cdot}456, 0{\cdot}2924, -0{\cdot}001927, -0{\cdot}4126].$$

If these rows are used in the above conditions, and y_1 and y_4 $(=\dot{y}_1)$ are then expressed in terms of the remaining coordinates, we find

$$y_1 = 0{\cdot}000807 y_2 + 1{\cdot}918 y_3 + 0{\cdot}001580 y_5 - 0{\cdot}1789 y_6,$$
$$y_4 = 0{\cdot}005896 y_2 + 1{\cdot}840 y_3 + 0{\cdot}001186 y_5 + 2{\cdot}023 y_6.$$

We may therefore write

$$\begin{bmatrix} y_1 \\ y_2 \\ y_3 \\ y_4 \\ y_5 \\ y_6 \end{bmatrix} = \begin{bmatrix} 0, & 0{\cdot}000807, & 1{\cdot}918, & 0, & 0{\cdot}001580, & -0{\cdot}1789 \\ 0, & 1, & 0, & 0, & 0, & 0 \\ 0, & 0, & 1, & 0, & 0, & 0 \\ 0, & 0{\cdot}005896, & 1{\cdot}840, & 0, & 0{\cdot}001186, & 2{\cdot}023 \\ 0, & 0, & 0, & 0, & 1, & 0 \\ 0, & 0, & 0, & 0, & 0, & 1 \end{bmatrix} \begin{bmatrix} y_1 \\ y_2 \\ y_3 \\ y_4 \\ y_5 \\ y_6 \end{bmatrix}.$$

Introducing this condition on the right-hand side of the equation $\lambda y = uy$, we obtain a modified equation $\lambda y = vy$, where

$$v = \begin{bmatrix} 0, & 0{\cdot}005896, & 1{\cdot}840, & 0, & 0{\cdot}001186, & 2{\cdot}023 \\ 0, & 0, & 0, & 0, & 1, & 0 \\ 0, & 0, & 0, & 0, & 0, & 1 \\ 0, & -0{\cdot}09652, & -15{\cdot}60, & 0, & -0{\cdot}01663, & 0{\cdot}4279 \\ 0, & -1{\cdot}561, & 274{\cdot}8, & 0, & -0{\cdot}9548, & -11{\cdot}52 \\ 0, & -0{\cdot}04678, & -7{\cdot}861, & 0, & -0{\cdot}01058, & -0{\cdot}6963 \end{bmatrix}.$$

This matrix possesses properties similar to those of u except that zero latent roots have been substituted for the original dominant roots $\mu \pm i\omega \equiv -0{\cdot}885 \pm i\,8{\cdot}443$. Hence the iterative process, when applied to v, would yield the subdominant latent root or roots of u and the associated mode of motion.

CHAPTER XI

DYNAMICAL SYSTEMS WITH SOLID FRICTION

11·1. Introduction. (*a*) *General Range of the Theory.* The theory* to be given relates solely to dynamical systems which, apart from the presence of solid friction, obey linear laws. The term *solid friction* is interpreted to mean a resistance between two bodies, due to tangential surface actions, which follows the simple idealised law that its magnitude has a constant value so long as relative motion between the bodies occurs. Whenever the forces tending to produce relative motion fall short of the foregoing value, the resistance adjusts itself to balance these forces, and no relative motion results. These simple laws, which are adopted as a mathematical convenience, take no account of the difference between static and dynamic friction, and thus only approximately represent true conditions.

The influence of friction on the oscillatory behaviour of a dynamical system is often somewhat unexpected. For instance, the introduction of friction may cause an otherwise stable system to develop maintained oscillations.

(*b*) *Experimental Illustration of the Influence of Friction.* Some wind tunnel experiments† carried out at the National Physical Laboratory in connection with tail flutter of an aeroplane illustrate some of the effects of solid friction. The model used had a rigid front fuselage but a flexible rear fuselage and tail unit, and it was suspended by a sling of wires from a turn-table locked to the roof girders of the wind tunnel. Occasionally the locking bolts of the turn-table worked slack under the tunnel vibration, with the result that the whole model became capable of angular displacement in yaw, but only under considerable frictional constraint. The degrees of freedom of the dynamical system, when tested under such conditions, can (for simplicity) here be taken as angular displacement of the rudder, lateral bending of the rear fuselage, and yawing of the whole model resisted by a considerable frictional moment. At moderate airspeeds the motion of the model was of a peculiar spasmodic type, which gave the impression that increasing and decreasing oscillations were occurring in succession.

* Chaps. XI and XII are based on investigations originally described in Refs. 38 and 39.

† See pp. 233 and 260 of Ref. 40.

The explanation is probably as follows. At the wind speeds which gave rise to the spasmodic motion, the binary system composed of the rudder and the fuselage was unstable, whereas the addition of the freedom in yaw rendered the system stable. A very small initial impulse was incurred, and this was sufficient to start growing rudder-fuselage oscillations. It was, however, too small to overcome the friction opposing yawing movement, so that meanwhile the model remained gripped in yaw. However, ultimately, a yawing moment due to the growing oscillations developed, which was sufficiently large to overcome the friction. All three degrees of freedom then cooperated in the motion, but since the system was then stable, the yawing moment eventually decreased and the model again became effectively gripped. The cycle of changes then repeated. Briefly, the phenomenon is attributable to the successive "sticking" and "unsticking" of a frictionally constrained member of the system.

A feature of importance in the illustration is the fact that the system would have been definitely stable had the friction been absent. Clearly then the state of stability of a system when friction is completely removed is no sure guide to the behaviour of that system when friction is present.

(c) *Ankylosis.* The effect just described as the "sticking" of a degree of freedom is known as *ankylosis*.* Whenever the system is moving without its full complement of degrees of freedom the motion is said to be *ankylotic*, and the particular generalised coordinates which are arrested are the *ankylosed coordinates*.

(d) *Types of Oscillation.* When solid friction is operative the oscillations which occur in a given degree of freedom can be of three different general types, namely:

(i) *Decaying oscillations*, in which the coordinate concerned either tends to a constant value or becomes permanently ankylosed.

(ii) *Unbounded oscillations*, in which the coordinate tends to attain indefinitely great values.

(iii) *Bounded oscillations*, which are defined to be any oscillations other than those of types (i) and (ii).

(e) *Linear and Non-linear Systems.* When friction is taken into account we have to deal with forces whose functional dependence on

* The term 'ankylosis' is used by Poincaré, Jeans and others to denote loss of one or more degrees of freedom in a dynamical system.

the generalised velocities is not linear. Strictly speaking, therefore, the problem must be classed (though as a very special case) under the general subject of non-linear dynamics. In the case of a linear system the values of the generalised coordinates and velocities at any instant of motion are all changed proportionally when the initial values are changed proportionally. In other words, so far as concerns similar disturbances, the motion may be said to be proportional to the magnitude of the initial disturbance. With a non-linear system the motion will depend in a very much more complicated way on the initial disturbance.

(*f*) *Effects of Very Large Disturbances.* If the system is linear in all respects apart from the friction, two general propositions are self-evident:

(i) No disturbance, however large, can produce unbounded oscillations when friction is present if these cannot occur when friction is absent. For even if increasing oscillations are possible, yet the growth necessarily ceases when the movements become so large that the frictional forces are insignificant in comparison with the other applied forces.

(ii) A sufficiently large disturbance (in general) produces unbounded oscillations when friction is present if such oscillations are possible with the frictionless system.

(*g*) *Effects of Arbitrary Disturbances.* The ultimate type of motion which results from a disturbance of arbitrary magnitude will depend upon the number of sets of bounded oscillations which are possible and upon the question as to whether these particular motions are stable when slightly disturbed. Suppose, for instance, that a very large disturbance gives rise to unbounded oscillations and that a small similar disturbance produces decaying oscillations leading to ankylosis and ultimately to complete rest of the system. Then it may be inferred that for one magnitude at least of the disturbance the oscillations must be bounded. If these are the only possible bounded oscillations, they must obviously be unstable, and they will therefore not be realisable in practice. However, more generally there may be an odd number of sets of bounded oscillations, alternately unstable and stable, the largest and smallest being unstable. In these cases, for a certain range of the initial disturbance, the resulting motions will tend to one or other of the stable sets of oscillations.

11·2. The Dynamical Equations. The m generalised coordinates q are supposed to be measured from the mean configuration, and the system is taken to be linear except for the friction. The dynamical equations are then of the type

$$f(D)\,q \equiv (AD^2 + BD + C)\,q = \rho, \qquad \ldots\ldots(1)$$

where ρ denotes the generalised frictional forces. When the friction is applied in a general manner the precise specification of ρ is complicated, and as a simplification attention will be restricted to systems in which each degree of freedom has its own independent frictional constraint. Thus, it is assumed that the typical generalised frictional force ρ_i has a constant magnitude R_i when $\dot{q}_i \neq 0$, and that the sign is such that the force always opposes the motion in q_i: hence $\rho_i = -R_i$ if $\dot{q}_i > 0$, and $\rho_i = +R_i$ if $\dot{q}_i < 0$. On the other hand, if $\dot{q}_i = 0$, the value of ρ_i lies between the limits $\pm R_i$, and is determined by the other circumstances of the motion.

The motion in any time interval may be such that continuous movements occur in all the degrees of freedom, or it may be such that throughout the interval one or more of the generalised coordinates are arrested by friction. In the first case the motion will be described as *complete* in the interval, and in the second case as *ankylotic*.* If a generalised velocity happens to vanish in the interval without becoming stationary, the motion will still be classed as complete. On the other hand, if the velocity which vanishes is also stationary, the motion at that stage will be viewed as ankylotic.

For brevity, any stage of the motion throughout which $\dot{q}_i$ is continuously positive, and never actually zero, will be described as an *up-stroke* in q_i: similarly, if $\dot{q}_i$ is continuously negative and never zero, the motion is a *down-stroke* in q_i. Hence $\rho_i = -R_i$ or $+R_i$ according as the stroke in q_i is upwards or downwards. The two instants at which a given stroke begins and ends will be referred to as the *terminal instants* for that stroke: when a distinction between the two terminal instants is necessary the one which relates to the beginning of the stroke will be called the *starting instant* and the other the *stopping instant*.

Throughout any time interval in which no generalised velocity vanishes the values of the forces ρ are constant both in sign and

* Note that the definition of ankylotic motion would require generalisation if the frictional constraints in the several degrees of freedom were not independent.

magnitude. The solution of the dynamical equations for such an interval will now be given. For simplicity it is assumed in the analysis that the n ($= 2m$) roots $\lambda_1, \lambda_2, \ldots, \lambda_n$ of $\Delta(\lambda) = 0$ are all distinct.

Write $$C^{-1}\rho = \theta, \qquad \ldots\ldots(2)$$

and let Θ denote the column formed from the m constants θ and from m ciphers, so that in the partitioned form

$$\Theta = \{\theta, 0\}. \qquad \ldots\ldots(3)$$

Also put $$y(t) = \{q(t), \dot{q}(t)\}. \qquad \ldots\ldots(4)$$

Then if $M(t)$, k, Λ and l are matrices as defined in §§ 5·10 and 6·4, and if τ denotes some datum instant in the interval, the direct matrix solution is readily verified to be

$$y(t) - \Theta = lM(t-\tau)\, l^{-1}\{y(\tau) - \Theta\}.$$

Alternatively, if n reducing variables $\alpha(t)$ (see example (iii) of § 6·4) and n quantities β are introduced such that

$$\alpha(t) = l^{-1}y(t), \qquad \ldots\ldots(5)$$

$$\beta = l^{-1}\Theta, \qquad \ldots\ldots(6)$$

the solution is expressible as

$$\alpha(t) - \beta = M(t-\tau)\{\alpha(\tau) - \beta\}. \qquad \ldots\ldots(7)$$

Matrices formed from particular columns or rows of the $(\overline{m}, n)$ modal matrix k will often be used in the sequel. The matrix $\{k_{ir}\}$ formed from the rth column of k (namely, the rth modal column) will as hitherto be written in the abbreviated form k_r. Similarly the matrix $\lfloor k_{pj} \rfloor$ formed from the pth row of k (namely, the pth *modal row**) will be denoted for brevity by $\mathfrak{k}_p$ (see also remarks in § 1·2 (c)).

11·3. Various Identities. Certain identities, which will be of use later, will now be obtained.

(a) *The Reducing Variables* $\alpha(t)$. The relation (11·2·5) gives immediately by definition of the matrix l

$$\{q(t), \dot{q}(t)\} = \{k, k\Lambda\}\, \alpha(t),$$

so that $$q(t) = k\alpha(t), \qquad \ldots\ldots(1)$$

$$\dot{q}(t) = k\Lambda\alpha(t). \qquad \ldots\ldots(2)$$

* Note that whereas the rth modal column k_r is naturally associated with the rth root λ_r of $\Delta(\lambda) = 0$, the pth modal row is associated with the pth generalised coordinate q_p. The modal rows must not be confused with the rows of the matrix κ defined by (3·6·8).

Again, differentiation of (2) yields $\ddot{q}(t) = k\Lambda\dot{\alpha}(t)$, while differentiation of (11·2·7) leads to

$$\dot{\alpha}(t) = \Lambda M(t-\tau)\{\alpha(\tau)-\beta\} = \Lambda\{\alpha(t)-\beta\}.$$

Accordingly,
$$\ddot{q}(t) = k\Lambda^2\{\alpha(t)-\beta\}. \qquad \text{......(3)}$$

When the expressions for q, $\dot{q}$ and $\ddot{q}$ given by (1), (2) and (3) are substituted in (11·2·1), the result is

$$(Ak\Lambda^2+Bk\Lambda+Ck)\,\alpha(t)-Ak\Lambda^2\beta = \rho.$$

But on account of the properties of the modal columns

$$Ak\Lambda^2+Bk\Lambda+Ck = 0.$$

Hence, writing $a \equiv A^{-1}$, we have

$$k\Lambda^2\beta = -a\rho. \qquad \text{......(4)}$$

Equation (3) may be therefore expressed as

$$\ddot{q}(t) = k\Lambda^2\alpha(t)+a\rho. \qquad \text{......(5)}$$

It is to be noted that since the discriminants of the kinetic energy are positive the principal diagonal elements of the matrix a are all positive.

The relations (1), (2) and (5) give the displacements, velocities and accelerations in terms of the reducing variables.

It will be seen later that the reducing variables and the quantities β usually occur multiplied by a modal constant. Formulae to aid the calculation of these products can be derived by use of the identity (6·5·8), which for the present second-order equations reduces to

$$kM(t-\tau)\,l^{-1} = \sum_{r=1}^{n} \frac{\underset{(1)}{F}(\lambda_r)}{\Delta(\lambda_r)}[\lambda_r A+B, A]\,e^{\lambda_r(t-\tau)}.$$

Since this is true for all values of t, the matrix coefficients of $e^{\lambda_r(t-\tau)}$ on the right and on the left can be identified. Hence

$$kE_r l^{-1} = \frac{\underset{(1)}{F}(\lambda_r)}{\Delta(\lambda_r)}[\lambda_r A+B, A], \qquad \text{......(6)}$$

in which E_r is the null square matrix of order n with a unit substituted in the rth principal diagonal place. On postmultiplication of both sides of (6) by $y(t)$ and application of (11·2·5) we readily obtain the identity

$$\alpha_r k_r = \frac{\underset{(1)}{F}(\lambda_r)}{\Delta(\lambda_r)}[\lambda_r A+B, A]\{q(t), \dot{q}(t)\}, \qquad \text{......(7)}$$

where k_r denotes the rth modal column. This can be written alternatively as

$$\alpha_r k_r = \frac{F(\lambda_r)}{\lambda_r \overset{(1)}{\Delta}(\lambda_r)} [-C, \lambda_r A] \{q(t), \dot{q}(t)\}. \qquad \text{......(8)}$$

(*b*) *The Quantities* β. From equations (11·2·3) and (11·2·6) it follows that $\{\theta, 0\} = \{k, k\Lambda\}\beta$, so that

$$k\beta = \theta, \qquad \text{......(9)}$$

$$k\Lambda\beta = 0. \qquad \text{......(10)}$$

Again, on postmultiplication of both sides of (6) by Θ and use of the relation $C\theta = \rho$, we readily find

$$\beta_r k_r = -\frac{F(\lambda_r)\rho}{\lambda_r \overset{(1)}{\Delta}(\lambda_r)}. \qquad \text{......(11)}$$

An alternative to (11) can be deduced from the relation

$$\Delta(\lambda) = f_{11}(\lambda) F_{11}(\lambda) + \ldots + f_{1s}(\lambda) F_{1s}(\lambda) + \ldots + f_{1m}(\lambda) F_{1m}(\lambda).$$

The only term on the right of this equation which involves the stiffness coefficient C_{1s} is evidently f_{1s}. Hence on total differentiation with respect to C_{1s} we have

$$\frac{d}{dC_{1s}} \Delta(\lambda) = \overset{(1)}{\Delta}(\lambda) \frac{\partial \lambda}{\partial C_{1s}} + \frac{\partial}{\partial C_{1s}} \Delta(\lambda) = \overset{(1)}{\Delta}(\lambda) \frac{\partial \lambda}{\partial C_{1s}} + F_{1s}(\lambda).$$

If in the variation considered λ is chosen to correspond to the root λ_r, so that $\Delta(\lambda_r) = 0$ and $\dfrac{d}{dC_{1s}} \Delta(\lambda_r) = 0$, we obtain

$$F_{1s}(\lambda_r) = -\overset{(1)}{\Delta}(\lambda_r) \frac{\partial \lambda_r}{\partial C_{1s}}.$$

We may accordingly write

$$F(\lambda_r) = -\overset{(1)}{\Delta}(\lambda_r) N(\lambda_r),$$

where

$$N(\lambda_r) \equiv \begin{bmatrix} \dfrac{\partial \lambda_r}{\partial C_{11}}, & \dfrac{\partial \lambda_r}{\partial C_{21}}, & \ldots, & \dfrac{\partial \lambda_r}{\partial C_{m1}} \\ \dfrac{\partial \lambda_r}{\partial C_{12}}, & \dfrac{\partial \lambda_r}{\partial C_{22}}, & \ldots, & \dfrac{\partial \lambda_r}{\partial C_{m2}} \\ \cdots & \cdots & \cdots & \cdots \\ \dfrac{\partial \lambda_r}{\partial C_{1m}}, & \dfrac{\partial \lambda_r}{\partial C_{2m}}, & \ldots, & \dfrac{\partial \lambda_r}{\partial C_{mm}} \end{bmatrix}$$

and on substitution in (11) this gives the identity

$$\lambda_r \beta_r k_r = N(\lambda_r)\rho. \qquad \text{......(12)}$$

By performing the differentiations with respect to the damping coefficients B_{ij} or the inertial coefficients A_{ij} instead of the stiffness coefficients, we can derive the relations

$$\frac{\partial \lambda_r}{\partial C_{ij}} = \frac{1}{\lambda_r}\frac{\partial \lambda_r}{\partial B_{ij}} = \frac{1}{\lambda_r^2}\frac{\partial \lambda_r}{\partial A_{ij}}. \qquad \text{......(13)}$$

The application of the preceding formulae to systems having only a single frictionally constrained coordinate will now be considered.

11·4. Complete Motion when only One Coordinate is Frictionally Constrained. The frictionally constrained coordinate will be assumed to be q_m. In this case all the frictional forces ρ_i are zero with the exception of ρ_m, and $\rho_m = \pm R_m$ according to the sense of the stroke. If ϕ denotes the column of m constants ϕ_i such that

$$C\phi = \{0, \ldots, 0, 1\},$$

then clearly $\theta = -R_m\phi$ for an up-stroke and $\theta = +R_m\phi$ for a down-stroke. Further, if Φ is defined similarly to Θ as the column of m values ϕ_i followed by m ciphers, and if γ is such that $\gamma = l^{-1}\Phi$, then $\beta = -R_m\gamma$ for an up-stroke and $\beta = +R_m\gamma$ for a down-stroke. Equations (11·3·11) and (11·3·12) may in this case be expressed as

$$\lambda_r\gamma_r k_r = \left\{\frac{\partial \lambda_r}{\partial C_{m1}}, \frac{\partial \lambda_r}{\partial C_{m2}}, \ldots, \frac{\partial \lambda_r}{\partial C_{mm}}\right\} = -\frac{1}{\overset{(1)}{-\Delta}(\lambda_r)}\{F_{m1}(\lambda_r), \ldots, F_{mm}(\lambda_r)\}, \qquad \text{......(1)}$$

while (11·3·9), (11·3·10), and (11·3·4) yield respectively

$$k\gamma = \phi, \qquad \text{......(2)}$$

$$k\Lambda\gamma = 0, \qquad \text{......(3)}$$

$$k\Lambda^2\gamma = -\{a_{im}\}. \qquad \text{......(4)}$$

Let the initial disturbance, assumed imposed at $t = \tau$, be given by the values $y(\tau)$ or $\alpha(\tau)$, and denote as t_0, t_1, etc. the successive terminal instants at which—while the complete motion is in progress—the velocity $\dot{q}_m$ vanishes. If $\dot{q}_m(\tau) \neq 0$, the sign of this velocity will fix the sign of ρ_m in the first stage of the motion. For the sake of definiteness we shall assume that $\dot{q}_m(\tau) > 0$, so that the first motion in q_m is upwards and $\rho_m = -R_m$ (see Fig. 11·4·1).

On application of (11·2·7) we have for the solution in the first time interval

$$\alpha(t) + R_m\gamma = M(t-\tau)\{\alpha(\tau) + R_m\gamma\}. \qquad \text{......(5)}$$

Hence the displacements, velocities, and accelerations at any time are given by

$$q(t) = k\alpha(t) = kM(t-\tau)\{\alpha(\tau)+R_m\gamma\}-R_m\phi,$$
$$\dot{q}(t) = k\Lambda\alpha(t) = k\Lambda M(t-\tau)\{\alpha(\tau)+R_m\gamma\},$$
$$\ddot{q}(t) = k\Lambda^2\{\alpha(t)+R_m\gamma\} = k\Lambda^2 M(t-\tau)\{\alpha(\tau)+R_m\gamma\}. \quad \text{......(6)}$$

In particular the velocity $\dot{q}_m$ vanishes when

$$\mathfrak{k}_m\Lambda M(t-\tau)\{\alpha(\tau)+R_m\gamma\} = 0,$$

where $\mathfrak{k}_m$ denotes the mth modal row. The stopping instant will be that root $t = t_0$ of this equation which exceeds, and lies nearest to, the value τ.

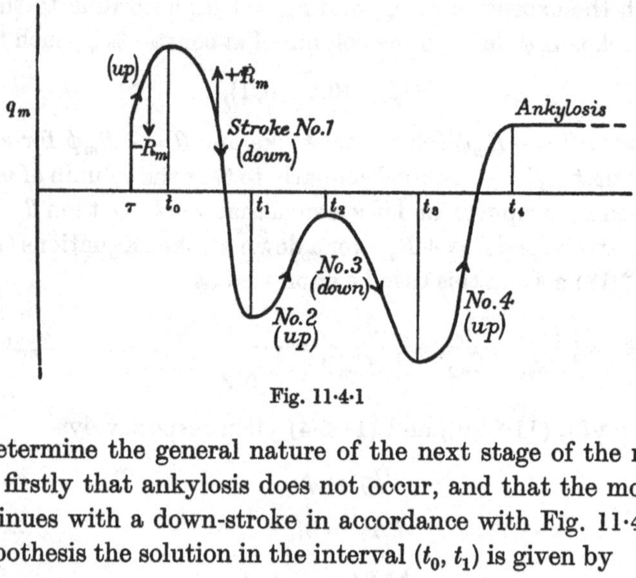

Fig. 11·4·1

To determine the general nature of the next stage of the motion, assume firstly that ankylosis does not occur, and that the motion in q_m continues with a down-stroke in accordance with Fig. 11·4·1. On this hypothesis the solution in the interval (t_0, t_1) is given by

$$\alpha(t)-R_m\gamma = M(t-t_0)\{\alpha(t_0)-R_m\gamma\}. \quad \text{......(7)}$$

This yields for the starting acceleration $\ddot{q}_m(t_0)$ the value

$$\mathfrak{k}_m\Lambda^2\{\alpha(t_0)-R_m\gamma\},$$

which, on application of (4), can be written

$$\mathfrak{k}_m\Lambda^2\alpha(t_0)+a_{mm}R_m. \quad \text{......(8)}$$

If this expression is negative, the motion in q_m obviously continues with a down-stroke as assumed.

Alternatively, suppose that an up-stroke in q_m ensues at time t_0. The solution in this case would be given by an equation similar to (5), but with τ replaced by t_0. The necessarily positive starting acceleration for the up-stroke in q_m would then be

$$\mathfrak{k}_m \Lambda^2\{\alpha(t_0)+R_m\gamma\}.$$

But on application of equation (6) this expression is seen to be equal to the stopping acceleration for the previous partial up-stroke in q_m, which is negative Hence the possibility of the stroke being upwards is excluded. We may thus conclude that if the expression (8) is negative the motion remains complete and continues with a down-stroke in q_m, but that otherwise the coordinate q_m remains arrested.

If ankylosis does not occur, the motion proceeds in accordance with (7) until q_m again becomes stationary. This occurs when

$$\mathfrak{k}_m \Lambda M(t-t_0)\{\alpha(t_0)-R_m\gamma\} = 0,$$

and the corresponding stopping instant t_1 is that root of this equation which exceeds, and lies nearest to, the value t_0. The reducing variables then have the values given by

$$\alpha(t_1)-R_m\gamma = M(t_1-t_0)\{\alpha(t_0)-R_m\gamma\},$$

and the condition that a second up-stroke in q_m is realised is readily shown to be

$$\mathfrak{k}_m \Lambda^2\alpha(t_1)-a_{mm}R_m > 0.$$

A continuation of this process formally determines the whole motion until ankylosis occurs. The sequences of relations which provide the step-by-step solution will now be summarised. The formulae are suited to the case where $\dot{q}_m(\tau)>0$, so that the first motion in q_m is upwards.*

(A) *Successive Stages of Motion.* After the first partial up-stroke in q_m the down-strokes and up-strokes in q_m occur in alternation. The motion in the typical down-stroke in q_m commencing at time t_{2r} is

$$\alpha(t)-R_m\gamma = M(t-t_{2r})\{\alpha(t_{2r})-R_m\gamma\}, \qquad \text{......(9)}$$

while that in the typical up-stroke in q_m commencing at time t_{2r+1} is

$$\alpha(t)+R_m\gamma = M(t-t_{2r+1})\{\alpha(t_{2r+1})+R_m\gamma\}. \qquad \text{......(10)}$$

(B) *Recurrence Relations for the Reducing Variables.* These may be written conveniently as

$$M(-t_0)\{\alpha(t_0)+R_m\gamma\} = M(-\tau)\{\alpha(\tau)+R_m\gamma\},$$
$$M(-t_1)\{\alpha(t_1)-R_m\gamma\} = M(-t_0)\{\alpha(t_0)-R_m\gamma\},$$
$$M(-t_2)\{\alpha(t_2)+R_m\gamma\} = M(-t_1)\{\alpha(t_1)+R_m\gamma\},$$

* The formulae can be adapted at once to the case where $\dot{q}_m(\tau)=0$.

and generally for $s \geqslant 1$

$$M(-t_s)\{\alpha(t_s)+(-1)^s R_m\gamma\} = M(-t_{s-1})\{\alpha(t_{s-1})+(-1)^s R_m\gamma\}. \qquad \text{......(11)}$$

Direct addition of these relations gives

$$(-1)^s\alpha(t_s)+R_m\gamma = (-1)^s M(t_s-\tau)\{\alpha(\tau)+R_m\gamma\}+2\Gamma(s)R_m\gamma, \qquad \text{......(12)}$$

where $\Gamma(s) \equiv M(t_s-t_{s-1})-M(t_s-t_{s-2})+\ldots+(-1)^{s-1}M(t_s-t_0)$.

(C) *The Terminal Equations.* These fix the terminal instants t_0, t_1, etc. at which the successive stationary values of q_m occur. The first equation of the sequence is explicitly

$$\sum_{r=1}^{n} \lambda_r k_{mr} e^{\lambda_r t_0}(\alpha_r(\tau)\,e^{-\lambda_r\tau}+R_m\gamma_r e^{-\lambda_r\tau}) = 0,$$

while for $s \geqslant 1$ the typical equation is $\dot{q}_m(t_s) = 0$, or

$$\mathfrak{k}_m\Lambda\alpha(t_s) = 0. \qquad \text{......(13)}$$

On substitution for $\alpha(t_s)$ from (12) this gives explicitly

$$\sum_{r=1}^{n} \lambda_r k_{mr} e^{\lambda_r t_s}((-1)^s\alpha_r(\tau)\,e^{-\lambda_r\tau}+R_m\gamma_r E(s,\lambda_r)) = 0, \qquad \text{......(14)}$$

where

$$E(s,\lambda_r) \equiv 2e^{-\lambda_r t_{s-1}}-2e^{-\lambda_r t_{s-2}}+\ldots+2(-1)^{s-1}e^{-\lambda_r t_0}+(-1)^s e^{-\lambda_r \tau}.$$

The terminal instant t_s must be that root of the sth terminal equation which exceeds, and lies nearest to, the value t_{s-1}. It may be noted that if the roots of $\Delta(\lambda) = 0$ are complex then each terminal equation necessarily has an infinite number of roots.*

(D) *Criterion for Ankylosis.* Ankylosis will occur (with the coordinate q_m arrested) at the first stopping instant t_s for which

$$(-1)^s\mathfrak{k}_m\Lambda^2\alpha(t_s)+a_{mm}R_m \geqslant 0. \qquad \text{......(15)}$$

This criterion is valid for $s \geqslant 0$.

Example

System with a Single Degree of Freedom.† With $m = 1$ the equation for complete motion is

$$A_{11}\ddot{q}_1+B_{11}\dot{q}_1+C_{11}q_1 = \mp R_1,$$

leading to a single constant $\phi_1 = 1/C_{11}$. The two roots λ are assumed

* If t_s is large the least damped exponential will predominate in equation (14). It readily follows that this equation necessarily has an infinite number of large real roots.

† Some aspects of this problem are discussed in Ref. 41.

to be complex, so that, say, $\lambda_1 = \mu + i\omega$ and $\lambda_2 = \mu - i\omega$. The modal matrix may be taken as $[1, 1]$, and we may accordingly write

$$l = \begin{bmatrix} 1 & 1 \\ \lambda_1 & \lambda_2 \end{bmatrix} \quad \text{and} \quad 2i\omega l^{-1} = \begin{bmatrix} -\lambda_2 & 1 \\ \lambda_1 & -1 \end{bmatrix}.$$

Hence
$$\{\alpha_1, \alpha_2\} = l^{-1}\{q_1, \dot{q}_1\} = \frac{i}{2\omega}\{\lambda_2 q_1 - \dot{q}_1, -\lambda_1 q_1 + \dot{q}_1\},$$

$$\{\gamma_1, \gamma_2\} = l^{-1}\{\phi_1, 0\} = \frac{i}{2\omega C_{11}}\{\lambda_2, -\lambda_1\}.$$

If the initial disturbance is assumed to be imposed at the starting instant t_0 of a down-stroke, then

$$\{\alpha_1(t_0), \alpha_2(t_0)\} = \frac{iq_1(t_0)}{2\omega}\{\lambda_2, -\lambda_1\},$$

and the motion in the first down-stroke is given by

$$\alpha(t) - R_m\gamma = M(t - t_0)\{\alpha(t_0) - R_m\gamma\}.$$

Further, the equation for the stopping instant t_1 is explicitly

$$\lambda_1(\alpha_1(t_0) - R_1\gamma_1)\, e^{(t_1 - t_0)\lambda_1} + \lambda_2(\alpha_2(t_0) - R_1\gamma_2)\, e^{(t_1 - t_0)\lambda_2} = 0.$$

This reduces to

$$-\frac{1}{\omega}\left(q_1(t_0) - \frac{R_1}{C_{11}}\right)(\mu^2 + \omega^2)\, e^{\mu(t_1 - t_0)} \sin \omega(t_1 - t_0) = 0,$$

of which the lowest root is $t_1 - t_0 = \pi/\omega$. Since this time interval is independent of the starting displacement $q_1(t_0)$, the time for any stroke is clearly $T = \pi/\omega$.

The formula (12), with the simplifications $\tau = t_0$ and $t_s - t_0 = sT$, may now be applied to determine the displacement at the end of the sth stroke. In the present case

$$M(sT) \equiv \begin{bmatrix} e^{\lambda_1 sT} & 0 \\ 0 & e^{\lambda_2 sT} \end{bmatrix} = (-1)^s e^{\mu sT} I,$$

$$\Gamma(s) \equiv M(T) - M(2T) + \ldots + (-1)^{s-1} M(sT)$$
$$= -e^{\mu T}\left(\frac{e^{\mu sT} - 1}{e^{\mu T} - 1}\right) I,$$

and equation (12) therefore reduces to

$$(-1)^s\{\alpha_1(t_s), \alpha_2(t_s)\} = \frac{i}{2\omega}\left(q_1(t_0)\, e^{\mu sT} - \frac{R_1}{C_{11}}(e^{\mu sT} - 1)\coth\frac{\mu T}{2}\right)\{\lambda_2, -\lambda_1\}.$$

Since
$$\begin{bmatrix} q_1 \\ \dot{q}_1 \end{bmatrix} = \begin{bmatrix} 1 & 1 \\ \lambda_1 & \lambda_2 \end{bmatrix}\begin{bmatrix} \alpha_1 \\ \alpha_2 \end{bmatrix},$$

we obtain finally

$$q_1(t_s)-(-1)^s\frac{R_1}{C_{11}}\coth\left(\frac{\mu\pi}{2\omega}\right)=(-1)^s\left(q_1(t_0)-\frac{R_1}{C_{11}}\coth\frac{\mu\pi}{2\omega}\right)e^{s\mu\pi/\omega}. \quad \text{......(16)}$$

The condition that the motion can actually begin is (see (15) with $t_s = t_0$)

$$\lambda_1^2\alpha_1(t_0)+\lambda_2^2\alpha_2(t_0)+a_{11}R_1<0,$$

which reduces to

$$q_1(t_0)>\frac{R_1}{C_{11}}.$$

If this inequality is satisfied, and if μ is positive, it is evident from (16) that the motion will grow to an indefinitely large amplitude.

It should be noted that with systems having more than a single degree of freedom, growing oscillations strictly analogous to the simple type (16) cannot occur. With such systems, oscillations having a constant semi-period are also (in general) necessarily steadily maintained (see example to § 11·6).

11·5. Illustrative Treatment for Ankylotic Motion. For the sake of simplicity attention will again be restricted to the case where only a single coordinate q_m has frictional constraint. The analysis is thus supplementary to that of § 11·4.

Suppose, then, that at some particular stopping instant t_s ankylosis occurs. Since the coordinate q_m now remains arrested, the dynamical equations for the special case under consideration become

$$\left.\begin{array}{l} f_{11}(D)\,q_1+\dots+f_{1,m-1}(D)\,q_{m-1}=-C_{1m}q_m(t_s), \\ \dots\dots\dots\dots\dots\dots\dots\dots\dots\dots\dots\dots \\ f_{m-1,1}(D)q_1+\dots+f_{m-1,m-1}(D)\,q_{m-1}=-C_{m-1,m}q_m(t_s), \end{array}\right\} \quad \text{......(1)}$$

while the value of the frictional force ρ_m at any stage of the ankylotic motion is given by

$$f_{m1}(D)\,q_1+\dots+f_{m,m-1}(D)\,q_{m-1}+C_{mm}q_m(t_s)=\rho_m.$$

To render the previous analysis immediately applicable to equations (1), without the introduction of new symbols, the convention will now be made that any symbol originally used in relation to the m degrees of freedom $q_1, q_2, \dots, q_m$ will, if shown in clarendon type, have a similar significance in relation to the $m-1$ degrees of freedom $q_1, q_2, \dots, q_{m-1}$. Thus $\boldsymbol{y}(t)$ denotes the column of the $m-1$ displacements $q_1, q_2, \dots, q_{m-1}$ and $m-1$ velocities $\dot{q}_1, \dot{q}_2, \dots, \dot{q}_{m-1}$ ($n-2$ quantities in all); while $\boldsymbol{M}(t)$ and $\boldsymbol{l}$ are square matrices of order $n-2$ involving the fundamental constants of the system (1) and the $n-2$ roots $\boldsymbol{\lambda}_1, \boldsymbol{\lambda}_2, \ldots, \boldsymbol{\lambda}_{n-2}$ of the appropriate determinantal equation.

The solution of equations (1) then is

$$\boldsymbol{y}(t) - \boldsymbol{\Theta} = \boldsymbol{l}\boldsymbol{M}(t-t_s)\,\boldsymbol{l}^{-1}\{\boldsymbol{y}(t_s) - \boldsymbol{\Theta}\}. \qquad \ldots\ldots(2)$$

The column matrix $\boldsymbol{\Theta}$ has $n-2$ elements, of which the last $m-1$ are ciphers and the first $m-1$ are the constants $\boldsymbol{\theta}$ defined by the relations

$$C_{11}\boldsymbol{\theta}_1 + \ldots + C_{1,m-1}\boldsymbol{\theta}_{m-1} = -C_{1m}q_m(t_s),$$

$$\cdots\cdots\cdots\cdots\cdots\cdots\cdots\cdots\cdots\cdots$$

$$C_{m-1,1}\boldsymbol{\theta}_1 + \ldots + C_{m-1,m-1}\boldsymbol{\theta}_{m-1} = -C_{m-1,m}q_m(t_s).$$

The motion will continue in $m-1$ degrees of freedom in accordance with (2) until $\rho_m > R_m$ or $< -R_m$.

11·6. Steady Oscillations when only One Coordinate is Frictionally Constrained. As a relatively simple application of the theory the conditions will now be examined under which the oscillations of a system having only a single frictionally constrained coordinate q_m are *steady*. Oscillations will be said to be steady when they are of the special type in which the displacements and velocities in all the degrees of freedom at the stopping instant of any stroke in q_m are exact reversals of the corresponding displacements and velocities at the starting instant of the stroke.

Suppose that a typical down-stroke in q_m commences at time $t = t_0$ and ends at time $t = t_1$. Then the solution during that stroke is given by

$$\alpha(t) - R_m\gamma = M(t-t_0)\{\alpha(t_0) - R_m\gamma\}, \qquad \ldots\ldots(1)$$

and the displacements and velocities at time $t = t_1$ are exact reversals of those at time $t = t_0$ provided $\alpha(t_1) = -\alpha(t_0)$, so that

$$-\alpha(t_0) - R_m\gamma = M(T)\{\alpha(t_0) - R_m\gamma\},$$

where $T = t_1 - t_0$.

The last equation may be written

$$[M(T) + I]\,\alpha(t_0) = R_m[M(T) - I]\,\gamma,$$

whence

$$\alpha_r(t_0) = R_m\gamma_r\frac{e^{T\lambda_r}-1}{e^{T\lambda_r}+1} \qquad \ldots\ldots(2)$$

for all values $r = 1, 2, \ldots, n$.

The condition that q_m shall be stationary at time t_1 is $\mathfrak{k}_m\Lambda\alpha(t_1) = 0$. Since by hypothesis $\alpha_r(t_1) = -\alpha_r(t_0)$, and $\alpha_r(t_0)$ is given by (2), this condition requires that

$$R_m\sum_{r=1}^{n}\lambda_r k_{mr}\gamma_r\frac{e^{T\lambda_r}-1}{e^{T\lambda_r}+1} = 0.$$

This equation fixes the semi-period T, and the appropriate initial conditions at time $t = t_0$ are then given by (2). On substitution for $\alpha_r(t_0)$ from (2) in (1), the solution during the typical down-stroke in q_m is explicitly

$$q_s(t)/R_m = \sum_{r=1}^{n} k_{sr}\alpha_r(t)/R_m = \sum_{r=1}^{n} k_{sr}\gamma_r f(t-t_0, \lambda_r),$$

$$\dot{q}_s(t)/R_m = \sum_{r=1}^{n} \lambda_r k_{sr}\alpha_r(t)/R_m = \sum_{r=1}^{n} \lambda_r k_{sr}\gamma_r f(t-t_0, \lambda_r),$$

where $s = 1, 2, \ldots, m$, and

$$f(t-t_0, \lambda_r) \equiv 1 - \frac{2e^{(t-t_0)\lambda_r}}{e^{T\lambda_r}+1}.$$

The acceleration $\ddot{q}_m(t)$ at any instant in the down-stroke is readily seen to be given by

$$\ddot{q}_m(t)/R_m = \sum_{r=1}^{n} \lambda_r^2 k_{mr}\gamma_r f(t-t_0, \lambda_r) + a_{mm}.$$

The conditions for steady oscillation will now be summarised.

(A) The semi-periodic time must be a real positive root T of the equation

$$R_m\Omega(T) \equiv R_m \sum_{r=1}^{n} \lambda_r k_{mr}\gamma_r \frac{e^{T\lambda_r}-1}{e^{T\lambda_r}+1} = 0. \qquad \text{......(3)}$$

(B) The necessary initial disturbance, which is assumed to occur at the starting instant $t = t_0$ of a down-stroke in q_m, is given by

$$\left.\begin{aligned} q_s(t_0)/R_m &= \sum_{r=1}^{n} k_{sr}\gamma_r \frac{e^{T\lambda_r}-1}{e^{T\lambda_r}+1}, \\ \dot{q}_s(t_0)/R_m &= \sum_{r=1}^{n} \lambda_r k_{sr}\gamma_r \frac{e^{T\lambda_r}-1}{e^{T\lambda_r}+1}, \end{aligned}\right\} \qquad \text{......(4)}$$

where $s = 1, 2, \ldots, m$, and R_m is to be taken definitely with the positive sign.

(C) The initial displacement $q_m(t_0)$ must be positive* and the initial acceleration $\ddot{q}_m(t_0)$ negative: thus

$$q_m(t_0)/R_m = \sum_{r=1}^{n} k_{mr}\gamma_r \frac{e^{T\lambda_r}-1}{e^{T\lambda_r}+1} > 0, \qquad \text{......(5)}$$

and

$$\ddot{q}_m(t_0)/R_m = \sum_{r=1}^{n} \lambda_r^2 k_{mr}\gamma_r \frac{e^{T\lambda_r}-1}{e^{T\lambda_r}+1} + a_{mm} < 0. \qquad \text{......(6)}$$

* If $q_m(t_0)$ were negative and $\ddot{q}_m(t_0)$ also negative, the first stroke would still be downwards (as postulated), but the displacement q_m at the end of that stroke would then necessarily be negative, and not the reverse of $q_m(t_0)$.

(D) The velocity $\dot{q}_m(t)$ must not vanish within the interval (t_0, t_0+T). Hence the equation

$$\sum_{r=1}^{n} \lambda_r k_{mr} \gamma_r \left(1 - \frac{2e^{(t-t_0)\lambda_r}}{e^{T\lambda_r}+1}\right) = 0 \qquad \ldots\ldots(7)$$

must have no positive real root (other than $t-t_0 = 0$) less than the selected root $t-t_0 = T$ of (3).

Example

Oscillations with a Constant Period are (in general) also Steady. If T is the given constant period, the successive terminal instants for the strokes in q_m may be taken as $t_0 = 0$, $t_1 = T$, ..., $t_s = sT$. The recurrence relations for α may be written

$$\alpha(T) - R_m\gamma = M(T)\{\alpha(0) - R_m\gamma\},$$
$$\alpha(2T) + R_m\gamma = M(T)\{\alpha(T) + R_m\gamma\},$$
$$\alpha(3T) - R_m\gamma = M(T)\{\alpha(2T) - R_m\gamma\},$$

and so on. Addition of the equations in successive pairs gives

$$\alpha(T) + \alpha(2T) = M(T)\{\alpha(0) + \alpha(T)\},$$
$$\alpha(2T) + \alpha(3T) = M(T)\{\alpha(T) + \alpha(2T)\} = M(2T)\{\alpha(0) + \alpha(T)\},$$

and generally

$$\alpha(sT) + \alpha(\overline{s+1}\,T) = M(sT)\{\alpha(0) + \alpha(T)\}.$$

Premultiplication of the last equation by $\mathfrak{k}_m\Lambda$, and use of the conditions $\dot{q}_m(sT) = \dot{q}_m(\overline{s+1}\,T) = 0$ yields the relation

$$0 = \sum_{r=1}^{n} e^{sT\lambda_r} A_r, \qquad \ldots\ldots(8)$$

where $A_r \equiv \lambda_r k_{mr}(\alpha_r(0) + \alpha_r(T))$. Taking (8) for any n consecutive values of s, we derive a set of n equations, which will be compatible either if $\alpha(0) + \alpha(T) = 0$ (in which case the motion is steady and T is a root of (3)), or if

$$\begin{vmatrix} 1 & 1 & \ldots & 1 \\ e^{T\lambda_1} & e^{T\lambda_2} & \ldots & e^{T\lambda_n} \\ \ldots & \ldots & \ldots & \ldots \\ e^{(n-1)T\lambda_1} & e^{(n-1)T\lambda_2} & \ldots & e^{(n-1)T\lambda_n} \end{vmatrix} = 0.$$

This last equation would require $T = \pi/\omega$, where $\mu \pm i\omega$ are any two conjugate complex roots of $\Delta(\lambda) = 0$, say λ_1 and λ_2; and equations (8) would then be satisfied if in addition

$$A_1 + A_2 = 0 \quad \text{and} \quad A_3 = A_4 = \ldots = A_n = 0.$$

It can be shown without difficulty that the particular value $T = \pi/\omega$ would in this case still have to be a root of (3), and unless $m = 1$ this would demand a special relationship between the dynamical coefficients of the system. Hence, if the periodic time is constant, and if the system has more than a single degree of freedom, the motion is in general also steady.

11·7. Discussion of the Conditions for Steady Oscillations. The nature of the roots of the "semi-period equation" (11·6·3) will be considered first. Denote by $\xi_r + i\eta_r$ the complex value of $\lambda_r k_{mr} \gamma_r$ corresponding to a complex root $\lambda_r = \mu_r + i\omega_r$. Then

$$\Omega(T) \equiv \sum_{r=1}^{n} (\xi_r + i\eta_r) F(\lambda_r T), \qquad \text{......(1)}$$

where $$F(\lambda T) \equiv \frac{e^{(\mu+i\omega)T} - 1}{e^{(\mu+i\omega)T} + 1} = \frac{e^{\mu T}\cos\omega T - 1 + ie^{\mu T}\sin\omega T}{e^{\mu T}\cos\omega T + 1 + ie^{\mu T}\sin\omega T}.$$

(i) *Case of Two Purely Imaginary Roots of* $\Delta(\lambda) = 0$. Suppose now that two, and not more than two, of the roots λ are purely imaginary, so that for this pair $\mu = 0$. Then for the one root (say $+i\omega$),

$$F(i\omega T) = (\cos\omega T - 1 + i\sin\omega T)/(\cos\omega T + 1 + i\sin\omega T). \qquad \text{......(2)}$$

If $\omega T = (2s+1)\pi - \epsilon$, where s is an integer and ϵ is small, equation (2) gives to first order $F(i\omega T) = 2i/\epsilon$. The two terms of (1) corresponding to the two roots $\pm i\omega$ combine at this value of T into $-4\eta/\epsilon$. Hence unless $\eta = 0$ the function $\Omega(T)/\eta$ changes from $-\infty$ to $+\infty$ on passage of T through each of the values $(2s+1)\pi/\omega$.* It follows that when two (and not more than two) of the roots λ are purely imaginary and the corresponding value of η is not zero, the semi-period equation has an infinite number of real roots T separated by the values π/ω, $3\pi/\omega$, $5\pi/\omega$, etc. A diagrammatic representation of the graph of $\Omega(T)$ for the case considered, with η positive, is curve No. 1 of Fig. 11·7·1: the possibility of any odd number of roots T between successive asymptotes is not excluded. If η is negative, the sense of approach to the asymptotes is the reverse of that shown in the diagram.

On application of the identity (11·4·4) it is readily shown that the slope of the graph $\Omega(T)$ at $T = 0$ has always the negative value $-a_{mm}/2$. It follows that when $\eta < 0$, an odd number of real roots T exists between the origin and the first asymptote.

* For brevity, the symbol T is here temporarily used in two senses; it signifies a root of the semi-period equation, and also the current variable associated with the function $\Omega(T)$.

(ii) *Case of Two Conjugate Roots of $\Delta(\lambda)=0$ with Real Parts Small.* Next suppose the real part of one pair of roots $\mu \pm i\omega$ to be numerically very small but not actually zero. Then for the root $\mu+i\omega$ the value of $F(\lambda T)$ corresponding to $\omega T=(2s+1)\pi-\epsilon$ is approximately

$$F(\lambda T)=2\left((2s+1)\frac{\mu\pi}{\omega}+i\epsilon\right)\Big/\left((2s+1)^2\frac{\mu^2\pi^2}{\omega^2}+\epsilon^2\right), \quad \ldots\ldots(3)$$

so that the terms in (1) corresponding to the two roots $\mu \pm i\omega$ contribute

$$(\xi+i\eta)F(\lambda T)+(\xi-i\eta)F(\lambda T)$$
$$=4\left((2s+1)\frac{\mu\pi\xi}{\omega}-\epsilon\eta\right)\Big/\left((2s+1)^2\frac{\mu^2\pi^2}{\omega^2}+\epsilon^2\right).$$

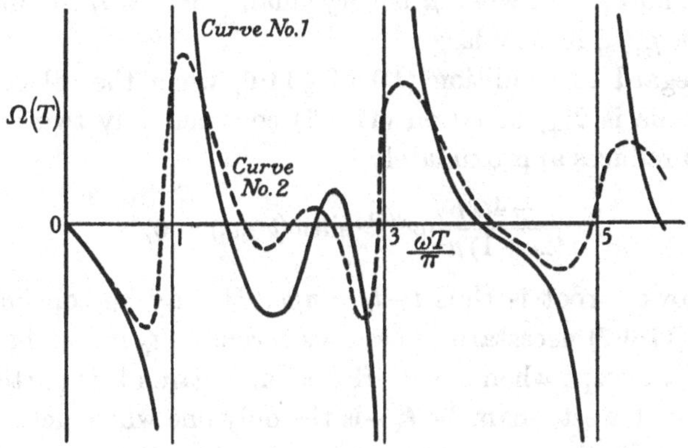

Fig. 11·7·1

This remains always finite, but changes from a large quantity of one sign to a large quantity of opposite sign as ϵ passes through the value $(2s+1)\mu\pi\xi/\omega\eta$. The graph of $\Omega(T)$ in the case where ξ, η and μ are all positive will therefore be as represented by curve No. 2 in Fig. 11·7·1. Hence a new group of real roots T has been gained, situated close to the values π/ω, $3\pi/\omega$, etc. If T_{s+1} denotes the $(s+1)$th of the new set of roots, then with $\epsilon=(2s+1)\mu\pi\xi/\omega\eta$ this root is approximately

$$T_{s+1}=(2s+1)\frac{\pi}{\omega}\left(1-\frac{\mu\xi}{\omega\eta}\right). \quad \ldots\ldots(4)$$

The corresponding value of $F(\lambda T_{s+1})$ given by (3) reduces to

$$F(\lambda T_{s+1})=\frac{2\omega\eta(\eta+i\xi)}{(2s+1)\mu\pi(\eta^2+\xi^2)},$$

and, if the important terms only are retained in (11·6·4), the initial displacement in q_m is approximately given by

$$q_m(t_0)/R_m = \frac{\xi+i\eta}{\mu+i\omega}F(\lambda T_{s+1}) + \frac{\xi-i\eta}{\mu-i\omega}F(\bar{\lambda}T_{s+1})$$
$$= 4\eta/(2s+1)\,\mu\pi.$$

Similarly, the initial acceleration is approximately given by

$$\ddot{q}_m(t_0)/R_m = -4\omega^2\eta/(2s+1)\,\mu\pi.$$

It follows that the conditions (C) of § 11·6 for a positive initial displacement and a negative initial acceleration are both satisfied if

$$\eta/\mu > 0. \qquad \text{......(5)}$$

It is to be noted that when μ is very small (and $\eta \neq 0$) the initial displacement $q_m(t_0)$ is very large.

With regard to condition (D) of § 11·6, when the selected semi-periodic time is T_{s+1} equation (11·6·7) contains only two important terms and reduces approximately to

$$\frac{-4\omega\eta}{(2s+1)\,\mu\pi}e^{\mu(t-t_0)}\sin\omega(t-t_0) = 0,$$

and the lowest root is thus $t-t_0 = \pi/\omega$. It may be concluded that equation (11·6·7) necessarily possesses a root $t-t_0$ lower than any of the set T_{s+1} except when $s = 0$. Hence, as regards this particular set of roots, the lowest—namely T_1—is the only one which actually leads to steady oscillations.

11·8. Stability of the Steady Oscillations. The stability of the steady oscillations when slightly disturbed can be investigated as follows. Let T denote the semi-periodic time for the undisturbed oscillations: thus T is assumed to be a root of (11·6·3). Also let A_r be the value of the typical reducing variable at the starting instant of any down-stroke in q_m in the steady oscillations, so that (see (11·6·2))

$$A_r = R_m\gamma_r\frac{e^{T\lambda_r}-1}{e^{T\lambda_r}+1}, \qquad \text{......(1)}$$

for $r = 1, 2, \ldots, n$.

For simplicity adopt $t_0 = 0$ as the datum starting instant and assume this to be appropriate to a down-stroke in q_m. Then in the undisturbed motion the successive terminal instants are 0, T, $2T$, ..., sT.

Now assume a small disturbance to occur at $t = 0$, and in the disturbed motion let the successive terminal instants for the strokes become $t_1 = T+\epsilon_1$, $t_2 = 2T+\epsilon_2$, ..., $t_s = sT+\epsilon_s$. Lastly, let $2R_m\gamma_r\delta_r$ be the increment of the typical initial reducing variable in the disturbance, so that for the disturbed motion

$$\alpha_r(t_0) = A_r + 2R_m\gamma_r\delta_r.$$

Only first-order terms in ϵ and δ will be retained in the analysis.

The formulae obtained in § 11·4 may now be used with the simplifications $\tau = t_0 = 0$. To illustrate the treatment, consider firstly the equation which gives the terminal instant t_1 in the disturbed motion (see (C) of § 11·4). This is in the present case

$$\sum_{r=1}^{n} \lambda_r k_{mr} e^{\lambda_r(T+\epsilon_1)}(-A_r - 2R_m\gamma_r\delta_r + R_m\gamma_r) = 0,$$

and on substitution for A_r from (1) and expansion of the exponential we have approximately

$$2R_m \sum_{r=1}^{n} \lambda_r k_{mr}\gamma_r e^{T\lambda_r}(1+\epsilon_1\lambda_r)\left(\frac{1}{e^{T\lambda_r}+1} - \delta_r\right) = 0.$$

The terms independent of ϵ and δ may be omitted in view of (11·6·3) and (11·4·3). If we write for brevity

$$e^{T\lambda_r} \equiv x_r, \quad \lambda_r^2 k_{mr}\gamma_r/(1+e^{T\lambda_r}) \equiv P_r, \quad \lambda_r k_{mr}\gamma_r\delta_r \equiv Q_r,$$

the equation gives to first order

$$\epsilon_1 \Sigma P_r x_r - \Sigma Q_r x_r = 0,$$

where the summations are taken for all the n roots λ_r.

A similar treatment of the more general equation (11·4·14) yields

$$\Sigma P_r x_r(\phi_s - \phi_{s-1}x_r + \phi_{s-2}x_r^2 + \ldots + (-1)^{s-2}\phi_2 x_r^{s-2} + (-1)^{s-1}\phi_1 x_r^{s-1}) + (-1)^s \Sigma Q_r x_r^s = 0,$$

in which $\phi_s \equiv \epsilon_s - \epsilon_{s-1}$ for $s > 1$ and $\phi_1 \equiv \epsilon_1$. The s relations which serve to determine the values of $\phi_1, \phi_2, \ldots, \phi_s$ are accordingly

$$\left.\begin{aligned}
\Sigma P_r x_r \phi_1 &= \Sigma Q_r x_r, \\
\Sigma P_r x_r(\phi_2 - x_r\phi_1) &= -\Sigma Q_r x_r^2, \\
\Sigma P_r x_r(\phi_3 - x_r\phi_2 + x_r^2\phi_1) &= \Sigma Q_r x_r^3, \\
&\cdots\cdots\cdots\cdots \\
\Sigma P_r x_r(\phi_s - x_r\phi_{s-1} + x_r^2\phi_{s-2} + \ldots & \\
+ (-1)^{s-2}x_r^{s-2}\phi_2 + (-1)^{s-1}x_r^{s-1}\phi_1) &= (-1)^{s-1}\Sigma Q_r x_r^s.
\end{aligned}\right\} \quad \ldots\ldots(2)$$

These can be solved by the following artifice. Let $z_1, z_2, \ldots, z_{n-1}$ denote the roots of the algebraic equation in z,

$$(z+x_1)(z+x_2)\ldots(z+x_n)\,\Sigma\frac{P_r x_r}{z+x_r}=0.$$

Then, for a general value of the variable z, we have

$$\frac{\Sigma\, Q_r x_r/(z+x_r)}{\Sigma\, P_r x_r/(z+x_r)}=\frac{(z+x_1)(z+x_2)\ldots(z+x_n)\,\Sigma\, Q_r x_r/(z+x_r)}{\text{Const.}\times(z-z_1)(z-z_2)\ldots(z-z_{n-1})}.$$

The numerator and the denominator of the rational fraction on the right are both of degree $n-1$ in z, and by resolution of the fraction into simple partial fractions we can therefore derive an identity of the form

$$\Sigma\frac{Q_r x_r}{z+x_r}=\left(E_0+\frac{E_1}{z-z_1}+\frac{E_2}{z-z_2}+\ldots+\frac{E_{n-1}}{z-z_{n-1}}\right)\Sigma\frac{P_r x_r}{z+x_r},\quad\ldots\ldots(3)$$

where $E_0, E_1, \ldots, E_{n-1}$ are constants which we shall not require to determine in the present discussion of the stability. Now if the modulus of z is assumed sufficiently great, the expressions on the left and on the right of (3) can legitimately be expanded in powers of $1/z$ to give

$$\sum_{r=1}^{n} Q_r x_r\left(\frac{1}{z}-\frac{x_r}{z^2}+\frac{x_r^2}{z^3}-\frac{x_r^3}{z^4}+\ldots\right)$$

$$=\left(E_0+\frac{1}{z}\sum_{i=1}^{n-1}E_i+\frac{1}{z^2}\sum_{i=1}^{n-1}E_i z_i+\ldots\right)\sum_{r=1}^{n}P_r x_r\left(\frac{1}{z}-\frac{x_r}{z^2}+\frac{x_r^2}{z^3}-\frac{x_r^3}{z^4}+\ldots\right).$$

Hence, on collecting together the coefficients of the separate powers of $1/z$ and equating the results to zero, we obtain the sequence of relations

$$\Sigma\, P_r x_r E_0=\Sigma\, Q_r x_r,$$

$$\Sigma\, P_r x_r\left(\sum_{i=1}^{n-1}E_i-x_r E_0\right)=-\Sigma\, Q_r x_r^2,$$

$$\Sigma\, P_r x_r\left(\sum_{i=1}^{n-1}E_i z_i-x_r\sum_{i=1}^{n-1}E_i+x_r^2 E_0\right)=\Sigma\, Q_r x_r^3,$$

and so on. A comparison with (2) shows immediately that $\phi_1=E_0$, and that for $s>1$

$$\phi_s\equiv\epsilon_s-\epsilon_{s-1}=\sum_{i=1}^{n-1}E_i z_i^{s-2}.\quad\ldots\ldots(4)$$

It may be concluded from (4) that if every modulus $|z_i|$ is less than unity, then ϕ_s tends to zero as s increases indefinitely; but that if any modulus exceeds unity, then ϕ_s tends to grow large. Now the time interval for the sth stroke in the disturbed motion is $T+\phi_s$. In the first case therefore the motion regains its original constant semi-period T, whereas in the second case the motion is unstable. The results of this investigation may be summarised as follows: *Let* $z_1, z_2, \ldots, z_{n-1}$ *denote the roots of the equation*

$$\sum_{r=1}^{n} \lambda_r^2 k_{mr} \gamma_r \frac{e^{T\lambda_r}}{1+e^{T\lambda_r}} \prod_{s \neq r} (z+e^{T\lambda_s}) = 0. \qquad \ldots\ldots(5)$$

Then the steady oscillations corresponding to the semi-periodic time T *will be stable or unstable according as every modulus* $|z_i| < 1$ *or as any modulus* $|z_i| > 1$.

Example

Case of Complex Pair of Roots with Real Part Small. A case of interest is that considered under heading (ii) of § 11·7, where the real part μ of one pair of conjugate roots $\lambda_1 = \mu + i\omega$ and $\lambda_2 = \mu - i\omega$ is numerically small. It was there shown that the lowest root of the series T_{s+1}, namely that root having the value $T_1 = \frac{\pi}{\omega}\left(1 - \frac{\mu\xi}{\omega\eta}\right)$, may be expected to lead to genuine steady oscillations provided that $\eta/\mu > 0$. If this value for T is adopted in (5), the equation contains only two important terms and reduces to

$$(z+e^{T_1\lambda_3})(z+e^{T_1\lambda_4})\ldots(z+e^{T_1\lambda_n})\,W = 0,$$

where $$W = (\mu+i\omega)(\xi+i\eta)\left(\frac{z+e^{T_1(\mu-i\omega)}}{1+e^{T_1(\mu+i\omega)}}\right)e^{T_1(\mu+i\omega)} + \text{conjugate}.$$

Now to first order $$e^{T_1(\mu+i\omega)} = -1 + \frac{\mu\pi}{\omega\eta}(-\eta+i\xi),$$

and $$\frac{e^{T_1(\mu+i\omega)}}{1+e^{T_1(\mu+i\omega)}} = \frac{i\omega\eta}{\mu\pi(\xi+i\eta)}.$$

Hence after some reduction we find

$$W = -\frac{2\omega^2\eta}{\mu\pi}\left(z - 1 - \frac{\mu\pi}{\omega}\right).$$

The approximate values of the roots z_i for the case considered are accordingly

$$z_1 = 1 + \frac{\mu\pi}{\omega}, \quad z_2 = -e^{\frac{\pi\lambda_3}{\omega}}, \quad \ldots, \quad z_{n-1} = -e^{\frac{\pi\lambda_n}{\omega}}. \qquad \ldots\ldots(6)$$

Consequently, if the real parts of all the roots λ other than λ_1 and λ_2 are negative, then the motion is stable or unstable according as $\mu < 0$ or > 0.

11·9. A Graphical Method for the Complete Motion of Binary Systems. The computation of the motion which follows any given disturbance is, in general, a matter of considerable difficulty. In the special case of a system having two coordinates q_1, q_2, of which q_2, say, is frictionally constrained, the solution can be greatly simplified by the use of a graphical treatment, provided that only rough results are required. Some remarks follow regarding the underlying principles of the method.

The problem for solution is the description of the complete motion which results from any given initial disturbance. For simplicity we shall assume that this disturbance is imposed at the starting instant t_0 of a down-stroke in q_2, so that initially $\dot{q}_2(t_0) = 0$. Attention will be restricted to the case in which the four roots of $\Delta(\lambda) = 0$ are complex. For the purpose of the graphical method it is convenient to denote these roots by

$$\lambda_1 = \mu + i\omega, \quad \lambda_2 = \mu - i\omega,$$
$$\lambda_3 = \mu' + i\omega', \quad \lambda_4 = \mu' - i\omega',$$

and to write

$$\lambda_1 k_{21} \alpha_1(t_s)/R_2 \equiv X_s + iY_s \equiv Z_s,$$
$$\lambda_2 k_{22} \alpha_2(t_s)/R_2 \equiv X_s - iY_s \equiv \bar{Z}_s,$$
$$\lambda_3 k_{23} \alpha_3(t_s)/R_2 \equiv X'_s + iY'_s \equiv Z'_s,$$
$$\lambda_4 k_{24} \alpha_4(t_s)/R_2 \equiv X'_s - iY'_s \equiv \bar{Z}'_s,$$

together with

$$\lambda_1 k_{21} \gamma_1 \equiv \xi + i\eta \equiv \zeta,$$
$$\lambda_2 k_{22} \gamma_2 \equiv \xi - i\eta \equiv \bar{\zeta},$$
$$\lambda_3 k_{23} \gamma_3 \equiv \xi' + i\eta' \equiv \zeta',$$
$$\lambda_4 k_{24} \gamma_4 \equiv \xi' - i\eta' \equiv \bar{\zeta}'.$$

Then if $\tau_s \equiv t_s - t_{s-1}$ denotes the time interval for the sth stroke in q_2, the recurrence relation (11·4·11) yields for the special system considered the four scalar equations

$$\left.\begin{aligned} Z_s + (-1)^s \zeta &= (Z_{s-1} + (-1)^s \zeta)\, e^{(\mu+i\omega)\tau_s}, \\ \bar{Z}_s + (-1)^s \bar{\zeta} &= (\bar{Z}_{s-1} + (-1)^s \bar{\zeta})\, e^{(\mu-i\omega)\tau_s}, \\ Z'_s + (-1)^s \zeta' &= (Z'_{s-1} + (-1)^s \zeta')\, e^{(\mu'+i\omega')\tau_s}, \\ \bar{Z}'_s + (-1)^s \bar{\zeta}' &= (\bar{Z}'_{s-1} + (-1)^s \bar{\zeta}')\, e^{(\mu'-i\omega')\tau_s}, \end{aligned}\right\} \qquad \text{......(1)}$$

while the condition (11·4·3) requires that

$$\xi + \xi' = 0. \qquad \text{......(2)}$$

Further, since $\dot{q}_2(t_s) = 0$ it follows from (11·4·13) that

$$X_s + X'_s = 0. \qquad \text{......(3)}$$

The foregoing relations may be interpreted geometrically by the use of two pairs of diagrams. Figs. 11·9·1 (*a*) and (*b*) represent the pair appropriate to the unaccented symbols, while two similar diagrams,

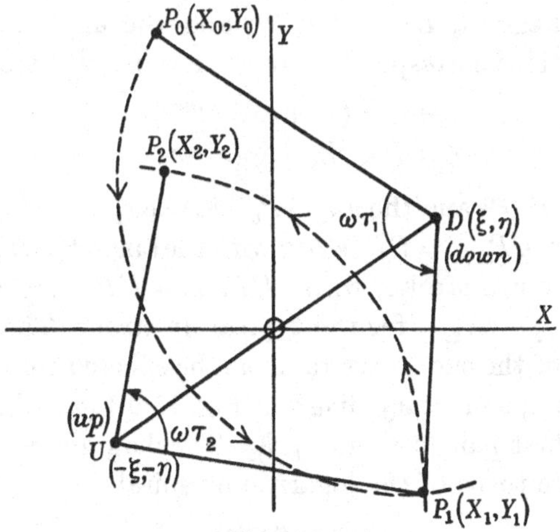

Fig. 11·9·1 (*a*)

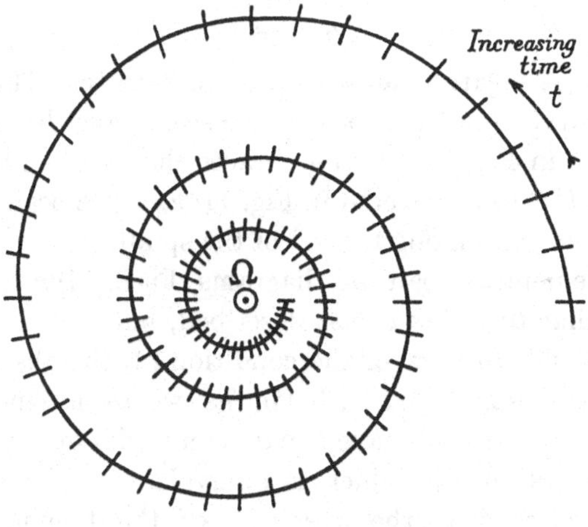

Fig. 11·9·1 (*b*)

referred to hereafter as Figs. 11·9·1 (*a′*) and (*b′*) but not actually drawn, would relate to the accented symbols. A description of Figs. 11·9·1 (*a*) and (*b*) follows.

In Fig. 11·9·1 (*a*) the points D and U have respectively the coordinates (ξ, η) and $(-\xi, -\eta)$, and are respectively marked "down" and "up"; while the points P_0, P_1, etc. have the coordinates (X_0, Y_0),

(X_1, Y_1), etc. Now the equations (1) corresponding to the first down-stroke are

$$Z_1 - \zeta = (Z_0 - \zeta)\, e^{(\mu+i\omega)\tau_1},$$
$$\bar{Z}_1 - \bar{\zeta} = (\bar{Z}_0 - \bar{\zeta})\, e^{(\mu-i\omega)\tau_1}.$$

Hence, in the diagram, $DP_1 = DP_0 e^{\mu\tau_1}$ and the angle $P_0DP_1 = \omega\tau_1$. Again, the equations corresponding to the ensuing up-stroke are

$$Z_2 + \zeta = (Z_1 + \zeta)\, e^{(\mu+i\omega)\tau_2},$$
$$\bar{Z}_2 + \bar{\zeta} = (\bar{Z}_1 + \bar{\zeta})\, e^{(\mu-i\omega)\tau_2},$$

so that $UP_2 = UP_1 e^{\mu\tau_2}$ and the angle $P_1UP_2 = \omega\tau_2$.

More generally, $DP_{2s+1} = DP_{2s} e^{\mu\tau_{2s+1}}$ with the angle $P_{2s}DP_{2s+1} = \omega\tau_{2s+1}$ for the typical down-stroke, while $UP_{2s+2} = UP_{2s+1} e^{\mu\tau_{2s+2}}$ with the angle $P_{2s+1}UP_{2s+2} = \omega\tau_{2s+2}$ for the typical up-stroke. These rotations and expansions of the successive radii can be effected conveniently by the use of the supplementary diagram Fig. 11·9·1 (*b*), which depends solely upon the first pair of roots λ_1, λ_2. The diagram consists of one or more complete turns of the logarithmic spiral

$$x + iy = e^{(\mu+i\omega)t}$$

or

$$r = e^{\mu t},$$
$$\theta = \omega t,$$

where t (time) is regarded as a variable parameter. The number of turns of the curve, and the choice of the scale, must be such that the radius vector r in Fig. 11·9·1 (*b*) embraces the range of values of the radii DP and UP to be covered in Fig. 11·9·1 (*a*). A scale for the time parameter t is marked along the arc of the spiral.

The complementary pair of diagrams Figs. 11·9·1 (*a'*) and (*b'*) would be similar to the two just described, but would relate to the accented symbols. In view of the condition (2) the abscissae for the homologous centres D, D' (or U, U') in the two displacement diagrams Figs. 11·9·1 (*a*) and (*a'*) will have equal magnitudes but opposite signs. Again, the condition (3)—which is equivalent to the sth terminal equation—requires that the abscissae of the homologous points P_s, P'_s shall also be equal and opposite.

The actual manipulation of the two pairs of diagrams may now be explained. For the two displacement diagrams Figs. 11·9·1 (*a*) and (*a'*) transparent graph paper is used; whereas the two spiral diagrams Figs. 11·9·1 (*b*) and (*b'*) are preferably prepared in ink on white unruled paper. It should be noted that the positions of the centres D, U and D', U' in the two displacement diagrams are fixed by the dynamical constants and are independent of the initial disturbance. The first step in the actual solution is to mark in Figs. 11·9·1 (*a*) and (*a*′) the positions P_0, P_0', corresponding to the chosen disturbance: since the disturbance is assumed to occur at the starting instant of a down-stroke in q_2, the abscissae of the initial points will be equal and opposite. To find the positions of P_1 and P_1' hold sheet 11·9·1 (*a*) superposed on sheet 11·9·1 (*b*) with D registered on the pole Ω of the first spiral and P_0 on the arc of the curve; similarly, hold sheet 11·9·1 (*a*′) superposed on sheet 11·9·1 (*b*′) with D' registered on Ω' and P_0' on the second spiral. Then travel continuously in the anticlockwise direction along the spirals to the first positions P_1, P_1' for which the increments of the parameter t are the same (τ_1) in the two diagrams, and the abscissae are again equal and opposite. To find P_2, P_2' from P_1, P_1', the procedure is similar, but the up-centres U, U', instead of the down-centres D, D', are in this case registered on the poles of the spirals. Two observers—one for each pair of diagrams—are necessary.

It is possible to obtain the full history of the complete motion resulting from an arbitrary disturbance by the use of the foregoing method. The velocities and displacements at any time t, not necessarily a terminal instant, can clearly be deduced by expressing the formulae (11·4·9) and (11·4·10) in terms of the current coordinates of the points $P(t)$, $P'(t)$ along the spirals. In particular, the value of the velocity $\dot{q}_2(t)/R_2$ at any stage is given by twice the sum of the current abscissae. A numerical illustration is given in example (iv) of § 12·9.

It should be noted that the condition (11·4·15) for ankylosis at the terminal instant t_s is expressible in the present modified notation as

$$2(-1)^s(\mu X_s - \omega Y_s + \mu' X_s' - \omega' Y_s') + A_{11}/p_0 \geqslant 0,$$

where $p_0 \equiv A_{11}A_{22} - A_{12}A_{21}$.

This inequality can if necessary be tested from time to time as the work proceeds. An indication of the occurrence of ankylosis is, however, provided by the graphical method itself. For, in any genuine down-stroke, the starting acceleration in the frictionally constrained coordinate must be negative, while in any genuine up-stroke it must be positive. At the starting instant itself the velocity is zero. Hence, if, for instance, a down-stroke is due, and if after one time step, or fraction of a step (say δt) the sum of the current abscissae of P and P' is negative, then $\delta\dot{q}_2/\delta t$ is negative and the stroke is realisable: whereas, if $\delta\dot{q}_2/\delta t$ is positive, ankylosis is indicated.

CHAPTER XII

ILLUSTRATIVE APPLICATIONS OF FRICTION THEORY TO FLUTTER PROBLEMS

12·1. Introductory. (*a*) *Flutter of Frictionless Aeroplane Structures.* In the simple theory of flutter* it is assumed that no solid friction is present in the aeroplane structure and that linear laws remain applicable throughout the motion. In this ideal case the ordinary

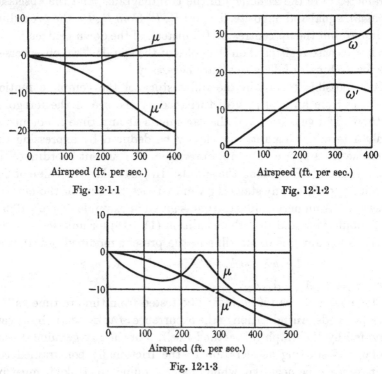

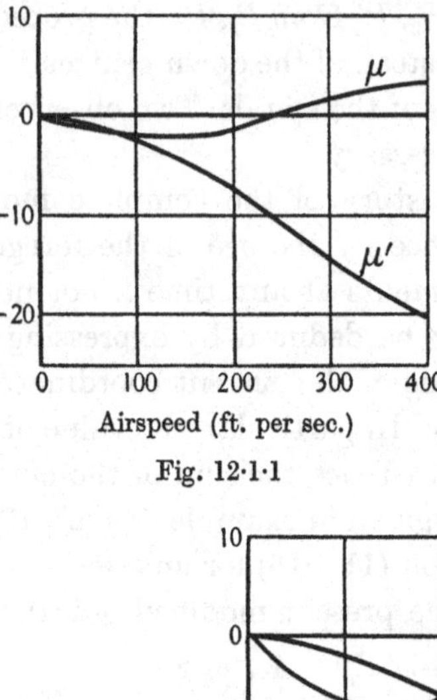

Fig. 12·1·1

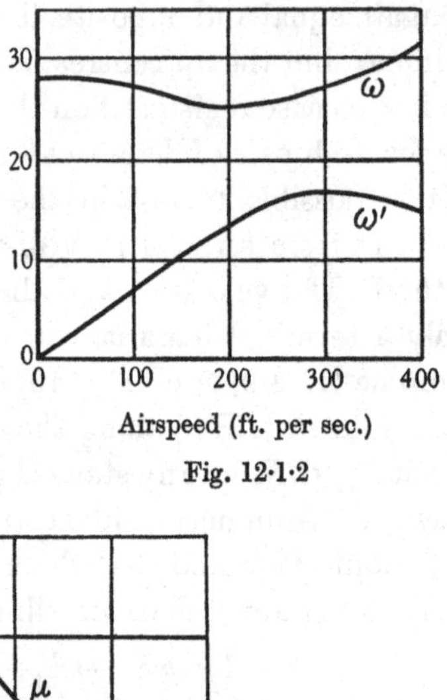

Fig. 12·1·2

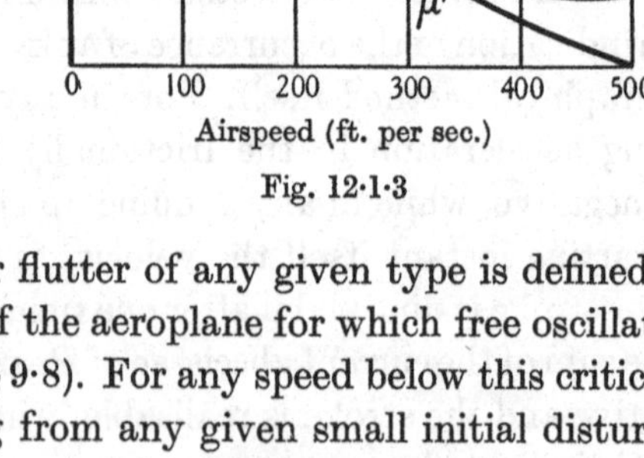

Fig. 12·1·3

critical speed for flutter of any given type is defined to be the lowest forward speed of the aeroplane for which free oscillations of that type are steady (see § 9·8). For any speed below this critical value the oscillations resulting from any given small initial disturbance eventually die away; at the actual critical speed the motion tends to become

* See for instance Ref. 30.

simply sinusoidal; while for all speeds over a certain range whose lower limit is the critical speed, oscillations occur which increase to an indefinitely large amplitude, however small the initial disturbance may be. The term critical speed will in the present chapter be used with the foregoing significance, and will, without further specific qualification, always refer to the ideal frictionless aeroplane. It is clear that at the critical speed for flutter the determinantal equation will have at least one pair of conjugate purely imaginary roots. It will be convenient to refer to one or other of these particular roots as the *critical root*.

The flutter characteristics of an aeroplane can be represented by means of two diagrams, showing respectively the variation of damping factor μ (the real part of $\lambda = \mu \pm i\omega$) and frequency $\omega/2\pi$ with airspeed, for the several constituents of the motion. Figs. 12·1·1 and 12·1·2 are examples of a pair of such diagrams.* They relate to the rudder-fuselage flutter of an aeroplane which is referred to as aeroplane No. 1 in the sequel. The curves marked μ and ω correspond to an oscillatory constituent which is damped for all speeds less than the critical value $V_c = 240$ feet per second, and grows indefinitely for all higher speeds. The second constituent μ', ω' remains a decaying oscillation throughout the range of speeds covered by the diagrams.†

It is possible for an aeroplane to be free from flutter throughout the range of its flying speeds, but to have a very small margin of stability in the vicinity of a particular flying speed. Fig. 12·1·3, which is purely diagrammatic, illustrates this condition.

(*b*) *Flutter in Practice.* In the practical sense "flutter" means an oscillation which grows, and finally either breaks the structure or remains bounded at some amplitude whose value is dependent upon the departures from linear laws. As pointed out in § 11·1 (*e*), solid friction introduces a very special departure from linearity.

(*c*) *Applications to be Considered.* The numerical examples to be given are restricted to tail oscillations involving angular displacement of the rudder and torsion of the fuselage. Moreover, they relate mainly to the simplest aspect of the theory—namely, the question of steady oscillation with either the rudder or the fuselage frictionally constrained.

* The curves drawn are based on data given in Chap. v of Ref. 42.

† At a speed of about $V = 580$ this constituent in fact degenerates into a pair of subsidences.

It will be useful to review briefly the conditions for the existence and stability of steady oscillations in these applications.

In accordance with the notation of § 11·7 we shall suppose the frictionally constrained coordinate (rudder or fuselage) to be q_2, and $\xi + i\eta$ to be the value of $\lambda_r k_{2r} \gamma_r$ for the critical speed, where λ_r is the critical root. If this root becomes $\mu + i\omega$ when the airspeed is slightly changed, then μ will be small and will be positive or negative according as the airspeed is above or below its critical value. It follows from equations (11·7·5) and (11·8·6) that if $\eta > 0$, then for a certain range of the speed above the critical steady oscillations will be possible, but will be unstable and therefore not realisable in practice. On the other hand for speeds below the critical, since $\mu < 0$, stable steady oscillations will be possible provided that $\eta < 0$. The particular oscillations here considered are those appropriate to the root T_1 of the semi-period equation which lies adjacent to the first asymptote (see Fig. 11·7·1).

Unfortunately, the numerical data available for the calculations are very scanty. The first example given is that of an actual aeroplane (aeroplane No. 1), for which, when solid friction acts either on the rudder or the fuselage, steady oscillations are theoretically possible only at speeds above the critical; as already explained, these oscillations are unstable. Aeroplane No. 2 is an artificial system, in the sense that its dynamical constants have been derived from those of aeroplane No. 1 by a transformation which changes the dynamical constants of the tail system, but leaves the roots of the determinantal equation unaltered at all speeds. Hence the tail flutter characteristics of the two aeroplanes are effectively identical. Nevertheless, when friction is present, the behaviour of aeroplane No. 2 is different from that of No. 1. On the modified aeroplane, if the friction acts on the rudder only, stable steady oscillations can occur at speeds below the critical. The graphical method of § 11·9 is applied to provide a description of the complete tail oscillations under representative initial disturbances. The final example relates to another artificial system (aeroplane No. 3) which is completely stable at all speeds when frictionless, but which is shown to admit bounded oscillations at a particular speed when friction is introduced.

A rather more detailed summary follows of the main conclusions for the three different aeroplanes considered.

(*d*) *Characteristics of Aeroplane No.* 1 (see Part I). The numerical data for this case are appropriate to a full scale aeroplane, the tail flutter characteristics of which have in fact been investigated by means of a model in a wind tunnel. From Fig. 12·1·1 it will be seen that the critical speed is about 240 feet per second.

The results when the friction acts on the rudder only are as follows:

(i) *Speeds below the critical.* The motion dies for all disturbances.

(ii) *Speeds above the critical.* Large disturbances produce flutter and small ones decaying motion. Steady oscillations (constant amplitude and frequency) could result from a special initial disturbance, which varies with the airspeed and is directly proportional to the limiting friction. The steady oscillations appropriate to any particular speed are unstable, so that any slight variation from the correct initial disturbance will give rise either to a decaying motion or to flutter. Figs. 12·3·2 and 12·4·1 show the steady oscillations for the airspeeds 260 feet per second and 500 feet per second, respectively.

When the frictional moment is applied to the fuselage only, the results are similar.

(*e*) *Characteristics of Aeroplane No.* 2 (see Part II). Aeroplanes Nos. 1 and 2 have the same damping-factor and frequency diagrams, and therefore effectively the same flutter characteristics. But their responses to disturbances at speeds below the critical are quite different in the case where the friction acts on the rudder. On aeroplane No. 1, if the airspeed is less than the critical value 240, the oscillations decay. This also occurs with aeroplane No. 2, provided the airspeed is less than about 215. On the other hand, for any speed within the range 215–240 two distinct sets of steady oscillations are possible, one (small amplitude) of unstable type and the other (large amplitude) of stable type. Consequently, very small disturbances produce motions which die; moderate ones give rise to oscillations which grow and finally tend to the large amplitude stable steady motion: still greater disturbances result in oscillations which decrease to the same stable motion.

The histories of the motion due to three different initial disturbances, for an airspeed of 230, are represented in Fig. 12·9·2. Curve *B* is the unstable (small amplitude) steady motion, the angular amplitude of the rudder in this motion being about 0·039 degree per foot-pound of (limiting) frictional hinge moment applied to the rudder. The corresponding amplitude for the stable steady oscillations (not actually shown in the diagram) is 0·278 degree per foot-pound of friction. Any disturbance similar to, but less than, that required to initiate oscillations B, gives rise to a motion which decays (e.g. oscillations A): whereas any similar disturbance greater than that corresponding to motion B produces oscillations which increase to a limiting amplitude of 0·278 degree per foot-pound of friction (e.g. oscillations C). A maximum amplitude increase of about 7 : 1 is thus to be expected at the particular airspeed considered.

The theory shows that the amplitude of the stable steady oscillations increases very rapidly as the critical speed is approached. On the other hand, the amplitude of the unstable set is not greatly affected. Hence the effect of the friction in the present instance virtually amounts to a slight, and somewhat indefinite, reduction of the critical speed.

(*f*) *Characteristics of Aeroplane No.* 3 (see Part III). This aeroplane is free from tail flutter at all airspeeds, but its margin of stability is low. The feature of interest is the fact that at an airspeed of 230, its oscillatory characteristics, when friction is present, are identical with those of aeroplane No. 2. This shows the possibility of bounded free oscillations occurring on an aeroplane which has no true critical speed.

PART I. AEROPLANE NO. 1

12·2. Numerical Data. The system here considered is the aeroplane of which the tail flutter characteristics are discussed in Chapter v of Ref. 42. The two degrees of freedom* correspond to angular displacement q_1 of the tail unit in torsion (measured positively when the starboard tailplane moves downwards) and angular displacement q_2 of the rudder (measured positively when the rudder trailing edge moves

Table 12·2·1

Dynamical Coefficients for Aeroplane No. 1

Fuselage (q_1)		Rudder (q_2)	
Coefficient	Value	Coefficient	Value
A_{11}	44·7	A_{21}	−1·15
B_{11}	1·77 V	B_{21}	0·041 V
C_{11}	33,700	C_{21}	0
A_{12}	−1·15	A_{22}	0·745
B_{12}	−0·186 V	B_{22}	0·034 V
C_{12}	−0·101 V^2	C_{22}	0·00358 V^2

* The coordinates q_1, q_2 correspond respectively to Ω and ξ of Chap. v of Ref. 42.

to port). The dynamical coefficients appropriate to the frictionless aeroplane are reproduced in Table 12·2·1 from Table 43 of Ref. 42: for simplicity the gravitational cross stiffness is neglected.

Except where otherwise stated angles are assumed to be measured in radians, moments in pounds feet, stiffnesses in pounds feet per radian, and airspeeds V in feet per second.

The determinantal equation for a general airspeed V is

$$p_0\lambda^4+p_1\lambda^3+p_2\lambda^2+p_3\lambda+p_4 = 0, \qquad \text{......(1)}$$

where

$$p_0 = 31{\cdot}979,$$
$$p_1 = 2{\cdot}6717V,$$
$$p_2 = 25106{\cdot}5+0{\cdot}111682V^2,$$
$$p_3 = 1145{\cdot}8V+0{\cdot}0104776V^3,$$
$$p_4 = 120{\cdot}646V^2.$$

Table 12·2·2 gives the calculated roots for a number of different airspeeds: the four roots are denoted by

$$\lambda_1 = \mu+i\omega, \quad \lambda_2 = \mu-i\omega, \quad \lambda_3 = \mu'+i\omega', \quad \lambda_4 = \mu'-i\omega'.$$

Table 12·2·2

Roots of Determinantal Equation for Aeroplane No. 1

(Also applicable to Aeroplane No. 2)

V	μ	ω	μ'	ω'
0	0	28·0195	0	0
100	−1·6932	27·3904	−2·4838	6·6277
200	−1·3826	25·5809	−6·9719	13·4658
230	−0·3590	25·4747	−9·2487	14·8973
239·807	0·0	25·5811	−10·0174	15·2049
260	0·6868	25·9749	−11·5477	15·6327
280	1·2684	26·5275	−12·9647	15·8515
300	1·7600	27·1810	−14·2929	15·9176
400	3·1912	31·2253	−19·9002	14·7198
500	3·7493	35·9271	−24·6357	10·7665

The critical speed for flutter is thus 239·8.

12·3. Steady Oscillations on Aeroplane No. 1 at $V = 260$. (Rudder Frictionally Constrained.) The calculations for this case will be explained in some detail. The dynamical coefficients, as deduced from Table 12·2·1 with $V = 260$, given in Table 12·3·1.

Further (see Table 12·2·2),

$$\lambda_1, \lambda_2 \equiv \mu \pm i\omega \quad = \quad 0{\cdot}6868 \pm 25{\cdot}9749i,$$
$$\lambda_3, \lambda_4 \equiv \mu' \pm i\omega' = -11{\cdot}5477 \pm 15{\cdot}6327i.$$

Table 12·3·1

Dynamical Constants for Aeroplane No. 1 at V = 260.

Fuselage (q_1)		Rudder (q_2)	
Coefficient	Value	Coefficient	Value
A_{11}	44·7	A_{21}	−1·15
B_{11}	460·2	B_{21}	10·66
C_{11}	33,700	C_{21}	0
A_{12}	−1·15	A_{22}	0·745
B_{12}	−48·36	B_{22}	8·84
C_{12}	−6827·6	C_{22}	242·008

The values of the constants $\lambda_r k_{2r} \gamma_r$ are given by (11·4·1), and the following formula is typical:

$$\lambda_1 k_{21} \gamma_1 = -\frac{F_{22}(\lambda_1)}{\overset{(1)}{\Delta}(\lambda_1)} = -\frac{1}{\overset{(1)}{\Delta}(\lambda_1)}(A_{11}\lambda_1^2 + B_{11}\lambda_1 + C_{11}). \quad \text{......(1)}$$

The results are

$$\lambda_1 k_{21} \gamma_1 \equiv \xi + i\eta \;\; = (7{\cdot}81516 + 9{\cdot}38083i)\, 10^{-3}, \quad \text{......(2)}$$

$$\lambda_3 k_{23} \gamma_3 \equiv \xi' + i\eta' = (-7{\cdot}81516 + 35{\cdot}23619i)\, 10^{-3}, \quad \text{......(3)}$$

while $\lambda_2 k_{22} \gamma_2$ and $\lambda_4 k_{24} \gamma_4$ are respectively the conjugates of (2) and (3). It should be noted that an immediate check of the accuracy is here provided by the condition (see (11·4·3))

$$\sum_{r=1}^{4} \lambda_r k_{2r} \gamma_r = 0. \quad \text{......(4)}$$

The next step is the solution of the semi-period equation (11·6·3), namely

$$\Omega(T) \equiv \sum_{r=1}^{4} \lambda_r k_{2r} \gamma_r \frac{e^{T\lambda_r} - 1}{e^{T\lambda_r} + 1} = 0. \quad \text{......(5)}$$

Now
$$\frac{e^{T\lambda_1} - 1}{e^{T\lambda_1} + 1} = \frac{\sinh T\mu + i \sin T\omega}{\cosh T\mu + \cos T\omega},$$

and the expression appropriate to the root λ_3 is similar. Hence (5) is reducible to

$$\tfrac{1}{2}\Omega(T) \equiv \frac{\xi \sinh T\mu - \eta \sin T\omega}{\cosh T\mu + \cos T\omega} + \frac{\xi' \sinh T\mu' - \eta' \sin T\omega'}{\cosh T\mu' + \cos T\omega'} = 0.$$

Curve No. 2 of Fig. 12·3·1 is part of the graph of $\Omega(T)$ plotted against T. The lowest root of the set discussed in § 11·7 works out, on close approximation, to $T_1 = 0{\cdot}11856$. Hence

$$\frac{e^{T_1\lambda_1}-1}{e^{T_1\lambda_1}+1} = 15{\cdot}56+11{\cdot}81i,$$

$$\frac{e^{T_1\lambda_3}-1}{e^{T_1\lambda_3}+1} = -1{\cdot}0136+0{\cdot}5293i.$$

The initial displacement $q_2(0)/R_2$ and initial acceleration $\ddot{q}_2(0)/R_2$ are next calculable by the formulae (11·6·5) and (11·6·6), namely

$$\frac{q_2(0)}{R_2} = \sum_{r=1}^{4} k_{2r}\gamma_r \frac{e^{T\lambda_r}-1}{e^{T\lambda_r}+1}, \qquad \text{......(6)}$$

$$\frac{\ddot{q}_2(0)}{R_2} = a_{22} + \sum_{r=1}^{4} \lambda_r^2 k_{2r}\gamma_r \frac{e^{T\lambda_r}-1}{e^{T\lambda_r}+1}. \qquad \text{......(7)}$$

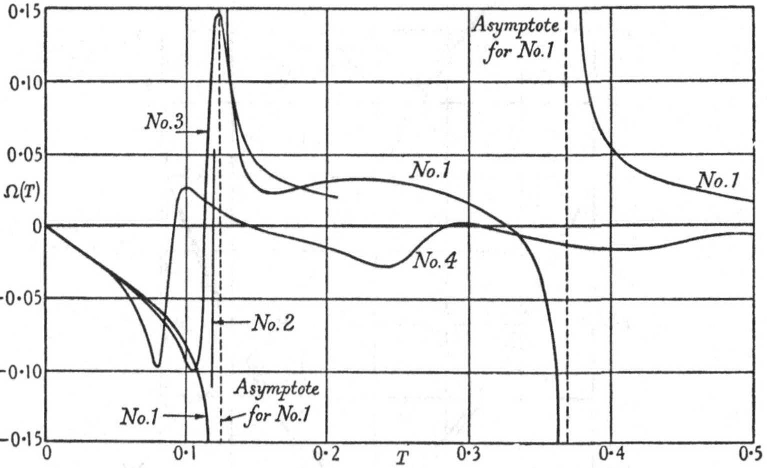

Fig. 12·3·1

From (2) and (3)

$$k_{21}\gamma_1 = (368{\cdot}847-291{\cdot}122i)\,10^{-6}, \qquad \text{......(8)}$$

$$k_{23}\gamma_3 = (1697{\cdot}198-753{\cdot}774i)\,10^{-6}, \qquad \text{......(9)}$$

$$\lambda_1^2 k_{21}\gamma_1 = -0{\cdot}238299+0{\cdot}209441i,$$

$$\lambda_3^2 k_{23}\gamma_3 = -0{\cdot}460581-0{\cdot}529072i.$$

Again $\qquad a_{22} \equiv A_{11}/p_0 = 1{\cdot}39779.$

At this stage several checks on the accuracy are possible. Firstly, from (11·4·2) and the definition of the constants ϕ it follows that

$$\phi_2 = C_{11}/p_4 = \sum_{r=1}^{4} k_{2r}\gamma_r.$$

The value of C_{11}/p_4 works out as $4132{\cdot}09 \times 10^{-6}$, which is in exact agreement with the value of $\sum_{r=1}^{4} k_{2r}\gamma_r$ given by equations (8) and (9).

Secondly (see (11·4·4)),

$$\sum_{r=1}^{4} \lambda_r^2 k_{2r}\gamma_r = -a_{22} = -A_{11}/p_0,$$

and the calculated values are $-a_{22} = -1{\cdot}39779$ and

$$\sum_{r=1}^{4} \lambda_r^2 k_{2r}\gamma_r = -1{\cdot}39776.$$

The numerical checks are thus satisfactory.

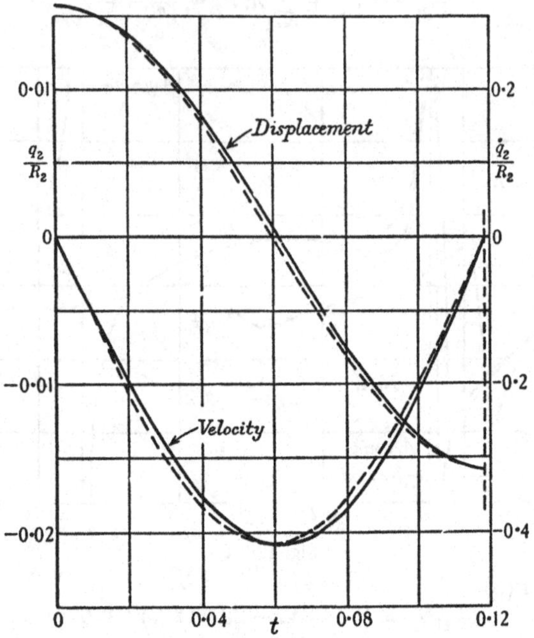

Fig. 12·3·2

From equations (6) and (7) we find that

$$q_2(0)/R_2 = +0{\cdot}015712,$$
$$\ddot{q}_2(0)/R_2 = -9{\cdot}4713.$$

Hence conditions (C) of § 11·6 are satisfied. Conditions (D) require that the velocity $\dot{q}_2(t)$ shall not vanish within the interval $t = 0$ to $t = T$. To decide this question we shall determine the actual motion of the rudder (q_2) during the first stroke. The displacement at any time t

during this stroke is

$$\frac{q_2(t)}{R_2} = \sum_{r=1}^{4} k_{2r}\gamma_r\left(1-\frac{2e^{t\lambda_r}}{1+e^{T\lambda_r}}\right),$$

where $T = 0{\cdot}11856$; while the velocity is given by the formula

$$\frac{\dot{q}_2(t)}{R_2} = -2\sum_{r=1}^{4} \lambda_r k_{2r}\gamma_r \frac{e^{t\lambda_r}}{1+e^{T\lambda_r}}.$$

The results, which are shown plotted in Fig. 12·3·2, show that all the required conditions are satisfied. The dotted curves correspond to a simple harmonic motion, and are added for comparison.

Equation (11·8·5) reduces in the present case to the cubic

$$17{\cdot}0109z^3 - 19{\cdot}9916z^2 + 2{\cdot}50234z - 1 = 0.$$

There is one real root $z_1 = 1{\cdot}0897$, and two conjugate complex roots having the modulus 0·2322. Since $|z_1| > 1$, the oscillations are unstable. It may be noted that the value of z_1 agrees quite well with the approximate value 1·083 given by the formula (11·8·6).

12·4. Steady Oscillations on Aeroplane No. 1 at Various Speeds. (Rudder Frictionally Constrained.)

A summary of results for other airspeeds will now be given. Table 12·4·1 gives the values of the constants $\lambda_1 k_{21}\gamma_1$ and $\lambda_3 k_{23}\gamma_3$.

Table 12·4·1

Values of $\lambda_1 k_{21}\gamma_1$ and $\lambda_3 k_{23}\gamma_3$ for Aeroplane No. 1 (Rudder Frictionally Constrained)

V	$\lambda_1 k_{21}\gamma_1 \times 10^3$	$\lambda_3 k_{23}\gamma_3 \times 10^3$
239·807 (critical)	$9{\cdot}6379 + 6{\cdot}8777i$	$-9{\cdot}6379 + 40{\cdot}7463i$
260	$7{\cdot}8152 + 9{\cdot}3808i$	$-7{\cdot}8152 + 35{\cdot}2362i$
300	$4{\cdot}0923 + 11{\cdot}2649i$	$-4{\cdot}0923 + 28{\cdot}7984i$
500	$-2{\cdot}5033 + 9{\cdot}8069i$	$+2{\cdot}5033 + 25{\cdot}5891i$

The graph of the function $\Omega(T)$ appropriate to the critical speed $V_c = 239{\cdot}8$ is curve No. 1 of Fig. 12·3·1, and it has asymptotes situated at the values $T = \pi/\omega$, $3\pi/\omega$, etc., where $\pi/\omega = 0{\cdot}12281$. The lowest non-zero root of the semi-period equation in this case is $T = 0{\cdot}32848$, but the oscillations corresponding to this root can be proved to be spurious, since $q_2(0)$ and $\ddot{q}_2(0)$ are both positive.

For a speed, say $V_c + \epsilon$, very close to the critical value, the lowest non-zero root lies adjacent to the first asymptote of curve No. 1 and its value is approximately

$$T = 0{\cdot}12281 - 0{\cdot}00023\epsilon.$$

In this case, if terms in $1/\epsilon$ only are retained,

$$\frac{e^{T\lambda_1}-1}{e^{T\lambda_1}+1}=\frac{250+230i}{\epsilon},$$

$$\frac{q_2(0)}{R_2}=\frac{0\cdot308}{\epsilon},$$

$$\frac{\ddot{q}_2(0)}{R_2}=-\frac{201\cdot5}{\epsilon}.$$

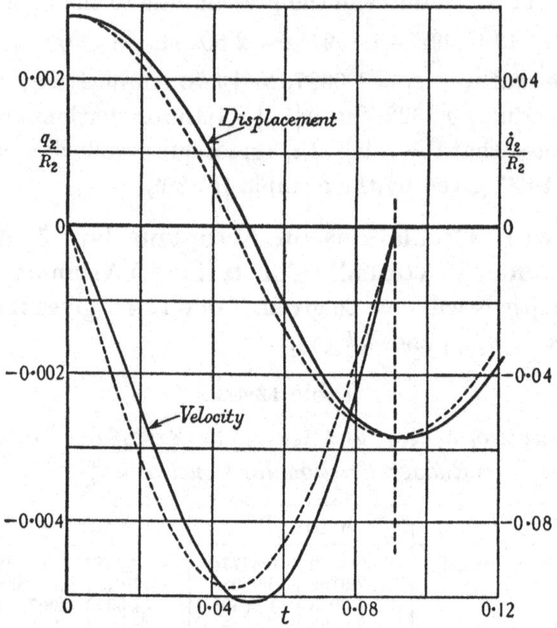

Fig. 12·4·1

Hence steady oscillations can occur if ϵ is positive (i.e. for speeds just above the critical). However, if ϵ is very small, the initial disturbance required will be correspondingly large, and the oscillations themselves will approximate closely to the sinusoidal type.

Curve No. 2 has already been discussed in § 12·3, while curves No. 3 and No. 4 relate respectively to $V = 300$ and $V = 500$. It may be noted that all the curves have the same slope $(-\frac{1}{2}a_{22})$ at the origin $T = 0$. The calculated values of $q_2(0)/R_2$ and of $\ddot{q}_2(0)/R_2$ for the lowest root T in the several cases considered are as follows.

Table 12·4·2

V	Lowest root T	$q_2(0)/R_2$	$\ddot{q}_2(0)/R_2$	Remarks
239·8 (critical)	0·32848	+0·005675	+1·4624	Condition (C) of § 11·6 violated
260	0·11856	+0·015712	−9·4713	Theoretically possible, but unstable
300	0·11340	+0·006801	−3·6527	
500	0·09111	+0·002851	−1·7573	

The first down-strokes in the rudder motion corresponding to the two cases $V = 260$ and $V = 500$ are shown compared against sinusoidal oscillations in Figs. 12·3·2 and 12·4·1. It will be seen that the deviation from the sine curve is quite appreciable in the case $V = 500$.

12·5. Steady Oscillations on Aeroplane No. 1. (Fuselage Frictionally Constrained.) The oscillations when the fuselage is frictionally constrained have not been investigated in detail. To render the formulae of § 11·6 directly applicable, the coordinate q_2 must in this case be chosen to refer to the fuselage and the coefficients in Table 12·2·1 must be correspondingly transposed. The value of $\lambda_1 k_{21} \gamma_1$ appropriate to the critical speed works out as

$$\xi + i\eta = (-128{\cdot}28 + 329{\cdot}55i) \times 10^{-6}.$$

Since $\eta > 0$, the steady oscillations occur at speeds above the critical.

Part II. Aeroplane No. 2

12·6. Numerical Data. The complete set of dynamical coefficients required for an application of the theory to rudder-fuselage oscillations has only been measured for the one aeroplane already considered. In order to provide an illustration of a system which exhibits steady oscillations below the critical speed, it has been necessary to adopt rather arbitrarily a new set of coefficients. These have been derived by postmultiplication of the inertial, damping, and stiffness matrices appropriate to aeroplane No. 1 by a non-singular matrix u of suitable constants. In this case, if $\boldsymbol{A}, \boldsymbol{B}, \boldsymbol{C}$ refer to the modified system, we shall have

$$\boldsymbol{A}\lambda^2 + \boldsymbol{B}\lambda + \boldsymbol{C} = (A\lambda^2 + B\lambda + C)\,u,$$

and the roots of the determinantal equation for each airspeed will accordingly be unchanged by the transformation. The elements of u must, of course, be chosen such that the symmetry of the inertial and elastic stiffness matrices is preserved and that the discriminants of the kinetic energy for the modified system are all positive. Moreover, in order that steady oscillations shall be possible on the modified system at speeds below the critical, we require $\eta < 0$. By an application of (12·3·1), it is easy to show that this last condition can be expressed as

$$(C_{11} - A_{11}\omega^2)(\mu'^2 + \omega'^2 - \omega^2) - 2\omega^2\mu' B_{11} < 0.$$

By trial a suitable postmultiplier is found to be

$$u = \begin{bmatrix} 1{\cdot}0 & 0 \\ -10{\cdot}0 & 7{\cdot}47826 \end{bmatrix}.$$

The dynamical coefficients for the modified system are summarised in Table 12·6·1.

Table 12·6·1

Dynamical Coefficients for Aeroplane No. 2

Fuselage (q_1)		Rudder (q_2)	
Coefficient	Value	Coefficient	Value
A_{11}	56·2	A_{21}	−8·60
B_{11}	3·63 V	B_{21}	−0·299 V
C_{11}	33,700 + 1·01 V^2	C_{21}	−0·0358 V^2
A_{12}	−8·60	A_{22}	5·57130
B_{12}	−1·39096 V	B_{22}	0·25426 V
C_{12}	−0·75530 V^2	C_{22}	0·026772 V^2

The coefficients of the new determinantal equation are the same as those for equation (12·2·1), multiplied by the constant factor 7·47826.

12·7. Steady Oscillations on Aeroplane No. 2. (Rudder Frictionally Constrained.) For the critical speed $V_c = 239{\cdot}8$ it is found that

$$\lambda_1 k_{21}\gamma_1 \equiv \xi + i\eta \;= (7{\cdot}88129 - 1{\cdot}41363i)\,10^{-3},$$
$$\lambda_3 k_{23}\gamma_3 \equiv \xi' + i\eta' = (-7{\cdot}88129 + 15{\cdot}29856i)\,10^{-3},$$

and since $\eta < 0$ stable steady oscillations are possible below the critical speed.

Curve No. 1 of Fig. 12·7·1 shows the graph of $\Omega(T)$ corresponding to the critical speed. The lowest non-zero root of the semi-period equation is $T = 0{\cdot}09608$, and this leads to the values $q_2(0)/R_2 = 0{\cdot}000551$ and $\ddot{q}_2(0)/R_2 = -0{\cdot}3474$. The corresponding steady oscillations can be shown to be unstable.

With $V = 230$, the calculated constants are

$$\left.\begin{aligned} \xi + i\eta &= (7{\cdot}51056 - 2{\cdot}10301i)\,10^{-3}, \\ \xi' + i\eta' &= (-7{\cdot}51056 + 15{\cdot}96533i)\,10^{-3}, \end{aligned}\right\} \qquad \text{......(1)}$$

and the semi-period equation is found to have for two of its roots the values $T_1 = 0{\cdot}09937$ and $T_2 = 0{\cdot}11607$ (see curve No. 2 of Fig. 12·7·1). The initial displacement and acceleration corresponding to the lower root T_1 are

$$q_2(0)/R_2 = 0{\cdot}000681 \quad \text{and} \quad \ddot{q}_2(0)/R_2 = -0{\cdot}4385;$$

while those appropriate to T_2 are

$$q_2(0)/R_2 = 0{\cdot}00485 \quad \text{and} \quad \ddot{q}_2(0)/R_2 = -3{\cdot}312.$$

In both cases condition (D) of § 11·6 is also satisfied.* An application of the criteria for stability shows that the first set of oscillations are unstable and the second set stable. This suggests that with aeroplane

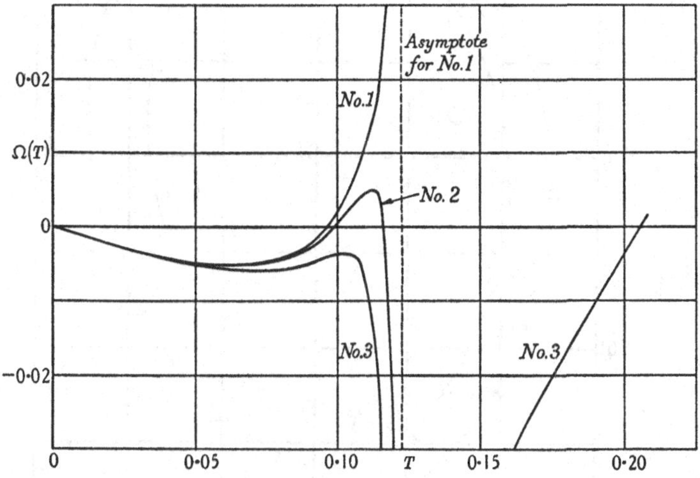

Fig. 12·7·1

No. 2, for a certain range of the airspeed below the critical speed, a moderately large disturbance may produce oscillations which grow to a relatively large constant limiting amplitude, but that a smaller disturbance will produce decaying oscillations leading ultimately to ankylosis. The limiting amplitude to be expected from growing oscillations will increase continuously the closer the speed approaches the critical value.

From Fig. 12·7·1 it is seen that at a speed as low as $V = 200$ (curve No. 3), the two real roots corresponding to T_1 and T_2 have disappeared, so that steady oscillations are no longer possible. Hence at this speed the oscillations due to any disturbance eventually die away.

* This is confirmed by the graphical method in § 12·9.

12·8. Steady Oscillations on Aeroplane No. 2. (Fuselage Frictionally Constrained.) The particular transformation used to derive the dynamical coefficients for aeroplane No. 2 from those of aeroplane No. 1 is such that the coefficients $\boldsymbol{A}_{22}$, $\boldsymbol{B}_{22}$ and $\boldsymbol{C}_{22}$ in Table 12·6·1

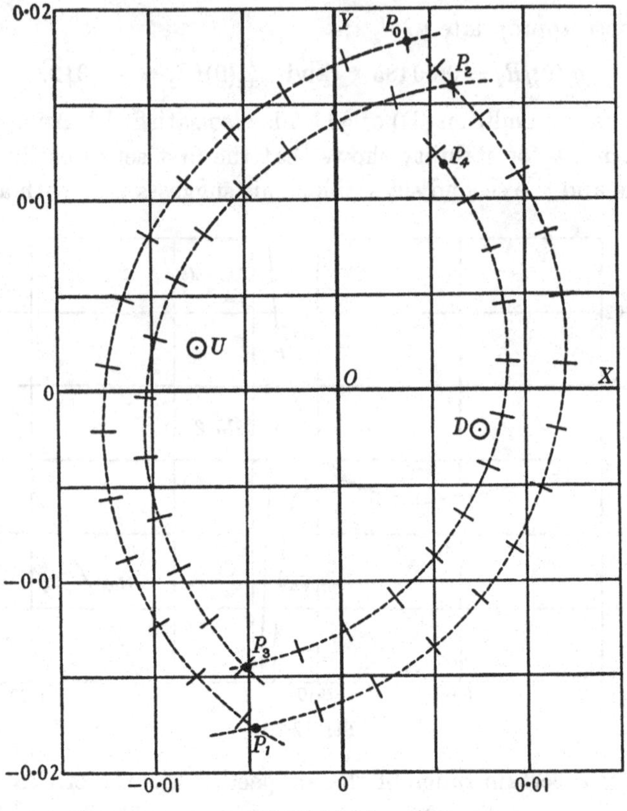

Fig. 12·9·1 (*a*)

are all proportional to the corresponding coefficients A_{22}, B_{22} and C_{22} in Table 12·2·1. Since also the roots of $\Delta(\lambda) = 0$ appropriate to the critical speed are the same in the two cases, it readily follows that the conclusions drawn in § 12·5 are equally applicable to aeroplane No. 2. Hence when the fuselage is frictionally constrained, the steady oscillations on aeroplane No. 2 occur above the critical speed, and are unstable.

12·9. Graphical Investigation of Complete Motion on Aeroplane No. 2 at $V = 230$. (Rudder Frictionally Constrained.) In § 11·9 a method is given for the graphical description of the complete motion of a binary system. This method will now be applied to aeroplane No. 2 for $V = 230$, which is about 10 feet per second below the critical speed. Throughout, the friction is assumed to be applied to the rudder only.

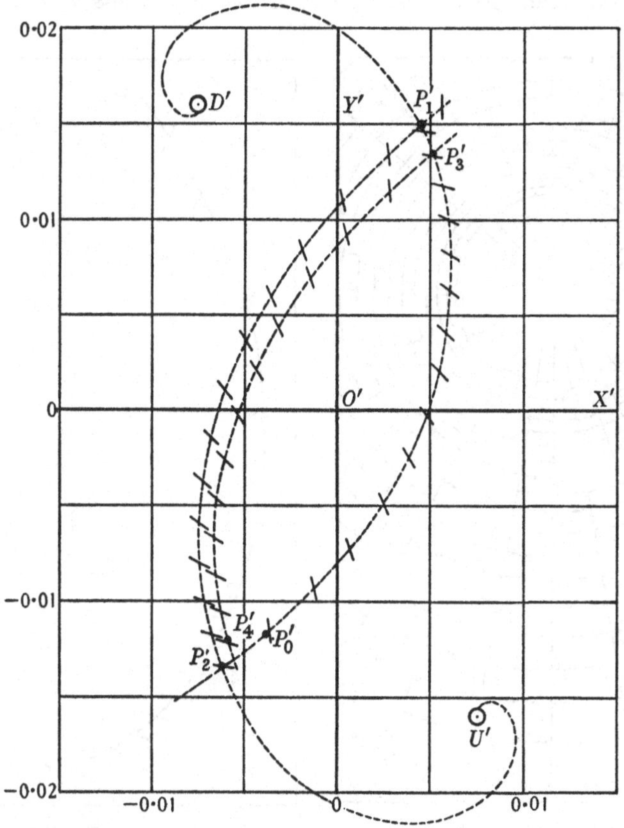

Fig. 12·9·1 (a′)

The appropriate values of $\lambda_r k_{2r} \gamma_r$ are (see (12·7·1))

$$\lambda_1 k_{21} \gamma_1 = (7{\cdot}51056 - 2{\cdot}10301i)\, 10^{-3},$$

$$\lambda_3 k_{23} \gamma_3 = (-7{\cdot}51056 + 15{\cdot}96533i)\, 10^{-3}.$$

Hence, in displacement diagram No. 1 (Fig. 12·9·1 (a), corresponding to Fig. 11·9·1 (a)), the down-centre D and up-centre U have, respectively, the coordinates $(+7{\cdot}51 \times 10^{-3},\ -2{\cdot}10 \times 10^{-3})$ and $(-7{\cdot}51 \times 10^{-3},\ +2{\cdot}10 \times 10^{-3})$; while in displacement diagram No. 2 (Fig. 12·9·1 (a′)) D' is the point $(-7{\cdot}51 \times 10^{-3},\ 15{\cdot}97 \times 10^{-3})$ and U' is $(+7{\cdot}51 \times 10^{-3},\ -15{\cdot}97 \times 10^{-3})$.

Spiral No. 1 (Fig. 12·9·1 (*b*), corresponding to Fig. 11·9·1 (*b*)) is the curve

$$r = e^{\mu\theta/\omega} = e^{-0.0140940\theta},$$

where θ is the angle of rotation of the radius r in radians, and $\theta = \omega t$. For convenience, the arc is divided into equal steps $\theta = 10$ degrees

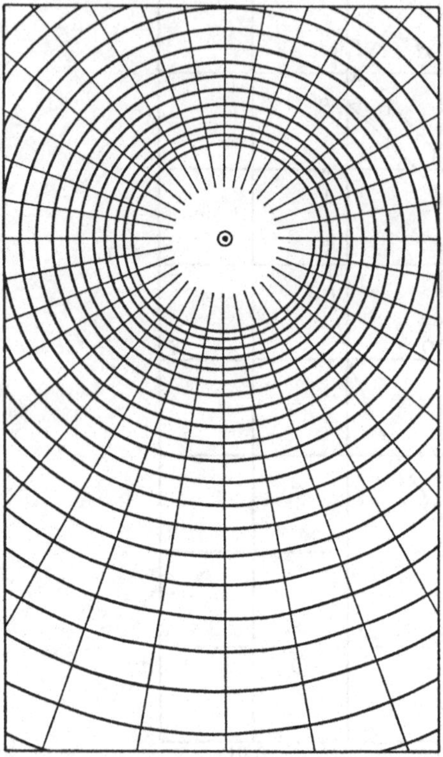

Fig. 12·9·1 (*b*)

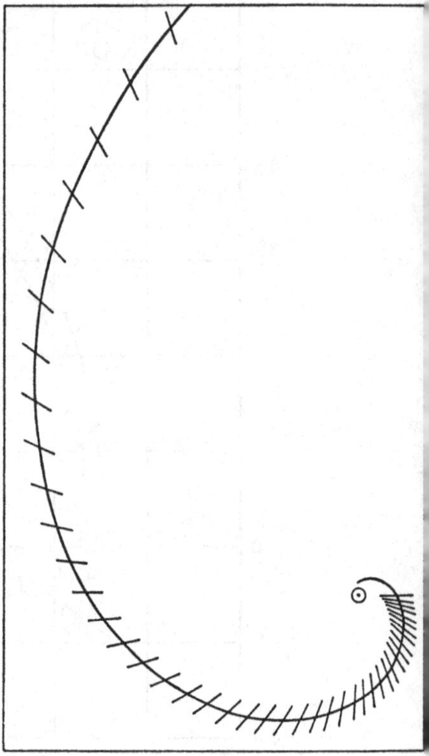

Fig. 12·9·1 (*b′*)

or $\pi/18$ radians, so that each step represents a time interval of $t = \pi/18\omega = 0.0068512$ sec. Spiral No. 2 (Fig. 12·9·1 (*b′*)) is the curve

$$r = e^{\mu'\theta'/\omega'} = e^{-0.620830\theta'},$$

where $\theta' = \omega' t$, and its arc is divided into steps representing the same time intervals as for No. 1. Thus $\theta'/\theta = \omega'/\omega$, so that a step in θ amounting to 10 degrees corresponds to a step in θ' amounting to 5·8479 degrees.

EXAMPLES

(i) *Steady Oscillations Corresponding to* $T_1=0{\cdot}09937$. This is one of the cases discussed in § 12·7. The relevant data are

$$\lambda_1 k_{21}\gamma_1 \frac{e^{T_1\lambda_1}-1}{e^{T_1\lambda_1}+1} = (5{\cdot}1738+24{\cdot}1772i)\,10^{-3},$$

$$\lambda_3 k_{23}\gamma_3 \frac{e^{T_1\lambda_3}-1}{e^{T_1\lambda_3}+1} = (-5{\cdot}1738-15{\cdot}7506i)\,10^{-3}.$$

Hence the positions of P_0, P_0' (which are fixed by the initial conditions for the first down-stroke) are

$$P_0 = (5{\cdot}17\times10^{-3},\;\; 24{\cdot}18\times10^{-3}),$$
$$P_0' = (-5{\cdot}17\times10^{-3},\;\; -15{\cdot}75\times10^{-3}).$$

These points are marked on the displacement diagrams Nos. 1 and 2, respectively, and the description of the motion is then begun. The first operation is to superpose displacement diagram No. 1 on spiral No. 1, with D and P_0 registered, respectively, on the pole and on the curve; the second pair of diagrams is superposed in a similar manner. The two observers (one for each pair of superposed diagrams) now follow the spirals with pointers, proceeding by equal time steps in the positive sense, so that in each case the radius vector DP is shrinking. In the early stages the abscissae of P, P' (initially numerically equal, although of opposite signs) begin to diverge; but after about 14·5 steps (i.e. about 145 degrees on diagram No. 1 and 85 degrees on diagram No. 2), they again become equal and opposite. At this stage the velocity again vanishes, and the first down-stroke is complete. It is found that the new positions of P and P' (namely P_1 and P_1') are such that O lies on and bisects P_0P_1, while similarly O' lies on and bisects $P_0'P_1'$. This shows that the motion is steady. Further, the angle of rotation θ amounts to 145 degrees, and this corresponds to a semi-period $T_1 = 0{\cdot}0993$ sec., in good agreement with the value obtained by direct calculation.

(ii) *Steady Oscillations Corresponding to* $T_2 = 0{\cdot}11607$. Here the points P_0, P_0' have, respectively, the coordinates

$$P_0 = (4{\cdot}11\times10^{-3},\;\; 82{\cdot}14\times10^{-3}),$$
$$P_0' = (-4{\cdot}11\times10^{-3},\;\; -19{\cdot}00\times10^{-3}).$$

The procedure is similar to that described for the previous case. It is found that the stroke is complete after 16·9 steps, and that the motion is steady. The value of the semi-period derived by this method is $T_2 = 0{\cdot}116$.

(iii) *Growing Oscillations following a Particular Disturbance.* The initial displacements and velocities are here all assumed to be 50 per cent in excess of those appropriate to the (unstable) steady oscillations considered in example (i).* Thus

$$P_0 = (7{\cdot}76 \times 10^{-3},\ 36{\cdot}27 \times 10^{-3}),$$
$$P_0' = (-7{\cdot}76 \times 10^{-3},\ -23{\cdot}62 \times 10^{-3}).$$

Table 12·9·1

Synopsis of Results for Example (iii) (*Growing Oscillations*)

Stroke no.	Type	No. of time steps	Time of stroke	Coordinates of Z_s				Coordinates of Z_s'			
				$X_s \times 10^3$		$Y_s \times 10^3$		$X_s' \times 10^3$		$Y_s' \times 10^3$	
				Initial	Final	Initial	Final	Initial	Final	Initial	Final
1	Down	15·7	$0{\cdot}107_5$	7·75	−7·2	36·25	−35·9	−7·75	7·2	−23·6	16·3
2	Up	16·15	$0{\cdot}110_5$	−7·2	4·0	−35·9	36·7	7·2	−4·0	16·3	−16·9
3	Down	15·6	0·107	4·0	−4·75	36·7	−37·5	−4·0	4·75	−16·9	17·5
4	Up	15·75	0·108	−4·75	4·75	−37·5	38·4	4·75	−4·75	17·5	−17·5
5	Down	15·7	$0{\cdot}107_5$	4·75	−4·75	38·4	−39·25	−4·75	4·75	−17·5	17·5
6	Up	15·85	$0{\cdot}108_5$	−4·75	4·75	−39·25	39·9	4·75	−4·75	17·5	−17·5
7	Down	15·85	$0{\cdot}108_5$	4·75	−4·75	39·9	−40·7	−4·75	4·75	−17·5	17·6
8	Up	15·9	0·109	−4·75	4·65	−40·7	41·5	4·75	−4·65	17·6	−17·6
9	Down	15·95	$0{\cdot}109_5$	4·65	−4·6	41·5	−42·5	−4·65	4·6	−17·6	17·6
10	Up	16·0	$0{\cdot}109_5$	−4·6	4·6	−42·5	43·7	4·6	−4·6	17·6	−17·5

The results obtained at the terminal instants of the first 10 complete strokes of the rudder are given in Table 12·9·1. To obtain a more detailed description of the motion it is necessary, as explained in § 11·9, to note the current positions of P and P' on the spirals for a number of the individual time steps. The value of the velocity $\dot{q}_2(t)/R_2$ at any stage is then twice the sum of the current abscissae, while the corresponding displacement $q_2(t)/R_2$ is twice the real part of $\dfrac{Z(t)}{\lambda_1} + \dfrac{Z'(t)}{\lambda_3}$. Alternatively, the displacement can be deduced by integration of the graph of the velocity. Curve C of Fig. 12·9·2 shows the complete motion determined by the first method for the first nine strokes of the rudder:

* Note that if the displacements and velocities in any initial disturbance are all increased proportionally, then the initial reducing variables will also all increase proportionally. Hence, in the present case the vectors OP_0 and $O'P_0'$ will be 50 per cent in excess of those appropriate to example (i).

the growth of the oscillations is obvious. However, this growth cannot continue indefinitely, and it is clear that the ultimate motion will be the stable steady oscillations discussed in example (ii). The initial angular displacement of the rudder is in the present case 0·0585 degree per foot pound of frictional moment R_2, while the maximum displacement is 0·278 degree per foot-pound. The ratio of initial to final amplitude is thus about 1 : 5.

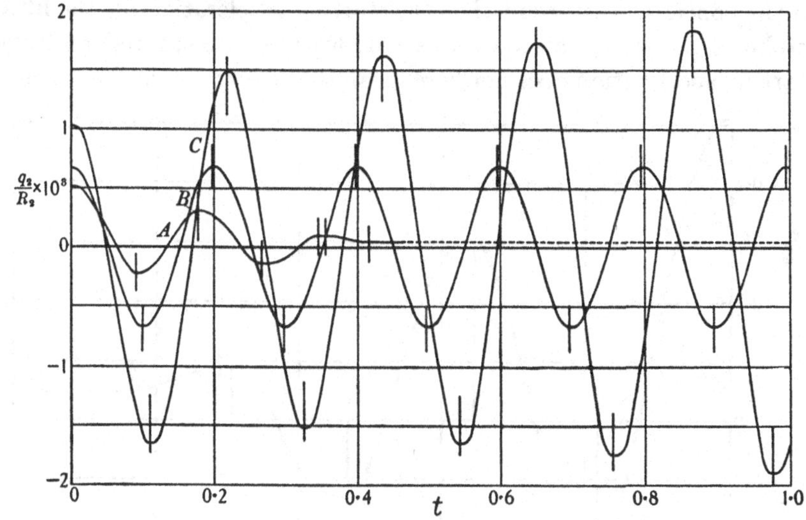

Fig. 12·9·2

(iv) *Decaying Oscillations leading to Ankylosis.* For this example the initial displacements and velocities are assumed to be only three-quarters of those appropriate to the (unstable) steady oscillations of example (i). The resulting motion is found to be complete for four strokes of the rudder; temporary ankylosis then occurs, and eventually permanent ankylosis.

The initial data for the first stage of the motion are

$$P_0 = (3{\cdot}9\times 10^{-3},\ 18{\cdot}1\times 10^{-3}),$$
$$P_0' = (-3{\cdot}9\times 10^{-3},\ -11{\cdot}8\times 10^{-3}),$$

corresponding to a rudder displacement $q_2(0)/R_2$ of about $0{\cdot}51\times 10^{-3}$. The displacements at the stopping instants of the first four strokes are found by the graphical method to be respectively

$$-0{\cdot}215\times 10^{-3},\quad 0{\cdot}305\times 10^{-3},\quad -0{\cdot}131\times 10^{-3},\quad \text{and}\quad 0{\cdot}087\times 10^{-3},$$

while the variation of velocity $\dot{q}_2/R_2$ during each stroke is determined by addition of the abscissae of P, P' for corresponding positions along the spirals. A graphical integration of the velocity diagram, controlled by the known values of the displacements at the terminal instants, yields the final graph of the rudder displacement (see Fig. 12·9·3) during the first four strokes.

An attempt to continue the graphical operations, with the down-centres D, D' registered on the poles of the spirals on the supposition that complete motion continues with a down-stroke, leads immediately to the contradictory result that the starting acceleration for the fifth stroke is positive. This indicates ankylosis. The measured positive starting acceleration does not here mean that complete motion ensues

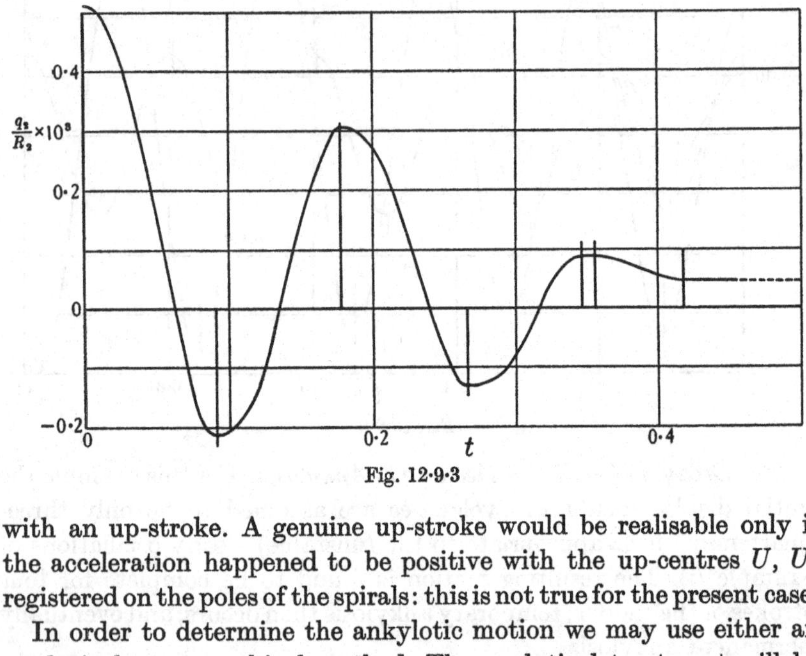

Fig. 12·9·3

with an up-stroke. A genuine up-stroke would be realisable only if the acceleration happened to be positive with the up-centres U, U' registered on the poles of the spirals: this is not true for the present case.

In order to determine the ankylotic motion we may use either an analytical or a graphical method. The analytical treatment will be illustrated here. The differential equation for the ankylotic motion is (see § 11·5)

$$A_{11}\ddot{q}_1 + B_{11}\dot{q}_1 + C_{11}q_1 = -C_{12}q_2(t_4), \qquad \text{......(1)}$$

where $q_2(t_4)$ is the constant value of the ankylosed rudder coordinate. Further, ankylosis will persist so long as the quantity

$$(A_{21}\ddot{q}_1 + B_{21}\dot{q}_1 + C_{21}q_1 + C_{22}q_2(t_4))/R_2 \qquad \text{......(2)}$$

is numerically less than unity.

The first step is to derive the initial displacements and velocities of the fuselage in the ankylotic motion from the corresponding known values of the reducing variables: for this purpose the formulae (11·3·1) and (11·3·2) can be used. The values of Z, Z' appropriate to the stopping instant t_4 of the fourth stroke in the complete motion are $Z_4 \times 10^3 = 5{\cdot}8 + 11{\cdot}8i$ and $Z'_4 \times 10^3 = -5{\cdot}8 - 12{\cdot}2i$. Hence

$$\lambda_1 k_{21} \alpha_1(t_4)/R_2 = (5{\cdot}8 + 11{\cdot}8i)\,10^{-3},$$
$$\lambda_3 k_{23} \alpha_3(t_4)/R_2 = (-5{\cdot}8 - 12{\cdot}2i)\,10^{-3}.$$

Also
$$\frac{k_{11}}{k_{21}} = -\frac{f_{12}(\lambda_1)}{f_{11}(\lambda_1)} = 0{\cdot}64057 - 0{\cdot}09874i,$$
$$\frac{k_{13}}{k_{23}} = -\frac{f_{12}(\lambda_3)}{f_{11}(\lambda_3)} = 0{\cdot}49702 + 0{\cdot}05452i.$$

The required initial values, which can now be calculated, are

$$q_1(t_4)/R_2 = 0{\cdot}058 \times 10^{-3} \quad \text{and} \quad \dot{q}_1(t_4)/R_2 = 5{\cdot}32 \times 10^{-3}.$$

Next, on substitution of the numerical data, equation (1) becomes

$$\ddot{q}_1 + 14{\cdot}856\dot{q}_1 + 1550{\cdot}34q_1 = 62 \times 10^{-3}R_2.$$

The solution appropriate to the initial values just determined is

$$10^3 \times q_1(t)/R_2 = 0{\cdot}040 - 0{\cdot}142e^{-7{\cdot}428t}\cos\phi,$$

where ϕ (assumed expressed in degrees) has the value $2215{\cdot}5t + 97{\cdot}3$, and $t = 0$ is temporarily adopted as the starting instant for the ankylotic motion. At $t = 0$ the quantity (2) has the value 0·5674; this confirms the occurrence of the ankylosis. However, at $t = 0{\cdot}0093$ the value is $+1{\cdot}0$, so that at this instant complete motion is resumed. The positive sign for (2) indicates that the ensuing stroke in q_2 will be downwards.

To proceed again by the graphical method for complete motion a preliminary conversion to the reducing variables is necessary. The formula appropriate to the root λ_1 may be written (see 11·3·8)

$$\{k_{11}\alpha_1, k_{21}\alpha_1\} = -\frac{\underset{(1)}{F}(\lambda_1)}{\lambda_1 \Delta(\lambda_1)}(Cq - \lambda_1 A\dot{q}),$$

and the numerical data actually required for the computation of $\lambda_1 k_{21} \alpha_1$ are
$$Cq - \lambda_1 A\dot{q} = 10^{-3}R_2\begin{bmatrix} 5493{\cdot}1 - 5824{\cdot}0i \\ -82{\cdot}52 + 891{\cdot}23i \end{bmatrix},$$
$$-f_{21}(\lambda_1) = -3710{\cdot}82 + 1594{\cdot}58i,$$
$$f_{11}(\lambda_1) = 5036{\cdot}94 + 20240{\cdot}78i,$$
$$-1/\underset{(1)}{\Delta}(\lambda_1) = (0{\cdot}11394 - 0{\cdot}08755i)\,10^{-6}.$$

Hence $$\lambda_1 k_{21} \alpha_1 / R_2 \equiv Z = (2{\cdot}469 + 11{\cdot}299i)\,10^{-3};$$

and a similar calculation gives

$$\lambda_3 k_{23} \alpha_3 / R_2 \equiv Z' = (-2{\cdot}469 - 9{\cdot}869i)\,10^{-3}.$$

These results fix the positions of the initial points in the displacement diagrams, and the next stroke can now be followed out in the usual manner. The velocity throughout the stroke is found to be negative, but always very small, and at the conclusion of the stroke permanent ankylosis is readily proved to occur.

The full history of the behaviour of the rudder is represented in Fig. 12·9·3.

Comparison between Types of Motion in Examples (i), (iii) *and* (iv). A comparison between three types of rudder motion on aeroplane No. 2 at $V = 230$ is given in Fig. 12·9·2. Graph A shows the decreasing motion just described in example (iv); graph B corresponds to the steady unstable oscillations, with the semi-period $T = 0{\cdot}0994$, considered in example (i); lastly graph C relates to the growing oscillations discussed in example (iii). The initial disturbances appropriate to the three types of motion can for brevity be referred to as $0{\cdot}75X$, X, and $1{\cdot}5X$ respectively. Here X denotes the complete disturbance required to produce the steady oscillations, and is specifically

$$\left.\begin{aligned} q_1(0)/R_2 &= 0{\cdot}0004865, & q_2(0)/R_2 &= 0{\cdot}000681, \\ \dot{q}_1(0)/R_2 &= 0{\cdot}007977, & \dot{q}_2(0)/R_2 &= 0. \end{aligned}\right\} \quad \ldots\ldots(3)$$

As already remarked, the motion C cannot grow indefinitely. It tends ultimately to the steady stable oscillations with the semi-period $T = 0{\cdot}116$.

PART III. AEROPLANE No. 3

12·10. Aeroplane No. 3. A final example is given to show that solid friction can produce bounded oscillations of a dynamical system, even when the frictionless system is always stable.

To derive a simple illustration we shall retain all the dynamical coefficients appropriate to aeroplane No. 2 (see Table 12·6·1) with the exception of the stiffness coefficient C_{11}. If the new stiffness coefficient is chosen to be $33{,}700 + 1{\cdot}01V^2 + a(V^2 - 230^2)$, and a is arbitrary, then evidently all the dynamical coefficients for the new system—which we shall call aeroplane No. 3—will agree with those for No. 2 when $V = 230$. Hence at this particular speed the oscillatory characteristics of the two aeroplanes, with or without friction, will be identical. In particular, the bounded oscillations already shown to be possible on aeroplane No. 2 will also be possible on aeroplane No. 3, irrespective of the value of a. It only remains to choose the constant a in such a way as to render aeroplane No. 3 stable at all airspeeds. A value for a which renders all the coefficients of the determinantal equation and the test function T_3 positive at all speeds is 0·6. The approximate values of the roots of $\Delta(\lambda) = 0$ for a few representative airspeeds are tabulated below.

Table 12·10·1

Roots of Determinantal Equation for Aeroplane No. 3

V	μ	ω	μ'	ω'
0	0	6·76	0	0
100	−0·59	12·40	−3·59	6·70
230	−0·36	25·47	−9·25	14·92
350	−0·32	38·16	−14·30	22·51
500	−0·33	54·21	−20·55	32·01

It will be noted that, although the aeroplane is completely free from flutter, yet it is only slightly stable at all speeds. Naturally no actual aeroplane would have such characteristics, but it is quite possible for an aeroplane which is immune from flutter to approach very close to flutter at particular flying speeds (see Fig. 12·1·3). From the examples already worked out it seems clear that at such speeds the development of much friction in the controls might sometimes result in the occurrence of bounded oscillations.

CHAPTER XIII

PITCHING OSCILLATIONS OF A FRICTIONALLY CONSTRAINED AEROFOIL

13·1. Preliminary Remarks. The theory given in Chapters XI and XII indicates that with particular aerodynamical systems possessing solid friction, steady oscillations can occur at airspeeds less than the critical speed for flutter. If the system has two degrees of freedom represented by the generalised coordinates q_1, q_2, and if q_2 only is subject to frictional constraint, then stable steady oscillations are possible below the critical speed provided that (see §§ 11·7, 11·8, and equations (11·3·13), (11·4·1))

$$\eta < 0, \qquad \text{......(1)}$$

where

$$\xi + i\eta \equiv \lambda_1 k_{21} \gamma_1 = \frac{\partial \lambda_1}{\partial C_{22}} = \frac{1}{\lambda_1}\frac{\partial \lambda_1}{\partial B_{22}}, \qquad \text{......(2)}$$

and λ_1 denotes the critical root.* If $\lambda_1 = i\omega$ at the critical speed, and if $\delta\mu$ is the increment of the real part of the root when the speed is slightly increased from the critical value, then the criterion (1) can be expressed as

$$-\omega\eta = \frac{\partial \mu}{\partial B_{22}} > 0. \qquad \text{......(3)}$$

This inequality has a simple physical significance. It requires that instability of the frictionless system shall develop if, while the airspeed is maintained at its critical value, the direct damping coefficient appropriate to the coordinate q_2 is given a small positive increment.

Steady oscillations of the type just referred to have been demonstrated in a wind tunnel† at the National Physical Laboratory. The system chosen for these experiments consisted of a single rigid aerofoil capable of pitching motion about two separate axes, and it was designed entirely from theoretical considerations. Part I of the present Chapter deals with some problems connected with the design of the apparatus, and Part II gives a brief description of the experiments.

* For the definition of critical root see § 12·1.

† Ref. 43.

PART I. THE TEST SYSTEM AND ITS DESIGN

13·2. Description of the Aerofoil System. The system as finally constructed is shown in Figs. 13·2·1, 13·2·2, 13·2·3. The two vertical axes about which pitching movements are allowed are AA and BB: these will be referred to as the "frame axis" and the "aerofoil axis",

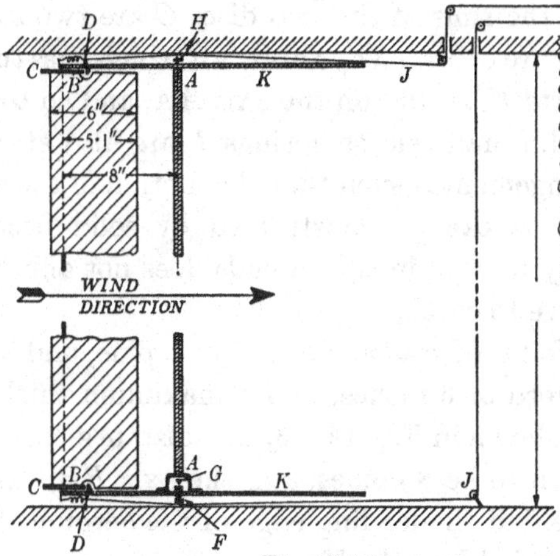

Fig. 13·2·1

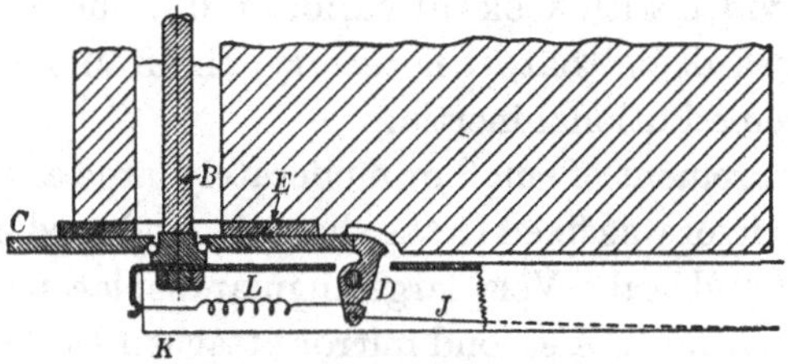

Fig. 13·2·2

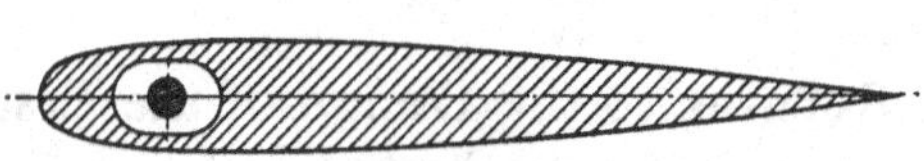

Fig. 13·2·3

respectively. The rigid frame $AABB$ consists of a steel rod AA of diameter 0·5 inch, two arms of square channel section K, and a steel rod BB of diameter 0·25 inch. The frame is pivoted about AA as an axis. At the bottom is a point and cup support F, with a locking point G carried on a small bridge; the top support is a simple journal bearing H. The rod BB carries a pair of ball-bearings which support a rigid wooden aerofoil of symmetrical section. Screwed to each end of the aerofoil is a metal plate E, and attached to each plate by metal screws is a circular steel disc C in the centre of which is the spherical ball-race. The attachment between C and E is such that it is possible to adjust the position of the aerofoil with respect to the axis BB; for this purpose the hole in the aerofoil, through which the rod BB passes, is of elongated section (Fig. 13·2·3).

Bearing against the rims of the two discs C are two steel levers D (Fig. 13·2·2). These are operated by threads J which pass through small holes in the supports F, H and on the axis AA, and so to the outside of the wind tunnel. Small release springs L are also attached to the levers D. The arrangement is such that the motion of the aerofoil with respect to the frame can be constrained by solid friction applied externally; but any tension in the threads does not affect the motion of the frame relative to earth.

The actual aerofoil used was made of yellow pine, and had a span of nearly 4 feet, a chord of 6 inches, and a maximum thickness of 0·75 inch: the profile is shown in Fig. 13·2·3. The distance between the axes AA, BB was chosen to be 8 inches, and the axis BB was at 0·9 inch (i.e. 0·15 chord) behind the leading edge of the aerofoil. Stability for the frame was provided by attaching springs in pairs to the arms K, which were provided with hooks at various radii. These arms were also extended backwards so that, if necessary, the frame could be cross-braced and given additional inertia.

For the measurement of small or moderately large angular displacements a small mirror was fixed to the shaft AA and used in conjunction with a spotlight and scale. Very large angular displacements were read directly on a quadrant. A second mirror attached to the aerofoil near the shaft BB was used in experiments with the frame locked.

13·3. Data Relating to the Design of the Test System. The preceding description applies to the test system as constructed, but the final design depended on much preliminary theoretical work. The aerofoil itself, and the position of the aerofoil axis BB at 0·9 inch behind the leading edge, were treated as definitely assigned from the outset. However, certain of the inertias and spring stiffnesses, and the position of the frame axis AA, were left free to be varied. The calculations relating to the aerodynamical derivatives and to the inertias will now be summarised.

(*a*) *The Aerodynamical Derivatives.* The derivatives appropriate to coupled pitching and normal oscillations are deduced from the theory of the accelerated motion of an aerofoil in Ref. 44. The generalised coordinates corresponding to these two degrees of freedom are θ (inclination in pitch) and z (linear displacement of the pitching axis normal to the chord). The senses of these coordinates are indicated in Fig. 13·3·1.

The derivatives given by the theory depend through a parameter on the frequency of oscillation. We shall see later that in relation to the application in view this frequency parameter, which is defined as $\omega c/V$ (where $\omega/2\pi$ is the frequency, c the chord, and V the airspeed), can be chosen at convenience: the value $\omega c/V = 0{\cdot}4$ will be adopted.

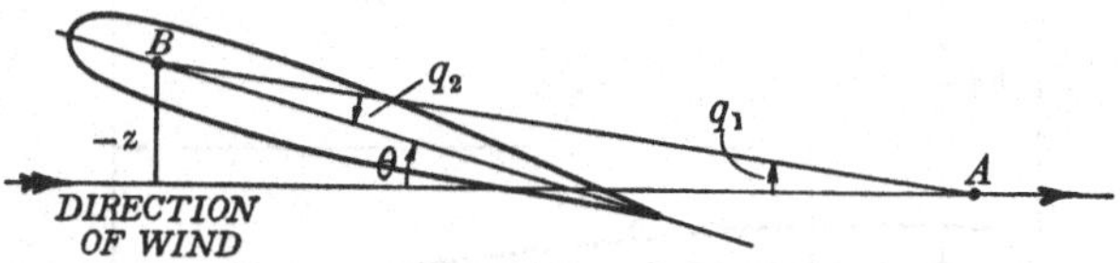

Fig. 13·3·1

If Z denotes the normal force and M the pitching moment about the aerofoil axis BB (prescribed at 0·15 chord behind the leading edge), the appropriate derivatives are as follows:*

Damping Matrix. $$\mathfrak{B} \equiv \begin{bmatrix} -Z_{\dot{z}} & -Z_{\dot{\theta}} \\ -M_{\dot{z}} & -M_{\dot{\theta}} \end{bmatrix} = 10^{-3}V\begin{bmatrix} 9{\cdot}34 & 1{\cdot}40 \\ 0{\cdot}467 & 0{\cdot}470 \end{bmatrix}.$$

Aerodynamic Stiffness Matrix.

$$\mathfrak{W} \equiv \begin{bmatrix} -Z_z & -Z_\theta \\ -M_z & -M_\theta \end{bmatrix} = 10^{-3}V^2\begin{bmatrix} 0 & 9{\cdot}34 \\ 0 & 0{\cdot}467 \end{bmatrix}.$$

For the investigation of the effects due to friction it is convenient to adopt for the generalised coordinates the angle of rotation q_1 about AA of the frame relative to earth and the angle of rotation q_2 about BB of the aerofoil relative to the frame (see Fig. 13·3·1). If l denotes the distance AB, these coordinates are connected with z, θ by the linear substitution

$$\begin{bmatrix} z \\ \theta \end{bmatrix} = \begin{bmatrix} -l & 0 \\ 1 & 1 \end{bmatrix}\begin{bmatrix} q_1 \\ q_2 \end{bmatrix},$$

* Since the aerofoil spanned the wind tunnel no corrections for aspect ratio were attempted. However, a slope of 2·7 was assumed for the lift coefficient curve instead of the usual theoretical value π.

or say $\{z, \theta\} = uq$. The aerodynamical damping and stiffness matrices B, W appropriate to the coordinates q_1, q_2 are then given by (compare example to § 9·3)

$$B = u'\mathfrak{B}u = 10^{-3}V\begin{bmatrix} -l & 1 \\ 0 & 1 \end{bmatrix}\begin{bmatrix} 9{\cdot}34 & 1{\cdot}40 \\ 0{\cdot}467 & 0{\cdot}470 \end{bmatrix}\begin{bmatrix} -l & 0 \\ 1 & 1 \end{bmatrix},$$

$$W = u'\mathfrak{W}u = 10^{-3}V^2\begin{bmatrix} -l & 1 \\ 0 & 1 \end{bmatrix}\begin{bmatrix} 0 & 9{\cdot}34 \\ 0 & 0{\cdot}467 \end{bmatrix}\begin{bmatrix} -l & 0 \\ 1 & 1 \end{bmatrix}.$$

In particular, if $l = \frac{2}{3}$ foot (as in Fig. 13·2·1), these yield

$$B = 10^{-3}V\begin{bmatrix} 3{\cdot}38 & -0{\cdot}46 \\ 0{\cdot}16 & 0{\cdot}47 \end{bmatrix} \quad \text{and} \quad W = 10^{-3}V^2\begin{bmatrix} -5{\cdot}76 & -5{\cdot}76 \\ 0{\cdot}467 & 0{\cdot}467 \end{bmatrix}.$$

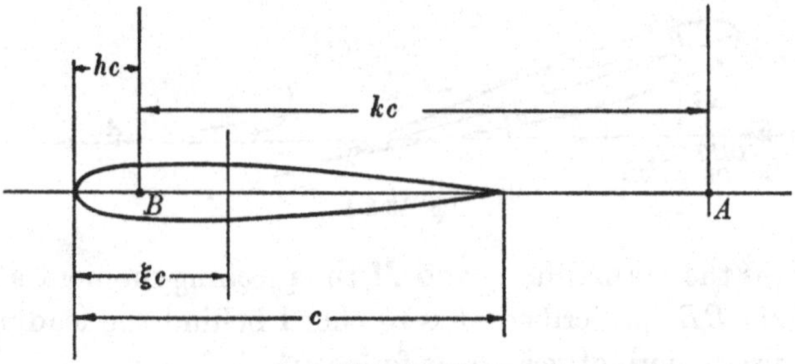

Fig. 13·3·2

(*b*) *The Inertias*. The contour adopted for the aerofoil is a blunt-nosed quartic oval.* The half-thickness t at a distance ξc behind the leading edge is

$$t = \frac{2\vartheta c}{3}(4\xi)^{\frac{1}{3}}(1-\xi),$$

where ϑ is the maximum thickness-chord ratio. Hence if s is the span, γ the material density, and $K = \frac{4}{3}\vartheta s c^4 \gamma (4)^{\frac{1}{3}}$, the inertias are given by (see Fig. 13·3·2)

$$A_{11} = K\int_0^1 \xi^{\frac{1}{3}}(1-\xi)(k+h-\xi)^2 \, d\xi,$$

$$P = A_{12} = A_{21} = K\int_0^1 \xi^{\frac{1}{3}}(1-\xi)(h-\xi)(k+h-\xi)\, d\xi,$$

$$A_{22} = K\int_0^1 \xi^{\frac{1}{3}}(1-\xi)(h-\xi)^2 \, d\xi,$$

in which hc, kc are respectively the distance between the leading edge and AA, and the distance between AA and BB.

* Defined in Ref. 45.

For the test system as finally constructed, $s = 4$ ft., $c = 0{\cdot}5$ ft., $\vartheta = 0{\cdot}125$, $\gamma = 0{\cdot}75$ slug/ft.^{3}, $h = 0{\cdot}15$, and $k = 1{\cdot}3333$. The numerical values of the inertias for the aerofoil alone are then*

$$A_{11} = 19{\cdot}1 \times 10^{-3} \text{ slug ft.}^2,$$
$$P = A_{12} = A_{21} = -3{\cdot}3 \times 10^{-3} \text{ slug ft.}^2,$$
$$A_{22} = 1{\cdot}8 \times 10^{-3} \text{ slug ft.}^2$$

For the complete system, the moment of inertia of the frame about the axis AA must be added to the value of A_{11} appropriate to the aerofoil only, while the values of P and A_{22} are also modified by the metal attachments to the aerofoil. Any addition of mass to the frame will increase A_{11} without altering P or A_{22}. Hence A_{11} is a separately variable parameter.

13·4. Graphical Interpretation of the Criterion for Steady Oscillations.

In connection with the design of the apparatus, two methods were developed for the discussion of the criterion (13·1·3). The first, a graphical method, is based on the use of a *test conic*,† and it will be illustrated with reference to the following particular binary system.

Table 13·4·1

Dynamical Coefficients for Illustrative Binary System No. 1

Coefficient	Value × 10^3	Coefficient	Value × 10^3
A_{11}	$35{\cdot}0$	$A_{21}\,(\equiv P)$	$4{\cdot}1$
B_{11}	$3{\cdot}74V$	B_{21}	$0{\cdot}703V$
C_{11}	$5{\cdot}14V^2 + 10^3\sigma_1$	C_{21}	$0{\cdot}467V^2$
$A_{12}\,(\equiv P)$	$4{\cdot}1$	A_{22}	$2{\cdot}3$
B_{12}	$1{\cdot}17V$	B_{22}	$(0{\cdot}47 + \alpha)V$
C_{12}	$5{\cdot}14V^2$	C_{22}	$0{\cdot}467V^2 + 10^3\sigma_2$

These dynamical constants were estimated for an aerofoil system identical with that described in § 13·2, except that the frame axis AA was situated 6 inches *forward* of the aerofoil axis BB. The coordinates q_1 and q_2 are as defined in § 13·3 and the frictionally constrained coordinate is assumed to be q_2. The spring stiffnesses σ_1, σ_2 are regarded as parameters temporarily left free for choice, while α in Table 13·4·1 is a parameter ultimately to be made zero.

* The *slug* is the unit of mass commonly adopted in aeronautics. A force of one pound applied to a mass of one slug produces an acceleration of one foot per second per second.

† For a fuller description of the use of test conics in relation to flutter problems the reader should consult Chap. III of Ref. 30.

If we write $$C_{11}\times 10^3 = 5{\cdot}14V^2+10^3\sigma_1 \equiv XV^2, \qquad \text{......(1)}$$
$$C_{22}\times 10^3 = 0{\cdot}467V^2+10^3\sigma_2 \equiv YV^2, \qquad \text{......(2)}$$
the coefficients of the determinantal equation are
$$\begin{aligned} 10^6 p_0 &= 63{\cdot}69, \\ 10^6 p_1/V &= 17{\cdot}3727+35{\cdot}0\alpha, \\ 10^6 p_2/V^2 &= 2{\cdot}3X+35{\cdot}0Y+3{\cdot}74\alpha-22{\cdot}05341, \\ 10^6 p_3/V^3 &= (0{\cdot}47+\alpha)X+3{\cdot}74Y-4{\cdot}15981, \\ 10^6 p_4/V^4 &= XY-2{\cdot}40038. \end{aligned}$$

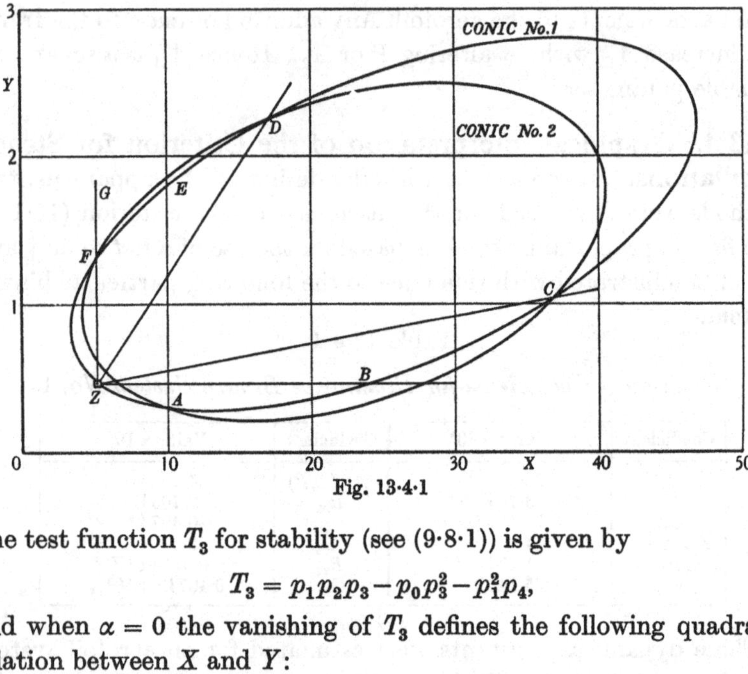

Fig. 13·4·1

The test function T_3 for stability (see (9·8·1)) is given by
$$T_3 = p_1p_2p_3-p_0p_3^2-p_1^2p_4,$$
and when $\alpha = 0$ the vanishing of T_3 defines the following quadratic relation between X and Y:
$$0 = X^2-9{\cdot}60548(2XY)+293{\cdot}629Y^2$$
$$-10{\cdot}3213(2X)-210{\cdot}211(2Y)+258{\cdot}154.$$
If X, Y are regarded as rectangular coordinates in a plane, this equation represents a *test conic*, which is shown as conic No. 1 in Fig. 13·4·1. Elimination of V^2 between (1) and (2) gives the *stiffness line*
$$\sigma_1(Y-0{\cdot}467) = \sigma_2(X-5{\cdot}14),$$
which always passes through the *stiffness point* $Z = (5{\cdot}14, 0{\cdot}467)$, and has the positive slope σ_2/σ_1. It is seen from the diagram that Z happens

to lie very nearly on the conic. The horizontal stiffness line ($\sigma_2 = 0$) is shown as ZB. When σ_1, σ_2 are known, the intersection of the corresponding stiffness line with the conic fixes the critical speed for the system.

The second curve (conic No. 2) shown in Fig. 13·4·1 represents the condition $\left(\frac{\partial T_3}{\partial \alpha}\right)_{\alpha=0} \equiv \frac{\partial T_3}{\partial B_{22}} = 0$, which reduces to

$$0 = X^2 - 5{\cdot}79196(2XY) + 255{\cdot}613Y^2 \\ - 14{\cdot}5159(2X) - 215{\cdot}904(2Y) + 326{\cdot}922.$$

It is easy to see that points lying inside curve No. 1 correspond to $T_3 < 0$ and hence to instability, and that points inside curve No. 2 satisfy the condition $\partial T_3/\partial B_{22} < 0$. Now consider any point on curve No. 1 which lies within curve No. 2. For such a point $T_3 = 0$ and $\partial T_3/\partial B_{22} < 0$, so that an increment in B_{22} leads to a negative value of T_3, i.e. to instability. Hence $\partial \mu/\partial B_{22} > 0$, which is the condition (13·1·3).

There are two separate arcs of curve No. 1, namely ABC and DEF, which lie inside curve No. 2 and which thus satisfy the foregoing condition. The slope of the stiffness line must, however, be positive. Hence only those portions of the two arcs for which $X > 5{\cdot}14$ and $Y > 0{\cdot}47$ yield permissible values of the stiffnesses. The horizontal line ZB excludes the portion AB of ABC, while a very small portion of DEF is excluded by the vertical line ZG. Stable steady oscillations will therefore occur at speeds below the critical when friction is present, provided the stiffnesses σ_1, σ_2 are such that the stiffness line lies within either of the angles BZC or DZG.

A system of the type discussed, with the frame axis AA forward of the aerofoil, was found to be open to certain practical objections and was not put into construction.

13·5. Alternative Treatment Based on the Use of Inertias as Parameters. The spring stiffness σ_2 is here assumed to be zero, and the stiffness σ_1 (and therefore C_{11}) and the inertias are left free for choice. The coefficients of the determinantal equation are

$$\left.\begin{aligned} p_0 &= A_{11}A_{22} - P^2, \\ p_1 &= A_{11}B_{22} + A_{22}B_{11} - P(B_{12} + B_{21}), \\ p_2 &= A_{11}C_{22} + A_{22}C_{11} - P(C_{12} + C_{21}) + B_{11}B_{22} - B_{12}B_{21}, \\ p_3 &= B_{22}C_{11} + B_{11}C_{22} - B_{12}C_{21} - B_{21}C_{12}, \\ p_4 &= C_{11}C_{22} - C_{12}C_{21}. \end{aligned}\right\} \quad \ldots\ldots(1)$$

At a critical speed, $\lambda_1 = i\omega$, and

$$p_0\omega^4 - p_2\omega^2 + p_4 = 0, \qquad \text{......(2)}$$

$$p_1\omega^2 - p_3 = 0. \qquad \text{......(3)}$$

On substitution of the expressions (1) for the coefficients in (2) and (3) two equations are obtained from which C_{11} can be eliminated. The result of this elimination, which is found to be independent of A_{11}, is

$$\begin{aligned} 0 = \omega^4\{P^2B_{22} - PA_{22}(B_{12}+B_{21}) + A_{22}^2B_{11}\} \\ + \omega^2[P\{C_{22}(B_{12}+B_{21}) - B_{22}(C_{12}+C_{21})\} \\ + A_{22}\{B_{12}C_{21} + B_{21}C_{12} - 2B_{11}C_{22}\}] \\ + \{\omega^2B_{22}(B_{11}B_{22} - B_{12}B_{21}) + B_{22}C_{12}C_{21} \\ - C_{22}(B_{12}C_{21} + B_{21}C_{12}) + B_{11}C_{22}^2\}. \qquad \text{......(4)} \end{aligned}$$

Since in this equation the four damping coefficients B_{ij} and the three stiffness coefficients C_{ij} are proportional to V and V^2 respectively, the equation effectively involves only the unknowns P, A_{22}, and the ratio ω/V. If the frequency parameter $\omega c/V$, and therefore ω/V, is arbitrarily assigned, a relation between P and A_{22} results.

Again, the formula (12·3·1) gives

$$\lambda_1 k_{21}\gamma_1 \equiv \xi + i\eta = -f_{11}(\lambda_1)/\overset{(1)}{\Delta}(\lambda_1),$$

and in the critical case, if $\lambda_1 = i\omega$, this can be reduced to

$$\xi + i\eta = (\overline{C_{11} - \omega^2A_{11}} + i\omega B_{11})/2(\omega^2p_1 + i\omega\,\overline{2\omega^2p_0 - p_2}). \qquad \text{...(5)}$$

The numerator in (5) can be expressed in a form independent of A_{11}, for on substitution from (1) in (3) we obtain

$$B_{22}(C_{11} - \omega^2A_{11}) = \omega^2(B_{11}A_{22} - P\overline{B_{12}+B_{21}}) - B_{11}C_{22} + B_{12}C_{21} + B_{21}C_{12}.$$

It follows that the expression (5) can be written

$$\xi + i\eta = \frac{\alpha + i\beta}{(aA_{11}+b) + i(cA_{11}+d)},$$

where α, β, a, b, c, and d are all independent of A_{11} and C_{11}. The condition (13·1·1) requires that

$$A_{11}(\alpha c - \beta a) > (\beta b - \alpha d). \qquad \text{......(6)}$$

The procedure is to assign at convenience the critical frequency parameter $\omega c/V$, to choose a series of values of P, and to determine a corresponding value of A_{22} from (4). These values of P and A_{22} are then used to determine α, β, a, b, c, and d. Any value of A_{11} which satisfies (6) and which exceeds P^2/A_{22} can then be assumed, and the system so defined will exhibit stable steady oscillations below the critical speed.

For example, suppose the dynamical coefficients to be as given in Table 13·5·1.

Table 13·5·1

Dynamical Coefficients for Illustrative Binary System No. 2

Coefficient	Value × 10^3	Coefficient	Value × 10^3
A_{11}	$A_{11}\times 10^3$	$A_{21}\,(\equiv P)$	$P\times 10^3$
B_{11}	$3{\cdot}38V$	B_{21}	$0{\cdot}16V$
C_{11}	$-5{\cdot}76V^2+10^3\sigma_1$	C_{21}	$0{\cdot}467V^2$
$A_{12}\,(\equiv P)$	$P\times 10^3$	A_{22}	$A_{22}\times 10^3$
B_{12}	$-0{\cdot}46V$	B_{22}	$0{\cdot}47V$
C_{12}	$-5{\cdot}76V^2$	C_{22}	$0{\cdot}467V^2$

The aerodynamical derivatives are here appropriate to the actual test system described in § 13·2. Since for that system $c = 0{\cdot}5$ and the frequency parameter $\omega c/V$ is assumed to be 0·4, we have $\omega = 0{\cdot}8V$.

If
$$C_{11}\times 10^3 = -5{\cdot}76V^2+10^3\sigma_1 \equiv XV^2, \qquad \ldots\ldots(7)$$
the formulae (1) yield

$$p_0 = A_{11}A_{22}-P^2,$$
$$10^3 p_1/V = 0{\cdot}47A_{11}+3{\cdot}38A_{22}+0{\cdot}30P,$$
$$10^3 p_2/V^2 = 0{\cdot}467A_{11}+A_{22}X+5{\cdot}293P+0{\cdot}0016622,$$
$$10^3 p_3/V^3 = 0{\cdot}47X+2{\cdot}71488,$$
$$10^6 p_4/V^4 = 0{\cdot}467X+2{\cdot}68992,$$

while equations (2) and (3) become

$$0{\cdot}4096p_0V^4-0{\cdot}64p_2V^2+p_4 = 0,$$
$$0{\cdot}64p_1V^2-p_3 = 0.$$

Equation (4), which is derived by substitution for the coefficients p and elimination of X, now becomes

$$0 = 409600P^2+261447A_{22}P+2945634A_{22}^2 + 3196{\cdot}745P-5846{\cdot}251A_{22}+1{\cdot}07144.$$

The values of A_{22} corresponding to a range of values of P, and the appropriate minimum values of A_{11} as calculated from equation (6), are given in Tables 13·5·2 and 13·5·3. In the case $P = 0$, only the greater value of A_{22} is used.

Table 13·5·2

$P \times 10^3$ (slug ft.^{2})	$A_{22} \times 10^3$ (slug ft.^{2})	
	First value	Second value
0	1·780419	0·204299
−2·0	2·636547	Negative
−4·0	2·936515	”
−6·0	2·909695	”

Table 13·5·3

$P \times 10^3$	0	−2·0	−4·0	−6·0
$\alpha \times V$	2·41813	5·54148	6·10508	5·16462
$\beta \times V$	2·704	2·704	2·704	2·704
a	0·6016	0·6016	0·6016	0·6016
$b \times 10^3$	7·70280	10·63875	11·16854	10·28450
c	1·07594	1·95262	2·25979	2·23233
$d \times 10^3$	−9·54798	−17·29053	−30·23657	−49·61866
$(\beta b - \alpha d) V \times 10^3$	43·917	124·582	214·796	284·071
$(\alpha c - \beta a) V$	0·9570	9·1937	12·1695	9·9024
Minimum $A_{11} \times 10^3$	45·041	13·551	17·650	28·687

From the results it is seen that quite a wide range of the inertial constants is permissible. A rough estimate of the inertias for the wooden aerofoil and some of its attachments indicated that the value $P = -4 \times 10^{-3}$, and the corresponding value of A_{22}, would be realisable in practice. The actual value of A_{11} was expected to exceed greatly the minimum required by Table 13·5·3, and a round figure of 40×10^{-3} was adopted for the further discussion of the test system.

13·6. Theoretical Behaviour of the Test System. In accordance with the conclusions drawn in § 13·5 the dynamical coefficients were chosen as follows:

Table 13·6·1

Dynamical Coefficients for Theoretical Test System

Coefficient	Value $\times 10^3$	Coefficient	Value $\times 10^3$
A_{11}	40·0	$A_{21} (\equiv P)$	−4·0
B_{11}	$3·38V$	B_{21}	$0·16V$
C_{11}	$10^3\sigma_1 - 5·76V^2$	C_{21}	$0·467V^2$
$A_{12} (\equiv P)$	−4·0	A_{22}	2·936515
B_{12}	$-0·46V$	B_{22}	$0·47V$
C_{12}	$-5·76V^2$	C_{22}	$0·467V^2$

(i) *Behaviour at a Critical Speed.* The critical value of X (see (13·5·7)) leading to simply sinusoidal oscillations of the frictionless system is readily found to be $X_c = 31{\cdot}7051$, and the critical speed V_c is then fixed in terms of σ_1 by the relation

$$10^3\sigma_1 = 37{\cdot}4651 V_c^2.$$

The coefficients of the determinantal equation for the same condition work out as

$$\begin{aligned} p_0 &= 101{\cdot}4606 \times 10^{-6}, \\ p_1 &= 27{\cdot}5254 V_c \times 10^{-6}, \\ p_2 &= 92{\cdot}2727 V_c^2 \times 10^{-6}, \\ p_3 &= 17{\cdot}6163 V_c^3 \times 10^{-6}, \\ p_4 &= 17{\cdot}4962 V_c^4 \times 10^{-6}. \end{aligned}$$

Further, the real and imaginary parts of the roots λ_r and of the constants $\lambda_r k_{2r} \gamma_r$ are given by

$$\begin{aligned} \mu &= 0, & \omega &= 0{\cdot}8 V_c, \\ \mu' &= -0{\cdot}135646 V_c, & \omega' &= 0{\cdot}501041 V_c, \\ V_c\xi &= 0{\cdot}077728 \times 10^3, & V_c\eta &= -0{\cdot}055964 \times 10^3, \\ V_c\xi' &= -0{\cdot}077728 \times 10^3, & V_c\eta' &= 0{\cdot}503823 \times 10^3. \end{aligned}$$

Lastly, the equation which determines the semi-period T is

$$0 = \tfrac{1}{2}\Omega(T) \equiv \frac{\xi \sinh T\mu - \eta \sin T\omega}{\cosh T\mu + \cos T\omega} + \frac{\xi' \sinh T\mu' - \eta' \sin T\omega'}{\cosh T\mu' + \cos T\omega'}.$$

It is found that $\Omega(T)$ vanishes when $TV_c = 3{\cdot}6563$, and that the graph of $\Omega(T)$ has an asymptote at $TV_c = 3{\cdot}9270$. The first value of TV_c corresponds to an unstable steady oscillation which, if disturbed, either dies to ankylosis or grows indefinitely.

(ii) *Behaviour at a Speed Less than the Critical.* If we next assume the airspeed to be reduced to $V = 0{\cdot}975 V_c$, the coefficients of the determinantal equation become

$$\begin{aligned} p_0 &= 101{\cdot}4606 \times 10^{-6}, \\ p_1 &= 26{\cdot}8373 V_c \times 10^{-6}, \\ p_2 &= 93{\cdot}1488 V_c^2 \times 10^{-6}, \\ p_3 &= 17{\cdot}1755 V_c^3 \times 10^{-6}, \\ p_4 &= 16{\cdot}6323 V_c^4 \times 10^{-6}, \end{aligned}$$

while

$$\begin{aligned} \mu &= -0{\cdot}008895 V_c, & \omega &= 0{\cdot}817539 V_c, \\ \mu' &= -0{\cdot}123360 V_c, & \omega' &= 0{\cdot}479605 V_c, \\ V_c\xi &= 0{\cdot}052411 \times 10^3, & V_c\eta &= -0{\cdot}051047 \times 10^3, \\ V_c\xi' &= -0{\cdot}052411 \times 10^3, & V_c\eta' &= 0{\cdot}510530 \times 10^3. \end{aligned}$$

In this case the lowest non-zero roots of the semi-period equation are given by $TV_c = 3{\cdot}6692$ and $TV_c = 3{\cdot}7714$. The first of these corresponds to the unstable oscillations referred to in the critical case, and the second (which develops from the asymptote) corresponds to stable steady oscillations. The characteristics of this stable motion will now be examined.

With $TV_c = 3{\cdot}7714$, it is found that

$$\frac{e^{T\lambda_1}-1}{e^{T\lambda_1}+1} = -14{\cdot}826+25{\cdot}764i,$$

$$\frac{e^{T\lambda_3}-1}{e^{T\lambda_3}+1} = -0{\cdot}55144+1{\cdot}11135i.$$

Moreover
$$V_c^2 k_{21}\gamma_1 = (-0{\cdot}063130-0{\cdot}063421i)\,10^3,$$
$$V_c^2 k_{23}\gamma_3 = (1{\cdot}024789-0{\cdot}154307i)\,10^3,$$
$$\lambda_1^2 k_{21}\gamma_1 = (0{\cdot}041267+0{\cdot}043302i)\,10^3,$$
$$\lambda_3^2 k_{23}\gamma_3 = (-0{\cdot}238388-0{\cdot}088116i)\,10^3,$$

while
$$a_{22} = A_{11}/p_0 = 0{\cdot}394242\times 10^3.$$

Hence (see (11·6·5) and (11·6·6))

$$q_2(0)/R = \sum_{r=1}^{4} k_{2r}\gamma_r \frac{e^{T\lambda_r}-1}{e^{T\lambda_r}+1} = 4{\cdot}3527\times 10^3/V_c^2,$$

$$\ddot{q}_2(0)/R = \sum_{r=1}^{4} \lambda_r^2 k_{2r}\gamma_r \frac{e^{T\lambda_r}-1}{e^{T\lambda_r}+1} + a_{22} = -2{\cdot}6019\times 10^3.$$

These expressions confirm that the oscillations are possible. Again, using equation (11·4·1), we find that

$$V_c^2 k_{11}\gamma_1 = (-0{\cdot}038914-0{\cdot}021948i)\,10^3,$$
$$V_c^2 k_{13}\gamma_3 = (0{\cdot}203520-0{\cdot}014215i)\,10^3,$$

and the motion of the frame is then given by

$$q_1(t)/R = \sum_{r=1}^{4} k_{1r}\gamma_r\left(1-\frac{2e^{t\lambda_r}}{1+e^{T\lambda_r}}\right).$$

A graphical examination shows that the coordinate q_1 lags in phase by 54° behind q_2, and that its maximum value is $2350R/V_c^2$.

If the critical speed is chosen to be 40 feet per second (corresponding to a spring stiffness σ_1 of 59·94 pounds feet per radian), the results obtained under the present heading are applicable at a speed of 39 feet per second. The amplitude of steady oscillations for the aerofoil then is 2·72 radians per foot pound of frictional couple, while that for the frame is 1·47.

PART II. EXPERIMENTAL INVESTIGATION

13·7. Preliminary Calibrations of the Actual Test System. The system as described in § 13·2 was mounted vertically for test in a wind tunnel. The following preliminary calibrations of the apparatus were made:

(i) *Frictional Couple.* The frictional couple applied to the aerofoil by the discs C and D (see Figs. 13·2·1 and 13·2·2) was varied in the wind tunnel experiments by the addition of masses to the scale-pan M outside the tunnel. A direct calibration was made of the couples required to overcome the "stick friction" and the "slip friction" due to a given mass in M: the frame was locked centrally in these tests. In the measurements of slip friction the aerofoil was given a small angular velocity, and the couple required to keep it just moving was determined. The slip friction was about 60 per cent of the stick friction.

The frictional couple appeared to be directly proportional to the outside load in the scale-pan, so that the weight of the latter balanced the tensions in the release springs L to within the limits of observation. The final result gave a slip frictional couple on the aerofoil of 0·0157 pound foot for 1 pound of outside load.

(ii) *Constants of Inertia.* These were determined by preliminary oscillation tests of the apparatus under gravity, with the axes AA and BB horizontal. The measured constants (in slug ft.^{2}) were

$$A_{22} = 2{\cdot}3 \times 10^{-3}, \quad P = -3{\cdot}0 \times 10^{-3}, \quad \text{and} \quad A_{11} = 41{\cdot}1 \times 10^{-3}.$$

(iii) *Spring Stiffnesses.* To provide stability four springs were attached to the frame, one pair to the top arm K and one pair to the bottom arm. A direct calibration with the four springs in place gave a stiffness σ_1 of 61·6 pounds feet per radian.

13·8. Observations of Frictional Oscillations. The observed critical speed, as indicated by the occurrence of definitely growing oscillations, was about 37 feet per second.* At speeds ranging from $V = 31$ to $V = 36$ steady oscillations were observed, the amplitudes of which could be increased by an increase of the frictional moment; and the frequency parameter of these oscillations had a constant value of 0·41. To this extent the experimental results accorded satisfactorily with the theory.

* If the measured values of the inertial constants and of σ_1 are substituted for the values given in Table 13·6·1, the calculated critical speed is almost exactly 37 feet per second.

The measured amplitudes of the frame during the steady oscillations are summarised in Table 13·8·1 and Fig. 13·8·1. For a given wind speed they should, according to the theory, increase linearly with the frictional moment, but this is only borne out by the experimental results for a restricted range of the friction. Within this range*, the measured increase of amplitude for an additional frictional moment of 0·01 pound foot is roughly 0·5 degree at $V = 35$. No exact comparison with theory is possible here, but it may be noted that for the system discussed in § 13·6 the calculated increase of amplitude is 0·84 degree at

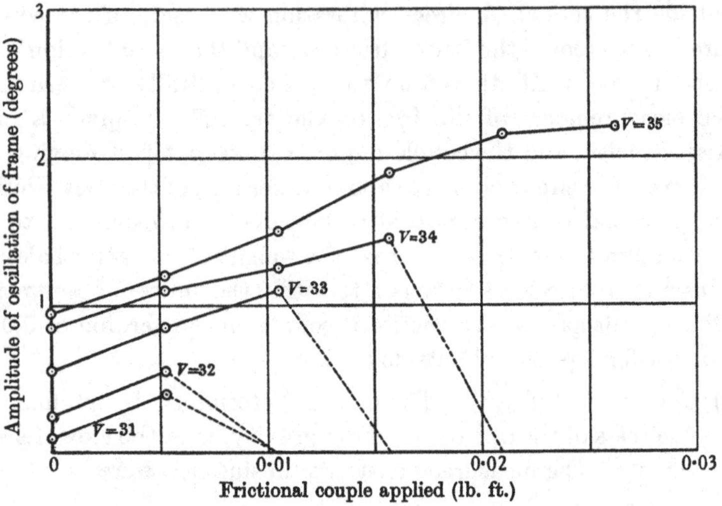

Fig. 13·8·1

a wind speed of 1 foot per second less than the critical speed. The behaviour of the actual system differed from that of the theoretical system in that its oscillations developed spontaneously from the small natural disturbances in the wind tunnel: the theory requires the initial disturbance to be at least comparable with the final motion.

Among the possible causes which might well account for the differences of behaviour just mentioned, reference may be made to the following:

(i) Departures of the actual aerodynamical forces from linear laws.†

* Beyond this range ankylosis occurred, as indicated by "A" in Table 13·8·1 and by dotted lines in Fig. 13·8·1.

† The non-linearity of these forces was to some extent confirmed by measurements of the aerodynamical stiffnesses for a static condition of the wing.

(ii) The presence of some friction in the ball-bearings, and increase of this internal friction with the amplitude of oscillation.

(iii) Departures from constancy of the externally applied friction during each stroke of the aerofoil.

A very close correlation between theory and experiment was, in fact, found impossible owing to the susceptibility of the actual test system to three other types of oscillation which were almost certainly attributable to the departures of the aerodynamical forces from linear laws. A brief description of these oscillations is given in § 13·9.

Table 13·8·1

Amplitudes of Frame during Steady Frictional Oscillations

V	Frictional Couple Applied (lb. ft.)					
	0	$5{\cdot}25\times10^{-3}$	$10{\cdot}5\times10^{-3}$	$15{\cdot}75\times10^{-3}$	21×10^{-3}	$26{\cdot}25\times10^{-3}$
31	0·1°	0·4°	A	—	—	—
32	0·25°	0·55°	A	—	—	—
33	0·55°	0·85°	1·1°	A	—	—
34	0·85°	1·1°	1·25°	1·45°	A	—
35	0·95°	1·2°	1·5°	1·9°	2·15°	2·2°
36	B	—	—	—	—	—

Remarks on Table 13·8·1. "A" indicates ankylosis. The method adopted was to increase the friction slowly, so that the amplitude could build up slowly to the steady value. If the friction was suddenly applied, the existing oscillation did not represent, in general, a sufficient disturbance to give rise to the new and larger oscillation; consequently the motion would die and the friction would then ankylose the aerofoil. Owing, presumably, to aerodynamical non-linearities, it was in practice impossible to build up the amplitude indefinitely, even slowly; and ankylosis would occur as indicated in the table.

"B" indicates the most curious of all the oscillations obtained. At $V=36$ and with zero frictional load, the oscillations grew to an amplitude of about 2·5°. Then, quite abruptly, the frequency dropped, and the amplitude immediately started to decrease. When it reached about 1°, the oscillation reverted to the frictional type, and the amplitude again increased, and so on. The result was a "hunting" oscillation which grew on one frequency and died on another. The decreasing oscillation was recognised to correspond to a sustained oscillation discussed in § 13·9, and there referred to as oscillation No. 1. Its existence was attributed to aerodynamical non-linearities.

13·9. Other Oscillations Exhibited by the Test System.

(i) *Oscillations with the Frame Locked.* Two types of sustained oscillation of the aerofoil were observed when the frame was locked in the central position. They will be referred to as oscillations Nos. 1 and 2. No. 1 was a smooth oscillation of half angle about 5 degrees, and characterised by a low value of the frequency parameter. No. 2 was a much larger oscillation, the half angle being about 20 degrees, and it was of a very jerky nature and profusely embroidered with overtones; the frequency parameter was appreciably higher than that of No. 1. To generate No. 2 it was always necessary to give the aerofoil a very large displacement (50 degrees or higher) and to allow it to swing down to the sustained state: the stability of the oscillation thus appeared to depend on the establishment of a turbulent oscillatory wake.

At low wind speeds, the equilibrium configuration was spontaneously unstable (i.e. growing oscillations would result from the natural disturbances present in the wind tunnel), and the oscillations increased to No. 1. Following a large initial disturbance, the motion decreased to No. 1, although there was a tendency for No. 2 to appear. Between $V = 10$ and $V = 24$ both oscillations were realisable; between $V = 24$ and $V = 30$ only No. 2 occurred. From $V = 30$ upward, all disturbances gave rise to decreasing oscillations, the equilibrium position being ultimately reached. The effect of friction was in all cases to reduce the amplitude, and ultimately, as the friction was increased, to cause ankylosis.

(ii) *Oscillations with the Frame Constrained by Springs.* In this condition it was possible to obtain three different types of maintained motion other than the frictional type. Up to $V = 25$ oscillations similar to Nos. 1 and 2 were again realisable. Oscillation No. 3 occurred at speeds above about 33 feet per second, and was very violent. Its amplitude was roughly 60°, and on several occasions it developed so suddenly that it broke the safety grab of the apparatus. Between $V = 30$ and $V = 37$ it could be produced by releasing the aerofoil from a large displacement. At $V = 37$, if the oscillation corresponding to the internal friction only was allowed to become sustained and if external friction was then applied impulsively, the motion jumped almost instantly to oscillation No. 3. At a speed just above 37 feet per second the spontaneous oscillations rose directly to No. 3. This was the critical speed for flutter.

LIST OF REFERENCES

R. & M. *denotes Reports and Memoranda of the Aeronautical Research Committee, published by H.M. Stationery Office.*

* *indicates unpublished report of the Aeronautical Research Committee.*

1. Bôcher, M. *Introduction to Higher Algebra.* New York (1933).
2. Turnbull, H. W. and Aitken, A. C. *An Introduction to the Theory of Canonical Matrices.* London (1932). (This includes numerous historical notes.)
3. Kelland, P. and Tait, P. G. *Introduction to Quaternions.* London, 3rd ed. (1904).
4. Smith, T. "Change of Variables in Laplace's and other Second-Order Differential Equations." *Proc. Phys. Soc.* Vol. 46, p. 344 (1934).
5. Ince, E. L. *Ordinary Differential Equations.* London (1927).
6. Baker, H. F. "On the Integration of Linear Differential Equations." *Proc. Lond. Math. Soc.* Vol. 35, p. 333 (1903).
7. Wedderburn, J. H. M. *Lectures on Matrices.* American Math. Soc. Colloquium Publications, Vol. 17 (1934). (This includes a bibliography of 549 references.)
8. Eddington, Sir Arthur. *New Pathways in Science.* Cambridge (1935).
9. Morris, J. "On a Simple Method for Solving Simultaneous Linear Equations by a Successive Approximation Process." *Journ. Roy. Aero. Soc.* Vol. 39, p. 349 (1935).
10. Duncan, W. J. and Collar, A. R. "A Method for the Solution of Oscillation Problems by Matrices." *Phil. Mag.* Series 7, Vol. 17, p. 865 (1934).
11. Duncan, W. J. and Collar, A. R. "Matrices Applied to the Motions of Damped Systems." *Phil. Mag.* Series 7, Vol. 19, p. 197 (1935).
12. Aitken, A. C. "On Bernoulli's Numerical Solution of Algebraic Equations." *Proc. Roy. Soc. Edinburgh,* Vol. 46, p. 289 (1927).
13. Routh, E. J. *Advanced Rigid Dynamics.* London, 6th ed. (1905).
14. Burnside, W. S. and Panton, A. W. *The Theory of Equations.* London, 9th ed. (1928).
15. Frazer, R. A. and Duncan, W. J. "On the Criteria for the Stability of Small Motions." *Proc. Roy. Soc.* Series A, Vol. 124, p. 642 (1929).
16. Jeffreys, H. *Operational Methods in Mathematical Physics.* Cambridge Mathematical Tracts, No. 23 (1927).
17. Bryant, L. W. and Williams, D. H. "The Application of the Method of Operators to the Calculation of the Disturbed Motion of an Aeroplane." R. & M. 1346 (1931).

18. Piaggio, H. T. H. *An Elementary Treatise on Differential Equations and their Applications*. London (1925).

19. Baker, H. F. "Note on the Integration of Linear Differential Equations." *Proc. Lond. Math. Soc.* Series 2, Vol. 2, p. 293 (1905).

20. Baker, H. F. "On Certain Linear Differential Equations of Astronomical Interest." *Phil. Trans. Roy. Soc.* Series A, Vol. 216, p. 129 (1916).

21. Frazer, R. A. "Disturbed Unsteady Motion, and the Numerical Solution of Linear Ordinary Differential Equations." Report T. 3179* (1931).

22. Darboux, M. G. "Sur les Systèmes Formés d'Equations Linéaires à une seule Variable Indépendante." *Comptes Rendus*, Vol. 90, p. 524 (1880).

23. Frazer, R. A., Jones, W. P. and Skan, S. W. "Approximations to Functions and to the Solutions of Differential Equations." R. & M. 1799 (1938).

24. Duncan, W. J. "Galerkin's Method in Mechanics and Differential Equations." R. & M. 1798 (1938).

25. Dinnik, A. N. "Galerkin's Method for Determining the Critical Strengths and Frequencies of Vibrations." *Aeronautical Engineering, U.S.S.R.* Vol. 9, No. 5, p. 99 (1935).

26. Grossman, E. P. "The Twist of a Monoplane Wing." *Transactions of the Central Aero-Hydrodynamical Institute, Moscow*, No. 253 (1936).

27. Bairstow, L. *Applied Aerodynamics*. London (1920).

28. Planck, M. (translated by Brose, H. L.). *General Mechanics*. London (1933).

29. Lamb, H. *Higher Mechanics*. Cambridge, 2nd ed. (1929).

30. Frazer, R. A. and Duncan, W. J. "The Flutter of Aeroplane Wings." R. & M. 1155 (1928).

31. Whittaker, E. T. *A Treatise on the Analytical Dynamics of Particles and Rigid Bodies*. Cambridge, 3rd ed. (1927).

32. Temple, G. and Bickley, W. G. *Rayleigh's Principle and its Application to Engineering*. Oxford (1933).

33. Küssner, H. G. "Augenblicklicher Entwicklungsstand der Frage des Flügelflatterns." *Luftfahrtforschung*, Vol. 12, No. 6, p. 193 (1935).

34. von Schlippe, B. "Zur Frage der Selbsterregten Flügelschwingungen." *Luftfahrtforschung*, Vol. 13, No. 2, p. 41 (1936).

35. Sezawa, K. and Kubo, S. "The Nature of the Deflection-Aileron Flutter of a Wing as Revealed through its Vibrational Frequencies." Reports of the Aeronautical Research Institute, Tôkyô Imperial University, Vol. 11, No. 140, p. 303 (1936).

36. Frazer, R. A. and Jones, W. P. "Forced Oscillations of Aeroplanes, with Special Reference to von Schlippe's Method of Predicting Critical Speeds for Flutter." R. & M. 1795 (1937).

37. Carter, B. C. "Dynamic Forces in Aircraft Engines." *Journ. Roy. Aero. Soc.* Vol. 31, p. 278 (1927).

38. FRAZER, R. A. and COLLAR, A. R. "Influence of Solid Friction on the Oscillations of Dynamical Systems." Report T. 3481* (1933).

39. FRAZER, R. A. and COLLAR, A. R. "Supplementary Theory Regarding the Influence of Solid Friction on the Oscillations of Dynamical Systems." Report T. 3481 (*a*)* (1934).

40. FRAZER, R. A., DUNCAN, W. J. and SCRUTON, C. "Tail Flutter of a Model of the 'Puss Moth' Monoplane." Appendix 44 to R. & M. 1699 (1937).

41. ROWELL, H. S. "Note on the Analysis of Damped Vibrations." *Phil. Mag.* Series 6, Vol. 44, p. 951 (1922).

42. FRAZER, R. A. and DUNCAN, W. J. "The Flutter of Monoplanes, Biplanes and Tail Units." R. & M. 1255 (1931).

43. COLLAR, A. R. "Experiments on the Sustained Pitching Oscillations of an Aerofoil." Report T. 3481 (*b*)* (1934).

44. DUNCAN, W. J. and COLLAR, A. R. "Resistance Derivatives of Flutter Theory." R. & M. 1500 (1933).

45. DUNCAN, W. J. and COLLAR, A. R. "The Present Position of the Investigation of Airscrew Flutter." R. & M. 1518 (1933).

Supplementary List of Publications

46. TURNBULL, H. W. *The Theory of Determinants, Matrices, and Invariants.* London (1928).

47. CULLIS, C. E. *Matrices and Determinoids,* Vols. I–III. Cambridge (1913–1925).

48. MUIR, SIR THOMAS. *The Theory of Determinants,* Vols. I–IV. London (1906–1923).

49. MUIR, SIR THOMAS. *Contributions to the History of Determinants,* 1900–1920. London (1930).

50. MUIR, SIR THOMAS. "The Literature of Cayleyan Matrices." *Trans. Roy. Soc. S. Africa,* Vol. 18, p. 219 (1930).

51. MACDUFFEE, C. C. *The Theory of Matrices.* Berlin (1933).

LIST OF AUTHORS CITED

Clarendon numbers indicate the serial order in the List of References on pages 399–401: other numbers refer to pages of the main text.

INDEX

The numbers refer to the pages.